P9-CIV-168

PRAISE FOR *REINVENTING FIRE*

"My friend Amory Lovins knows that the most important question of the 21st century is the 'how' question—how we turn good ideas into working solutions. *Reinventing Fire* is a wise, detailed, and comprehensive blueprint for gathering the best existing technologies for energy use and putting them to work right now to create jobs, end our dependence on climate-changing fossil fuels, and unleash the enormous economic potential of the coming energy revolution."

—PRESIDENT BILL CLINTON

"Amory Lovins and his team of extraordinary professionals provide an analytically sound, detailed, coherent plan for transforming our national use and supply of energy—and for saving $5 trillion in the process! *Reinventing Fire* is a towering work, a page-turning tour de force of compelling wisdom that deserves a permanent place on the desk—nay, in the mind—of whoever holds the chair in the Oval Office."

—ROBERT C. MCFARLANE, NATIONAL SECURITY ADVISOR TO PRESIDENT REAGAN; COFOUNDER AND COCHAIR OF THE UNITED STATES ENERGY SECURITY COUNCIL

"If you wanted to bring America happiness and prosperity, *and* address unemployment, government gridlock, and climate change, *and* create meaning in a world rife with contradictory views and ideologies, you can do one thing: read *Reinventing Fire* . . . and then see to it that it is read by every decision maker in the land. This is a stunning work of enormous dimension. *Reinventing Fire* outlines an eminently practical path to a durable and meaningful future by reimagining how we use and produce the lifeblood of civilization—energy in its myriad forms."

—PAUL HAWKEN, AUTHOR OF *BLESSED UNREST*; COAUTHOR OF *NATURAL CAPITALISM*

"A brilliant, thorough, innovative plan for a complete and profitable restructuring over the next four decades of how we use energy for transport, electricity, buildings, and industry. RMI's new fire will . . . help us see our way out of the massive problems caused by our dependence on oil and coal."

—R. JAMES WOOLSEY, VENTURE PARTNER, LUX CAPITAL; FORMER DIRECTOR OF THE U.S. CENTRAL INTELLIGENCE AGENCY; CHAIRMAN, FOUNDATION FOR DEFENSE OF DEMOCRACIES

"*Reinventing Fire* crackles with fresh perspectives and compelling insights about our energy past, present, and future. Drawing on the logic of economics, physics, geology, national security, and just plain common sense, Lovins and his colleagues blaze a trail toward an energy future that is cleaner, cheaper, and safer. A 'must read' book for business leaders, policymakers, environmentalists, academics, and anyone else who cares about our planet's future and our nation's prosperity."

—DAN ESTY, DIRECTOR, CENTER FOR BUSINESS AND THE ENVIRONMENT AT YALE UNIVERSITY, AND AUTHOR OF *GREEN TO GOLD*

"*Reinventing Fire*'s business-led energy revolution is driven by new revenue, lower risk, and striking competitive advantage. In real estate as in other sectors, that's a powerful strategy for success. Smart developers will follow it to their benefit; laggards will ignore it at their peril."

—GERALD D. HINES, FOUNDER AND CHAIRMAN, HINES

"A must-read 'new baseline' analysis for innovators and policy makers."

—BILL JOY, PARTNER, KLEINER PERKINS CAUFIELD & BYERS; COFOUNDER, SUN MICROSYSTEMS

REINVENTING FIRE®

BOLD BUSINESS SOLUTIONS FOR THE NEW ENERGY ERA

AMORY B. LOVINS AND
ROCKY MOUNTAIN INSTITUTE

FOREWORDS BY
MARVIN ODUM, PRESIDENT, SHELL OIL COMPANY
JOHN W. ROWE, CHAIRMAN AND CEO, EXELON CORPORATION

Chelsea Green Publishing
White River Junction, Vermont

PROJECT MANAGER: Patricia Stone
DEVELOPMENTAL EDITOR: Joni Praded
COPY EDITOR: Nancy Ringer
PROOFREADER: Eileen Clawson
INDEXER: Peggy Holloway
DESIGNER: Maureen Forys, Happenstance Type-O-Rama

Printed in the United States of America

First printing September, 2011

10 9 8 7 6 5 4 3 2 1   11 12 13 14 15

Chelsea Green Publishing is committed to preserving ancient forests and natural resources. We elected to print this title on FSC®-certified paper containing at least 10% postconsumer recycled fiber, processed chlorine-free. As a result, for this printing, we have saved:

32 Trees
14,809 Gallons of Wastewater
13 million BTUs
939 Pounds of Solid Waste
3,284 Pounds of Greenhouse Gases

Chelsea Green Publishing made this paper choice because we are a member of the Green Press Initiative, a nonprofit program dedicated to supporting authors, publishers, and suppliers in their efforts to reduce their use of fiber obtained from endangered forests. For more information, visit www.greenpressinitiative.org.

Environmental impact estimates were made using the Environmental Defense Paper Calculator. For more information visit: www.papercalculator.org.

Our Commitment to Green Publishing

Chelsea Green sees publishing as a tool for cultural change and ecological stewardship. We strive to align our book manufacturing practices with our editorial mission and to reduce the impact of our business enterprise in the environment. We print our books and catalogs on chlorine-free recycled paper, using vegetable-based inks whenever possible. This book may cost slightly more because it was printed on paper that contains recycled fiber, and we hope you'll agree that it's worth it. Chelsea Green is a member of the Green Press Initiative (www.greenpressinitiative.org), a nonprofit coalition of publishers, manufacturers, and authors working to protect the world's endangered forests and conserve natural resources. *Reinventing Fire* was printed on FSC®-certified paper supplied by QuadGraphics that contains at least 10% postconsumer recycled fiber.

LIBRARY OF CONGRESS CATALOGING-IN-PUBLICATION DATA
Lovins, Amory B., 1947–
Reinventing fire : bold business solutions for the new energy era / Amory B. Lovins and Rocky Mountain Institute.
p. cm.
Includes bibliographical references and index.
ISBN 978-1-60358-371-8 (hbk.)—ISBN 978-1-60358-372-5 (ebook)
1. Energy development—United States. 2. Renewable energy sources—United States. 3. Energy consumption—United States. 4. Energy policy—United States. 5. Industries—Energy consumption—United States. 6. Industries—Energy conservation—United States. I. Title.

HD9502.U52L674 2011
333.790973—dc23

2011029947

Chelsea Green Publishing Company
Post Office Box 428
White River Junction, VT 05001
(802) 295-6300
www.chelseagreen.com

Dedicated to two friends and mentors
who informed and inspired this work:

Ray C. Anderson (1934–2011)

Lee Schipper (1947–2011)

There will be no end to the good they have done.

Caminante, no hay camino
Se hace camino al andar

Walker, there is no path
The path is made by walking

—Antonio Machado (1875–1939)

ROCKY MOUNTAIN INSTITUTE'S *REINVENTING FIRE* PROJECT TEAM

Senior author: Amory B. Lovins

Coauthors: Mathias Bell; Lionel Bony; Albert Chan; Stephen Doig, PhD; Nathan J. Glasgow; Lena Hansen; Virginia Lacy; Eric Maurer; Jesse Morris; James Newcomb; Greg Rucks; Caroline Traube

Project manager: Lionel Bony (research and writing); Eric Maurer (editing and production)

Modelers: Mark Dyson; Mark Gately

Researchers: Josh Agenbroad; Sarah Bahan; Dan Gorman; Ryan Matley; Natalie Mims; Jay Tankersley; Katherine Wang, PE

Also: Michael Bendewald; Steve Brauneis; Aaron Buys; Betsy Cannon; Bennett Cohen; Andrew Dietrich; Amanda Gonzalez; Chris Hart; Leah Kuritzky; Nicole LeClaire; Angie Lee; Davis Lindsey; Luisa Lombera; Matt Mattila; Miriam Morris; Brendan O'Donnell; Tyler Ruggles; Anna Shpitsberg; Roy Torbert; Brendan Trimboli; Molly Ward

Production: Betsy Ronan Herzog; Carrie Jordan; Clay Stranger

Senior reviewers: Robert Hutchinson; Brad Mushovic; Michael Potts

Contributors External to RMI

Editor: John Carey

Contributing author: Jason Denner

Researchers: Will Clift; Emily Grubert; Darrin Magee, PhD; Glenn Mercer; Scott Muldavin; Eric Wanless

Senior reviewers: E. Kyle Datta; Jonathan G. Koomey, PhD

Peer reviewers: Please see pages 253–254

CONTENTS

About This Book ix

Preface xi

Foreword by Marvin Odum (President, Shell Oil Company) xv

Foreword by John W. Rowe (Chairman and CEO, Exelon Corporation) xvi

1. DEFOSSILIZING FUELS xx

The True Cost of Oil Addiction 3

Oil and Insecurity 5

Coal's Hidden Costs 6

Turning the Supertanker 8

Lighting the New Fire 9

No Miracles Required 12

2. TRANSPORTATION: FITTER VEHICLES, SMARTER USE 14

Designing and Building Autos Differently 17

Using Autos More Productively 42

The Rest of the Story: Beyond Automobiles 49

Powering Vehicles with Cleaner Energy 62

Conclusion: Better Mobility at Lower Cost without Oil 69

3. BUILDINGS: DESIGNS FOR BETTER LIVING 76

Understanding Today's Building Quagmire 82

The Efficiency Revolution: What's Profitable and What's Possible 86

The Conundrum and the Challenge 103

Solving the Efficiency Puzzle 105

Conclusion: More Comfort, More Productivity, Less Energy, Stronger Economics 117

4. INDUSTRY: REMAKING HOW WE MAKE THINGS 122
How the Industrial Jungle Drives U.S. Energy Demand 128
Viewing Industry through the Efficiency Lens 133
How Much More Productive Can Industry Become? 144
Transforming the Industrial Jungle 155
Conclusion: Competitiveness through Radical Energy Productivity 158

5. ELECTRICITY: REPOWERING PROSPERITY 164
Imagining the Next Electricity System 167
Maintain: The Elusiveness of "Business-as-Usual" 171
Migrate: The Conventional Approach to "Carbon-Free" Electricity 180
Renew: Tapping Nature's Inexhaustible Energy Sources 187
Transform: A Seismic Shift in Scale 202
Four Cases, One Broad Direction 211
How Do We Get There from Here? 217
Conclusion: The Path Forward 222

6. MANY CHOICES, ONE FUTURE 226
Looking Back from 2050 229
How Can Reinventing Fire Evolve Smoothly? 232
How Do We Seize the 2050 Prize? 246
Kindling the New Fire 250

Acknowledgments 252

About the Authors 258

About Rocky Mountain Institute 263

Other Publications by Rocky Mountain Institute 265

Figure Credits 266

Notes 267

References 289

Index 322

ABOUT THIS BOOK

Reinventing Fire offers a roadmap for navigating the United States' economy through the end of the fossil-fuel era. It is the fruit of several years' work by dozens of scientists, engineers, architects, economists, business experts, and other practitioners at Rocky Mountain Institute (RMI)—an independent, nonprofit think-and-do tank that drives the efficient and restorative use of resources by transforming design, busting barriers, and spreading innovation (see page 263). The Reinventing Fire synthesis and this book, summarizing its findings for U.S. business and other leaders, have both been extensively reviewed by outside experts (see pages 253–254).

RMI launched its Reinventing Fire initiative to answer two questions: Could the United States realistically stop using oil and coal by 2050? And could such a vast transition toward efficient use and renewable energy be led by business for durable advantage? The answer to both questions proved to be yes.

Those answers emerged from combining conventionally accepted data about future energy needs and production with reasonably conservative projections of what our energy future could look like in 40 years if businesses adopted currently available technologies at normal rates of return, and if policy shifts removed some of the current barriers to adopting energy and design innovations. Surprisingly, our analysis did not depend on pricing carbon or any other externality; it required no act of Congress nor any new national taxes, subsidies, or mandates; and it cost $5 trillion (in 2010 net present value[1]) *less* than business-as-usual, creating big profit opportunities.

Reinventing Fire's narrative centers around two big stories—oil and electricity. Burning oil and fueling power plants (nearly half with coal) each release over two-fifths of America's and the world's fossil carbon. Nearly three-fourths of our oil fuels mobility, while nearly three-fourths of our electricity runs buildings (the rest powers industry). Thus efficient *transportation*, *buildings*, and *industry* are the keys to saving oil and coal, as well as much natural gas that can substitute for both, and these shifts in turn make possible major shifts in how *electricity* is made. Chapters 2 through 5 explore these four sectors, reflecting what the RMI team has learned through three decades of strategic and technical collaboration with leading firms in all four sectors around the world. This global experience suggests that all the technologies and many of the U.S.-oriented policy suggestions in this book should be widely adoptable or adaptable elsewhere.

We wanted to make *Reinventing Fire* uncluttered and easy to read, yet keep its detailed technical analyses transparent, scrutable, credible, and documented. You'll find our basic references at the back of this book, and supporting methodological and technical material at www.reinventingfire.com, which is also a portal to RMI's own initiatives for implementing the Reinventing Fire strategy in the four key sectors. A few basic conventions deserve mention here.

Throughout, we express value in 2009 U.S. dollars (except where noted), and we choose technologies that make money at hurdle rates appropriate to each sector's risk-reward tolerance. Since automakers' decisions are driven by relatively shortsighted auto buyers, for instance, we assume their fuel savings must pay back within three years. For efficiency gains in buildings, our hurdle rate is 7%/y real, reflecting longer time horizons. In the even riskier and more fiercely competitive industrial sector, we screen efficiency gains at 12%/y real. For electricity generation, we use 5.7%/y real, reflecting many shareholder-owned utilities' weighted-average cost of capital. Throughout, we present-value the societal value of costs and benefits back to 2010 at the social (3%/y real) discount rate prescribed by the U.S. Office of Management and Budget for valuing the federal government's investments in energy efficiency. We assume the same robust economic growth and activity levels as the U.S. Energy Information Administration's 2010 Reference Case forecast to 2035, extrapolated to 2050, and use that baseline to compare our proposed alternatives.

Readers are invited to send any proposed improvements or corrections to rfsuggestions@rmi.org to help us refine later editions, and to send queries to rf@rmi.org.

PREFACE

Imagine fuel without fear. No climate change. No oil spills, dead coal miners, dirty air, devastated lands, lost wildlife. No energy poverty. No oil-fed wars, tyrannies, or terrorists. Nothing to run out. Nothing to cut off. Nothing to worry about. Just energy abundance, benign and affordable, for all, for ever.

That richer, fairer, cooler, safer world is possible, practical, even profitable—because saving and replacing fossil fuels now works better and costs no more than buying and burning them.

We just need a new fire.

The old fire nurtured our ancestors for the past few million years. As glaciers retreated and woolly mammoths roamed, shaggy fur-clad humans warmed their families and cooked their food at wood-fueled hearths. Later some humans gathered lumps of coal from beaches and outcrops, scooped up oil from natural seeps, and, in China 2,400 years ago, drilled down nearly a mile for natural gas and some liquid hydrocarbons, delivered in bamboo pipes.[2] But nearly all energy in the world came from wood, sun, wind, water, draft animals, and brute-force human toil. Life was short and hard. Winters were cold. Nights were dark.

To varying degrees, nearly half of our fellow human beings still live in that medieval world. One and a half billion have no electricity; they inhabit the vast dark spaces on the satellite photos of Earth at night. Three billion cook over smoky wood, dung, or charcoal fires. But for the more fortunate four billion of us, over the past two centuries fossil fuels have changed everything. Just as fire made us fully human and agriculture made possible cities and states, fossil fuels made us modern. They transformed energy from a preoccupation with personal scavenging to a ubiquitous commodity continuously delivered by extraordinary specialists, esoteric attainments, unthinkably huge machines, the world's largest corporations, and the world's vastest industry.

That industry, invisible to most of us, has become immeasurably skillful and powerful. It delves miles beneath continents and oceans. It inverts mountains. It smoothly delivers sophisticated energy carriers like gasoline, diesel, jet fuel, natural gas, and electricity to our buildings, vehicles, and factories. It is the foundation of our wealth, the bulwark of our might, the unseen metabolic engine of our modern life. Whenever we drive a car, flip a switch, or heat a house, we enjoy its widely affordable potency, convenience, versatility, and reliability. Without fossil fuels, or a similarly capable alternative, most of us would quickly start to experience the struggle for survival that only the world's poorest still suffer daily.

Yet this enabler of our civilization, this magic elixir that has so enriched and extended the lives of billions, has also begun, ever less subtly, to make our lives more fearful, insecure, costly, destructive, and dangerous. Its growing costs and risks erode, and at times may even seem to exceed, its manifest benefits. It puts asthma in our children's lungs and mercury in their lunchbox tuna. Its occasional mishaps can shatter economies. Its wealth and power buy politicians and dictate to governments. It drives many of the world's rivalries, corruptions, despotisms, and wars. It is changing the composition of our planet's atmosphere faster than it has changed at any time in about the past 60 million years.

In short, our rich legacy of fossil fuels is starting to undermine the very security it built. Military leaders, among our society's most farsighted risk managers, are worried. In February 2010, the lead feature article in *Joint Force Quarterly*, the magazine of the chairman of the Joint Chiefs of Staff, began:

> *Energy is the lifeblood of modern societies and a pillar of America's prowess and prosperity. Yet energy is also a major source of global instability, conflict, pollution, and risk. Many of the gravest threats to national security are intimately intertwined with energy, including oil supply interruptions, oil-funded terrorism, oil-fed conflict and instability, nuclear proliferation, domestic critical infrastructure vulnerabilities, and climate change (which changes everything).*[3]

A year later, the chairman responded with a call to energy action for security and prosperity—a mission the Pentagon is increasingly helping to lead.[4]

Another threat, too, hangs over the global energy system: the ultimate certainty of fossil fuels' physical and economic depletion. Only its timing is in question. Despite prodigious technological progress in finding and extracting fossil-fuel deposits—exploration geologists now enjoy the digital equivalent of X-ray eyes—the round earth is not getting any bigger. The easy oil is rapidly dwindling and concentrating in fewer countries; the easy coal has only decades left; the huge deposits of U.S. natural gas, trapped in dense shale rock, that are now starting to be exploited are contained in bubbles finer than a human hair. As economists (and some geologists) start to understand how oil-reserve data were widely misinterpreted or misreported,[5] opinions of fossil-fuel abundance are shifting rapidly. In late 2010, the International Energy Agency said world crude-oil output had already peaked in 2006;[6] the Pentagon's Joint Forces Command warned that surplus capacity could disappear by 2012 and urged readiness for an oil-free military by 2040.[7] The same story is emerging even for coal, long thought too abundant to survey carefully.[8] Whether from the perspective of economic geology, affordability, security, or side effects, the Age of Fossil Fuels, viewed in the longer sweep of human civilization, is just a blip about two centuries long.

The fossil-fuel party is drawing to a close. It's time for something completely different.

What might that new fire look like?

The old fire was dug from below. The new fire flows from above. The old fire was scarce. The new fire is bountiful. The old fire was local. The new fire is everywhere. The old fire was transient. The new fire is permanent. And except for a little biofuel, biogas, and biomass, all grown in ways that sustain and endure, the new fire is flameless—providing all the convenient and dependable services of the old fire but with no combustion.

That sounds daunting. Yet, as the reflective Republican secretary of health, education, and welfare John Gardner said when he joined President Lyndon Johnson's Cabinet in 1965, "what we have before us are some breathtaking opportunities, disguised as insoluble problems."

The problems of fossil fuels are not necessary, either technologically or economically. We can avoid them in ways that tend to *reduce* energy costs—because technological progress has quietly been making fossil fuels obsolete.

About 78% of all human activity[9] is fueled by digging up and burning the rotted remains of primeval swamps. But today we have alternatives more modern than sucking up and burning decayed muck hundreds of millions of years old. The same ingenuity and entrepreneurship that now scrape the bottom of the barrel from the ends of the earth can instead energize and enhance our own lives, and enrich the lives of the world's teeming billions, at little or no extra cost and often—even pretty generally—at a profit.

In fact, the new fire will enrich society by many trillions of net dollars in cold, hard cash. These pages will explain how—and what you can do to capture your piece of that once-in-a-civilization opportunity. For at root this is a story not of energy and fear but of energy and hope; not of restrictions and mandates but of choices and enterprise; not of danger and impoverishment but of security and wealth creation.

The new fire described here combines two elements: it uses energy very efficiently, and it gets that energy from diverse and mainly dispersed renewable sources. But this twin transition to efficiency and renewables, already under way and accelerating, isn't just about the old "what"—technology—and the old "how"—public policy. Technology and public policy are important and rich with innovation, so we'll have a lot to say about them: existing ones needing adoption, emerging ones needing refinement, on-the-horizon ones needing development. But they are less than half the story. Today's energy transition is also, often even more, about the new "what"—integrative design that combines technologies in unexpected ways—and the new "how"—novel business models and competitive strategies. In each of these four areas, important innovations are converging to create perhaps the biggest flood of disruptive opportunities ever seen, with effects as pervasive as those of the Information Age but even more fundamental.

These four tools for energy transformation total far more than the sum of their parts. Together, as we'll see, they can create the greatest business opportunity of our time—indeed, of all time. The human species has begun the most important infrastructure shift in its history, melding energy with information technology and new ideas, blending technical with social breakthroughs, creating one astonishment after another and then merging them into still more. And this crucial next decade, even the next few years, is when most of the big bets will be placed, the foundations of their success laid, their outcomes set into motion to unfold into the midcentury.

Business-as-usual is no longer an option: Too much is changing too quickly. The new energy era is already rising up all around us. We must gaze piercingly, understand deeply, plan humbly, and act bravely.

This book outlines how we can grasp the shift to the new fire, speed it, integrate it, and help steer it along advantageous paths to prudent destinations. This shift will not be easy, but it can be easier than not making it. Safer, too. Because with these striking opportunities come equally stark uncertainties, risks, and dangers. Business leaders should ask themselves questions like these:

- How would your business work without oil, on just a few weeks' or days' notice?
- What would your firm do if the lights didn't turn on tomorrow morning—or next year?
- Do you understand the implications of vastly higher energy prices and price volatility for your company, your customers, and your suppliers?
- How can you win by *eliminating* your energy operating costs—before your competitors do?
- What pieces of the multi-trillion-dollar new energy economy do *you* intend to get?

Such questions are not fanciful. They reflect both the opportunities this book describes and the existential risks, already visible, that insidiously grow every day we fail to act. Judicious daring is the opportunity and responsibility of every leader with resources and the imagination to invest them differently. And what we most need now is not just management but leadership.

Today, one energy and business system is dying, and another struggling to be born. In the midst of such turbulence, standing firmly in one place can seem comfortable. There are always many beguiling reasons not to change. T. S. Eliot wrote, "Between the idea / And the reality / Between the motion / And the act / Falls the Shadow." Yet the ground is shifting. We cannot just stand here. Unquestionably, between today's and tomorrow's energy systems loom inflection points where immense fortunes will be made and lost. Which will be your legacy?

When your shareholders in this decade, and later your grandchildren in your retirement, ask what you did to meet humanity's supreme energy challenge, how will you answer?

And when you're at work, how steeply do you discount your great-granddaughter's future?

My colleagues and I hope this book will help guide, support, inspire, and speed you in creating and exploiting your important piece of the new energy solution for us all.

—Amory B. Lovins
Old Snowmass, Colorado
August 2011

FOREWORD

BY MARVIN ODUM

The idea of "Reinventing Fire" appeals to the mind and to the imagination as much as it applies to the current energy landscape. I commend Amory Lovins for broaching the idea that energy—in its most primitive form and as a fundamental component of the human existence—*can* be reinvented, and then advocating that it *must* be.

Over the past several years, the energy arena has undergone dramatic changes—both in terms of the demand expressed around the world and the sources available to meet that demand, like the technological innovations that have produced vast new supplies of natural gas in North America, for example.

However, our current energy system is not without an array of hidden costs and disadvantages—and there is room for exponentially more change than we have already seen. Global energy demand is going to *continue* to grow. Developments in oil and gas, along with growth in alternatives, will provide us with more energy, but not enough to meet future demand on their own. So we are left with a "zone of uncertainty"—a gap that must be filled in a way still to be identified.

I have always believed that no source should be excluded from the future energy mix—including fossil fuels. As the source of the majority of society's energy, fossil fuels have driven the growth and success of modern society, stimulating economic activity and enabling prosperity and higher standards of living for millions of people around the world. That is why by 2012, my company will actually produce more natural gas than oil—and why we're the world's largest distributor of biofuels and also have a significant wind business. As a global society, we need it all.

In *Reinventing Fire,* Amory offers compelling arguments and perspectives on the challenges and resistance that often come with change—as well as the rewards that are possible. His vision is an important stimulus for honest dialogue, the urgency of which grows more pressing all the time.

Our biggest impediment to "Reinventing Fire" may be our own economic and technological inertia. As in physics, it takes energy to steer something away from its present course. And the good news is that a few things are already moving in this new direction and building momentum.

—Marvin Odum
President, Shell Oil Company
Director, Upstream Americas
June 2011

FOREWORD

BY JOHN W. ROWE

In *Reinventing Fire,* Amory Lovins and Rocky Mountain Institute envision a world where our energy supply—both for transportation and electricity—is transformed. It is a world that no longer relies on fossil fuels. This new energy paradigm produces very little pollution and minimal greenhouse gases, and oil spills and coal mine disasters are a thing of the past. *Reinventing Fire* makes a business case for why this transition can be both profitable for businesses and affordable for consumers. The book, like all of Lovins's work, will spur debate and help to change the way we think about our energy future.

Burning fossil fuels is not free, and it is imperative that as a nation we address this problem head on. The science of climate change is definitive as spelled out by many reports by the National Academy of Sciences. But carbon is not the only harm that comes from burning fossil fuels. Burning coal emits sulfur dioxide, nitrogen dioxide, particulates, mercury, arsenic, lead, hydrochloric acid and other acid gases, dioxins, and the other toxins that are harmful to human health. According to the National Research Council, just three of these pollutants cause $62 billion per year in damages. In addition to reducing pollution, moving away from fossil fuel creates national security and economic benefits. We cannot sustain our dependence on them for all these reasons, and this book presents one pathway to eliminate our reliance on these harmful sources of energy.

Reinventing Fire outlines a possible future for each sector of the economy—transportation, buildings, industry, and electricity. Lovins has been a visionary in the transportation sector, having seen the need for and development of hybrid electric vehicles long before they became en vogue. In the following pages, he calls for the next generation of vehicles—lighter, more efficient ones that run on cleaner fuels like electricity.

Improving building efficiency is a key element of Lovins's plan to get to a fossil-fuel-free world. We at Exelon have seen the benefits of increasing the efficiency of our buildings first hand. Since 2007 we have reduced the energy consumption in our buildings by 25.2% over 2001 levels. We have done so through behavioral changes, retrofitting lighting and HVAC systems, and utilizing the LEED criteria when renovating our spaces. Exelon currently has 10 LEED-certified facilities, including our renovated headquarters in Chicago. A number of the investments we have made delivered a payback in less than one year, and our corporate headquarters renovation resulted in an energy savings of more than 40%. Efficiency, as the book suggests, does pay.

As *Reinventing Fire* recognizes, the old model of electricity generation is changing due to new developments in technology. This book outlines one possible future in 2050 that is dominated by renewables, both large-scale and distributed generation, efficiency, smart grids, and microgrids. This is one possible future, but there are many others. New nuclear and coal with carbon capture and sequestration are currently uneconomical at today's market prices and will likely be for the next decade—but they may eventually become economical as conditions change. Efficiency is an untapped resource in any scenario. Renewables will play an ever-increasing role in our energy mix, but without significant advances in storage, they cannot do it alone. Natural gas, an abundant, inexpensive, domestic resource, will play a key role as the bridge to whatever energy future prevails. For the next decade it is the most economical way to meet our environmental, economic, and security challenges.

Reinventing Fire is a valuable contribution to the debate about what our energy future should look like. It will drive new thinking and conversations in and among businesses, nongovernmental organizations, and policy makers. Whether the future as outlined by Lovins and his colleagues comes to fruition remains to be seen. What is certain is that the energy world as we know it must and will change.

—John W. Rowe
Chairman and CEO, Exelon Corporation
June 2011

CHAPTER 1

DEFOSSILIZING FUELS

FIG. 1-1.

We learn by going where we have to go.

—Theodore Roethke (1908–1963)

You cannot travel the path until you have become the path.

—Gautama Buddha (ca. fifth century BCE)

Our country has met many great tests. Some have imposed extreme hardship and sacrifice. Others have demanded only resolve, ingenuity, and clarity of purpose. Such is the case with energy today.

—National Energy Policy Development Group, 2001

INTRODUCTION

Think about how much our world has changed in just the past two hundred years. We can fly around the globe in less than two days, buy Chilean grapes at a local supermarket, enjoy similar amenities whether we choose to work in skyscrapers or in rural cabins, build mammoth machines and tiny chips, and use a handheld device to connect instantaneously with people in virtually every distant corner of the planet. All these advances have sprung from humankind's restless imagination and ingenuity—and from one crucial enabler: fossil fuels.

Before the fossil-fuel era, people could only harvest energy on the spot. They pumped water with windmills, plowed fields with draft animals, cooked with wood, and explored the seas with the wind in their sails. None of these sources could provide portable energy on demand. But coal and oil and natural gas were different. They store in concentrated form immense amounts of ancient sunlight accumulated over vast areas and geological periods. Tapping into these deposits of plants and animals that grew tens or hundreds of *millions* of years ago allows us to grab and use amounts of energy unimaginable to our ancestors. It's like being able to live a life of wealth by withdrawing vast sums of money from a gargantuan bank account that took eons to accumulate.

Fueled by those ancient deposits, ordinary people in industrialized countries now use up to 100 times more energy than their predecessors did before the burgeoning use of coal in the 1700s, the birth of commercial oil in 1859, and the gas pipelines of the early 1920s. Each of us, in effect, has been able to harness the energy equivalent of several hundred human workers, not to mention the hundreds of horses under the hoods of our cars.

The effect has been profound. Fossil fuels made possible James Watt's steam engine and the Industrial Revolution, the growth of cities, the Internet Age—indeed modern civilization. Between 1800 and 2000, "total worldwide energy use grew by 80–90-fold, the most revolutionary process in human history since domestication" of plants and animals, writes Georgetown University historian John R. McNeill.[10] Adopting fossil fuels, he adds, was "one of the three or four most crucial 'choices' in the history of our species, and more than anything else has shaped the tumultuous relationship between human society and the ecosystems on which it depends." The modern age represents such a break from past planetary history that many scientists say we now live in a new human-driven "Anthropocene Era."

But now it's time for another historic break from the past. In this next revolution, we need to stop using fossil fuels. This book will show how to eliminate oil and coal completely by 2050, with less risk and less cost to society than business-as-usual. Natural gas use can be moderated and ultimately phased out too.

Why would we want to take such a drastic step? Why give up on gasoline stations, coal-fired power plants, and the other fossil-fueled founts of energy that have made our lives so much easier and wealthier than our great-grandparents' or even our grandparents'? Why not just keep pumping up the oil, drilling the gas, mining the coal—in all, the four-odd cubic miles a year of that magic carboniferous stuff that keeps the global economy humming? Why go to the trouble of Reinventing Fire?

The short answer: to create wealth (trillions of dollars' worth, as we'll see in the following chapters), manage risk, capture opportunity and choice, and expand innovation and jobs. But the broader business and social reasons for farsightedly achieving displacement before depletion are compelling. Beyond opportunities for profit, they include correcting structural weaknesses in our economy and threats to our health and our way of life. Indeed, the list of reasons to wean ourselves from fossil fuels is long and strong. Let's start with basic economics.

THE TRUE COST OF OIL ADDICTION

The U.S. has built its economy on cheap oil and coal. At late 2010 prices, gasoline was cheaper per gallon than milk, orange juice, and domestic bottled water. This low price is a tribute to the oil industry's remarkable skill, not just in its complex operations and technologies but also in the politics of keeping subsidies high and fuel taxes low. In 2006, federal subsidies to oil and gas, mostly oil, totaled about $39 billion,[11] shifting several dollars a barrel (abbreviated "bbl") from pump prices to income taxes and budget deficits financed by borrowing abroad. Subsidies to oil-*using* systems are even bigger, estimated in 1998 at $111 billion a year for autos alone, equivalent to $16/bbl.[12]

But this seemingly cheap fuel has been a dangerous illusion. The low price covers just a fraction of the total costs society actually pays to mine and burn fossil fuels. When all those costs are included, the true price soars to well above the cost of renewable alternatives. The surprise, though, is that, as we'll see, practically all energy efficiency initiatives and many renewables are already cost-competitive even *without* counting those hidden costs—and indeed our analysis values them all at zero, a conservatively low estimate. Switching from fossil fuels is often justified today just on head-to-head price competition, and renewables are widening that gap. But switching also avoids the hidden costs, and they are large.

Consider that the low U.S. price of gasoline, about half to one-third what's normal in other industrialized countries, has helped to create a pervasive pattern of inefficient vehicles and settlement patterns that maximize driving, causing a massive treasure transfer from America to oil-exporting nations. Of America's $0.9 trillion[13] oil bill in 2008, $388 billion went abroad. Some of this money paid for state-sponsored violence, weapons of mass destruction, and terrorism. This wealth transfer also worsens U.S. trade deficits, weakens the dollar, and boosts oil prices even higher as sellers try to protect their purchasing power. It's equivalent to a roughly 2% tax on the whole economy, without the revenues.[14] Since 1975, America's oil imports have sucked well over $3 trillion of cash out of other expenditures and investments. Much of it has never returned. Dr. David Greene of Oak Ridge National Laboratory graphs this wealth transfer and its costs, including the economic damage (called "dislocation losses") from oil price shocks, in figure 1-2.[15]

The economics are worsened by the pricing power of the Organization of Petroleum Exporting Countries (OPEC). The OPEC cartel can charge above-free-market prices so long as America's oil addiction helps keep markets tight. In 2000, Greene calculated that over the past three decades this had probably cost Americans more than a year's worth of GDP and depressed total cumulative real GDP by about 10%.[16]

Price shocks alone have cut the GDP growth rate by about half a percentage point. High oil prices have preceded every recession since 1973[17] and put mobility industries (such as automakers, trucking companies, airlines, and tourism) at special risk. At least until the past decade, the effects of oil shocks were amplified because the inflation they helped trigger typically caused the Federal Reserve to raise interest rates, choking off economic expansion.

Oil-price volatility would still rattle the U.S. economy even if all oil were domestic, because oil prices would still be set in the world market. And derivatives markets tell us what that volatility is worth—what you'd have to pay a trader to bear your price risk. For example, price volatility over the next five years was recently expected to cost $40 a barrel for crude oil and $124 a barrel ($2.95 a gallon) for gasoline. These costs of volatility respectively add 40% and 47% to those commodities' spot prices five years out[18] —even more over the much longer period appropriate for big energy investments. Thus 2010 U.S. gasoline consumption alone bore a hidden volatility cost (over just five years) of $0.41 trillion, or well over a half-trillion dollars for total oil consumption.[19] That microeconomic cost is borne directly by oil users as a business risk and appears to be mainly additional to the macroeconomic costs shown in figure 1-2.

In total, U.S. oil dependence's economic cost just in 2008 was on the order of a trillion dollars *beyond the cost of the oil itself*. The only escape is to stop using oil.

FIG. 1-2. Estimated direct costs of oil dependence to the United States, 1970-2008[20]

OIL AND INSECURITY

Looming even larger are added national-security costs. Our petrodollars, notes former CIA director R. James Woolsey, have effectively funded both sides of recent U.S. wars. The United States has diverse interests at stake in the Middle East, of course, but surely it wouldn't have sent a half-million troops to Kuwait in 1991 if Kuwait just grew broccoli. And as Alan Greenspan said in 2007, the Iraq War "is largely about oil." That war has already cost more than 4,400 U.S. lives, plus one to several trillion borrowed dollars.[21]

The costs mount even without war. In the 1990s, the U.S. paid two to three times as much to maintain forces poised to intervene in the Persian Gulf as it paid to buy oil from the Persian Gulf.[22] In 2000, if oil imports from the Gulf had been charged with those costs, they'd have been priced $77/bbl higher, or 2.7 times that year's price of Saudi crude.[23] Those costs count only the Gulf-focused Central Command, yet *every* Combatant Command has oil-protection duties, whether fighting tanker-hijacking pirates off Somalia, instability in Latin America, or militants blowing up pipelines in Faroffistan. In 2010, a Princeton study[24] pegged the cost of U.S. forces just in the Persian Gulf in just one year (2007) at half a trillion dollars, or about three-fourths of the nation's total military expenditures.[25] That's similar to the peak expenditure rate for the Cold War. It is also about ten times what the U.S. typically pays for all the oil it imports from the Persian Gulf.

So the economic costs of oil dependence, plus U.S. military expenditures for Persian Gulf forces (and minus the cost of the oil itself), total roughly $1.5 trillion a year, or 12% of GDP—far more than our total energy bill. This hidden surcharge, paid not at the pump but through business risks, taxes, and deficits, exceeds the per-gallon U.S. price of gasoline. It also rivals or exceeds the gasoline

THE NATURE OF THE U.S. OIL PROBLEM

The United States still has both coal and natural gas in relative abundance, but no longer oil: U.S. oil output is back down to pre-1950 levels. With 4.5% of the world's population, the U.S. produces about 23% of world GDP, and it uses 22% of world oil but supplies only 11% and owns just 2%, so we can't drill our way out of depletion. Indeed, after one and a half centuries of skilled and nearly ubiquitous drilling, U.S. oil resources are now so depleted that a new domestic barrel consistently costs more than an imported barrel. This leaves only three ways for a market economy to avert rising import dependence:

- *Protectionism* distorts oil prices by taxing foreign oil or subsidizing domestic oil (the latter also retards efficient use and substitution). This violates free-market and free-trade principles, sacrifices competitiveness and economic efficiency, and illogically supposes that the solution to domestic depletion is to deplete faster—a policy that conservationist and retired Army major David R. Brower called "Strength Through Exhaustion."
- *Trade* simply buys oil unsentimentally from the lowest bidder, as the U.S. did for 49% of its 2010 use and as most other countries do—both those good at earning money to buy oil, like Germany and Japan, and the rest, like most developing countries. Trade can be economically efficient and is generally a sound concept; it underpins the global economy on which the U.S. depends. But when a resource is vulnerable or unreliable, trade makes everyone share shortages and price spikes as well as surpluses. Trade also creates dependent relationships and imposes political and diplomatic constraints that a superpower may find onerous.
- *Substitution* replaces oil with more efficient use or alternative supplies whenever they're cheaper. These domestic substitutes offer the advantages of protectionism without its drawbacks, and the advantages of trade without its vulnerabilities. Since substitution is cheaper and less risky than buying oil in the world market, it's the focus of chapter 2's exploration of how to achieve mobility without oil.

taxes that most industrial countries' citizens pay to their national treasuries, while Americans pay their version mostly to oil suppliers (Canada, Mexico, OPEC, and others), military contractors, and foreign lenders.

Meanwhile, the whole oil supply chain is astonishingly vulnerable. For example, failing to foil just one of the steady stream of terrorist plots against key Saudi oil facilities could crash the global economy. Two-thirds of Saudi oil flows through one processing plant and two terminals, both already repeatedly attacked. Yet oil chokepoints are spread around the world, including in the United States. A Pentagon study found that a handful of people in one evening could cut off three-fourths of the oil and gas supplies to the eastern U.S. without even leaving Louisiana.[26] Drilling for more oil at home (see The Nature of the U.S. Oil Problem sidebar) could simply shift oil insecurity from well-known instabilities and chokepoints abroad[27] to equally brittle domestic infrastructure, such as what Woolsey called the "frighteningly insecure" Trans-Alaska Pipeline.[28]

Want even more reasons? Dependence on oil gives disproportionate power to exporters like Iran. It hurts America's reputation around the world (since many foreign-policy issues look like they're just about oil, eroding U.S. moral authority). It hinders the spread of democracy: 67% of oil reserves are in "unfree" nations, 25% "partly free," and only 7% "free," as rated by Freedom House.[29] With few exceptions—above all Canada, which supplied 37% of U.S. oil imports in 2010—countries that depend heavily on oil revenue tend to have more corruption,[30] autocratic governments, repression, inequity, and excessive militarization. *The Economist* in 2005 called this "the oil curse."[31] Quite a few oil-endowed countries are unstable: Algeria, Chechnya,[32] Indonesia, Iraq, Nigeria, Sudan, and Yemen all harbor Islamic extremists. And don't forget the environmental and health costs of oil spills.

COAL'S HIDDEN COSTS

Oil isn't the only fossil fuel with large hidden costs. Coal fires the power stations that generate 45% of U.S. and 41% of world electricity. Burning coal emits sulfur and nitrogen oxides (causing acid rain), particulates, mercury, and other toxic metals. Coal ash from power plants pollutes streams. Mining coal injures and kills workers and inverts landscapes. Such hidden costs of U.S. coal-fired electricity total $180 to $530 billion per year. Properly charging that on our electric bills, rather than to our health and our kids, would double or triple the price of coal-fired electricity.[33] Up to half those hidden costs come from conventional air pollution—more than the National Academy of Sciences' estimate of coal's climate cost. And as we'll learn in chapter 5, the transportation and burning of coal is not an entirely secure system, and the grid that transmits its electricity across the country is alarmingly vulnerable to physical attack, cyberattack, and solar storms. But coal's gravest threat to national security may emerge from those power plants' smokestacks.

Burning coal and oil each emit about two-fifths of U.S. and global fossil carbon,[34] incurring the risks of climate change. Military leaders worry that climate change's possible droughts, floods, famines, disease spread, geopolitical shifts, mass migrations, and other likely effects could trigger and amplify serious instability, conflict, and humanitarian-relief missions. Defense Secretary Robert Gates, Chairman Mike Mullen of the Joint Chiefs of Staff, and other military leaders therefore include climate change among the Pentagon's core security concerns.[35] In January 2011, Admiral Mullen commented at the front of *Joint Force Quarterly*, his magazine for military commanders:

> *Near the polar cap, waterways are opening that we could not have imagined a few years ago, rewriting the geopolitical map of the world. Rising sea levels could lead to mass migrations*

similar to what we have seen in Pakistan's recent flooding. Climate shifts could drastically reduce the arable land [available that is] needed to feed a burgeoning population as we have seen in parts of Africa. As glaciers melt and shrink at a faster rate, crucial water supplies may diminish further in parts of Asia. This impending scarcity of resources compounded by an influx of refugees if coastal lands disappear not only could produce a humanitarian crisis, but also could generate conditions that could lead to failed states and make populations more vulnerable to radicalization. These troubling challenges highlight the systemic implications—and multiple-order effects—inherent in energy security and climate change.[36]

Besides, we'll have to stop burning oil and coal—and ultimately even cleaner natural gas at some point—because we can't withdraw from nature's vast deposits forever. Humanity has already withdrawn about one-third of the original fossil-fuel bank balance, and those once-in-a-civilization withdrawals are accelerating: half their total has occurred just since 1985.

While there's a raging debate over whether we've reached or passed the peak of oil production (fig. 1-3), there's no doubt that the cost of finding and lifting new oil is rising, despite ever-better technologies, as those huge reserves dwindle. And whatever its pace, depletion will concentrate oil ownership, raise tensions, boost prices, and

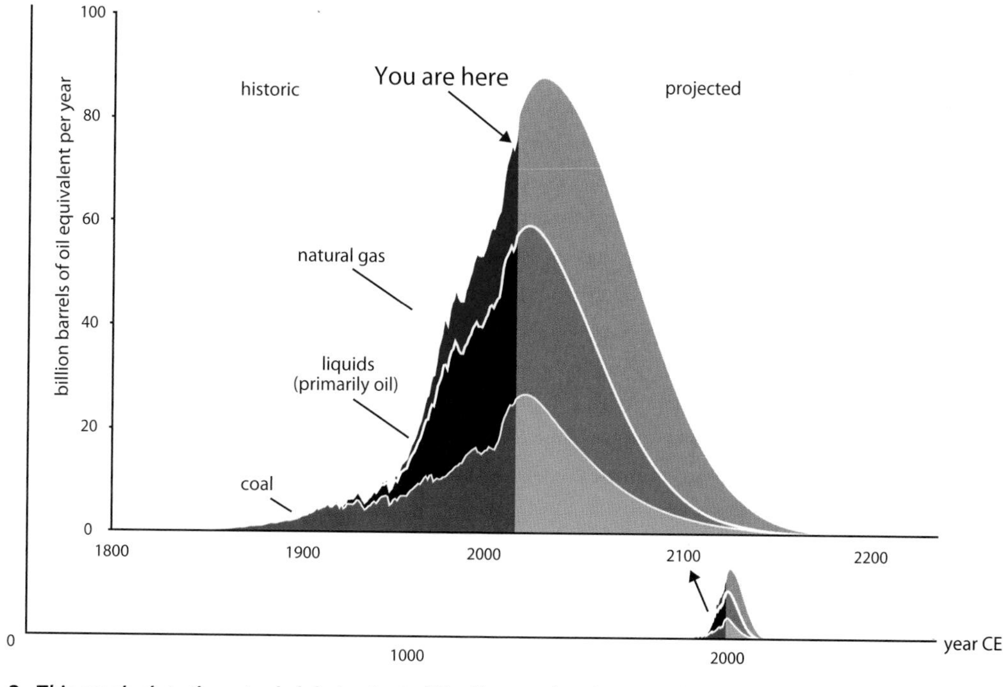

FIG. 1-3. This graph plots the actual global output of the three major classes of hydrocarbons through 2009, then projects the remaining amounts of each believed likely to be recovered if there are no aboveground constraints.[37] The historic data are accurate but the smoothed illustrative projections are quite approximate, reflecting leading resource experts' knowledge in early 2011 but subject to many uncertainties. The projections include unconventional resources such as shale gas, heavy oil, tar sands, and shale oil, but not methane hydrates, potential Arctic and Antarctic resources, or Alaskan North Slope and central Siberian coal.

increase price volatility. If you believe oil production has peaked or is about to, getting off oil is a practical and profitable way to reduce those risks; if not, it's negative-premium insurance, prudently hedging your bets while cutting you costs.

Figure 1-3 also shows how coal depletion, long assumed to be centuries off, may arrive unexpectedly soon. Coal resources had long been assessed with little or no attention to their exploitation cost. Recent reassessments of coal's economic geology are more sobering, suggesting that "peak coal" will occur within decades even in such coal-rich countries as the U.S. and China. Physical depletion could take much longer, but the cheap coal is going fast.

Obviously, weaning ourselves from fossil fuels isn't easy. Every American president since Richard Nixon has vowed to break the country's addiction to imported oil, yet U.S. oil-import dependence hit 60% in 2005 before falling back to 49% in 2010. U.S. coal use also rose from 15 quadrillion BTU in 1980 to nearly 21 quadrillion in 2010. Its mining is deeply embedded in some regions' way of life, and a huge fleet of power stations burns it.

TURNING THE SUPERTANKER

Though most trends are in the wrong direction, encouraging examples prove that reducing or eliminating fossil-fuel use is possible. For instance, the U.S. and others responded to the oil price shocks of the 1970s with a spate of efficiency gains that halved OPEC's sales in just eight years and "broke its pricing power for a decade," Greene found.[38] (Specifically, during 1977–1985, U.S. GDP grew 27%, oil use fell 17%, oil imports fell 50%, and oil imports from the Persian Gulf fell 87%.) Similarly, as we'll see later, coal's share of electricity generation plummeted starting in 2009 as utilities switched to cheaper natural gas, while efficiency and renewables further eroded its market.

Or consider Denmark. A Danish home built in 2008 uses half the energy per unit of floorspace of one built before 1977. The Danish economy grew by two-thirds during 1980–2009, while energy consumption returned to its 1980 level and carbon emissions fell 21%.[39] All new power plants, too, are either renewable or generators of both electricity and useful heat. Of all Danish electricity, 53% is cogenerated and 30% is renewable.[40] The average Dane releases 52% less carbon than the average American. Yet Danes have an excellent quality of life, with the most reliable electricity in Europe, at some of the lowest pretax prices.[41] And in 2011, Denmark's conservative government announced nearly self-financing, and over time highly advantageous, plans to get the country completely off fossil fuels by 2050 by boosting efficiency and switching to renewables.[42]

Motivating 21st-century Danes to reverse their 19th-century switch to fossil fuels are the same forces that are motivating people everywhere. People want their energy services to be secure against being accidentally or deliberately cut off; affordable even in hard times; stably priced; clean and safe; fair, not unduly disadvantaging others; and modern, continuously improving through innovation. Many people are realizing that fossil fuels are no longer the only or the best way to achieve these goals. Business-as-usual with fossil fuel still works, but its costs and risk are increasing, and so are opportunities for change.

Consider autos. A 20-mile round-trip in an average new two-ton auto to buy a gallon of milk burns a gallon of gasoline probably costing less than the milk. The gallon of milk could have come from a nearby cow that did two hours' work yesterday to eat and digest nine dry pounds of grass. But the gallon of gasoline—pumped up, refined, and perhaps hauled halfway around the world—was formed over eons from a *quarter-million* pounds of primeval plants, only a tiny fraction of which formed oil, got geologically trapped, and ultimately was extracted and processed. Thus when your auto burns a gallon of gasoline, it's consuming over 60 times its own weight in ancient plants—17 million times more plant weight per gallon than the cow ate to make the gallon of milk. Why so much? Not just because copious ancient pond scum forms little recoverable oil, but also because the auto's basic physics make it so inefficient that, as we'll see in the next chapter, a mere fraction of a percent of its fuel energy moves the driver. How long can such autos compete in a world of scarcer oil?

Looking at the bigger picture, CODA Automotive's then CEO Kevin Czinger said the backbone of the U.S. economy used to be locally made cars fueled with relatively locally produced oil. That industry sparked America's great industrial growth. But now we often drive imported cars fueled with imported oil, so "every time we buy a car we're exporting $15,000 of capital, paying for it with borrowed money, and running it on foreign energy sources. We've gone from autos' being a middle-class-making machine to a middle-class-destroying machine."[43] Of course, the more we make efficient and competitive cars at home, the less we'll need to import both cars and oil, and the more our dollars and jobs will stay home.

Competitive strategy similarly demands moving beyond wastefully used electricity and the coal burned to make it. Radical efficiency can save much or most of the electricity used in buildings and industry. And renewables have already captured half the global power-plant market,[44] threatening U.S. coal in its last big use. Judicious choices can speed all these trends, benefiting the economy, the environment, and our lives.

The tipping point for shifting from fossil fuels to the new fire has arrived. And the new fire can work worldwide. It can be scaled up quickly in many diverse conditions and cultures.

LIGHTING THE NEW FIRE

This book explains how business—motivated by enduring advantage, supported by civil society, sped by effective policy—can advantageously achieve the ambitious transition beyond oil and coal by 2050, and later beyond natural gas too. New technologies, and new ways of combining them, can wring severalfold more work from the same amount of energy. Those efficiency gains then allow renewable energy sources, equally enabled by modern information technology, to be deployed faster. The transition will create new industries with vast potential for jobs, profits, and better, cheaper, more robust services.

These new energy futures promise total economic costs comparable to or less than those of the old fossil-fueled approach, with a cost advantage increasing over time. Meanwhile, far more importantly than any modest cost differences, they already offer far lower risks to national security, the economy, the environment, and public health, and a rapidly expanding range of choices.

Unfortunately, many barriers lurk in the transition to the new fire, which explain why it's now only just beginning. But as we'll see, clever policies and business strategies can help surmount those barriers, and business logic favors getting in early.

We will explore these themes and ideas in the four sectors of the economy that burn fossil fuels—transportation, buildings, industry, and electricity generation—and see how they intertwine into a whole new structure for the energy system, creating a remarkable range of public and private benefits. We can best illustrate the possible paths to the future by looking first at how these four sectors use fossil fuels, and how the United States used all forms of energy in 2010 (fig. 1-4).

These energy flows show only how fuels were burned for energy, and not how they were used as raw materials to make plastics, chemicals, lubricants, asphalt, fertilizers, and other materials. (When Russian chemist D. I. Mendeleyev said in 1877 that crude oil was far too precious to burn, he was right.) The U.S. Energy Information Administration expects all these combustion uses to grow, as figure 1-5 shows, under its standard assumption of existing laws, rules, and (mostly) technologies.[45]

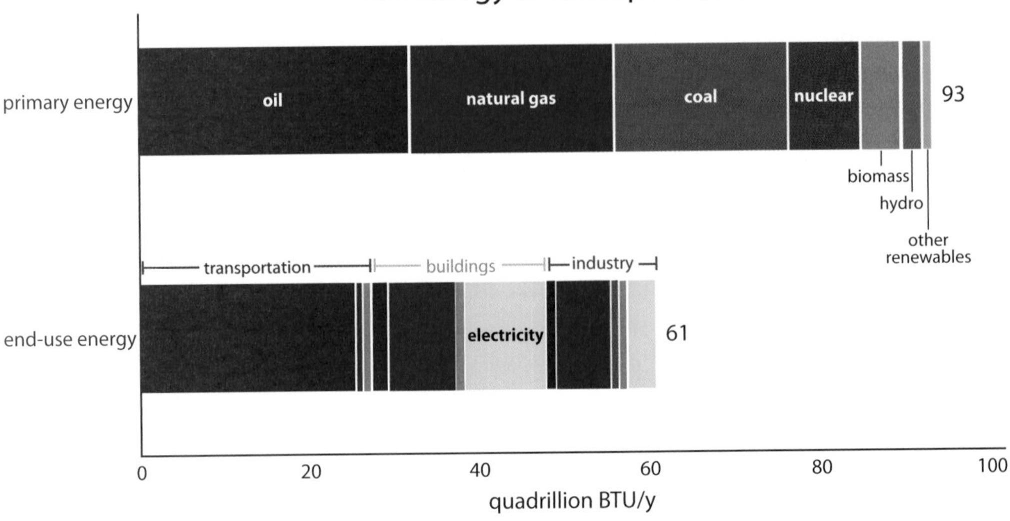

FIG. 1-4. U.S. energy use in 2010, measured in the U.S. in "quads" or quadrillion BTU (million billion British thermal units) and in the rest of the world as a 5.5% larger number of EJ (exajoules, or billion billion joules)[46]

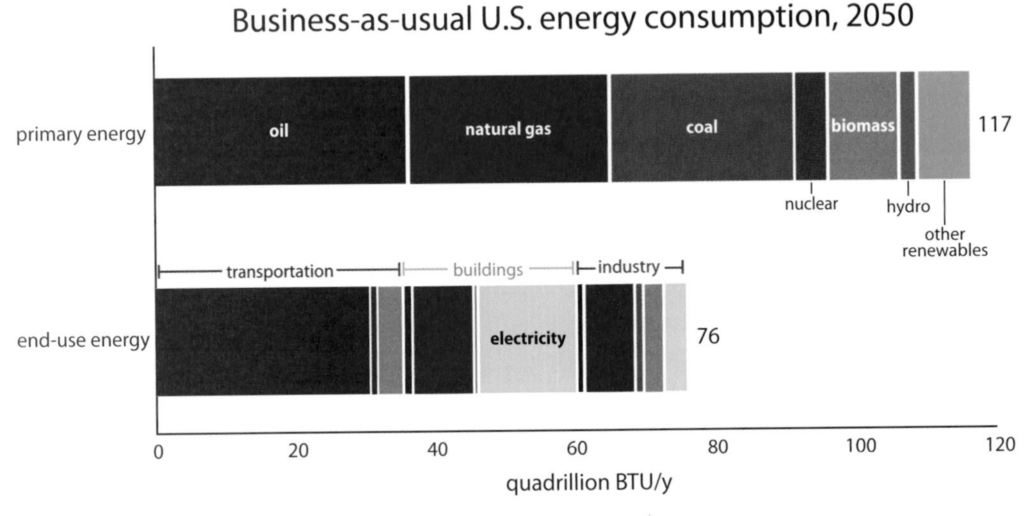

FIG. 1-5. RMI's extrapolation to 2050 from the Energy Information Administration's 2010 forecast of U.S. energy supply and use to 2035.[47]

Yet trend is not destiny. Our energy future could be startlingly different. These uses could shrink and the fuels that drive them shift (fig. 1-6), while providing exactly the same amenities—access, comfort, and industrial production, for instance—and using the same or better vehicles, buildings, appliances, and other equipment as in the 2050 base-case forecast.

While our energy economy is complex and the uses of energy are myriad, we can move away from fossil fuel by focusing on three simple principles: reduce use, modulate demand, and optimize supply. These principles are not new, but they must be applied holistically, with consistent passion and relentless patience.

Do more with less. The cheapest, best "source" of energy is needing less of it in the first place by converting, delivering, and using it more efficiently. By 2009, wringing more work from each barrel, ton, therm, or kilowatt-hour had become the biggest energy source in the U.S. economy. In that year, it fueled half of all economic activity and provided 174% more energy services than oil-burning did. For businesses, this principle, called "raising energy productivity," is a veritable gold mine. It delivers the same or better services every day at lower cost, while also reducing the risk of energy price spikes or supply failures. We can then use less energy but get even more work out of it. Almost unnoticed, the U.S. economy actually did that in nine of the past 36 years by raising energy productivity even faster than GDP grew. The following chapters show how we could do it *every* year—both by wringing more work from our energy and by using more productively the services it provides.

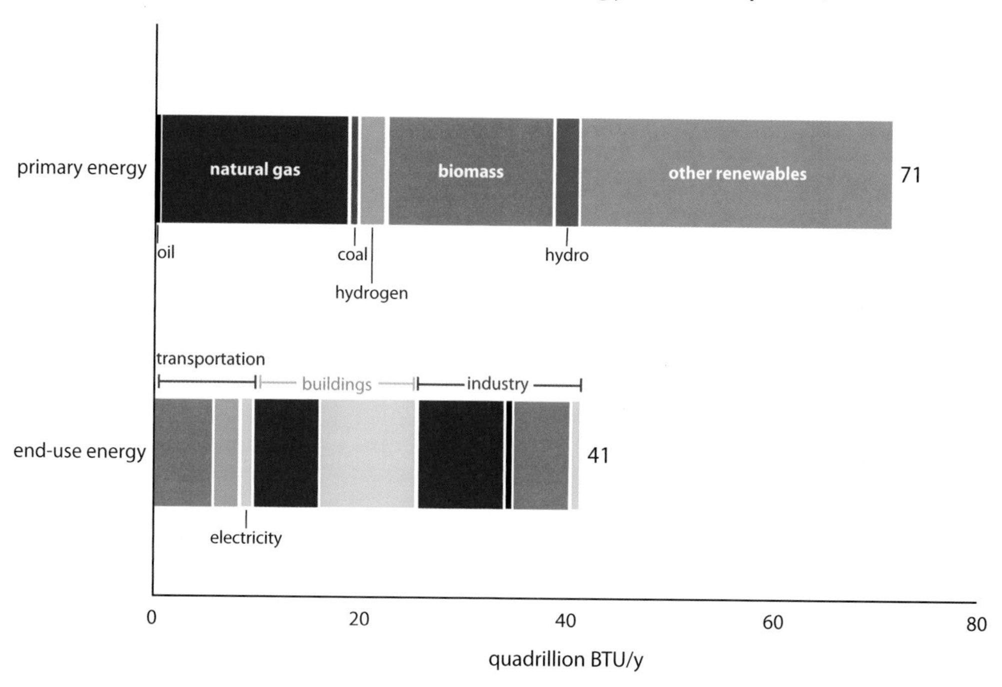

FIG. 1-6. The following chapters will show how we can run the same 2050 economy as in figure 1-5, but with half the delivered energy, with less risk, and for $5 trillion less (in 2010 net present value).[48]

Modulate demand. We have been trained to believe that our economy and our way of life depend on a ceaseless, constant, ever-growing and never-ending supply of electricity. That paradigm is rooted in 19th- and 20th-century thinking, from a time when factories were steeped in mass production and large central generators sought "baseload" customers to utilize their capital every hour of the year. Today, new technologies, smart controls, and IT-enabled services empower us to adjust our energy demands to match more closely and advantageously a wide range of supply technologies. The upshot for businesses is that they can use electricity when it is cheapest and, without disruption or inconvenience, reduce their use and costs when electricity is most precious. For commercial building owners this may mean storing energy in the cool of the night and releasing it in the heat of the day. For other customers it may mean charging an electric vehicle when prices are lowest. Even aluminum producers have begun to modify demand in response to price signals—a practice unheard of a decade ago.

Optimize supply. Once we have dramatically reduced our demand and learned how to control and modify it (all without diminishing or compromising services), we can expand our choices about how to meet it. The old fire left few choices, all of which chiefly involved burning one or another abundant fuel for heat or electricity. Today and in the near future we can choose from an ever-expanding array of options ranging from fuel-switching (coal to gas) to windpower to solar photovoltaics and someday fuel cells. These new options give business new ways to control energy risks, often at ever lower and more stable costs. The rest of us can become both consumers and producers of energy—"prosumers"—as rapid innovation expands our choices too.

NO MIRACLES REQUIRED

The next four chapters explain how to optimize, execute, and combine these three principles in all four energy-using sectors of the economy—transportation, buildings, industry, and electricity generation. But our story is about more than just reducing each sector's fossil fuel use. What's especially exciting is how in each sector, those three principles launch self-reinforcing cycles that bring more gains in the other sectors. In just one example, switching to electric cars to eliminate oil use could also make the electricity grid more efficient and resilient, speeding the electricity sector's path to replacing fossil fire with renewable energy. Such leapfrogs are challenging, but with business leadership and policy support, they're also realistic, already under way, and strikingly rewarding.

If you think things must always remain as they are, remember that history tells a different story. Energy sources have been coming and going for millennia. Renewable energy technologies such as wind and solar have become widespread several times in the past few millennia, only to be displaced by a glut of apparently cheap fuel, which then dwindled, making room for the renewables to be rediscovered.[49]

But some sources go away and never come back, simply because their time has passed. No matter how dominant their role—even if they bestride the earth like a colossus—there comes a time when competitors overtake them. That's been happening for both oil and coal. The U.S. stopped directly using coal for transportation by 1920 and for buildings by 1960. Industrial coal use has halved in the past 40 years, U.S. coal output peaked in energy terms in 1998, and coal's last big use, for power generation, probably peaked in 2007. Now the costs of using oil and coal are climbing, even as the price of renewables drops inexorably. The curves are already crossing. The endgames of oil and coal have already begun.

When upstart competitors knock off the dominant source of energy, the end can be swift and cruel for the established industry. Consider how one type of oil—a versatile, convenient, ubiquitous oil—was snuffed out way back in the mid-19th century.[50] In 1850, most American houses were lit by whale-oil lamps. Today we think of whaling fleets only when we read *Moby Dick* or visit a whaling museum, but in 1850, whaling was America's fifth-biggest industry. The demand for whale oil was so great that whales were getting shy and scarce. But rising whale-oil prices brought competition, mainly from kerosene and gas, both at that time synthesized from coal. Entrepreneurs started selling cheap kits to convert lamps from whale oil to coal oil, repaying their cost within months.

By 1859, when Drake struck oil in Pennsylvania, creating another source of kerosene, more than five-sixths of whale oil's lighting market had been taken over by these competitors, in less than a decade. The inattentive whalers were astounded to find they had run out of customers before they ran out of whales. The whaling industry was reduced to begging for federal subsidies on national-security grounds—and soon American whaling was history. The remnant whale populations had been saved by technological innovators and profit-maximizing capitalists. And within another few decades, kerosene lighting was history too, replaced by Edison's 1879 electric light.

In November 1973, as economies reeled under the shock of the Arab oil embargo, a Texas A&M economics professor named Phil Gramm recalled the whale-oil history in a *Wall Street Journal* op-ed. Don't worry about oil price shocks and shortages, he said. Markets would clear, innovators would invent, and oil, like any other commodity, would in due course be saved or displaced. He was right. By 2009, America was making a dollar of real GDP using 60% less oil, 50% less total energy, 63% less directly used natural gas, and 20% less electricity than in 1975. But this journey toward using energy in better, cheaper ways has only just begun.

It has happened despite general indifference, gridlocked national policy, 26 years of stagnant efficiency standards for automobiles, and 48 states' rewarding utilities for selling more electricity and natural gas. Imagine what we can do together once we pay attention to the trillions of dollars of savings and potential profits lying on the table. So with the whale-oil history fresh in our minds, let's start with the vital function that uses 71% of America's oil—transportation. Right now, oil is the lifeblood of the most mobile society in the history of the world. But we can reinvent the fire that now burns in our internal-combustion engines as well as in our buildings and factories, eliminating oil burning by 2050—all while saving about $5 trillion.

CHAPTER 2

TRANSPORTATION: FITTER VEHICLES, SMARTER USE

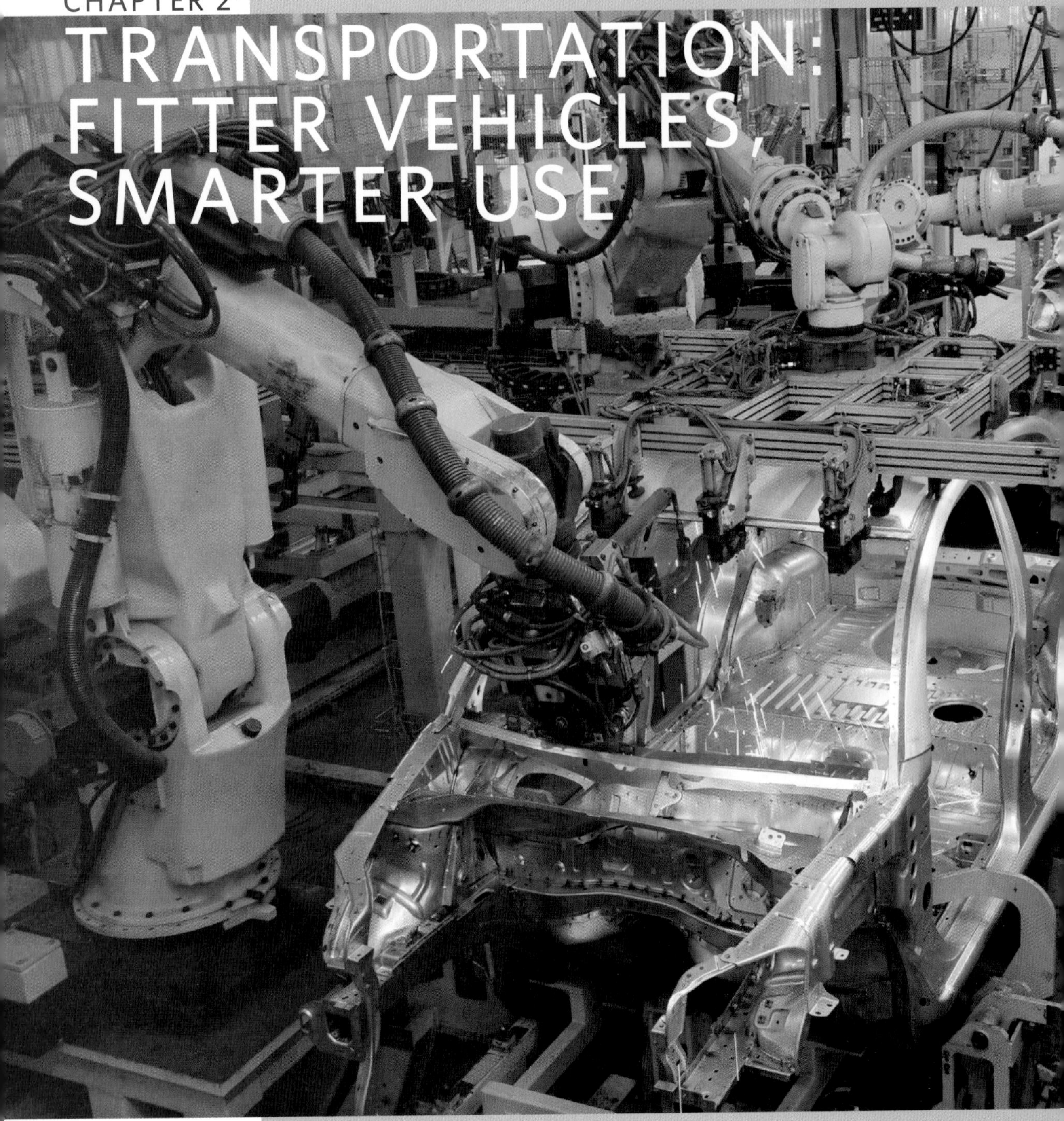

FIG. 2-1.

CHAPTER HIGHLIGHTS

- **THE GOAL.** In 2050, superefficient autos, trucks, and planes, far more productively used, need three-fourths less fuel and no oil and have less life-cycle cost than the vehicles of today. Yet they provide 90% more automobile-miles, 118% more truck-miles, and 61% more airplane seat-miles with uncompromised convenience, safety, and performance.
- **THE BUSINESS OPPORTUNITY.** Radical efficiency enables alternative propulsion and fuels, transforming vehicle manufacturing for breakthrough competitive advantage, while customers save money and mobility options expand. Society eliminates oil dependence, reducing many security and business risks.
- **THE BOTTOM LINE.** Oil not needed saves $3.8 trillion (in 2010 net preset value).
- **BUSINESS SECTORS THAT CAN PROFIT.** Vehicle manufacturers and suppliers, chemical and electronics industries, fleet operators, entrepreneurs, real-estate developers, electric utilities, farmers and foresters.
- **POLICY ENABLERS.** Innovative state, regional, or federal policies can remove barriers to buying superefficient vehicles and using them in smarter ways—without new federal taxes, subsidies, mandates, or laws (with a minor exception about truck weight limits[51]).

Whatever you can do, or dream you can, begin it: Boldness has genius, power, and magic in it.

—JOHN ANSTER, 1835, LOOSELY PARAPHRASING GOETHE'S FAUST, 214–30

The United States burns 13 million barrels of oil a day driving to work, shuttling kids to soccer games, hauling cargo, jetting to meetings and vacations, and keeping its vast transportation system humming. The costs of this use are huge and often hidden. They include oil spills, air pollution, climate risks, and a billion dollars a day plucked from Americans' pockets to buy petroleum from other countries, some of them unfriendly.

Yet burning that oil is simply unnecessary, and the money is largely misspent. We can imagine a world where spacious, peppy, ultrasafe autos sip fuel at 125–240 miles per gallon, but they need no gasoline; where heavy trucks haul goods along the interstates using a third the fuel they do now, but they need no diesel fuel; where planes use severalfold less fuel, but they need no oil either.

This isn't just a dream. It's a clear pathway that requires no technological miracles, only the continued development and adoption of innovations already well under way. Going down this pathway is not just an option but an imperative, because the transportation sector is in one of those rare periods of transformation. Now is the time when smart, light vehicles could take over the roads and skies, radically changing the world's biggest businesses.

The combination of better vehicles and smarter use would bring huge benefits to society—and to the companies that lead the way. Transportation is now America's number two consumer cost after housing, totaling $740 billion in 2009—17.6% of household expenditures. Yet the 13 million barrels of oil (fig. 2-2), roughly a billion dollars' worth, that keep America moving each day are mostly wasted.

Eliminating that waste is a multi-trillion-dollar business opportunity—not just for companies that manufacture vehicles, but also for suppliers innovating new materials and processes, and for investors who get in early and wisely.

If we do this right, we'll make America stronger and safer by keeping that billion-dollar-a-day oil-import cost at home. We'll be less buffeted by volatile oil prices and less anxious to defend access to oil. And if America, which put the world on wheels and wings, sets the pace and pattern for transport innovations in the next stage of global development, we could help head off the nightmare of an auto-choked but oil-starved world.

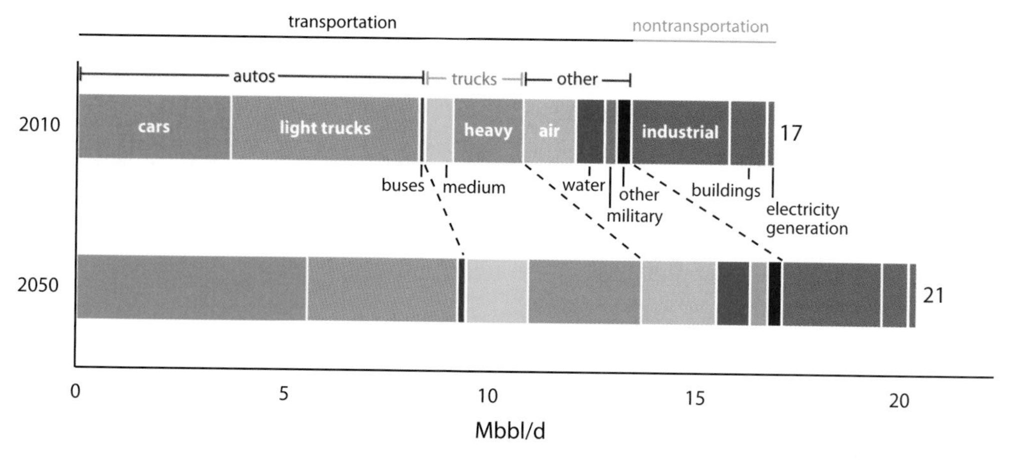

FIG. 2-2. U.S. oil use in 2010 and the U.S. Energy Information Administration's 2035 projection extrapolated to 2050 (our base case). (Only uses that burn oil are shown—not uses of oil as a raw material.) In 2010, transportation used 71% of U.S. oil and was 94% oil-fueled; the rest was 3% biofuels and 3% natural gas to run gas pipelines. Later chapters describe how to eliminate oil's nontransportation uses.[52]

This isn't to say that creating a better, oil-free world of transportation will be easy. For most auto buyers, and until lately some truck operators, fuel efficiency and fuel costs have historically been minor considerations. Up-front costs loom large for individuals, while individual benefits can be small: the fuel savings from switching even to an all-electric auto, though important over years, would barely buy a daily latte, inspiring little sense of urgency. Individual transportation modes often present advantages over public transit, as do road- and air-based shipping over rail- and water-based shipping. And the current pattern of sprawl and traffic congestion is firmly entrenched, thanks to subsidies, mandates, and Americans' own choices. That's why this transformation needs and merits both an extra initial nudge and fairer competition.

The ever graver consequences of our oil addiction leave us little choice. We must design and use our vehicles differently, transforming industries in the process. We can do it, because the technological path is clear, the business case is compelling, and we know new ways to use carefully crafted and light-handed public policies to bust barriers. If we do succeed, the benefits—to customers, to companies, and to society as a whole—will be vast and enduring.

DESIGNING AND BUILDING AUTOS DIFFERENTLY

The key innovation behind this transformation will be a shift to ultralight but ultrastrong autobodies, made of advanced materials. Not only will these bodies be simpler and cheaper to manufacture, they will also trigger snowballing weight savings. With drastically lighter platforms, propulsion systems can be smaller, lighter, cheaper, more efficient, and, for autos, electrified. Several major automakers (and airplane manufacturers) are already adopting or seriously considering this gamechanging strategy. And if the world is going to wean itself from fossil fuels, the rest of the world's vehicle makers need to adopt this strategy as well—or risk falling far behind.

TRANSPORTATION-SECTOR TERMINOLOGY

Mbbl/d is the abbreviation for "millions of barrels [of oil] per day," a common U.S. unit of oil production or use. An oil barrel contains 42 U.S. gallons. **Cost of saved energy (CSE)** is the cost of saving a unit of energy, directly comparable to the avoided cost of the saved energy.

The term "**autos**" is used in this book to refer to all **light-duty vehicles**, which comprise cars, light trucks (sport-utility vehicles [SUVs], pickup trucks, and vans), and **crossover** vehicles (SUVs with sedan attributes), with a gross vehicle weight up to 10,000 pounds (4,537 kg).

A **powertrain** generates an auto's propulsion and delivers it to the surface of the road. A **drivetrain** or **"driveline"** connects the source of torque (like an engine) to the driving axles.

Battery-electric vehicles (BEVs) are powered entirely by electricity. **Fuel-cell vehicles (FCVs)** are powered by a hydrogen fuel cell and electric motors. **Plug-in hybrid electric vehicles (PHEVs)** are powered by both an internal combustion engine **(ICE)** and batteries that can be recharged from an electrical outlet. **Electric vehicles (EVs)** comprise the three previous categories but do not include the popular **hybrid-electric vehicles** that use both a fueled engine and electric motor(s) but don't ever plug in.

A **ton-mile** is equivalent to a ton (in this book, 2,000 pounds) of freight moved one mile. A **seat-mile** refers to one commercial airline seat flown one mile and is used to measure performance standards in the aviation industry. The seat may be occupied or empty. **Vehicle-miles traveled (VMT)** refers to the total number of miles traveled by a vehicle over a given period of time, typically one year.

Physics at Work

By 2011, policies setting vehicle fuel consumption standards at roughly 30 mpg (miles per [U.S.] gallon) for 2016 were starting to shift the market, and 54.5 mpg standards—around 39 on the road—were agreed for 2025. Yet technology has far outrun policy: attractive 125–240 mpg autos can be achieved within a decade, with multi-trillion-dollar net benefits to society. How? The answer begins with the simple physics of automobiles.

Consider these two facts of automotive physics:

1. Less than 0.5% of the energy in the fuel of a typical modern auto actually moves the driver. Six-sevenths of the fuel energy is lost in the propulsion system, during idling while the vehicle is stopped or braking, and to run accessories like air-conditioning and lights. More than half of the remaining one-seventh of the fuel energy that reaches the wheels heats the air that the auto pushes aside or heats the tires and road. Only the last 5% of the fuel energy accelerates the auto. Depending on the type and size of vehicle and the weight of the driver, only about one-twentieth of the mass being accelerated is the driver—so only about 0.3%, and at most 0.5%, of the fuel energy accelerates the driver.

2. An auto's weight is responsible for more than two-thirds of the energy needed to move it. Heavy autos have more inertia, needing more force to accelerate. They also have more rolling resistance because more weight is pushing down on the tires, which therefore lose more energy. As a result, the energy needed to move the auto, called its "tractive load," increases about in proportion to its weight. Heavier autos also need proportionally more powerful engines for the same acceleration. U.S. autos' big engines use only 8% of their power in typical highway driving or just 5% in the city—and this mismatch to normal driving requirements halves their average efficiency.

A LITTLE AUTOMOTIVE HISTORY

The automotive industry is enormous and complex. Building autos is a $1.6-trillion-a-year global enterprise, producing every five seconds or so a shiny two-ton machine with more than 14,000 parts, mostly from a global web of suppliers. The automobile runs extremely reliably for 15 years in all kinds of harsh conditions, costs less per pound than a McDonald's quarter-pound hamburger, and meets conflicting requirements with immense skill honed over 120 years. Making major changes will be very hard.

However, radical change can occur quickly. America's changeovers from horses to cars, from bare automotive tailpipes to catalytic converters, and from steam to diesel-electric locomotives all went from 10% to 90% adoption (in the stock of devices in use, not merely in new units sold) in only 12 years. Airbags went from zero to 100% of the new vehicle market in seven years. Henry Ford sold 2.5 million Model Ts between 1908 and 1916 even though in 1908 the United States had almost no paved roads and affordable personal motorcars were so inconceivable that, as Ford quipped, if he'd asked his customers what they wanted, they'd have said, "Faster horses."

The industry can be responsive—if asked. Detroit proved its ability to respond to customer demand for efficiency during the years 1975–1985 (fig. 2-3), responding to President Gerald Ford's efficiency standards and fuel prices by raising rated efficiency 62%. The average new car shed nearly half a ton and, while becoming safer, far cleaner, and no less peppy, drove 1% fewer miles on 20% fewer gallons. Of that fuel saving, 96% came from smarter design, and 4% from smaller size. But the lost weight was more than regained by 2005. In the past decade, U.S. cars gained weight twice as fast as people did.

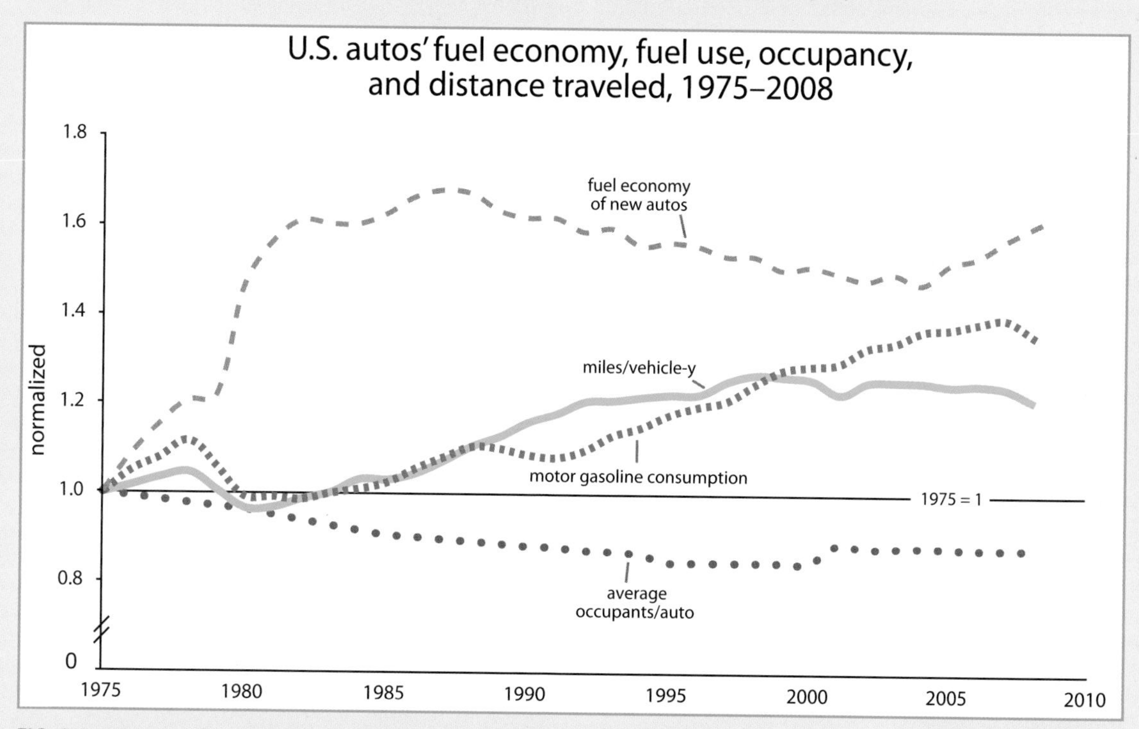

FIG. 2-3. Historic U.S. automotive fuel economy, vehicle-miles traveled, load factor, and fuel consumption. Automakers' big rise in fuel economy was driven mainly by 1975 federal CAFE (Corporate Average Fuel Economy) standards that took effect in 1978, and secondarily by fuel prices.[53]

From 1980 to 2004, automobile efficiency languished while the rest of the economy went to the moon, developed atomic power, and built the Internet.

The causes of the 30-year stagnation in vehicle efficiency are well understood. The combination of abundant oil supplies and efficiency's success crashed the world oil price in 1985–1986. In the U.S., heavy subsidies and light taxes helped keep it low for decades, ensuring that its price was one-half to one-third the price in almost all other countries. This boosted driving and sprawl and, as we'll see later, disadvantaged U.S. automakers against foreign ones. In the quarter century prior to 2008, global automaking's aftertax profits averaged just 1.26% of revenues; in the U.S., those profits were only 0.37% and more volatile. (Many brands were sustained by very few models, and by financing cars they sold at a loss.) As profit-starved Detroit kept on innovating, wringing more power from smaller engines while achieving remarkable improvements in emissions and safety, companies found they could make more money marketing acceleration, weight, and sheer size rather than efficiency. In 1998, a single factory, Ford's Michigan Truck Plant in Wayne, Michigan, earned $3.7 billion churning out giant 12 mpg Ford Expeditions and other SUVs, making it "the most profitable factory of any industry in the world," wrote Keith Bradsher in *High and Mighty*. To keep the SUV cash cow rolling, the auto industry's lobbyists blocked every effort to increase fuel economy standards (which by law were supposed to keep up with cost-effective technological advances). By 2008, new U.S. autos averaged a miserable 23 mpg on the road. No wonder America's best-selling vehicle in 2008, the Ford F150 pickup truck, got fewer miles per gallon than the groundbreaking Model T had a century earlier. But dependence on those highly profitable light trucks was risky, because their sales depended partly on oil prices staying low, while oil prices have actually been random since 1859. Inefficient autos in turn heightened pressure on world oil markets, making oil shocks more likely.

The industry is already responding to new conditions. Sure enough, the gasoline price spike of 2008, coinciding with recession and collapsing finance, led to a dramatic (if partly temporary) shift in customer preference from SUVs to more efficient vehicles. By 2010, the most popular SUV, for the first time in a quarter century, was outsold by a car—a Japanese compact—and GM couldn't even sell its Hummer business. The 2008–2009 Great Recession also sent U.S. auto sales plunging from nearly 17 million in 2005 to 10.4 million in 2009, helping push the U.S. industry to the brink of collapse. Concerns about climate change grew too. These trends, along with the waning clout of the crippled U.S. industry, emboldened Washington to enact the first higher fuel economy standards for cars in 35 years, raising new autos' minimum in 2016 to about 29.5 mpg—about where Europe was in 2008.

By 2011, efficiency was selling briskly and becoming good business. The emergence of plug-in hybrid and battery-electric vehicles such as the Chevrolet Volt, Tesla Roadster, and Nissan Leaf signalled a further shift in industry priorities.

These insights have a profound consequence: Making a conventional auto very light and smoothing its journey through the air and over the road has enormous leverage for saving fuel. By avoiding losses from tank to wheels, each unit of energy saved at the wheels saves seven units at the tank. A more efficient propulsion system wins no such leverage.

The logical goal, therefore, is achieving vehicle "fitness"—designing out weight, aerodynamic drag, and rolling resistance. Once autos are extremely light and efficient, *then* you can focus on the powertrain and change how autos are propelled and fueled.

Vehicle fitness is not news to automakers. They had compelling reasons not to pursue it seriously during the past few decades, but new conditions (see A Little Automotive History sidebar) are making the old rationale obsolete.

Boosting Efficiency, Step by Step: The Low-Hanging Fruit

Some improvements in automotive fitness are so straightforward with existing technology that they are considered, even now, the industry's quickest win.

Weight. The autos that will ultimately free us from oil will take the imperative for lightness as far as possible. Clever engineers are already working on the design, materials, and manufacturing innovations that will create ultralightweight autos with safety comparable to or better than today's heavier autos (as explained below). Automakers are taking the initial steps down that path, using conventional materials (see the There's a Lot of Life Left in Metals sidebar) and standard design and manufacturing techniques to wring weight out of existing vehicles, and doing so at little or no additional cost: a survey of 2010's new autos shows that across all models, lighter autos aren't priced higher.[54] Henry Ford said, "Weight may be desirable in a steam roller but nowhere else." In 2011, Ford CEO Alan Mulally called [light] weight "absolutely critical" and, reported Bloomberg, made lightweighting "the foundation of Ford's plan to meet rising fuel and safety mandates without scrapping the pickups and SUVs that generate most of the company's profits." Meanwhile, Nissan, Toyota, and Chinese automakers announced big weight cuts. Audi's aluminum concept version of its TT Roadster body got 35% lighter, twice as rigid, and a lot sportier—part of Audi's strategy to balance heavier electric propulsion components with lighter bodies.

Aerodynamics. Additional efficiency gains come from reducing the drag coefficient, the frontal area, or both. Smoothing the flow of air around an auto needn't constrain styling (it's often invisible, since much drag comes from airflow under the auto). One major automaker recently found that it could cut a popular model's aerodynamic drag by about 30%, which would boost fuel economy by 14%, at an extra manufacturing cost of around $100. Across all 2010 U.S. autos, there's no correlation between price and aerodynamic drag coefficient, which varies by nearly half among vehicles at any price level.

THERE'S A LOT OF LIFE LEFT IN METALS

The winner of the mainstream class of Progressive Insurance's 2010 Automotive X Prize for 100+ mpg designs, Edison2's Very Light Car, was made largely from steel and aluminum.

Nearly all modern autos are made mainly of steel, a strong, cheap, versatile material whose shaping technologies are exquisitely refined, and that is not nearly as heavy as it used to be. A study from a consortium of 35 steel producers shows that autobody structures could be made 25% lighter using advanced steels and manufacturing techniques, at no extra cost. How? Steel sheets can be made with varying thickness, putting strength only where needed. Hydraulic fluid can be used to shape metal in dies (a process called hydroforming), allowing larger and more complex shapes that add strength without weight. A follow-on project—"The Future Steel Vehicle 2020"—suggests that weight reductions of up to 35% are possible with steel.

Steel faces a challenge from aluminum, which is only one-third as dense for comparable strength. While aluminum is about five times as expensive per pound (thus about 1.5 times more expensive per part) and can be trickier to form and join, the metal is increasingly being used. Over the past three decades, the aluminum content of vehicles has increased from 2% to 8% as part of an effort to curb weight. Magnesium and even titanium are also increasingly used.

Rolling resistance. Another way to boost efficiency is by adding modern low-rolling-resistance tires. Again, highly efficient tires don't generally cost more. Shifting from the least to the most efficient in a common size boosts fuel economy by 8–12% but needn't cost more nor sacrifice performance, durability, or safety. Rolling resistance accounts for 9% of the world's oil use, worth a half-billion dollars a day, so innovation to cut that waste will continue.

Combined effects. These gains add up. Straightforward reductions in weight, drag, and tire losses could together boost fuel economy by about 50% with no electrification—not even the conventional hybrid drive now in millions of autos—at an attractive price. Watch your showrooms. Today, with rising oil prices, oil insecurity, and climate concerns, taking these evolutionary steps to boost efficiency sounds like a great idea. It is. But it captures only a part of the prize and won't get us off oil. We can be far bolder. Some automakers already sell autos with carbon-fiber composite parts or powered by electric motors. As we'll see next, these innovations are about to converge, not just to make a better, more efficient auto but to support the companies' long-term survival and success. So how can we make this revolutionary leap?

Revolutionary Autos: The Vision

The path to the answer began around 1992 behind the guarded doors of Lockheed Martin's legendary Skunk Works advanced research and development facility in Palmdale, California, where David F. Taggart, with a team of visionary engineers, led the development of an advanced airframe for the F-35 Joint Strike Fighter (JSF). It was 95% super-costly carbon-fiber composites, and hence one-third lighter than the benchmark 72%-metal JSF production design—yet it was two-thirds *cheaper.*[55]

How was that possible? Taggart's engineers began with a clean sheet of paper, reinventing the plane from scratch as a primarily composite airframe tailored for affordable manufacturing. For instance, novel snap-together joints would self-align large, complex composite shapes for bonding—a whole new way to make high-performance airplanes.

Taggart couldn't find a military customer for his radical fighter plane design, so he left Skunk Works, then joined Rocky Mountain Institute's Hypercar® Center in late 1998 to see if he could do for autos what he'd done for planes. In 2000, he and engineer David Cramer moved to England to build a development team with English and German Tier One auto-engineering firms expert in race cars, light structures, and advanced powertrains.

The team set out to design a midsize crossover sport-utility vehicle (SUV) that met a list of seemingly irreconcilable requirements: be as practical as a Ford Explorer, carrying five adults and their cargo in comfort and safety, with the driving dynamics of a BMW X5, at least three times the Explorer's fuel economy, and a mass-production extra price repaid by the first few years' savings at U.S. fuel prices.

Industry-standard performance, structural, and financial models and subsystem prototypes showed that the Hypercar team achieved these goals—in nine months and for a few million dollars. How did they do it? With integrative design.

Taggart and Cramer organized their people Skunk Works–style. The core team, initially seven engineers each leading a key vehicle subsystem, sat around the same table. Taggart deliberately set no requirements for each of these major systems, thus forcing the engineers to design the whole vehicle together from scratch. The only requirements were at the whole-vehicle level, so no subsystem could be optimized at the expense of another. They started from the wheels and worked back to the engine, giving each part exactly the needed size and strength. Perhaps most importantly, Taggart's engineers took to heart the central lesson from auto physics, relentlessly striving for lighter weight.

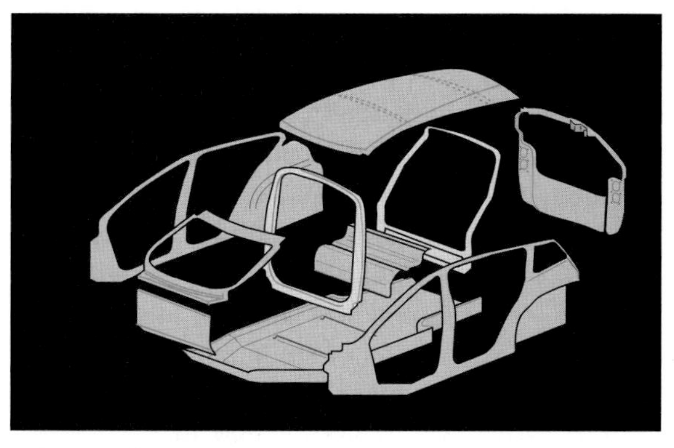

FIG. 2-4. The Hypercar (2000) SUV design's airframe-inspired ultralight carbon-fiber-composite body (left)—suspended from rings, not built up from a tub—and a full-scale physical mockup of the complete virtual design (right). It won the 2003 World Technology Award.

They designed the composite structure in a novel way[56] that enhanced crashworthiness and eased assembly, while putting strength and stiffness only where they were needed. At each key design milestone, leveraging further efficiencies in both cost and weight made the vehicle still lighter and cheaper. The final body had just 14 main parts (fig. 2-4, left), each liftable without a hoist, and designed for mass-production techniques devised specifically for making carbon-fiber autos, including parts that snap precisely into position for bonding.

The designers integrated parts and functions so that many parts each did multiple tasks, substituted software for hardware, replaced mechanical and hydraulic with electrical and electronic components, and trimmed superfluous features. For example, Taggart's team made the interior and trim 72% lighter by exposing the body structures to the interior and making their components simultaneously vibration-damping, crash-absorbing, heat-insulating, good-looking, and therefore fewer.

With the size of a Ford Edge, the final Hypercar SUV design (fig. 2-4, right) was 53% lighter and simulated to be 3.6 times more efficient on gasoline (6.3 times on hydrogen) than the most comparable steel SUV, the 2000 Audi Allroad 2.7T. The Hypercar SUV could cruise at 55 mph on the same power to the wheels that a normal SUV uses on a hot afternoon to run its air conditioner. (The Hypercar design had an air conditioner too, but with seven times normal efficiency.) Hypercar, Inc., was unable to raise production capital in the late-2000 capital-market crash, but the design[57] continued to influence industry thinking in ways that are now moving toward the market. Both Toyota (fig. 2-5) and Volkswagen (fig. 2-6) have

FIG. 2-5. Toyota's 2007 1/X concept car—a carbon-fiber-monocoque four-seat plug-in hybrid with half the fuel use and one-third the weight of an equally spacious Prius. It weighs just 926 pounds (and would weigh only 880 pounds if it were an ordinary hybrid) and reportedly gets about 108 mpg on the European test cycle. Its half-liter flex-fuel backup engine is tucked under the rear seat.

FIG. 2-6. Volkswagen's 230-mpg-gasoline-equivalent XL1 concept car (2011)—a carbon-fiber two-seat plug-in hybrid. Its 0.8-liter 48 hp diesel engine is hybridized with a 27 hp electric motor, and its drag coefficient is an industry-leading 0.186. The car has a top speed of 99 mph and 0–62 mph time of 11.9 seconds, weighs just 1,752 pounds, and is slated to enter limited production in 2013.

shown Hypercar-class carbon-fiber sedans, the latter—like the BMW and Audi sedans mentioned below—with announced imminent production intent.

Revolutionary+ Vehicles: Key Enablers

The key enablers to vehicles of such breakthrough efficiency, which we'll call "Revolutionary," are (1) integrative, whole-system design optimized for (2) ultralight materials, particularly advanced composites. Adding (3) an electrified powertrain creates what we'll call the "Revolutionary+" auto—the key to getting autos off oil by 2050.

Each of the three successive Revolutionary+ enablers unlocks the effectiveness of the next, yielding benefits that multiply. The first two, integrative design and advanced materials, not only save energy directly but also make it more practical and affordable to move to the third—superefficient electrified powertrains (fig. 2-7).

Revolutionary+ Enabler 1: Whole-System Design

This requires new ways of thinking. Changing a design, Taggart explains, is like stretching a rubber band. The farther you stretch it from its current norm, the greater the resistance. Stretch too far from the comfort zone and the rubber band breaks. Breakthrough design therefore requires shifting to what engineers call a whole new "design space" with its own brand-new rubber band. If technology can't yet reliably deliver the performance you need, you can stretch back toward today's norms, but as the technology matures, the rubber band will relax toward your goal and pull you into the future.

Instead of assuming that an auto needs all the traditional parts and designing each part separately, why not think of the design as one integrated whole? The biggest benefits emerge when engineers repeatedly revise the whole design to exploit each gain they just made in its parts. This recursive "design cycle" exploits how lightness snowballs. The less weight you have, the less weight you need.

A lighter auto needs less power, so its powertrain can be smaller and simpler. That makes the auto even lighter, so in the next design go-round, the engine can be made yet smaller and lighter. The weight savings multiply with each component, from brakes to suspension parts, and with each turn around the design cycle. Parts and systems can even disappear entirely: put an electric motor in each wheel, for instance, and suddenly there's no need for a transmission, clutch, driveshaft, axles, universal joints, or differentials. Their disappearance in turn triggers still more weight savings. Lightness multiplies.

Revolutionary+ Enabler 2: Advanced Composites

By combining two or more materials with complementary properties, composites can maximize the

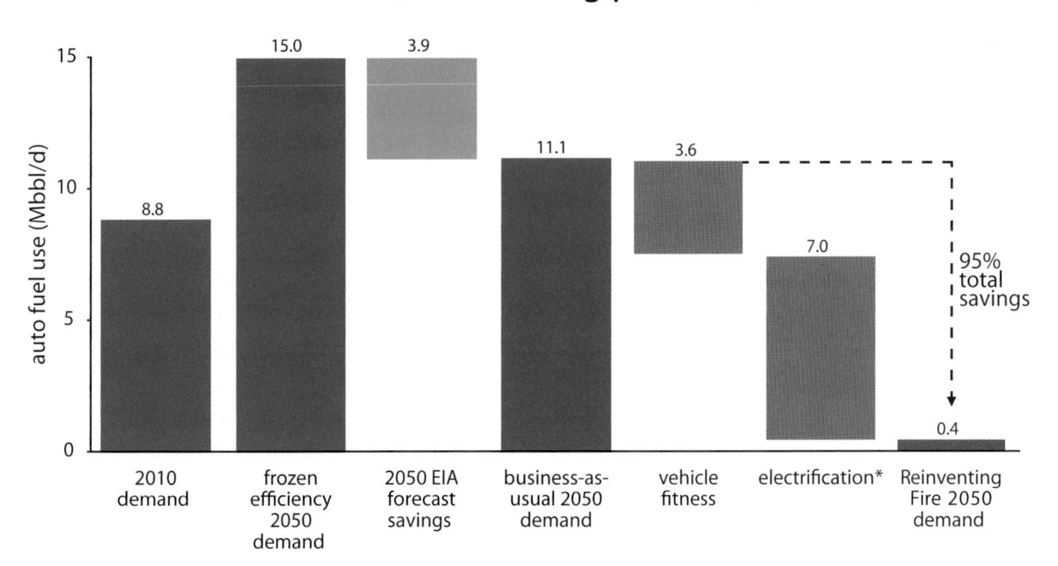

FIG. 2-7. Vehicle fitness (via integrative design and advanced materials) reduces normally projected automotive fuel consumption by one-third while enabling efficient and affordable electrification, the source of even larger fuel savings and the most crucial step toward getting U.S. autos completely off oil. The oil use remaining in 2050 to run surviving gasoline-engined vehicles is displaced by biofuels.[58]

benefits of each material. Like wood—cellulose fibers in a lignin matrix—carbon-fiber composites embed long strands of carbon atoms, with excellent tensile strength, in a tough plastic resin to yield a new material stronger and stiffer than steel but only one-third as dense. Modern methods can make it reparable and recyclable. It doesn't rust or fatigue. It could allow the chemical industry to muscle in on metalmakers. And such ultralighting of U.S. autos could cut up to two-thirds of their weight and half their fuel usage, make their electrification affordable, and thus ultimately save an amount of annual oil nearly comparable to finding a Saudi Arabia under Detroit.

Revolutionary+ vehicles will combine a mixture of materials. Composites aren't appropriate for all applications: advanced and even standard versions of today's conventional metals are sure to play a significant role in lightweight autos, just as they do in Boeing's half-composite 787. Designing with composites creates a new realm of lightweight possibilities, enabling Revolutionary+ vehicles and bringing drivers far lower fuel costs for the same or better performance. However, the currently high price of ultralight materials, their traditionally slow manufacturing processes, and the investment needed to retool factories all challenge automakers to manufacture such featherweights affordably. Are there ways to unlock lightweight materials' benefits without breaking the bank?

Mass-Producing Composite Structures

Building Formula One carbon-fiber autobodies is like crafting a handmade Italian suit. Bundles, sheets, tapes, or woven cloth made of stiff, finer-than-hair carbon fibers are laid by hand in precise patterns aligned for maximum strength. This "layup" is embedded in a costly thermoset resin like epoxy and then baked in big ovens to cure—a finicky process that takes hours per part. Much of the expensive fiber gets trimmed away as scrap. No wonder the automakers long considered carbon fiber prohibitive: they'd need a process roughly a thousandfold higher-volume and lower-cost, and a hundredfold faster.

A typical 250,000-units-a-year auto plant must make a vehicle every two minutes or less, for two main reasons. First, only half of a typical auto's retail price is manufacturing cost; the other half is fixed overhead. If production volume drops, overhead costs per auto rise and profits plunge. Second, production must keep pace with the $0.3 billion paint shop, whose huge scale spreads the cost of controlling its air pollution and protecting workers. Automakers would therefore need composite parts made in a minute, not hours.

But what if you could invent a rapid automated layup process and switch from thermosets to thermoplastics—tougher, cheaper, needing no curing, and quickly reshapable simply by melting, molding, and cooling? David Cramer and other Hypercar engineers tried that and it worked. Hypercar developed automated equipment that if scaled and matured could meet automotive speed and cost goals, and it became Fiberforge Corporation. Now its third-generation equipment is making high-performance composite parts for aerospace, military, and other customers, in competition with such firms as Electroimpact, Forest-Liné, Ingersoll Machine Tools, MTorres, and MAG Cincinnati.

Automakers are also partnering to develop their own large-scale manufacturing processes, already at several-minute cycle times—initially with thermosets but moving toward thermoplastics. Toray, the world's largest carbon-fiber supplier, announced the day before Toyota showed the 1/X—a clear signal of both firms' strategic intent—a $0.3 billion factory to "mass-produce carbon-fiber auto parts for Toyota," then added Honda, Nissan, Subaru, Daimler, and others, seeking "billion-dollar automotive sales." Arch-rival Teijin announced a sub-minute thermoplastic forming process. Toray and U.S. rival Zoltek each opened an automotive advanced-composite application center; so, in 2010, did the Japanese government, to accelerate private-sector composite technology. A half-dozen automakers know that, as VW said in 2011, large-scale manufacture of carbon-fiber automotive structures, "simply not viable" in 2002, "is now possible." The *Wall Street Journal*, surveying lightweighting progress from Lamborghini to Land Rover, concluded: "Deftly trimming the weight from cars . . . will be vital to the competitiveness of every auto maker. . . . Soon, a luxury car made only of steel and plastic could be as déclassé as a cinder-block sized cellphone."[59] And meanwhile, Toyota and Honda have entered the carbon-fiber airplane business, doubtless aiming to cross-pollinate new materials skills back to their core automotive businesses.

Transforming Automaking

Though at least two firms' technologies can already achieve one-minute cycle times, that may still be slower than steel-stamping. But plants can compensate for any initially longer composite cycle times by setting up parallel lines in the floorspace previously needed for one steel-based production line. The new equipment's drastically lower cost and size could shift production economics profoundly, since costlier but fewer parts with far cheaper assembly can help offset the costlier materials.

Consider first that composites can reduce by approximately tenfold the 100 to 200 parts needed for a typical autobody.[60] The roughly $0.3 billion tooling cost to stamp them would fall far more, because molding each part takes a single dieset with composites, versus about four progressive diesets to stamp steel. Fewer parts also mean fewer assembly stations and fewer robots. Lighter parts mean less powerful, less costly equipment. Bonded or induction-welded joints can replace thousands of spot welds.

Composites may also prove able to slash auto plants' biggest investment and toughest operation—the paint shop. A shiny, flawless Class A finish costs about $400 per auto. With composites, it may be possible to use "paint-in-mold" techniques to prime or color a part while forming it, greatly simplifying the paint shop or eliminating it altogether.

Studies show that Revolutionary autos' manufacturing fixed costs can be reduced by 80% (fig. 2-8), cutting total manufacturing cost by about 35% in a 250,000-a-year plant.[61] Other savings in variable nonmaterial costs, such as factory energy, are a useful bonus. Manufacturing with composites may thus shift the longstanding automaking business model toward lower investment, smaller plants, faster product cycles, and hence a more diverse, agile, and rapidly evolving product portfolio—all helpful in managing uncertainty.

Right now, the raw material for a carbon-fiber composite is 15–30 times more expensive per pound than steel. So an unfinished autobody made of composite parts 60% lighter than their steel equivalents would be roughly 300 pounds lighter and $1,000 to $3,000 costlier. But only 4–8% of the manufacturing cost of a typical steel car is the steel, so shaping, finishing, and all the rest of the car's manufacture drive far more cost. The composites' material-cost premium gets partly offset by simpler manufacturing, smaller powertrain, snowballing weight savings that make other parts smaller, and such valuable performance benefits as more stiffness, better handling and ride quality, faster acceleration, and fuel savings.

Making Carbon Fiber Cheaper

Composite autobodies' extra materials cost is wide-ranging because carbon fiber comes in different types with differing prices. In non-cosmetic applications, substituting cheaper, often recycled carbon fiber scrap with strength comparable to that of high-grade fiber can save up to one third of material cost. Further savings result from placing carbon fiber only in areas where its exceptional properties are required and filling the remaining space with lighter, cheaper glass fiber or core material. Carbon fiber's price will also fall as the industry first equilibrates and then matures.

Carbon fiber was long a boutique product with global tonnage comparable to U.S. gourmet chocolate sales. Its price then soared with sudden demand for making airplanes and wind-turbine blades (as well as spiking oil prices), but it will ease—even without cheap new precursors—as suppliers catch up and move down the learning curve[62] to compellingly low prices.[63]

About half today's production cost is for precursor material—96% of which is polyacrylonitrile made from oil (propylene) or natural gas (propane), both of which have volatile prices. Carbon-fiber manufacturers are starting to make their own precursors and expect to cut their costs by about 20%. But much cheaper precursors are emerging. Their strands of carbon atoms are commonplace; the trick is removing the other elements and forming the remaining carbon skeleton into long, pure strands. Solve those problems, and carbon fiber could be made from biomaterials like plant fibers, or even from recycled plastic trash. Oak Ridge National Laboratory (ORNL) believes these alternatives could potentially cut carbon-fiber costs by up to 90%, matching or even beating steel prices on a direct dollar-per-pound comparison[64]—not that anyone buys autos by the pound.

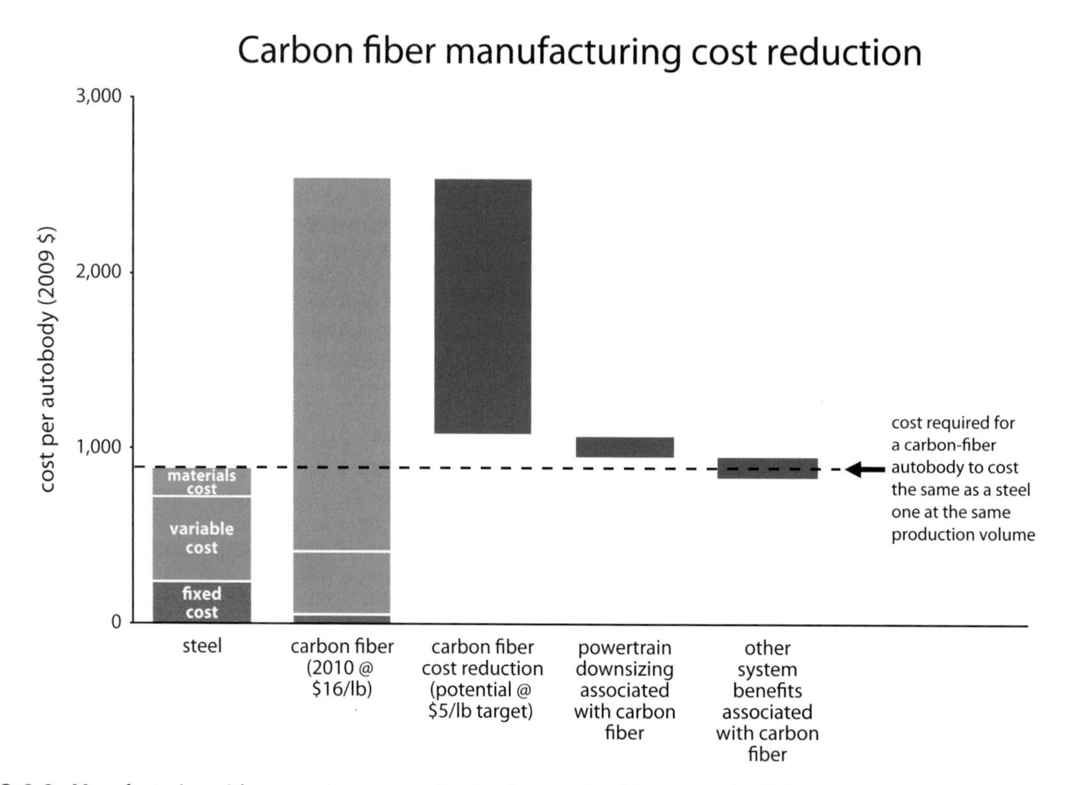

FIG. 2-8. Manufacturing with composites can cut the fixed costs of making an autobody by 80% and its variable non-materials costs by 25%. Both reductions are currently overshadowed by raw material costs, but those are likely to fall as the composites industry matures and are partly offset by other factors (as described in the text).[65]

Ultralight but Ultrastrong

What about the safety of composites-based vehicles? Until recently, the prevailing view in the U.S. auto industry was that efficient autos are small, unsafe, sluggish, costly, or otherwise so undesirable that customers would buy them only if the government required or subsidized them. However, the physics of autos shows that light weight and efficiency can actually mean spacious, safer (see Crash Safety with Composites sidebar), peppier, and cost-competitive. People will buy such autos because they're *better*, not just because they're more efficient, much as most of us switched from vinyl phonograph records to CDs and then to iPods.

Revolutionary+ Enabler 3: Electrified Powertrains

So far we've seen the alluring benefits of designing autos to be as light as possible, maximizing their overall vehicle fitness while maintaining or improving safety. We've also identified a path to reducing the cost of manufacturing these ultralight autos so that they offer a compelling business case.

Yet there's still a trump card to be played. While vehicle fitness alone accounts for a third of autos' 2050 fuel reduction, it more importantly enables the essential element that finally liberates us from oil. That final ingredient is powertrain

CRASH SAFETY WITH COMPOSITES

The perception that safety requires weight stems mainly from flawed studies done between 1977 and 2004 by the National Highway Traffic Safety Administration (NHTSA). Analyzing U.S. crashes, NHTSA concluded that making autos 100 pounds lighter would kill an extra 400–1,300 Americans per year.[66] NHTSA therefore encouraged heavier autos by letting them be less efficient. At the same time, big autos got more popular and profitable, while materials and design choices made big autos heavy. The combined effect was a frenzied "mass arms race" in which, trying to protect your kids, you drive an Expedition, your neighbor drives a Hummer, and the guy down the street drives an 18-wheeler.

But NHTSA's analysis had erred by conflating weight with size. Careful reanalysis of the data showed that making all autos 100 pounds lighter would *save* 1,500 lives, because what increases crash safety is not weight but size.[67] Buyers were right about larger autos being safer—they have more crush space to absorb impacts. But larger autos needn't be heavy. NHTSA has since switched to regulating autos by size, not weight. The goal, explains University of Michigan physics professor Marc Ross, should be to make "heavy vehicles lighter (but not smaller) and . . . lighter cars larger (but not heavier)."[68] We can thus make autos big, and hence comfortable and protective, without making them heavy, and hence hostile and inefficient. By decoupling size from weight, ultralight materials can save lives, oil, and money simultaneously.[69]

At any weight, using any materials, design for safety is vital: it's why, across all cars on U.S. roads, observed crash death rates vary by about threefold between different models of the same weight. But lighter, stronger materials magnify the design opportunity. Aluminum absorbs about twice as much crash energy per pound as steel, while carbon-fiber composites are up to six times better than aluminum. Such materials' strength, combined with good design, helps explain why Formula One race-car drivers usually suffer only minor injuries in horrific 200 mph crashes. It also explains how ORNL's and Hypercar's designs halved vehicle weight without reducing safety. And it explains why your sports safety helmet is probably made of carbon fiber, not steel.

Also, as part of a clean-sheet design approach, integrating the latest active safety features could enhance safety while allowing further weight savings. MIT's *Reinventing the Automobile* shows how features based on wireless technology and electronic sensing could help make accidents less likely and less severe even with smaller autos.[70]

A fleet of lightweight, well-designed autos would be a triple play on safety. They'd protect their own drivers and passengers better. They'd cause less damage if they hit other vehicles, structures, or pedestrians, thus protecting others better. And they'd protect us all against the dangers of buying and burning oil.

electrification, which unlocks a further 63% reduction of 2050 fuel consumption (fig. 2-7).

An electric motor both drives the wheels and acts as a generator to convert unwanted motion back into useful electricity. That electric motor can be run on battery power or fuel cells, or it can be augmented with a small onboard fueled engine, as in hybrid and plug-in hybrid autos. Batteries store far less energy per pound than gasoline but convert it more efficiently into motion, helping to justify their higher cost.

Do we really need to go to electric propulsion? After all, several innovations are under way with internal combustion engines (ICEs) that could boost efficiency by as much as 50% (see New Engine Technologies sidebar).

Despite such advances, some simple math shows that electrification is a better answer. If the best ICE technology were combined with 30% lighter vehicle weight, U.S. autos would still burn about six million barrels of biofuels a day (8.5 times 2010 production), far exceeding the country's projected supply of inedible biofeedstocks (see Pumping Biofuels on page 65). Those biofuels would be better used by aviation and heavy trucking, for which electric power is not a viable option.

The saga of JB Straubel further illuminates the advantages of electric propulsion. Earning Stanford engineering degrees, he built a gasoline reformer and fuel cell (they worked and didn't blow up) and then electrified his Porsche. He worked with ultralight airplane pioneer Burt

NEW ENGINE TECHNOLOGIES

In the 1960s, Israeli engineering genius Eddie Sturman designed digitally controlled valves for NASA's huge rocket engines. His valves' energy frugality helped the crippled *Apollo 13* return to Earth. Now his Colorado firm and its larger collaborators are trying to revolutionize the diesel engine, operating the valves with tiny electric actuators instead of mechanically driven camshafts.

A diesel engine's piston compresses the air in a heavy metal cylinder to extremely high pressures, making the air so hot that sprayed-in fuel oil explodes, pushing back the piston to turn a crankshaft. Modern diesel engines in autos and trucks get peak efficiencies in the 40s of percent, versus the 30s for nondiesel car engines.

Fast, small, light, and cheap, Sturman's retrofittable valves permit very precise fuel and air injection under closed-loop digital control. This could boost the efficiency of a diesel engine by half (to around 60%), increase torque by more than half, and make the engine at least one-third smaller and lighter, more than a tenth cheaper, and able to burn any fuel so cleanly it would need no emission-control equipment.

Digital valves also permit unusual event sequences. How about first injecting the fuel into leftover exhaust gas, so it's vaporized and mixed for free, then adding the air through a separate digital valve? Or switching the engine on the fly between two-, four-, six-, and eight-stroke operation? Or better yet, since the piston has the least leverage when trying to push the crankshaft from the top of its stroke, how about eliminating the crankshaft? Simple: Let the back of the piston directly compress hydraulic fluid into a separate vessel, then later turn that stored pressure back into mechanical work exactly when and at the force required. This system, Sturman believes, could push the engine well past 60% efficiency and offer many advantages for propulsion, as well as for stationary engines.

Another brilliant engine innovation to watch is the opposed piston-opposed cylinder (OPOC) design developed for DARPA (Defense Advanced Research Projects Agency, a military research agency) to make very light and efficient portable generators. OPOC is very compact because it balances and integrates the forces of two aligned pistons moving in opposite directions. EcoMotors International, led by Don Runkle (formerly GM's head of advanced engineering), is commercializing it. The firm claims its engine is 30% lighter, 75% smaller, and 50% more efficient than today's state-of-the-art turbocharged diesel engines. It needs no valves. An electricity-generating version could even eliminate the crankshaft by using magnet-containing free pistons, nonmagnetic cylinder walls, and surrounding copper coils.

Rutan and then designed hybrid powertrains at Rosen Motors. When PayPal cofounder and commercial-space-rocket mogul Elon Musk decided to build a breakthrough battery-electric car, founding Tesla Motors in Silicon Valley, Straubel was the perfect choice to lead the design.

Straubel already knew that electric motors are lighter, smaller, cheaper, quieter, cleaner, more rugged and reliable, and severalfold more efficient than modern fueled engines, as well as enabling sizzling acceleration. And electric drive can automatically recover for storage and reuse (accelerating the auto) up to about 70% of the energy otherwise wasted as heat by the brakes.

Could autos fully capture those advantages? Straubel's first task was to design a powerful but affordable battery. He bought consumer-electronics lithium cells that look like drugstore AAs—but slightly older models to cut cost. He figured out how to arrange 6,831 of them safely in flatpacks that Tesla now sells to other automakers. He married a light, low-drag Lotus body with an advanced watermelon-size motor and inverter previously used in GM's EV-1 electric car. The resulting Roadster EV costs $109,000 to buy but one cent per mile to run—and its acceleration is on par with that of the world's fastest sports cars.

But a problem appeared: the car's original two-speed gearbox kept breaking. Two top suppliers couldn't meet both the acceleration and top-speed specifications. Straubel's novel solution: eliminate the gearbox and let his *electrical* engineers solve the problem by wringing more torque from the electric motor and helping it shed heat better. The result was 40 more horsepower, 10-mile longer range, 14-pound lighter weight, less noise and maintenance, less warranty cost, and lower manufacturing costs.

This story offers an important lesson for automakers. The world of power electronics, microchips, software, and systems integration is at a far earlier stage than 120-odd-year-old engine-and-gears designs, so it offers far more scope for innovation, scaling, and cost reduction (see Electric-Powertrain Learning Curves sidebar). Early-generation, low-volume electric traction systems can compete *today* against mechanical systems

ELECTRIC-POWERTRAIN LEARNING CURVES

Our manufacturing-cost learning curves for battery packs and fuel-cell systems are based on a 2007 MIT study on electric powertrains (Kromer and Heywood 2007) and informed by extensive industry data.[71] The cost decrease is very steep for the first half-million vehicles, supporting rapid takeoff to higher volume.

Lithium batteries are similarly getting cheaper as more are made, but batteries' underlying science and technology are driven less by new automotive markets than by consumer-electronics buyers' willingness to pay a premium for longer operating time from smaller batteries. To get a marketing edge, electronics manufacturers therefore pay suppliers about twice the price for a battery that safely packs the same energy into half the volume. The battery-maker's incentive is even bigger because such an innovation typically sells twice as well, quadrupling revenues. Battery packs for propelling autos have very different requirements (cycle life and depth, temperature, ruggedness, safety, et cetera), but fundamentally they're made from many small batteries. In time, the basic battery innovations driven by the consumer-electronics market do tend to trickle down from cellphones and laptops to autos.

Much government effort now goes into making batteries cheaper, and used to go into making vehicular fuel cells cheaper. Yet investing R&D effort in vehicle fitness (which got about 100-fold less in U.S. research budgets through 2010) will yield the same result with less cost, time, and risk. First making batteries *fewer* (fit vehicles need two- to threefold less energy per mile) makes them affordable; that sells more electric autos; that volume makes batteries cheaper. Smaller *and* cheaper batteries then hit the jackpot.

that sweated out most potential cost reductions decades ago. Imagine how far electric traction will jump ahead with greater experience and higher volume. And the advantage may as easily go to small, agile firms as to big, rich ones.

Contrary to some recent reports, electrification won't be constrained by critical materials (see The Rare-Earth Conundrum sidebar); rather, they're vibrant business opportunities to displace scarce elements and to use them more productively, durably, and recoverably.

The Transition to Revolutionary+ Autos

As we have seen so far, incremental efficiency gains can apply to any auto. The real automotive magic—and the best hope for eliminating oil—happens when electric traction is combined with Revolutionary vehicle fitness, making any advanced powertrain more affordable and providing the range buyers have come to expect. So how can automakers best make this transition to Revolutionary+ vehicles? Producing these designs at scale is the next and most difficult step, one that no auto manufacturer has yet taken.

As with any breakthrough technology, however, there are "first movers" who have already begun the transition. Some begin with a substitution phase in which a few standard parts on an existing auto model are replaced with lightweight composite parts, enabling a manufacturer to build analysis and design prowess for composites while working out the raw-material supply chain and gaining a head start on tooling. This advances all four elements of automotive innovation—plant, people, product, and process—with due deliberate speed but not all at once, which would create undue risk.

The experience thus gained can then be applied to manufacturing an all-new, clean-sheet, integrative design that takes full advantage of advanced materials. BMW, for instance, announced in 2010

THE RARE-EARTH CONUNDRUM: ARE CRITICAL MATERIALS FOR ELECTRIFIED VEHICLES REALLY IN SHORT SUPPLY?

Much has lately been written about supposed shortages of critical materials for electric vehicles (and for renewable energy and even energy efficiency), notably "rare earth" elements. The U.S. Department of Energy has set up a special group to examine these issues, though the Pentagon found they're not important to national security. On closer examination, such serious critical-materials issues aren't likely, especially for autos.[72] (The U.S. Geological Survey in 2010 reported 1,300 years' worth of U.S. rare-earth deposits.[73])

Lithium, currently the best battery material, is relatively abundant and is readily recoverable from old batteries, much as 97% of the lead in today's auto batteries is recovered for reuse: you must often turn in your old battery to buy a new one or reclaim a deposit. Rare-earth elements like neodymium, mined mainly in China and currently in the midst of a market bubble[74] hyped by stock promoters, are part of the recoverable superstrong magnets in certain compact and powerful electric motors and generators: a Prius's main motor contains nearly half a pound of neodymium and dysprosium, and the world makes about 50,000 tonnes of such magnets per year. (Another and commoner rare-earth element, lanthanum, is in recyclable nickel-metal-hydride batteries, but virtually all automakers, now including Toyota, have switched new models to lighter-weight lithium batteries.) However, there's no necessity nor good reason to use permanent-magnet motors. Induction (asynchronous) motors like Tesla's have no magnets, nor do switched-reluctance motors that match or beat permanent-magnet motors in all respects including cost.[75] And of course the need for all these special materials remains small—both before Revolutionary+ autos, because electrification's cost severely limits its market, and after, because their tractive loads and powertrains are two to three times smaller.

a $748 million investment to mass-produce what it described as "the world's first volume-produced vehicle with a passenger cell made of carbon" (fig. 2-9)—and in 2011, confirmed that "The smaller battery pack required by a lighter car offsets the cost of the carbon fiber body."[76]

Audi announced production of a carbon-fiber electrified auto for 2012, a year before VW's and BMW's releases. These three companies are presumably counting on strong initial sales to early adopters to increase their market share as production expansion and the gains from mutually reinforcing technologies make these models more affordable. There are also signs that Japan's automakers have significant, though shrouded, ultralighting efforts under way.

So far, such first movers as Audi, BMW, Volkswagen, and Toyota have pursued the compound benefits of lightweighting plus electrification. But what about introducing electrification *before* full vehicle fitness? Chevrolet's Volt and Nissan's Leaf did just that, pioneering emerging powertrain technologies in 2010. The Volt, a midsize plug-in hybrid, goes 35 miles on electric power alone before a gasoline engine kicks in and generates enough electricity to carry it more than 300 additional miles on a single charge. But it weighs in at a hefty 3,781 pounds. The Leaf, a midsize battery-electric vehicle, goes 73 miles on a charge, at 99 mpg equivalent, but it's also relatively heavy at 3,366 pounds. Meanwhile Honda, with a long lightweighting heritage, broke new ground with

FIG. 2-9. July 2011 preliminary version of BMW's i3 (originally Megacity) carbon-and-aluminum battery-electric car announced in 2010 for 2013 mass production—the same year as VW's XL-1 (fig. 2-6). BMW's 2,756-lb, 4-seat flat-floor city hatchback is agile, compact, airy, and spacious. Its range is about 100 miles without an optional "REx" range-extending gasoline engine so small it fits alongside the drive-motor above the rear axle. The 170-hp electric motor delivers 184 lb-ft from a standstill and accelerates 0–62 mph in 7.9 seconds, rivaling a BMW 120i, though top speed is governed to 93 mph. The carbon-fiber passenger cell, with replaceable plastic exterior panels, is more crashworthy and durable than steel but half the weight. Initial production is reportedly planned for 30,000 a year but is rapidly scalable upwards.

its FCX Clarity fuel-cell car—a hydrogen-fueled sedan with a 240-mile range—but it weighs 3,582 pounds.

These cars are already impressive. But think how much more amazing they would be at half or one-third the weight, if their manufacturers also made the investments in advanced materials and clean-sheet design to turn them into Revolutionary+ vehicles. For example, Bright Automotive's nominally 80 mpg commercial van (fig. 2-10) is aluminum-intensive yet weighs less with a ton of payload than its competitors weigh empty. Even that partial gain in fitness eliminated 40% of the costly batteries needed to make it a plug-in hybrid. That in turn made its business case compelling to fleet buyers (who take a longer view on fuel savings than do private auto buyers), without the subsidy that all other plug-in hybrids currently need. A future carbon-fiber general-market version could then attract individual car buyers.

Making electrified vehicles light would cost more for structural materials but less for batteries. A fit Nissan Leaf, for example, could save $3,000 in battery costs for the same range and cut the charge time on a standard home outlet from 20 to just over 13 hours; or it could keep the current battery pack while increasing range 50%. Revolutionary fitness could similarly extend the range or reduce the price of fuel-cell electric vehicles like Honda's FCX Clarity.

Seasoned automakers that have become first movers are mainly interested in a stepwise manufacturing transition to reduce risk and ease losses of legacy investments in equipment and tooling. But start-ups like Tesla and rapidly emerging Asian competitors (see New Asian Competitors sidebar) can adopt the latest manufacturing technology from scratch. Start-ups nonetheless face their own barriers linked mainly to economies of scale. They start at low production volumes not to minimize risk

FIG. 2-10. Bright Automotive—a 2009 RMI spinoff that in 2010 entered a strategic partnership with General Motors—showed in 2009 a driving prototype of this commercial utility/service/delivery van, the Bright IDEA, with 3- to 12-fold higher fuel economy depending on driving cycle. It carries 5 cubic meters of cargo, two people, and their fold-down front-seat office for 30 miles electric-only or 430 miles total, and gets 100 mpg on a 50-mile-a-day urban route. This vehicle segment is 7% of U.S. auto sales but uses about 20% of their fuel.

but rather because of limited capacity and high barriers to market entry. Like incumbent manufacturers, they sell initially to an early-adopter market willing to pay for new technology, hoping then to descend learning curves, cut prices, and broaden sales.

But is becoming a first mover the only way to get a piece of this emerging market? Not if followers move swiftly. "Fast followers" are mindful that often "the pioneers get the arrows, the settlers get the land." They believe there may be little intellectual property advantage for first movers, since manufacturing innovation could lie within the supply chain, so all competitors would ultimately pay to license it. If not—if first movers own the intellectual property themselves—fast followers believe it can be affordably licensed from them. In either case, the strategy is to offset the brand value of their pioneering competitors by starting further down the learning curve and piggybacking on an increasingly commoditized supply chain, perhaps aided by better understanding of the market.

Whether first mover or fast follower, existing manufacturer or start-up, the first and most important step is to establish the goal by *designing* an ultralight vehicle that takes full advantage of advanced materials to enable a smaller, cheaper powertrain. The sooner the better: Integrative design could require major organizational change, which isn't easily emulated. Those with

NEW ASIAN COMPETITORS

The California phenomenon of smart, hungry, unknown engineers tinkering in the garage, hatching the next Apple or HP or Xerox, still lives on, but now it's happening worldwide, from Shenzhen to Bangalore and São Paulo to St. Petersburg.

Despite China's many challenges, its industry is dynamic, capable, and supported by nearly free and almost unlimited state capital and determined central policy. Automaking is now a pillar of the nation's growth strategy. The same intensity and drive that created the world's largest-ever construction boom to serve an increasingly affluent and urbanized 1.4 billion people is joining with Chinese leaders' aversion to the oil trap, commitment to electrification and fuel cells, strong interest in advanced lightweight materials, and strategic goal of becoming a formidable advanced-vehicle exporter. China plans a carbon-fiber plant as big as Toray, the world's top producer. In 2010, China became the world's largest auto maker and buyer; Zhejiang Geely bought Volvo; and GM sold half of its India operations to its Chinese 51% partner, Shanghai Automotive, which will now coinvest in GM's Indian expansion. China's auto market, after a sevenfold surge from 2000 to 2009, shows few signs of flagging. And the pattern of turning Western partnership into competition, as in high-speed rail and wind turbines, may well repeat with road vehicles, especially lightweight electric ones.[77] As journalist Thomas L. Friedman recently told a U.S. audience, "The bad news is we'll buy all this stuff from China; the good news is it'll cost less than your tennis shoes." The more China exercises its potentially vast market power, the faster it will drive—even lead—the global automotive transformation.

India is another emerging force in the world automotive market. Capable, aggressive conglomerates like Mahindra and Tata have already snapped up Land Rover and Jaguar. In 2009 Tata launched the Nano, a decent four-seat family car more efficient than a Prius, and in late 2010 priced at $2,900—less than half Ford's fast-selling $7,700 Figo four-seat subcompact hatchback. It's crammed with clever, deeply frugal "Gandhian engineering." Indian quality has improved at least as fast as Korea's: seven years ago, Tata exported 20,000 cars to Britain under the MG Rover badge. India's auto market in 2009 was still only one-sixth that of China but growing briskly. India's 1.2 billion people, including an educated elite as populous as France, have vast innovation potential that's already transformed industries from prosthetics to software. India still lags behind China in overall development and coherent central policy, but the country could be ahead of China in other key institutional factors. It'll be quite a horserace—not to mention Brazil, Korea, maybe Russia, and more.

an established Revolutionary+ design will be best positioned to produce it quickly, whether to win or defend market share.

The Enabling Role of New Policies

The nirvana of autos that use little or no oil is in sight. Mutually reinforcing technologies, economies of scale, and manufacturing innovation, in both raw materials and finished products, will make Revolutionary+ autos affordable; the question is when. In 2030, both battery-electric and fuel-cell vehicles would still be priced a few thousand dollars higher than the Energy Information Administration projects for comparable business-as-usual models, but their higher price would be more than repaid from fuel savings within three years. By 2050, ever-cheaper batteries and fuel cells would bring the vehicles' price down to $29,000—about $500 more than business-as-usual autos (fig. 2-11). Attractive, safe, sporty, fuel-sipping, and eminently affordable, such autos should fly out of showrooms.

But that still leaves a huge problem. How do we jump-start the rapid development of Revolutionary+ vehicles today, so that their prices will drop to economically compelling levels by 2030? Their early price premium would be so high that they'd sell only to a high-end niche market of early adopters.

Fortunately, there are ways to ensure the widespread adoption of Revolutionary+ autos by 2050. Smart policies can unlock and accelerate this transition by changing buyers' price signals to favor advanced-technology vehicles, and speeding retooling to make them. Non-fiscal, light-handed policies—fresher and more effective than fuel taxes or CAFE standards—could boost innovation and speed retooling even better than standards without picking technology winners, mandating specific solutions, or raising subsidies or taxes. Then government can steer the transition while

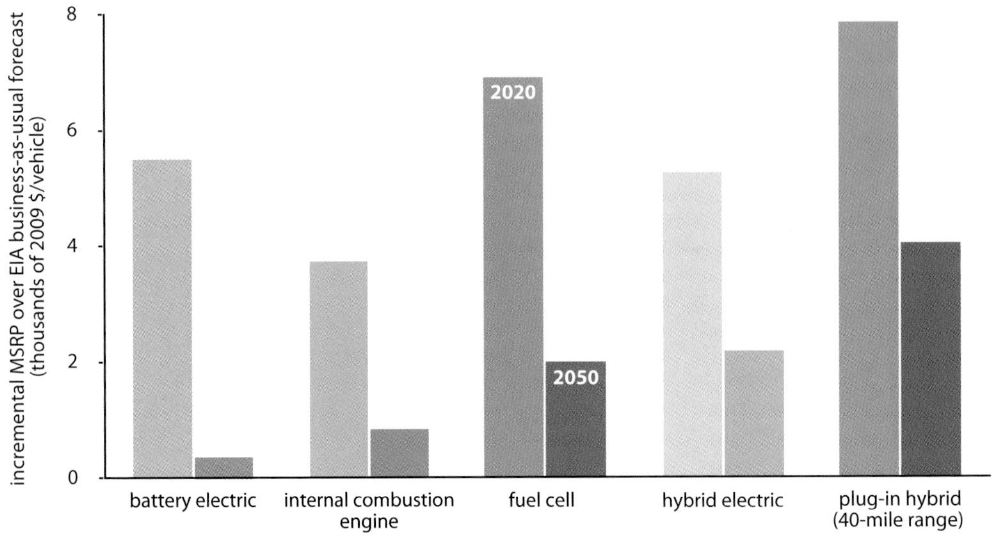

FIG. 2-11. By 2050, the sticker-price premium for Revolutionary vehicles, compared to EIA's projected business-as-usual autos, would drop dramatically.[78]

free enterprise does the rowing. To cross the finish line first, one adjustment to the tiller has unique strategic value: the "feebate."

Feebates

Feebates make efficient autos cheaper to buy and inefficient autos costlier.[79] Buy a fuel hog and you'd pay an up-front fee, right on the price sticker, that climbs as its fuel economy declines. But choose a fuel sipper instead and you'd get a rebate funded by others' fees: the more efficient the auto, the bigger the rebate. Crucially, this is not a wealth transfer scheme. Feebates offer buyers an incentive to buy more of a good thing for themselves and for society—efficient autos—and less of a bad one—inefficient autos. Nor is it a tax: just choose an efficient model and you get an on-the-spot rebate. The Treasury's revenues don't change. And setting feebates separately for each size class would reward you for buying a more efficient model of the size you want.

Feebates provide a powerful price signal that influences auto-buying decisions at the instant they're made. Feebates also maintain a continuous incentive for automakers to innovate. In contrast, government standards may stagnate for decades and give automakers no incentive to beat the standard, while gasoline price spikes only temporarily and unpredictably change consumers' preferences.

Feebates work. The biggest program, in France (fig. 2-12), began at the start of 2008. It charges up to €2,600 or gives out up to €1,000 across seven classes of auto efficiency. The result? Market share for the more efficient models nearly doubled. Market share for fuel hogs fell threefold. While high fuel taxes and CO_2 standards also helped, feebates tripled the speed of efficiency gains, through periods of both high and low fuel prices. Countries with high fuel taxes and CO_2 standards, but without feebates, did not see the same shift, nor did they achieve the same rate of carbon emission

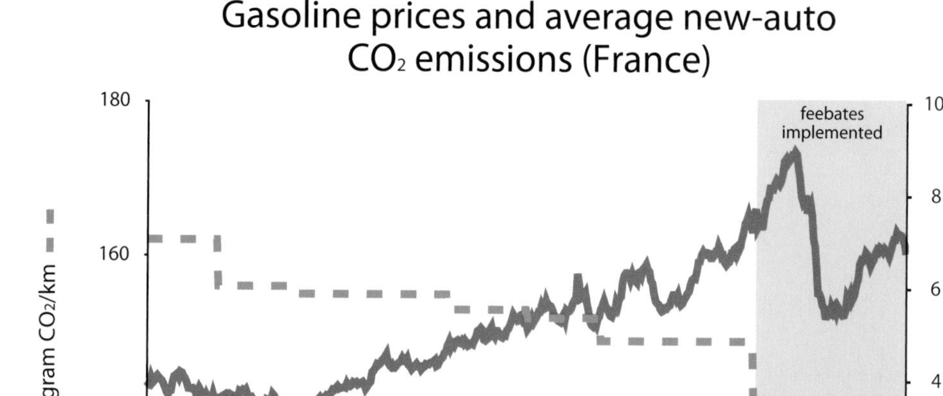

FIG. 2-12. France's spectacularly successful feebates tripled the speed of auto efficiency gains, even when fuel prices were low. Details are adjusted annually. Other feebate programs have succeeded in Denmark, Norway, Holland, and Austria.[80]

reduction. The French program was such a howling success that the rebates totaled far more than the fees, running up a €710 million deficit in 2010. A revenue-neutral design, adjusted each year, would avoid such deficits.

So are U.S. feebates politically feasible? Inside-the-Beltway mutter says no, but the evidence says yes. If a federal feebate bill (introduced in 2009) didn't pass, states or regions could fill the gap: California and its 16 state partners on auto-efficiency rules are two-fifths of the U.S. auto market, enough to swing the whole market. The California legislature passed a feebate by a seven-to-one margin in 1980 (though outgoing governor George Deukmejian pocket-vetoed it because automakers, who hadn't been properly engaged in its design, were uncertain or divided). Now the state is considering a feebate again, and a late-2009 politically balanced survey of 3,000 California households found 76% support.[81] Some automakers, too, already believe well-crafted feebates would help them make more money at less risk and speed innovation. Dealers could support feebates to boost sales and margins. Such industry support, plus national-security and environmental constituencies, could be politically potent.

So what could feebates achieve? Large rebates and avoided fees—totaling perhaps up to $4,000–$5,000 per auto, comparable to manufacturers' SUV rebates of the mid-2000s, and lower than the current $7,500 credit for electrified vehicles—would trigger a virtuous cycle. People would buy enough Revolutionary+ autos to propel automakers down the three mutually reinforcing learning curves that together can achieve three-year or shorter paybacks before 2030 (fig. 2-13). Feebates could even be conditional on electrification in order to speed the journey beyond oil, replacing

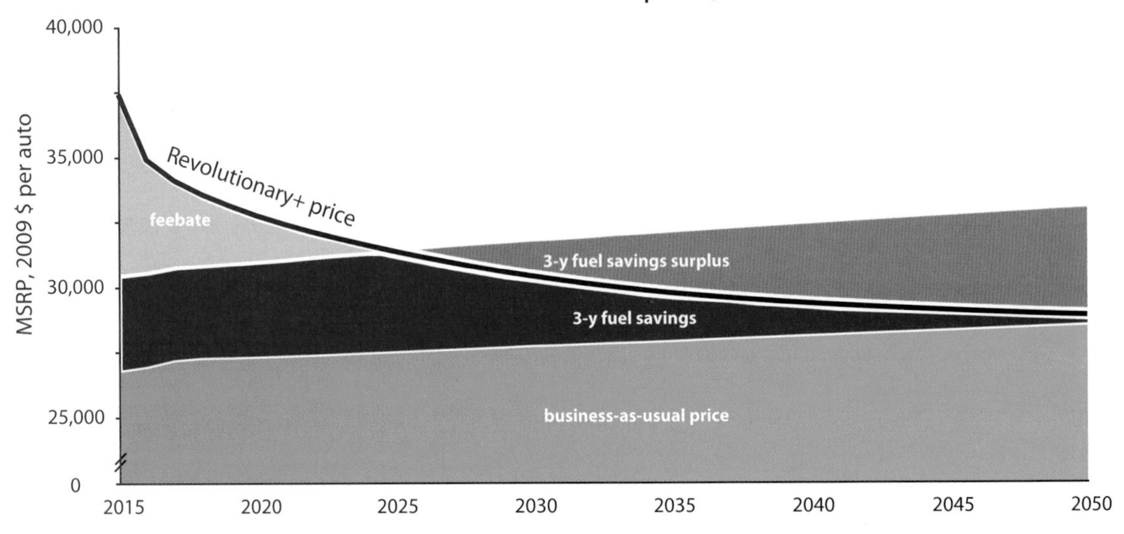

FIG. 2-13. The first three years' total cost of ownership[82] starts higher for Revolutionary+ autos. An initial revenue-neutral feebate covers the premium as up-front cost falls with production volume. By about 2030, fuel savings pay back that premium in three years, so no feebate is needed. Soon thereafter, the autos are creating a societal surplus totaling $2 trillion in 2010 net present value (assuming government-forecast gasoline prices). The curve for fuel-cell autos is very similar, lagging by a few years the battery-electrics shown here.[83]

federal tax credits with self-financing feebates and eliminating current perverse incentives that favor big batteries over fitter autos.

The autos then would rapidly become cheaper, further accelerating their success, which would stimulate more innovation and bring even lower prices. As a result, feebates could be phased out entirely by 2030, when Revolutionary+ autos will pay back in about three years at reasonable fuel prices. Further improvements would then continue, cutting autos' mobility-fuel needs 84% by 2050.

Could we do even better? Besides feebates, there's a host of clever ideas that could further offset the initial price premium of Revolutionary+ autos, in turn accelerating manufacturers' retooling to build them, and sooner achieving their benefits.

Fleet Procurement

Fleets, including rental fleets, employ about 7% of U.S. autos and light trucks, drive their vehicles about twice as much as private owners, and resell them about twice as fast, so they can strongly influence the whole auto market. If governments or big fleet owners, perhaps encouraged by feebates, bought the most efficient vehicles available that are cost-effective on a life-cycle basis (as federal rules require), that investment could jump-start learning curves, driving down everyone's costs quickly. This would be a boon for manufacturers because fleets have such strong purchasing power. Just three officials in the U.S. General Services Administration, the Department of Defense, and the Postal Service together control more than 650,000 autos and buy about 70,000 annually.[84] If they could commit to future purchases of highly efficient autos and all commercial fleets followed suit, this would slash manufacturers' risk and accelerate Revolutionary+ retooling.[85] And when big buyers are prepared to commit to buys at certain specifications and prices, a proven "golden carrot" program that shares this intention with automakers would encourage them to offer such vehicles without worrying about whether they'll have timely customers.

New York City's taxi fleet is currently one-third hybrid-electric. Mayor Bloomberg's plan to mandate the conversion of the entire fleet to hybrids was struck down by the U.S. Supreme Court because city governments can't regulate emissions and efficiency—only Washington can. Efforts are under way to revise the federal law on which the ruling was based, so local officials can enact laws to improve emissions and efficiency. But meanwhile, absent federal regulations, they can adopt policies that reasonably influence (but don't mandate) fleet choices—for example, by focusing just on life-cycle cost to minimize taxi fares.[86] Cities and states remain free to mandate changes in public fleets too, so Massachusetts governor Mitt Romney in 2004 ordered 5,600 hybrids for state use and ordered state agencies to buy vehicles with a 20 mpg rating or higher.

Cash for Clunkers

Another approach: speed the retirement of old, inefficient vehicles to save oil and speed turnover of the automotive stock. A recent example is the "cash for clunkers" Consumer Assistance to Recycle and Save Act of 2009. CARS got nearly 700,000 "clunkers" off the road and replaced them at an unexpectedly high average gain of 9.1 mpg. The $3 billion allocated was committed in probably the first week of the program, which boosted GDP by an estimated $3.8–$6.8 billion and created or saved 60,000 jobs. The program could have been improved by pegging the size of the award to the efficiency gain; by paying for clunkers without requiring a new replacement be bought; by encouraging carsharing and other modes of mobility; and by financing low-income Americans' purchase of efficient autos so they can afford to drive to work. Making efficient autos affordable to these households, when combined with accelerated scrappage of inefficient autos, would also offer Detroit a new million-autos-a-year market from customers who otherwise couldn't qualify to buy a new auto.[87]

Affordable Government Financing

There's already a long and successful history of using federal dollars to jump-start innovation and new industries, from microchips to the Global Positioning System. Now, taxpayer dollars could be loaned to automakers, with appropriate accountability and safeguards, to convert or build production capacity and retrain workers for Revolutionary+ vehicles, as is already being done for ordinary autos' efficiency improvements.[88]

Prize Competitions

Lindbergh flew the Atlantic to win a prize. Longitude became measurable because of a prize. The $10 million Progressive Automotive X Prize motivated private teams to build safe, affordable, two- and four-seat cars that achieved at least the equivalent of 100 mpg with a 200-mile range. Future prizes, private or public, could be larger and based on autos sold. These policies all have the same effect: by boosting demand, guaranteeing purchases, or funding development, they help automakers spread production cost and reduce their sales risk for new superefficient vehicles.

Modernizing Old Policies That Hobble U.S. Automakers

Finally, to compete fairly and fully in tough global markets, U.S. automakers and suppliers need not just these new policies but also changes in some old ones (see U.S. Automakers' Three Handicaps sidebar) that make fast followership riskier for them than for their competitors. They're talented enough to overcome that risk by extraordinary effort—but why make their path so much harder?

The Risks and Rewards

For the whole value chain from suppliers through automakers to dealers (see Implications for Auto

U.S. AUTOMAKERS' THREE HANDICAPS

The Big Three automakers, besides being in an inherently tough industry,[89] labor under three handicaps imposed by public policy and national politics. First, U.S. auto efficiency standards and offerings still lag those abroad. Ten leading countries beat the efficiency of U.S. autos now on the road by an average of about 30% and aim to stay ahead. The new 54.5 U.S. standard for 2025 is one-sixth weaker than Europe's for 2020, and may be leapfrogged again—even by China, whose current standards flunk most U.S. SUVs.

Second, historically cheap gasoline makes America the only automaking country where such inefficient cars can be affordably fueled. European autos are about 30% more efficient, and driven 60+% less per capita,[90] not because their designers are more capable, but largely because a gallon there is taxed to cost $6–$9, not $2–$4. This affects auto design. Saving a pound by shifting from steel to aluminum costs about a dollar and saves about a gallon of gasoline over 12 years' driving, so if that gallon is worth, say, $3, the lightweighting pays back in four years, longer than most buyers want. U.S. automakers' unusually cheap domestic gasoline makes domestic buyers less eager than foreign ones to buy efficient autos. This mismatch between home and export markets' preferences is a competitive weakness in a global industry.

Third, gridlocked federal policy hasn't helped. For more than two decades, oil companies have called for stiffer auto-efficiency standards, and automakers for higher gasoline taxes. Many environmentalists want both; many politicians want neither. These titanic lobbies fought each other to a debilitating standoff until a brief interlude of federal policy coherence, and a wave of reform as the automakers surveyed the wreckage in 2008–2009, began to break the logjam. Yet the same conflict between domestic and foreign market expectations for this global industry is now being replayed in climate policy.

Dealers sidebar), a new automotive strategy is emerging: (1) Using feebates to vault initial price barriers, quickly introduce Revolutionary vehicle fitness, and use it to enable electrification. Such Revolutionary+ autos tightly integrate three game-changing technologies—advanced ultralight materials, their rapid structural manufacturing, and electric powertrains. (2) Drive and exploit the rapidly falling costs of all three to build volume, gain share, and cut costs even more. While legacy manufacturers are wringing pennies out of the nearly flat learning curves of century-old steel stamping and engines, these three learning curves are fresh and steep, savings thousands of dollars per car—and all three powerfully reinforce each other.

This strategy could be as transformational as jumping from tiny refinements in mechanical typewriters to the dramatic Moore's Law–driven gains in computers. IT and electronics are now America's biggest industrial sector; typewriter makers are gone. The CEO of BMW, pushing the carbon-fiber-and-electrified auto frontier, gets it: his speeches announce that his firm does not intend to be a typewriter maker.

For the companies in the highly competitive auto industry, however, the challenges are both exhilarating and terrifying. Executives must decide which of many types of vehicles to build, what materials and manufacturing processes to use, and how quickly to invest in revolutionary advances. They must make these choices without knowing whether the coming decades will bring recessions or prosperity, spiky or stable oil prices, high or low interest rates, helpful or crippling regulations, even war or peace. The industry has long lead times: typically four years' research and eight years' development to design and start mass-producing a new vehicle, followed by roughly eight-year cycles of cosmetic freshening, reskinning, restyling, reengineering, and redesign. Wrong bets can be fatal. Look how many automakers, from Duesenberg to Hudson to Nash, have vanished, or how close GM and Chrysler came to disappearing in 2008.

Fundamental to a durable automotive sector, therefore, is a strategy of systematic derisking by cutting capital intensity, lead times, oil dependence, borrowing needs, complexity, inflexibility,

IMPLICATIONS FOR AUTO DEALERS

The shift to Revolutionary+ autos is important to auto dealers, who as a whole are a powerful force that provides a tenth of many American communities' sales-tax revenues. During 1999–2009, the average U.S. dealership made only about $50 net profit selling each new auto, or $40,000 per year, but cleared $94,000 on used-auto sales and $279,000 on service and parts. Ultrareliable, extra-durable, radically simplified autos could threaten this model. Even oil changes could become a tale to tell the grandkids.

However, new software and hardware upgrade opportunities would abound, creating new businesses that could rival providers of smartphone apps. Dealers could become the center for customizing increasingly software-based autos and for add-ons, ranging from extra range and pep modules to integrated entertainment and security options. Dealers have always found ways to exploit their customer relationships and skills when technologies changed. Mr. Goodwrench is becoming Ms. Goodchip, but life goes on.

The smarter dealers are already eager to get their hands on Revolutionary+ autos because of their powerful potential to create value and engage customers. Dealers could well find that early adoption and expertise could make them as popular as iPhone distributors with precious stocks of the latest hot model. When oil prices spiked in 2008, high demand and short supply drove up the prices of hybrids like the Toyota Prius and Ford Escape, often to thousands of dollars above list price. Prius had earlier doubled dealers' average margins for a considerable time. Supply-constrained introduction of Revolutionary+ autos could well repeat this happy (for dealers) history, but repeatedly and in short model cycles more akin to those of consumer electronics.

and societal impacts (especially carbon emissions). Revolutionary+ vehicles, once mature, could potentially do *all* these things.

Of course, they'd introduce new, nontrivial risks of their own. Manufacturers will need to change cherished perceptions of value, selling light weight and acceleration instead of heft and horsepower, and beating their own legacy products before competitors do. The perception of a safety problem can kill a model or tarnish an entire brand. Yet with Revolutionary+ autos, consumers will be asked to believe in the safety of a number of new technologies. Will they be convinced, for instance, that ultralight autos are safe on roads filled with 18-wheelers? Or that the featherweight vehicles made of the same light, super-tough, noncorroding materials now familiar from sporting goods will last for 15 years or perhaps much longer? Even clever marketing campaigns may not be enough to overturn deeply held beliefs that only weight brings safety and durability.

So there's no question that leading on this path to an oil-free future is risky. But here's the surprising twist: lagging can be still riskier. The cheapest and fastest ways to save oil and carbon, and to meet automakers' other seemingly conflicting requirements, are also the best ways to manage business risks and exploit new business opportunities.

That's because Revolutionary+ automobiles are potentially simpler, cleaner, higher performance, safer, more reliable, and more durable than today's autos. They permit mass customization because most functionality is in software—they're more like computers with wheels than cars with chips. They enable production with shorter cycles and more flexible scale. They offer more potential for further cost reduction and simplification as even better materials, manufacturing methods, and powertrain components emerge and converge. They make cheaper *any* of the four innovative powertrains now extant—battery-electric, fuel cell, plug-in hybrid, or advanced-biofuel advanced-engine hybrid—driving vibrant competition and rapid improvement in all four (and perhaps others not yet thought of). They need less capital investment. And the learning curves behind their three advanced technologies will give pioneering companies lower manufacturing costs than slower competitors, bringing the first movers and fast followers most of the spoils and the laggards most of the spills. This makes incrementalism the *high*-risk strategy.

Consider how Toyota's boldly accelerated 1997 Japan launch of Prius is still challenging competitors 14 years later to catch up to the company's overwhelming dominance of the hybrid marketplace. Being a pioneer let Toyota "green" and pep up its luxury models and hybridize Camry, one of the world's most popular sedans. Nissan now seems to be seeking similarly to capture the market lead for battery-electric vehicles. The first automaker to jump to Revolutionary+ autos will move much faster by exploiting three simultaneous and synergistic technologies, not just one—the hybrid powertrain—as Toyota did.

The risk of lagging is equally great for automakers' suppliers. A switch to electric traction would make obsolete nearly 30% of sales in Japan's $430-billion-a-year automotive supply chain.[91] But as Hiroshi Tsuda, a former Suzuki president, says: "This is not a crisis. It's a big opportunity." As always when technologies undergo tectonic shifts, innovators find ways to stay ahead—and to reap even greater profits. That will be true not only for autos but for all other vehicles.

USING AUTOS MORE PRODUCTIVELY

As we've seen, better designs and materials can enormously increase the efficiency of automobiles. Now we come to the second big part of the efficiency story: *using* autos more productively. We can eliminate the need for many trips entirely, and we can use vehicles in smarter ways, improving access to places or goods with fewer, shorter, or faster trips. America's real cost of driving (see The Real Cost of Driving sidebar) makes this an economic as well as a national-security imperative.

First, though, we must explode a deeply held myth—that efforts to reduce travel inevitably take away cherished freedoms, choices, and mobility. This myth is so powerful because such fears are real. After all, one effective way to take autos off the road is to simply decree that you can't drive. China, for instance, has yanked hundreds of thousands of autos off Beijing's streets to cut pollution. No wonder taking steps to reduce miles traveled raises the ominous specter of Big Brother or intrusive government.

The efforts we advocate don't curtail freedom or choice. On the contrary, they'd provide wider choice, greater freedom, and more mobility through diverse alternatives to individual autos. But will those alternatives still be attractive as we adopt Revolutionary+ vehicles that drastically reduce the cost of driving? (There are compelling reasons why this won't materially increase driving and cause fuel use to "rebound."[92])

Changing how we use autos doesn't just save oil; it also creates new business opportunities. And it can change a core function from chore to joy. Imagine waking up and having options: Do I feel like jumping into an electric sports car from my carsharing program, and getting on a road that is

THE REAL COST OF DRIVING

Most Americans have only one real transportation option and one option for fueling it; you can "choose" among several vendors of almost-identical gasoline. Due to our lack of other mobility choices, we're powerless when gasoline prices soar. Shrinking government coffers can't sustain our aging transportation infrastructure, let alone significantly expand it. Road accidents, though declining, still kill about as many Americans as breast cancer or diabetes and injure five million a year. Throw in pollution and climate change, and the hidden societal costs of oil-powered U.S. automobiles add up to $820 billion a year (fig. 2-14).[93] That analysis seriously understates energy security costs that by themselves are probably upwards of a half-trillion dollars per year (chapter 1), 40% of which is attributable to fueling autos.

These results are not necessarily surprising. Decades of dedicated road and parking-lot building with little or no proper pricing, competition, information, or opportunity have created socialism for drivers and free enterprise for most other modes of transportation—not a level playing field with honest prices and open choices. But we can fix these distortions by applying some smart market thinking to changing how, when, where, and why we use our vehicles. Along the way, we'll save millions of barrels of oil every day, adding to the savings from Revolutionary+ autos.

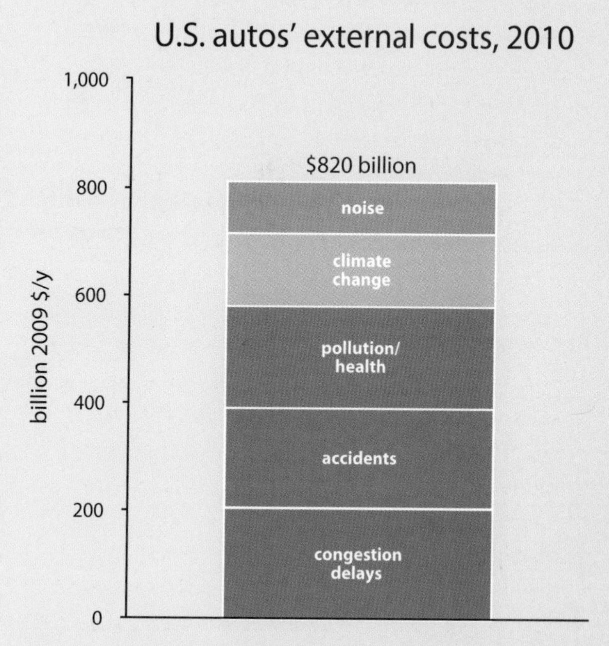

Fig. 2-14. The hidden costs of U.S. automobiles, all paid not at the pump but in ill health, delay, and loss of well-being. Not included are costs to national security and costs of roads, parking, policing, et cetera, paid through general taxation.[94]

fast and clear of congestion because of intelligent pricing and better management? Will I fund and enliven my ride by accessing a social network and offering to share it with a friend or stranger headed the same way? Or should I swipe my universal transit card, jump on a readily available bus, get some work done on the free Wi-Fi, then take a bikeshare for the last few blocks to get some exercise? Or do I want to work from home and attend the day's important meetings via my computer's or smartphone's virtual-presence features?

Providing all these options won't always be easy. Americans cherish the freedom to hop into their private autos and drive wherever they please. To some, trying to increase the use of buses or carpools or adding tolls to roads seems like intrusive government at its worst (though building those taxpayer-funded roads was a major imposition, especially on displaced residents and nondrivers). So the key is showing people that these wider choices improve their lives. We can make better use of existing roads, save time, build better communities, and expand options in everything from types of transport to insurance policies. Moreover, we can create a more equitable society where the poor, young, elderly, and disabled enjoy better access.

So how can we achieve these benefits? The most obvious solution is reducing the need to commute to work. Why fight traffic and burn fuel if you can do the same work from home? Telecommuting reduces an average worker's miles traveled by about 40%, saves more energy in avoided office space than the extra energy it uses in the home, and can bring important side benefits, such as more family time and improved morale, retention, and productivity—by 81% among British Telecommunications' telecommuting employees.[95]

Beyond telecommuting, the solutions fall into four main categories (fig. 2-15): innovative pricing, alternative commuting, smart growth, and system-wide transportation efficiency improvements. The basic idea is to make driving and parking bear their true costs at the time of use, foster

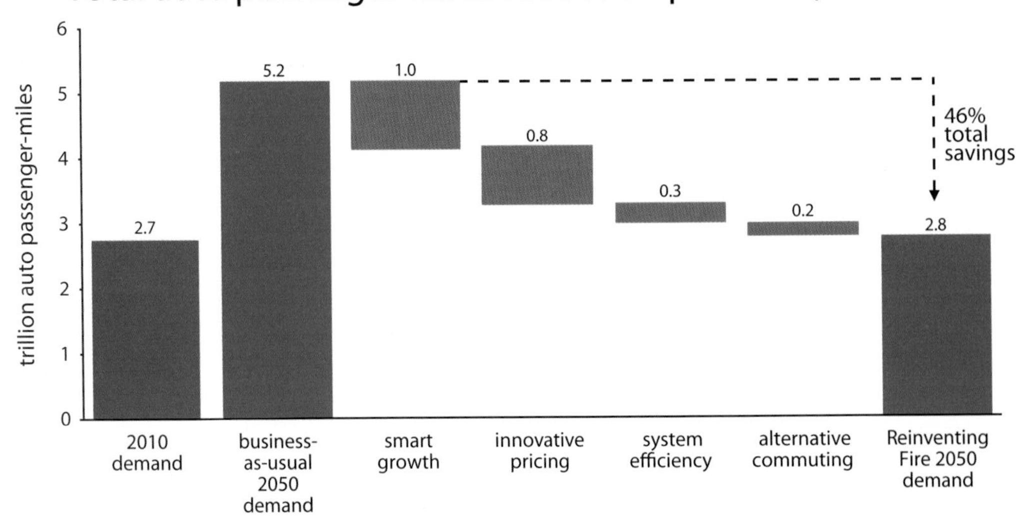

FIG. 2-15. Four ways of using autos more productively can provide the same or better access services at lower cost with 46–84% less driving in 2050.[96]

genuine competition between different modes of transportation (or between transportation and its substitutes), and integrate land-use choices with personal mobility options. Together, based on empirically observed U.S. performance in specific implementation experiments, *these opportunities can by themselves, without making vehicles any more efficient, save 46–84% of U.S. automotive fuel* by allowing us to travel fewer passenger-miles to do the same tasks, thus offsetting the entire projected growth in passenger-miles to 2050.[97]

Innovative Pricing

Reducing vehicle-miles traveled (VMT) can lead to another reboundlike effect: induced demand. If highways become less congested because more people are carpooling or not driving to work, why wouldn't we drive more to take advantage of a newly available, uncongested roadway? The solution is well-designed pricing to achieve the balancing act of decongesting roadways *and* discouraging driving during peak travel hours.

We must do this anyway, because as we adopt efficient autos that use far less fuel, little or none of it gasoline, the fuel-tax revenues that keep up our roads and bridges will dwindle and ultimately vanish. (It would be hard to tax car-charging electricity—a ubiquitous and fungible commodity—or hydrogen; even biofuels can be homebrewed.) The federal gasoline tax isn't indexed for inflation and hasn't risen since 1993 despite enormous increases in the cost and scale of highway construction and maintenance. Real highway spending per mile traveled has shrunk by nearly 50% since the Highway Trust Fund was established in the 1950s, and scary deficits loom: $400 billion by 2015, $2.3 trillion by 2035. So we must pay for our driving infrastructure differently—but how?

Driving imposes social costs like oil dependence and pollution. Autos take up space. Both driving and autos contribute to congestion. Weight also causes road wear, bridge fatigue, and fuel waste (hence pollution).[98] But probably a lot more of driving's societal cost comes from driving than from autos, so it'd be smart to fill the road-funding gap mainly by charging for vehicle-miles driven, by switching from a gasoline tax to a VMT tax.[99]

In 2005, Oregon tested this fundamental change in the way people pay to drive. GPS units recorded how many miles volunteers' autos drove. At the pump, instead of paying the state gasoline tax, the volunteers paid a fee (higher in peak travel periods) based on how many miles they'd driven. Compared to a control group, these Oregon drivers reduced their total mileage by 15% and their rush-hour mileage by 22%.[100]

Suspicions of Big Brother snooping could be eased by offering an array of choices, such as using odometer readings at regular inspection visits to record vehicle-miles, installing on-board diagnostic logs (OBDs), ensuring that GPS location data would be kept private, or simply using GPS, as Oregon did, to store and transmit only mileage, not position.

Pricing mechanisms that reduce peak driving also avoid the need for costly new infrastructure so we can focus on fixing the roads we already have. And congestion pricing can work even without a VMT tax. Just charge drivers a fee to drive when roads are clogged. After Singapore started charging to drive downtown during rush hour, the number of autos coming into town dropped 44% (and solo trips plunged 60%), speeding traffic by 20%. London's fee, now a hefty $16 per day, has cut average traffic inside the central city 15%, sped it up 30%, and greatly expanded bus and bicycle ridership—a special boon to low-income citizens.

Insurance companies are also starting to use a creative new approach that rewards both driving less and using efficient vehicles: pay-as-you-drive (PAYD) insurance. Paying casualty insurance by the mile makes the premium roughly proportional to the risk, so low-mileage drivers needn't subsidize high-mileage drivers. PAYD insurance has been observed to cut driving about 8%[101] and accidents even more.[102]

Still another way to reduce driving harnesses market-savvy parking policies. A typical city has three times as many parking spaces as autos, yet cruising for a spot is thought to cause a third of major cities' downtown traffic and even more congestion. All those spaces eat up a vast area of land—81% of the Los Angeles Central Business District, 31% of San Francisco, and 18% of New York. Where charged, parking fees rarely cover the true cost of building and maintaining each space, which totals tens of thousands of dollars per spot (just building the parking at Los Angeles' Disney Concert Hall cost $50,000 per space). In all, parking gets $151 billion worth of annual subsidies—perhaps the greatest single cause of excessive urban driving.[103] Yet much of that overbuilt parking capacity is required by zoning and building rules mandating as much parking as drivers might conceivably use if it were all free.

The alternative? Frankfurt, Germany, actually forbids developers of workplaces to provide parking. Britain plans to tax firms that provide free or below-market employee parking. Metro Sydney taxes nonresidential parking spaces and uses the revenue to fund transit improvements. In Tokyo, you can't even buy an auto without proving you own or rent a parking place.

Such policies would be a hard sell in the U.S., where cheap parking is viewed as almost a sacred right. But we can, as companies in smog-prone parts of California are required to do, pay employees the fair market value of the parking at their companies' lots and charge them that rate when they do park there. Workers can use the allowance to pay for parking, or they can leave their autos at home, get to work by other means, and pocket the money. That's an incentive for real competition and wider, smarter choices.

Charging drivers for the costs their driving incurs and imposes is fundamentally fairer than socializing their costs to all taxpayers, a third of whom, though they benefit from road haulage, are too old, young, poor, or infirm to drive themselves—a potentially potent coalition once they realize how they're subsidizing drivers. They'll argue that drivers should get what they pay for—but also pay for what they get. And if the gasoline-tax system isn't fixed, the already-broke Highway Trust Fund will become unable to keep America's traffic moving. Inaction is not an option.

Alternative Commuting

About 77% of U.S. job commuting is by single-person auto. Almost all autos are meant to carry at least four adults, yet for daily commutes, single-occupant drivers outnumber more than tenfold the *combined* total of all Americans who carpool, ride public transit, walk, bike, or telecommute. That lone driver may be the traditional American way, but it doesn't have to be our future. Coaxing more people into each auto, sharing autos, or eliminating trips can together reduce work-related VMT by 6–12%.

Carpooling and Ridesharing

The most direct way to reduce VMT is carpooling. For many years, metropolitan areas from Washington DC to Los Angeles and San Francisco have offered special highway lanes for high-occupancy vehicles (HOVs). Some regions have seen considerable success. In San Francisco, about 3,000 three-person carpools spontaneously form daily at East Bay pickup points, saving about $30 million a year in fuel, time, and transit subsidies. Most riders take the bus home. Similar "slug lines," like taxi lines for carpoolers, form in Washington DC and Houston. In the District, two million rides per year save more than two million gallons of gasoline.

Such ridesharing could get a boost from social networking. New applications with names like

Avego and NuRide link drivers with riders, eliminating the uncertainty of depending on slug lines. Even without such tools, ridesharing programs typically displace 5–15% of single-auto commutes—more when incentives are offered, which in the Puget Sound area attracted 10–30% of commute trips into vanpooling.

Carsharing

Ten years ago, Robin Chase and Antje Danielson were sitting in a Cambridge café. Danielson had just returned from Berlin, where she'd seen a branded shared car on the street. The women launched Zipcar—now the largest U.S. commercial carsharing company, with three-fourths of the market.

The business model was simple: Americans now each pay about $8,000 per year for a vehicle that stands idle 96% of the time. For people who don't need an auto to get to work, this makes little sense. People could save money (and Zipcar could make money) if they paid for an auto just when they needed one. As German climatologist Hans-Joachim Schellnhuber quips, "Buying a car to get mobility is like buying a three-star restaurant to get a good meal."

Zipcar now owns more than 8,000 autos in 30 diverse models, parked in 1,200 locations in cities and university towns across the United States, Canada, and the United Kingdom. Its members pay an annual fee plus an hourly rate, fuel and insurance included. Zipcar says each of its autos takes 15–20 personally owned autos off the road. Similar programs in Europe have resulted in a 30–70% drop in VMT. *The Economist* in 2010 reported carsharing models emerging in over a thousand cities worldwide; U.S. membership is projected to approach 4.5 million and revenue $3 billion by 2016.

A new take on the carsharing business model even allows individual auto owners to get in on the action via person-to-person ("P2P") business models. In Boston and San Francisco, owners can register their vehicles with companies like Spride and RelayRides that rent out their vehicles by the hour, day, or week.

Why limit such sharing schemes to autos? In some bike-friendly major cities in Europe, 30–40% of commuters walk or bike. Paris has had enormous success with its Vélib' (*vélo libre*) free public bicycles, now numbering 17,000 in 1,200 self-service stations throughout the city center. Washington DC recently expanded its own bikeshare program, which charges an annual fee and a small usage fee. Having bikes available can remove one of the major disincentives to use transit—that it doesn't go exactly where you want.

Getting people out of their autos sometimes requires new laws and policies. Some businesses can't eliminate costly parking places, for example, because zoning laws demand a certain number of spaces. Auto insurance policies often don't cover ridesharers or P2P (except in California), raising the financial risk when accidents happen. And states and cities impose auto rental taxes on carsharing, increasing the costs of the service.[104]

Smart Growth

For decades, Atlanta's inhabitants pursued the American Dream—moving to a nice house in the suburbs. But the resulting sprawl has dimmed the dream. Atlantans drive more miles than most other Americans and suffer crippling congestion. But in 1999, one developer in the sprawling city chose to build a dense community of residences, shops, and offices on an abandoned steel-mill site in the heart of Atlanta, rather than scattering them across three suburbs. The 130-acre Atlantic Station offered homes for 10,000 people, employment for 30,000, recreational opportunities for millions, and easy access to public transit. For auto-choked Atlanta, the results have been revelatory: area VMT dropped by 30%.[105]

Shifting 60% of new U.S. growth to become more "smart" and compact, like Atlantic Station, would save as much fuel as a 28% rise in the efficiency of new vehicles by 2020. Far from raising costs (to pay for transit), these smart-growth developments typically *de*crease buyers' costs, and later their tax bills. New Jersey found that each new homeowner in a sprawl development pays about $10,000 more for extra roads and extended infrastructure.[106] Most compact development avoids those costs, increases household savings rates, enhances real-estate values, holds value much better in market slumps, and boosts developers' profits.[107]

Smart growth also greatly improves quality of life. "Most people believe the alternative to autos is better transit—in truth, it's better neighborhoods," explains Alan Durning of the Sightline Group. Such neighborhoods, he adds, make the automobile "an accessory of life rather than its central organizing principle." This rebuilds community by reversing decades of what architect Andres Duany calls "meeting our neighbors only through windshields."

It's easy to see why smart growth commonly cuts the length and number of trips by auto by half, and in the best recent designs by three-fourths.[108] Such developments are dense, putting the dry cleaner, bagel shop, gym, or even office a short walk or bike ride away (while also facilitating local delivery and online shopping for heavy or bulky items[109]). They're typically built in underused urban neighborhoods, bringing people nearer jobs and kids nearer schools. They avoid the repetitive "dead-worm" cul-de-sacs of many suburbs, where visiting a neighbor on the next street may mean jumping in an auto and driving to a main road before dipping back into the neighborhood. They can also improve public health through exercise by making walking and biking safe again. And they make buses and light-rail systems more effective, especially in communities like Arlington, Virginia, where dense mixed-use development is encouraged to cluster around Metro stops.

Pricier urban housing can often *decrease* a household's *total* costs.[110] Why? Because suburban transportation costs—gasoline, congestion, accidents, et cetera—are higher. Traditionally, housing is deemed "affordable" if it consumes no more than 30% of income; on that basis, 69% of U.S. communities have affordable housing. But including transportation costs cuts the affordable fraction of communities to 39%.

That's why Fannie Mae offers—if you can get one—"locationally efficient mortgages" with easier qualification for households near work or transit, reflecting their better cash flow and lower default risk. How much lower? Natural Resources Defense Council scientist Dr. David Goldstein, who pioneered that concept, notes that an average location-inefficient homeowner, over a 30-year mortgage life, pays about $300,000 for car commuting and $75,000 for home utilities—in all, twice the $175,000 median price of the house. An energy-retrofitted, locationally efficient home saves at least 63% of that total, or one-third more than the price of the house. No wonder high-driving regions have lately had by far the highest mortgage default rates while smart-growth, compact, transit-served areas have had the lowest. The differences in default rate were as high as 40-fold after controlling for other variables like credit scores and income. If this were more widely appreciated, smart growth might help inoculate our economy against another mortgage-induced financial meltdown, protecting capital markets and cutting interest rates.[111]

Together, then, smart growth is important not just to the developers whose property values, margins, absorption, and appreciation it enhances, but to all business. Workers who spend less time commuting, arrive less stressed, can better balance work with family life, and have more time with their kids are more valuable and productive.

Businesses in a smart-growth area can better recruit and retain the best workers yet needn't pay them so much to offset their high commuting costs. Less traffic and perhaps more exercise also mean safer and healthier people, cutting health-care costs and sick time. Smart growth is just another part of smart business.

Boosting Efficiency in Transportation Systems

What's the most frustrating part of taking public transport? Often it's not knowing when the next bus or train will arrive.

Plenty of entrepreneurs are stepping into this breach. Transit systems like the Massachusetts Bay Transit Authority already broadcast real-time bus locations. NextBus uses GPS to provide accurate vehicle arrival and departure information and real-time maps to any passenger with Internet access. More innovation is brewing. Anna Jaffe's Mobi team at MIT is working on an ambitious integration: you enter your destination into your smartphone (which already knows where you are), and up pops a list and map of all the ways to get there—transit, ridesharing, Zipcar, free or rental bikes, whatever—with their cost, location, and real-time-savvy estimated arrival time.

Such "intelligent transportation systems" (ITS) can make traffic flow more smoothly by controlling traffic lights to match changing conditions, advising drivers about hazards or jams ahead, using ramp meters to smoothly insert autos into the traffic, or charging tolls on the fly electronically, to name just a few. Taking all these steps would cut fuel use by 5% and prevent 308 million person-hours of delay per year, worth $6.5 billion.[112]

That leaves one more "system" that can be tweaked to save fuel—drivers themselves. Changing *how* we drive can significantly boost efficiency in any auto, on any road.

Just bringing tire pressures up to recommended levels and using improved synthetic engine oil would cut U.S. gasoline use by 1–3%. Once on the road, the key to fuel efficiency is a light touch on the accelerator and brakes, coasting up to red lights or traffic jams, avoiding jackrabbit starts (except in many hybrids where brisk acceleration can *save* fuel), and driving slower. Hybrid autos, and some nonhybrids, also turn off their engines automatically when stopped. We can remove unused heavy junk from the trunk, open windows at low speeds instead of using air-conditioning, and put shades in the windshield when parked to keep auto interiors cooler.

The most effective way to encourage these changes in driver behavior, it turns out, is to give drivers more information about how they're doing. Real-time mpg indicators on the dashboard can turn driving into a contest to see who can get the best mileage. A British government report estimates that efforts to promote "eco-driving" can save 10–15% of fuel in the long run—and the cost is nearly zero.

Together, all these ways to use autos more productively can save a 2010 net present value of about $0.4 trillion, deliver the same or better access to where we want to be, and improve the quality of our lives and the strength of our families and communities. Far from losing convenient access, we'll improve it. We'll unclog traffic, save time and tension, cut pollution and noise, save lives now lost to traffic accidents, reclaim land from roads and parking lots, reduce tax burdens—and take another giant step toward getting off oil.

THE REST OF THE STORY: BEYOND AUTOMOBILES

Automobiles use 60% of U.S. transportation's oil—by far the biggest leverage point in the transportation sector's energy use. But what about all the other vehicles we rely on? What opportunities lie in their smarter design and use?

Heavy Trucks

Jimmy Ray is a straight-talking, charismatic veteran of the U.S. trucking industry with decades of experience maintaining, driving, and managing Class 8 trucks.[113] He runs Mesilla Valley Transportation, a large, highly successful, locally owned freight service provider based in western Texas and southern New Mexico. One secret to his success: efficiency. The big rigs in Jimmy's 800-tractor fleet are sleeker than typical 18-wheelers, since two-thirds of the energy needed to move trucks over the road is caused by aerodynamic drag. Many of his trucks sport wide tires in place of the usual two adjacent narrow ones, further reducing aerodynamic drag and cutting the rolling resistance that accounts for the remaining third of tractive load. Jimmy also offers quarterly rewards to the most efficient drivers (and every year, the most efficient driver wins a Harley-Davidson motorcycle). U.S. heavy trucks are officially projected to average 7.8 mpg by 2050. Yet in 2010, forty years early, Jimmy's fleet averaged 8.5 mpg.

His innovations are but the tip of an iceberg of technical, operational, and logistical improvements for trucks that can save about a tenth of U.S. oil, help insulate the trucking industry and its customers from high oil prices, and keep the U.S. economy humming. Moving goods, not people, burns upwards of 28% of the fuel used for transportation. Despite a sophisticated industry whose trucks use highly efficient diesel engines, most of that fuel is wasted. This doesn't have to happen. Focusing on design and operational changes can cut total U.S. big-rig diesel fuel consumption 41% by 2050 despite 88% more heavy-truck-miles.[114]

As with autos, the solution starts with physics. Modern aerodynamic improvements (fig. 2-16) and today's better, wider tires can take the typical 2010 Class 8 tractor-trailer rigs dramatically beyond their average of 6 miles per gallon of diesel fuel.

Another tweak connects fewer axles to the engine, cutting hundreds of pounds in driveshafts, gears, and differentials. The unpowered "tag axles" tag along behind the powered ones. Saved weight is usually taken up in more payload—a valuable feature.[115] Saved drag often permits an engine that is one size smaller, and hence lighter and cheaper. Electronic active suspensions can recover some of the energy of bouncing over bumps and potholes. Combining all these technologies could yield a competitively priced 8.9 mpg truck.[116]

FIG. 2-16. The Daimler Innovation Truck (left) and Renault Radiance (right) illustrate aerodynamic progress, including fairings, underbody panels, and side-view cameras replacing mirrors.

Adding some simple hardware brings another leap. To understand why, look at the typical truck stop, where dozens of trucks are parked with their engines running to power the air conditioner, lights, and electronics during drivers' mandated breaks (which must last 10 hours after 11 hours of driving). These idling engines eat up 12% of typical heavy trucks' fuel. Two-thirds of this idling waste can be saved with an auxiliary power unit (APU) such as a small diesel generator, fuel cell, or battery. Plugging into an electrified parking space (EPS), as moored ships and parked airplanes do, saves all of it.

A huge South Bronx truck depot handling most of greater New York's produce offers a glimpse of what's possible. Community activist Majora Carter persuaded the operators to double the overnight parking fee from $10 to $20, but also to include an EPS to eliminate idling. Drivers come out ahead by saving more than $10 worth of fuel plus engine wear and tear, while the neighborhood benefits from big drops in its extraordinary levels of diesel-particulate-induced asthma.

Such straightforward design improvements would save 1.7 million barrels of diesel fuel each day (fig. 2-17). Even assuming 2009's low diesel price of $2.47 a gallon, this translates to savings of $64 billion per year.

Fuel savings per ton-mile of cargo improve further when trucks hitch a second or third long trailer behind the first on highways, on which most cargo moves. Such mammoth rigs are controversial because of their size and weight, which rises from 80,000 to as much as 120,000 pounds. But the number of axles rises even more, from five to nine, so weight per axle—the key to road wear—*decreases* by one-sixth. Smarter links between trailers and active electronic safety aids

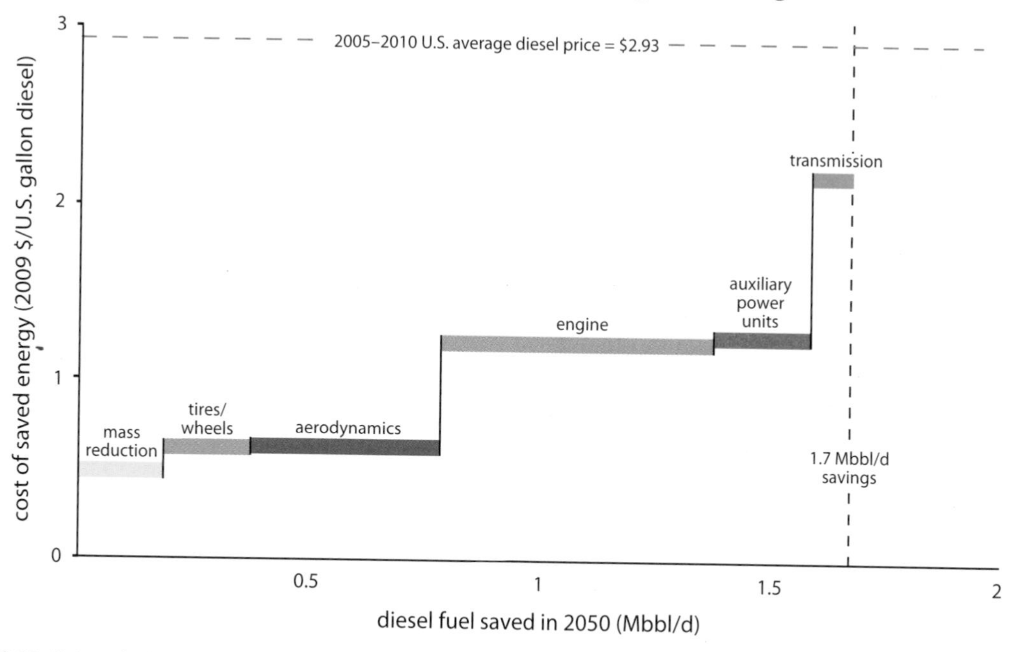

FIG. 2-17. Better design of single-trailer Class 8 trucks could save the U.S. 1.7 million barrels of diesel fuel per day at a fraction of the cost of diesel fuel.[117]

make the long trailers of such turnpike doubles, as they're known, inherently more stable than today's short double trailers, and much safer than today's triple trailers. Turnpike doubles can boost nationwide heavy truck efficiency by 7% (beyond the design changes cited in figure 2-17) because of their doubled cargo capacity. Shifting to these long combination vehicles can also cut congestion by reducing the total number of trucks on the road and cut cost by saving tractors and skilled drivers (of whom there's a worrisome shortage).[118]

To reap these benefits fully, the allowable trailer length should be increased nationwide—as some states have already done—from 53 to 59 feet. Furthermore, the maximum allowable gross vehicle weight rating (GVWR) should be increased to 97,000 pounds (which is still less than Britain's current limit of 110,000 pounds).[119]

With such big, technologically advanced rigs humming along the highways, the U.S. could haul its cargo with an amazing 41% less fuel than today's 6 mpg average. Further improvements could come from more efficient truck refrigeration and advanced emissions controls, superefficient engines, hybrids on some routes, and closer attention to auxiliary and accessory loads.

The even better news is that all these reductions would be relatively cheap: a fit, aerodynamic truck with wide single tires and a superefficient engine would pay back in just over three years with diesel at $3.00 a gallon. As Jimmy Ray likes to say, "We get our tires for free"—that is, the small incremental cost of superefficient tires pays back almost immediately.

Overcoming Obstacles to Heavy-Truck Innovation

The bad news is that this dramatic cut in trucking fuel faces more hurdles than just the cost of the technologies. Higher maintenance costs, such as for add-on aerodynamic features, sometimes cut into the fuel savings. Tractors and trailers are usually made and often owned by different companies, making it hard to integrate the design of the whole rig to achieve the best aerodynamics.

Industry standards, vertical integration, new business models that share the savings, and the demands of big customers can help solve these problems. For instance, in just five years Walmart was able to reduce by 60% its heavy-truck fuel use per ton-mile from its 2005 level (handily beating its 2005 goal of a 50% reduction by 2015) and is looking for more gains.[120] With one of the world's largest civilian heavy-truck fleets—6,400 tractors—Walmart could compel truck, and indeed tractor and trailer, manufacturers to work together to boost efficiencies.

Another problem is the structure of the trucking business. Only about 20% of truck ton-miles get moved by the private fleets of companies like Walmart, where increased fuel efficiency makes clear economic sense. Most of the rest is moved by for-hire firms that, in turn, rely on the services of more than 540,000 independent contractors. The companies hiring these truckers have little incentive to invest in efficiency because the drivers buy their own fuel. And while independent truckers typically drive older, inefficient rigs and would benefit greatly from fuel efficiency, they simply can't afford fuel-saving technologies or newer, more-efficient trucks.

However, clever entrepreneurs are figuring out how to get fuel-saving improvements into the hands of the small operators. Jon Gustafson at Cascade Sierra Solutions has set up a series of trucker-to-trucker kiosks at West Coast truck stops to advise on technologies and best operational practices. In 2004, his colleague Sharon Banks launched "Everybody Wins," a program that finances APUs with lease-to-own contracts along with grants and tax credits. It has since financed 350 mainly small operators' upgrades, saving more than 0.7 million gallons of fuel per year and much pollution.

More broadly, capitalizing the fuel savings of doubled- and tripled-efficiency heavy trucks into a lease could enable small operators to buy a new rig before big competitors who first buy those breakthrough trucks have put them out of business—a great opportunity for leasing and financial companies.[121]

Trucks can also save fuel when drivers are trained in efficiency. Highway cruising at 65 mph instead of 70 boosts fuel economy by more than 8%. Route optimization can more than make up for the slower speed and help operators deliver on or ahead of schedule. Still lower speeds further reduce air drag, since drag increases as the *cube* of speed.

Picking the right gear (out of as many as 18 on a big rig) saves as much as 10% of fuel, so driver training or dashboard up- and downshift lights help. So can innovative cruise control systems: Daimler's new Predictive Cruise Control plans optimal gearshifts a mile in advance.

Jimmy Ray trains his drivers to accelerate slowly and cruise efficiently with the aid of electronic controls and displays that show real-time mpg. He also limits their speed with a 63-mph governor and monitors their performance with an onboard tracking system. The result? A 6% gain in mpg.

These reforms aren't always easy to implement. Drivers paid by the haul have a powerful incentive to drive faster. Shippers and carriers are often not well coordinated, even internally. But we can take effective steps. Paying drivers by the hour would encourage efficiency and could save more fuel than the extra labor cost. So would regulations, electronic logs, GPS uplinks, and speed governors that keep drivers from driving too long and too fast. Compound trailers can be loaded in tandem, or their drivers paid to wait for loading—it's cheaper than needing an extra tractor and its driver for the second trailer's whole route.

Using Trucks More Productively

We needn't stop with doubled-efficiency trucks. We can also be much smarter about how we use trucks in the first place. We can reduce the number and length of trips, or figure out how to ship fewer goods, or make sure our trucks are full.

Here are the key areas for improvement.

Logistics

Both independent truckers and big fleets abhor "empty" miles—trips made with no cargo (which currently typically make up 10–28% of a heavy trucking fleet's total miles).[122] The solution is consolidating shipments across carriers, shippers, and platforms via third-party logistics (3PL) firms, IT companies, and initiatives like the Empty Miles Service. This boost in truck productivity per mile saves 5–15% of heavy-truck fuel at a profit. One European retailer raised the average load per truck from 85% to 93% of full capacity, saving 10,000 outbound loads per year. Carrefour's Demeter Environment and Logistics Club helps companies coordinate their shipments to reduce backhauls; one of its retailers has even reported that only 5% of its fleet's miles are empty.

Fewer Miles

Another way to cut fuel and cost is to drive more efficient routes. Coca-Cola's Simply Orange Juice shipments used to stop at Minute Maid's Florida distribution center to pick up juice for delivery. Now the product travels straight from factory to regional distribution centers. Eliminating the extra stop saves 144,000 gallons of diesel fuel each year while stretching product shelf life by up to six days.

Making products or growing food closer to customers can also shorten routes. Certain soft-drink producers are moving toward in-store or home-fountain production; Walmart makes bottled water at its distribution centers, not a remote factory. (Better yet: use tapwater.) Some medium- and light-duty commercial fleets are consolidating home deliveries into secure, insulated giant mailboxes, further reducing hauls.

Less Wasted Space

Surprisingly, empty air still makes up the biggest volume shipped in many firms' supply chains. IKEA, famous for making its bulky furniture products collapsible, even employs "air hunters" who design products to stack more densely and completely fill each truck while balancing denser with fluffier cargoes.[123] Big retailers are pushing their suppliers to cut packaging, redesign products to fit more in a truck, and fine-tune pallet designs to minimize empty space at the top of the truck.

Fewer Tons

We can also save large amounts of fuel by shipping less cargo. Rapid innovation and improvement are shrinking countless products, from industrial equipment to consumer goods like music players. Look at Proctor & Gamble's concentrated detergents. The large volume of water formerly in the bottle, which had to be shipped from afar, is instead added from the user's tap. Products are lasting longer and being designed for repair, remanufacturing, reuse, and recycling. Superfluous packaging is being driven out of the market by higher hydrocarbon and fiber prices, consumer rejection, and pressure from companies that don't want to pay for unneeded packaging, let alone twice—once to buy it and once more to discard it. And those bulky marine and rail shipping containers themselves are becoming radically lighter (even before advanced composites) and more foldable, making them far handier to store and return: by folding to one-fourth its volume in just four steps, Holland Container Innovations' container can save up to 25% of shipping lines' operational cost.[124]

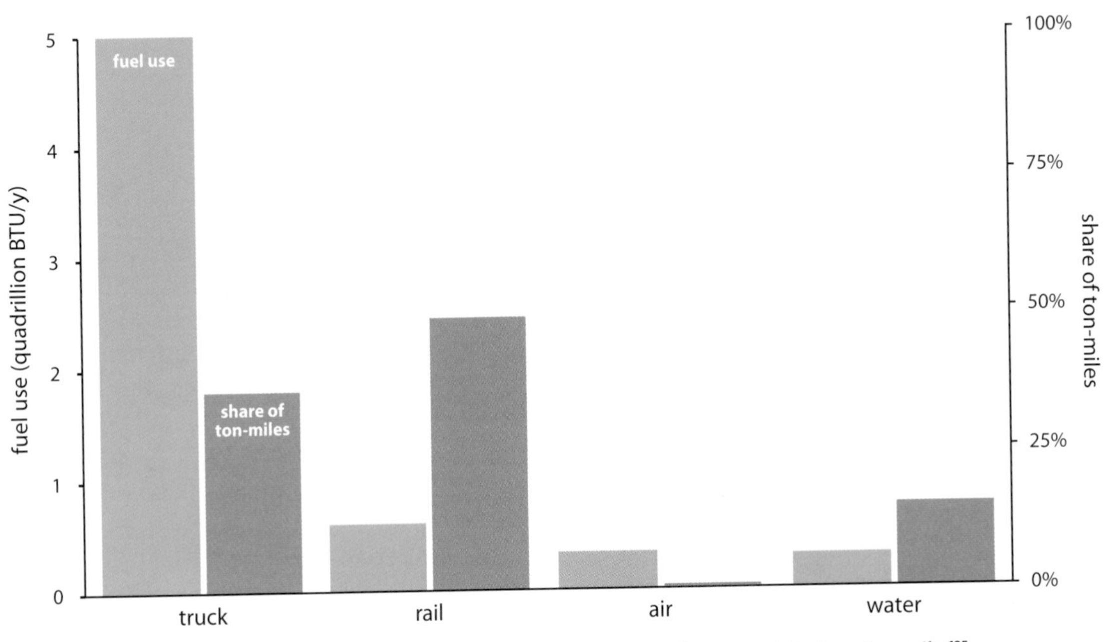

FIG. 2-18. Trucks haul less U.S. freight than railways but use more than twice as much fuel per ton-mile.[125]

Intermodal Freight: Merging Truck, Sea, and Rail Transport

The two oldest means of moving heavy goods—ships and trains—are still the most energy-efficient. Rail moves 49% of U.S. freight but uses only 9% of freight-sector fuel (fig. 2-18). Shifting from truck to rail nearly halves cost per ton-mile and cuts fuel use by nearly fivefold. The trick is to use each mode to do what it does best.

Both rail and ships also beat trucks in capacity, cost, and safety. This permits major gains from shipping goods most of the distance by rail or ship, shifting them to or from trucks for a few dozen miles on each end—so-called intermodal transport. The net energy saving, plus packaging improvements and reduced need to haul oil and coal, can total 33% of total heavy-truck fuel[126]—*beyond* the major potential for hauling fewer tons, cubes, and miles.

Intermodal transport has become so popular that it yields 21% of U.S. railway revenues—the second-biggest segment after coal hauling (which used 48% of 2009 U.S. ton-miles). As chapters 4 and 5 will describe, coal burning could be eliminated, effectively doubling rail capacity. Greater rail capacity could drastically reduce road congestion and deterioration, saving more truck and auto fuel too.

Shifting one-third of our freight to intermodality won't happen overnight. Both rail and ships are usually slower than trucks and can't deliver point-to-point. Current rail infrastructure is stressed and aging, while seaports have their own infrastructure challenges and often impose high harbor maintenance taxes.

However, advances in rail and port design are boosting intermodalism. Thruport terminals reduce the number of interchanges for rail-based freight from 13 steps to one. They improve terminal efficiency 97% by eliminating unneeded drayage between rail terminals, allowing crane transfers from rail to rail, and eliminating the need to store freight overnight outside the terminal. And some ports are shifting from diesel equipment to all-electric to save cost and pollution.

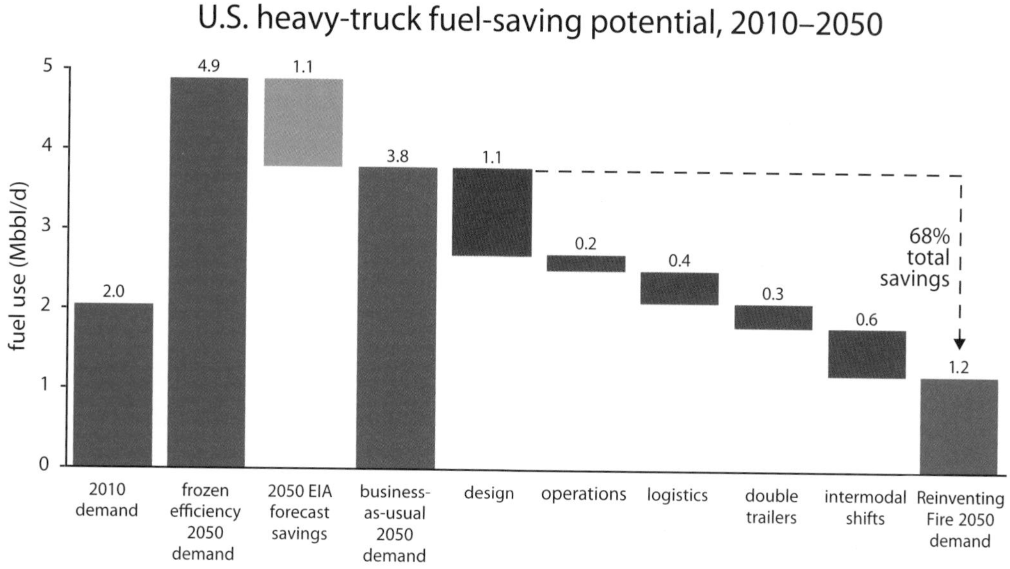

U.S. heavy-truck fuel-saving potential, 2010–2050

FIG. 2-19. A portfolio approach to enhanced efficiency in the U.S. domestic freight sector cuts the 2050 need for heavy-truck fuel to just 1.2 Mbbl/d—none of which need be oil.[127]

Combined Truck Savings

The Energy Information Administration predicts that Class 8 trucks will improve from 6.1 mpg in 2010 to 7 mpg by 2035, which we extrapolate to 7.8 mpg by 2050—29% better than 2010. But figure 2-19 shows how combining improvements in design (including idle reduction), improved operations and logistics, long compound rigs, and intermodalism can cut 2050 fuel use to one-third of projected demand.

Next, let's look to America's third-biggest oil user—in the skies.

Airplanes

Thanks to the marvel of flight, we can leave the U.S. and meet with Chinese clients within a day's time, pop over to Paris for a romantic weekend, or travel around the world in two days. But flying burns fuel—lots of it. Even after decades of improvements in airplane design and airline logistics, which slashed the fuel burned per seat-mile by 82% from 1958 to 2010, U.S. air transportation of people and freight uses 1.3 million barrels of oil per day and is projected to climb to 1.8 million by 2050. But efficiency continues to improve, thanks to many of the same technological advances that make Revolutionary+ vehicles possible—and to bold moves by companies like Boeing, Airbus, and their smaller competitors.

Faced with a critical development decision in 2000, Boeing initially favored a high-speed jet it called the Sonic Cruiser, but airline customers wanted efficiency. Even without changing the plane's basic form, this bore risks. Weight would clearly be the key: each pound removed from a typical midsize jet saves 124 pounds of fuel a year,[128] with a 30-year present value approaching a thousand dollars. The lightweight, high-performance design Boeing envisaged and airlines wanted would have to comprise 50% carbon fiber composites by weight (80% by volume). It'd also need to electrify more systems and depend on a global supply chain. But Boeing plunged ahead with what became the 787 Dreamliner. Some of the development risks proved all too real: design- and production-related delays of over three years diverted focus from the company's next all-new airplane, probably a replacement for the mainstay 737 to compete with new offerings from Airbus and China. But the risk seems to have been worth it: the 787 became the fastest-selling new jetliner ever, giving Boeing a leg up on archrival Airbus. It's expected to use 20% less fuel than a comparable 767-300, proving the market appetite for efficiency just as Prius did for autos.

Moreover, Boeing is turning that technological leapfrog into a breakthrough competitive strategy, using its head start to move technologies developed for the 787 rapidly into existing platforms. The coming 747-8, for example, uses a 787-style composite wing with the same supercritical airfoil shape to cut high-speed drag and, like the 787, electrifies energy-sapping pneumatic systems. The updated 747-8 will use 16% less fuel than its predecessor.

As airplane makers come down the learning curve of making advanced composite structures, they've discovered that the new manufacturing methods, when combined with lean principles gleaned from the Japanese automotive industry in the 1990s, are actually cheaper than those for making metal airplanes.[129] Now automakers stand to benefit in turn from composite airplanes' process advances.[130]

Next-Generation Efficiency

Are further efficiency gains possible? Beyond weight reduction enabled by advanced materials, the most critical airplane performance levers are engine efficiency, aerodynamics, and integrative design that maximizes snowballing weight savings.

The next jump in efficiency, therefore, will come from new designs that improve aerodynamics while making the engines and airframe work better together. Three such state-of-the-art designs offer the potential to reduce fuel use by 59–80%. The first (fig. 2-20, at left) braces the wing with a strut or a truss, making it much longer, lighter, and thinner. A thin wing smooths airflow for dramatically higher lift-to-drag ratio. The second, a tailless design, integrates a single aft engine into the body (fig. 2-20, in the middle). The third design (fig. 2-20, at right) was probably inspired by natural gliders like the Javan cucumber's five-inch seeds that can glide hundreds of yards.[131] The aft engines are more efficient and cut cabin noise. Such a carbon-fiber plane could save more than half the fuel used by today's most efficient aircraft. Composite construction also potentially enables "compliant" structures, allowing the seamless wing to flex and morph into optimal shapes for different flight modes, replacing today's hinged flaps for another 5–12% in fuel savings.[132]

So what's holding up this aeronautic revolution? With blended wing body (BWB) designs, manufacturers and airlines would need attractive ways to fit passengers and cargo into the novel shape. Airports, too, would need to change gate geometries (as some have for the giant A380). But the biggest obstacle is that most airlines lack the capital to replace their fleets, and turnover times tend to be long.

For standard jet transportation, the fastest route to short-term fuel savings is speeding the adoption of more efficient planes like the 787, in turn accelerating manufacturers' innovation and retooling. New policies, such as a scrappage program similar to "cash for clunkers" linked similarly to secured federal loan guarantees for buying superefficient new airplanes, could vault the capital hurdle. So could landing fees that rise with higher noise, emissions, and fuel use. Such policies are already being tried in England, Germany, Switzerland, and Sweden. The charges range from a 6% discount on landing fees for the cleanest aircraft at Basel (and rebates for low emissions at Heathrow and Gatwick—think feebates) to a 40% surcharge for the dirtiest aircraft at Zürich, in a steeply graduated fee structure meant to speed the next generation of advanced aircraft.

The full suite of efficient airplanes, if adopted at a feasible rate as they become available, could cut projected 2050 civil-aviation fuel use 47%—ultimately up to 70%, or more than tripled efficiency, with full adoption. Just as with long-haul trucks, we won't eliminate the need for liquid fuel for airplanes. But we can eliminate the need for oil-based airplane fuel by switching to biofuels now: the oils of many plants, from algae to halophytes, can be processed into jet-quality fuel that many civilian and military users are adopting, as we'll see below.

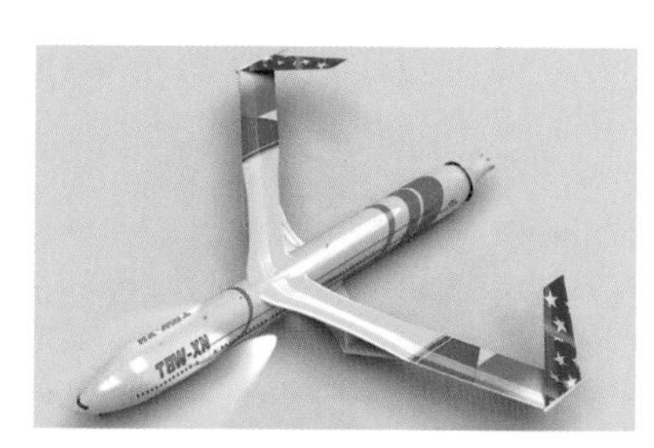

Fig. 2-20. From left: Boeing's SUGAR Volt electric-battery/gas-turbine hybrid propulsion system with a strut-braced wing, saving 70% of fuel; NASA's truss-braced wing design with a buried single rear propulsor, saving 60–80%; and MIT's H Series blended wing body (BWB) concept with podded, actively controlled boundary-layer-inlet propulsion, saving 59%.

Displacing Flights and Conserving Flight Time

Why travel to China to close a deal if you can still look your partner in the eye across the table using advanced teleconferencing? Even though high-end systems like Cisco's, HP's, and others' cost $300,000 each, software giant SAP found that they can pay back in just one year in reduced travel costs. Moore's Law is starting to bring vivid telepresence to the desktop. Using such teleconferencing nationwide could reduce business air travel by 12%, or 2.5% of total air travel.

When we do need to fly, finding ways to cut the 1.3 million barrels of jet fuel consumed every day is just good business. For airlines and airfreight companies, fuel is the single largest expense. Shaving off a few thousand gallons here or there can mean the difference between profits and red ink.

It's no surprise, then, that airlines are constantly searching for efficiency improvements. Planes are taxiing slower, for instance, using one engine instead of two or four. They're carrying lightweight catering carts, cruising slightly slower, gliding to direct landings without fuel-hungry maneuvers, and using new avionics with advances in air traffic management to chart the fastest routes and carry less excess fuel. These little steps add up to hundreds of millions of dollars saved.

The vast network of airports and airplane routes can also be made more efficient. In the current system, powerful major airlines hold near-monopolies on some airports, using them as hubs to connect many of their other flights so they can control more market share and charge higher prices. An exception is Southwest Airlines, which has achieved consistent profitability—in an industry that hasn't cumulatively broken even since the Wright brothers—largely by adopting direct point-to-point routing.[133] While Southwest is primarily in the business of domestic flights with single-aisle airplanes, previously discussed advances in efficiency would allow airplanes in all classes to fly farther more cheaply, enabling airlines to cover more direct routes that could expand the proven point-to-point model to international scale.[134]

A shift to point-to-point routing would nevertheless face barriers. Allocating airports' gates and slots through periodic auctions, rather than letting "fortress hub" monopolists keep hoarding them, would create a more level playing field and help shift the balance from hub-and-spokes business models to the more fuel- and capital-efficient point-to-point models.

Not only do short flights between regional hubs consume much more fuel per seat-mile than long flights due to less time spent in efficient cruising flight, but they also add to air, ramp, and terminal congestion. A more efficient alternative to these trips along dense corridors is high-speed rail (HSR). As trains have gotten faster and regional plane travel slower due to traffic and security-related delays, rail can beat air in door-to-door travel time at longer and longer distances. One study predicts that adopting current HSR technology would save the U.S. 29 million automobile trips and nearly 500,000 flights per year. HSR yields other benefits too: reduced airport-related road congestion, mixed-use development opportunities around main train stations, expanded labor markets from convenient and affordable medium-distance transport options, and higher business productivity through quicker travel.[135]

Combined Savings for Airplanes

Air travel's global importance makes continued growth in business and leisure air travel seem certain. But we can cut its fuel intensity even faster. Capturing available and likely use and design-related aviation improvements would cut 2050 demand 54% (fig. 2-21) despite 61% more seat-miles. That would leave 0.8 Mbbl/d of remaining demand, which could be met by advanced "drop-in" biofuels[136] at a lower cost than fossil-derived jet kerosene as soon as 2020, and ultimately by liquid hydrogen if desired.

Trains, Boats, and Other Vehicles

Americans don't use oil just to drive our autos, fly to business meetings or vacation spots, fuel our military mobility, and haul big truckloads of cargo. Gasoline and diesel also fuel the motorboats that take us fishing, the buses that cart our kids to school, the delivery trucks that bring our packages, and the trains and pipelines that carry everything from coal to hydrocarbons and chemicals. These miscellaneous uses add up: in 2010, they totaled 1.7 million barrels of oil a day.

But the same design principles and technologies that can wean our autos from fossil fuel can also slash the oil needed for these other civilian vehicles by a remarkable four-fifths by 2050. Today's 10 mpg urban delivery trucks average 30 cents per mile for fuel. Replacing these with lightweight, aerodynamic, battery-electric or plug-in-hybrid versions, recharged at the depot, drops that fuel cost to just 2.5 cents. FedEx's E700 hybrid delivery vehicles increased fuel economy by 36%, while UPS has considered using a hydraulic hybrid to save 60–70%.[137]

Buses, too, benefit from lightweight design, halved air drag, better tires, and electrification. Hybrids are especially useful for typical stop-and-start driving in urban areas. GM already makes hybrid buses that boost mpg by up to 55%, and cities from Albuquerque to New York have begun rolling out not just hybrids but also efficient biofueled buses. All these vehicles are also candidates for natural gas, which powers 15–20% of the world's new transit buses and garbage trucks. And bus rapid transit—the "surface subway" system pioneered in Curitiba, Brazil, providing subwaylike capacities at a tenth the cost of even surface light rail—is now found in more than 80 cities, chiefly in South America, but headed for Los Angeles. Bogotá's BRT system was built in three years and carrying a million riders a day by year six. Even costlier kinds of transit

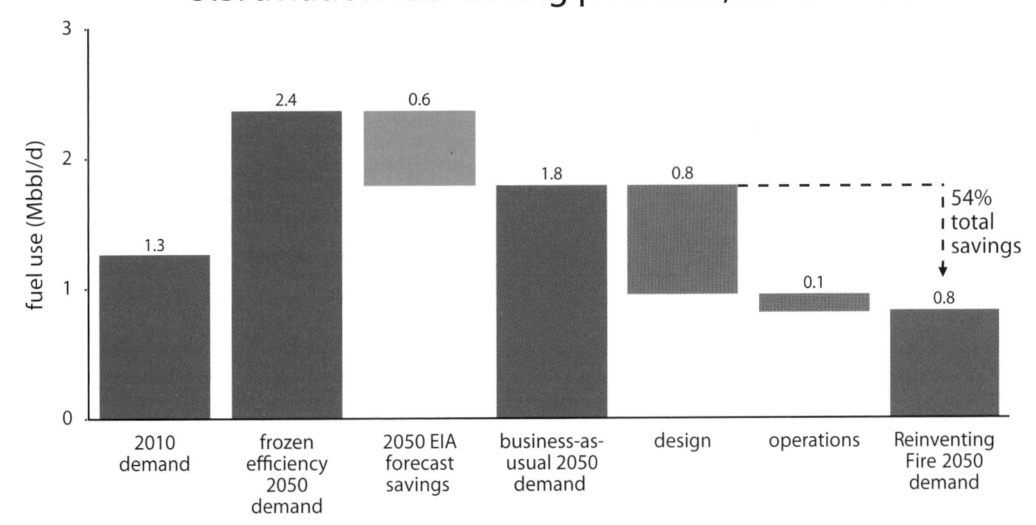

FIG. 2-21. Together, design and use improvements can cut 2050 jet-fuel demand by 54%.[138]

systems are far cheaper than automobile-based systems.

A new category of ultralight rail, like the CyberTran invented at the Idaho National Laboratory, shows promise of manyfold reductions in light-rail system costs, with easier installation and greater versatility. Its ingeniously light vehicle multiplies weight and cost savings in its truss-mounted overhead rail system. There's also plenty of room for improvement in conventional light trains. In the 1990s, Danish State Railways (DSB) developed the Copenhagen S-train with 46% lower weight per seat than the 1986 model. One of the most elegant light-rail energy-saving ideas is Victorian: humped track that decelerates arriving and accelerates leaving trains in London Underground stations.

Hauling goods by rail has already been an efficiency success story. Since 1980, U.S. rail freight has doubled while fuel use has barely increased, due to computerized throttle controls, hybrid-electric drive, efficient diesel engines, and idle control (the average Canadian locomotive was found in 1984 to be idling about 54–83% of the time). But further improvements are possible: fuel cells, widespread electrification, better aerodynamics, further reductions in idling, and regenerative braking. Norfolk Southern has tested a prototype 1,500 hp switching locomotive that's all battery-powered, can run 24 hours on a charge, and reportedly costs the same to build as a normal diesel locomotive.[139] Chinese bullet-train engineers have even invented a way for passengers to enter and leave without the train stopping, via a "connector cabin" that the train drops off at each station while picking up a new one.

Ships are an efficient way to move freight, but they still consume about 1% of U.S. transportation fuel. Now data on the fuel efficiency of nearly every large oceangoing vessel, collected by the nonprofit Carbon War Room,[140] help smart shippers pick the most efficient vessels. Low-cost upgrades can also raise the efficiency of each ship by 20–50%. Merely using low-friction paint the next time a ship needs repainting boosts its efficiency by 9%. Some fleet owners have already experimented with several kinds of modern sails to displace oil in their freighters: SkySails quotes a 35% fuel saving.

Military-Led Design Efficiencies

America's number one airline is not a civilian company but the U.S. Department of Defense (DoD)—the world's largest single buyer both of oil (three-fourths of which goes to its thousands of airplanes) and of renewable energy. DoD directly burns some 0.36 million barrels a day, 1.9% of U.S. oil use, plus whatever its contractors use. That's not a huge amount of oil—it could all come from two Gulf of Mexico platforms—but *delivering* it to thirsty vehicles in war zones is enormously costly in blood, treasure, and weakened combat capability. More than a thousand American servicemembers died in convoy attacks during the past decade, hauling mainly fuel. Just the monetary cost of delivering a gallon in Afghanistan averages $25–$45. In remote outposts, where winter resupply can be a 45-day struggle, delivery can cost up to tenfold more. The delivered cost of fuel thus totals about 20–36% of the total budgetary cost of the Afghanistan deployment.

Logistics—hauling things around, mostly fuel—uses about half the Pentagon's people and a third of its budget, so saving oil in combat could save tens of billions of dollars a year, free up whole divisions of logisticians and fuel guards for combat, and eliminate grave vulnerabilities. In 2010 alone, heating and cooling inefficient U.S. military structures in Iraq and Afghanistan cost $20 *billion*. Just spraying $95 million worth of foam insulation on tents in Iraq, inefficiently air-conditioned by electricity from 10% efficient oil-fired generators, is saving about $1 billion a year and taking 11,000 fuel trucks off the road. In Afghanistan, where foam hasn't yet been added, its payback is 51 days in a big base and 3 days in a remote one—and construction cost goes *down* because the foam costs less than the power, heating, and cooling equipment it avoids.[141] Such fuel savings also protect, multiply, and enable military forces.

Until 2010, the Pentagon hadn't counted the huge costs and risks of fuel delivery when buying the things that used the fuel. Now it values saved fuel at its full *delivered* cost, ranging from many to hundreds of times higher.[142] For contractors developing innovative military vehicles, this is lucrative news. The race is on to find and scale competitive advantages in energy-efficient design. For those squarely in the civilian sector, that race can spur domestic progress.

Those in or serving the airline industry, for instance, will want to follow the development of ultralight airframes, advanced engines and aerodynamics, and military blended-wing-body heavy aircraft, which can carry twice the weight twice as far using about 80–89% less fuel per ton-mile, or unmanned aircraft that can loiter for 50 hours using 97% less fuel (fig. 2-22a). Other proposals include a tiltrotor aircraft with tripled speed but five to six times greater range and fuel efficiency than legacy platforms (fig. 2-22b) and an airship that can float above 20,000 feet using far less fuel than normal heavy-lift dirigibles and can quietly deliver 20 tons, potentially hundreds of tons, of cargo with no ground infrastructure (fig. 2-22c). Cruise lines will want to emulate proposed ship retrofits that could save up to one-sixth of the Navy's nonaviation fuel—or consider new electric actuators that, if substituted for six hydraulic systems in one aircraft carrier, could save 1.4 million pounds, 61,000 square feet, 500 personnel,

FIG. 2-22. Four proposed military platforms with exceptional energy efficiency and combat effectiveness (see text), clockwise from upper left.

and \$20–\$25 million a year. And a new family of ultralight armored vehicles (fig. 2-22d) promises five-star protection from roadside bombs and agility like a great pickup truck's, all with lower cost, weight, and fuel use than a HMMWV ("Humvee").

Military R&D has long created new industries that reshape our entire economy, from the Internet and Global Positioning System to microchips and jet engines. DoD's keen new interest in the fitness and fuel-frugality of its land, sea, and air platforms seems bound to spill over into civilian vehicles, accelerating the national journey beyond oil and perhaps creating DoD's biggest-ever national-security win.

That, in turn, can transform our Armed Forces' risks and responsibilities. Light, agile platforms can go farther, faster, and longer, adding revolutionary combat capabilities—and speeding the oil savings that make them less likely to be needed.[143] End America's dependence on oil, especially imported oil, and it's no longer so important to protect unstable oil-producing nations and supply routes. Our sons and daughters have twice gone to fight in the Persian Gulf in half-mile-per-gallon tanks and 17-feet-per-gallon-equivalent aircraft carriers, in part because back home, we were still driving 15-mile-per-gallon SUVs. A worthy tribute to their sacrifice would be building military prowess on the fuel efficiency that makes such sacrifices less necessary, eases global tensions, undermines tyrants, and ultimately turns Persian Gulf intervention into "Mission Unnecessary."

POWERING VEHICLES WITH CLEANER ENERGY

Even if we realize all the potential for efficient design and use across the whole range of vehicles, we still need to fuel or power those vehicles by better and cheaper means than oil. Fortunately, four options—electricity (which will also become far cleaner), hydrogen, natural gas, and advanced biofuels—offer ample choices and robust competition.

Getting a Charge

When your gasoline gauge drops near empty today, you rarely have to worry about where your next tank of fuel is coming from. The U.S. has more than 110,000 gas stations, along with nearly 150 refineries and the intricate intricately linked pipelines and truck delivery routes that keep them supplied.

But what happens when you're in your electric auto, driving, say, from Los Angeles to San Francisco along Highway 1? As you marvel at the Big Sur scenery, you notice that you're so low on power that you might not make it to your destination, especially over some of the steep hills that lie ahead. And there's no place to get a charge.

Welcome to a new worry for the age of the electric auto: "range anxiety."[144] It happened to the very first buyer of the Nissan Leaf, Olivier Chalouhi. After all the hoopla of picking up his auto in Petaluma in December 2010 and driving for photo ops near the Golden Gate Bridge, Chalouhi saw he had just 37 miles left before it ran out of juice. His trip home was 37 miles. Chalouhi had to stop at City Hall for a charge before he could safely be on his way. It's no wonder, then, that when 85% of U.S. respondents to a recent Nielsen survey said they'd buy an electric auto, they strongly preferred plug-in hybrids, backed up by an onboard engine, to pure battery cars.[145]

The lesson: Getting off oil by moving to Revolutionary+ autos is not just a matter of developing the autos. To the extent they get their electricity only from onboard batteries, it also requires infrastructure to recharge them, preferably from renewables.

Fortunately, the physical infrastructure for recharging is rapidly maturing, based on an industry-standard smart plug. Companies like

AeroVironment, Inc., have also developed not just charging stations for homes and businesses but also charging technology that promises to cut charging time from many hours to a nearly full charge in a fraction of an hour. AeroVironment has teamed up with NRG Energy, a Princeton-based utility, to create a whole electric "ecosystem" in Houston. NRG plans to invest $10 million in a network of more than 50 fast-charging stations along major freeways, in business districts, at shopping malls, and in workplace parking areas. Rather than being at the whim of gasoline prices, NRG customers lock in subscription rates for charging plans. California is also rapidly building a fast-charging network and battery-swapping stations throughout the state. Other major players in this field include Coulomb Technologies, GE, Schneider Electric, and Better Place.[146]

One cost-effective way to build electric-auto infrastructure is to include the installation of charging equipment in the normal course of working on roads or building new parking garages, an idea promoted by RMI's Project Get Ready (PGR) that has spread to 16 cities. One PGR partner city, Vancouver, British Columbia, has led the world in requiring new mixed-unit dwellings to install electric conduits for future charging stations. Some major merchants even plan to add charging stations in their parking lots—often powered by solar cells so as not to raise the grid's daytime peak load—and to offer charging to attract customers.

Charging infrastructure could also bring an unexpected benefit to electric-auto owners—and to the whole electricity system. Plugged-in automobiles could sell their stored electricity back to the grid when it's most valuable, such as when utilities are struggling to meet downtown demand on hot summer afternoons. Thus Americans' second-biggest household asset could earn money during some of the 96% of the time that it's parked. And as we'll see in chapter 5, utilities' ability to draw on parked autos' "distributed storage" and to inform or control their charging times to match the grid's needs could be very valuable to the electricity system. Today's standard charging plugs have two-way communication to ensure that charging doesn't unduly burden the grid at peak periods, pricing matches scarcity, and drivers selling power back to the grid get paid for their electricity and their slight battery-life degradation.

Building a new infrastructure won't be cheap. By some estimates, each new electric vehicle will require about 1.1 charging stations—though 80% will be at homes and paid for (costing about $1,500) by the auto buyer. The other 20% will be a mix of workplace and public charging, ranging from $2,000 for a basic unit to tens of thousands of dollars for a direct-current fast-charging station. Some areas will need utility distribution upgrades, especially if on-peak charging isn't surcharged. A new financial infrastructure will enable users to pay by credit card, as easily as visiting an ATM.

But compared to building railroads or new highways, adding an electricity infrastructure is relatively easy. Electricity is already ubiquitous. As chapter 5 shows, we have enough of it. In most cases, hooking up an electric auto is no more difficult than buying and installing a new appliance.

Gassing Up with Hydrogen

A competing source of the electricity for electrified autos is fuel cells. Remember the high-school chemistry experiment where an electric current splits water into hydrogen and oxygen? Fuel cells do that backward, chemically reacting hydrogen with oxygen (from air) to make electricity, pure water, heat, and nothing else. There's no combustion. Fuel cells are compact, efficient, extremely reliable, costly if handmade, but competitive if mass-produced. Their hydrogen is stored at 5,000 psi pressure in ultrastrong, ultrasafe 1990s-vintage carbon-fiber tanks, refueled just like the compressed natural gas (CNG) discussed below for heavy trucks (which could use compressed hydrogen instead).

Several official studies have found, and some policymakers believe, that hydrogen-powered autos are impractical. But all those studies assumed unfit vehicles, and hence unaffordably big fuel cells and impossibly bulky hydrogen tanks. Vehicle fitness solves those problems[147] without a breakthrough in storage technology, making hydrogen a technically feasible, economically competitive option that's at least as safe as gasoline, and depending on its source, a reasonably or completely clean and climate-safe fuel for autos.[148]

Automotive hydrogen would initially be made at the filling station from natural gas, using efficient miniature "reformers" already developed and emitting two to three times less CO_2 per mile than gasoline cars emit today. Climate-safe biofuels could be reformed too if renewable electricity or direct use of sunlight to split water doesn't ultimately become even cheaper. Neither cost nor timing is problematic: Deutsche Shell said a decade ago it could sell hydrogen at all its German filling stations in about two years[149]—as fast as Portugal just built its national electric-car recharging network.

The supposedly intractable "chicken-and-egg" problem of hydrogen infrastructure—no auto sales without it, but no infrastructure without customers—was solved in 1999.[150] GM and independent experts even found that nationwide implementation would cost less than sustaining equivalent oil-fueling capacity.[151] A 2010 McKinsey study confirmed that hydrogen production and fueling infrastructure cost only about 5% as much as the vehicles they support.[152] When asked whether fuel-cell cars will come to market in 10, 20, or 50 years, the general manager of Lexus in the U.S. replied simply, "It will be far sooner than you think."[153]

Frigid liquid hydrogen is also feasible for airplanes. Though bulky, it has 2.8 times jet fuel's energy per pound. The U.S. Air Force and major airplane makers have established the feasibility and safety of such "cryoplanes," and Boeing has developed the "Phantom Eye," a spy plane whose hydrogen fuel keeps it aloft 60% longer. Boeing also successfully flight-tested a hydrogen-fuel-cell-powered two-seat airplane in 2008. The liquid hydrogen would even allow highly efficient, lightweight superconducting electric motors turning modern propellers—a recipe for long-run efficiency perhaps beyond the tripling already available from advanced airplanes,[154] and potentially extendable with onboard ultralight solar cells.

Putting Natural Gas on the Road

Unlike autos, even very efficient long-haul trucks can't yet be cost-effectively electrified. Normally they'd use biodiesel, with a longer-term option of hydrogen fuel cells. But in the near and medium term, switching from diesel to natural gas would save money and cut trucks' greenhouse gas emissions 20–30%. Such trucks could even become nearly fossil-fuel-free by using "renewable natural gas" from landfills, wastewater treatment plants, and livestock manures.

The technology is well established: more than 12 million vehicles around the world now use CNG, which in 2010 was on average 42% cheaper than U.S. diesel fuel per unit of energy contained. CNG has four times the volume of diesel fuel, so it's impractically bulky for today's long-distance heavy trucks, but it's well suited to trucks two to three times more efficient, shrinking the tanks correspondingly for the same range. Lightweight integrated tanks may soon help even less-efficient trucks to carry more CNG.

Another solution is liquefied natural gas (LNG), costlier but 2.4 times less bulky than CNG and increasingly cost-competitive with diesel fuel. Converting a long-haul truck to LNG is estimated to cost about $70,000. With natural gas costing

$0.75 less than an equivalent diesel gallon (as it does today in California), an average long-haul truck can recoup the conversion cost in about five years. At a $1.50 difference, costs can be recovered in just two years.[155]

The barriers? One concern is over safety.[156] LNG must be kept in vacuum-insulated tanks at –261°F. If released, it can shatter materials like steel and create a ground-hugging layer of supercold but highly flammable gas. New compact composite tanks developed by BMW and others can reduce cost and improve safety, but the gas could still be deliberately released to cause a ground-level firestorm worse than from propane or gasoline, whose tank trucks are already of homeland-security concern.

Another barrier is scarce infrastructure. Tens of thousands more natural-gas filling stations will be needed if natural-gas vehicles are to displace much diesel fuel, and LNG has even higher infrastructure costs than CNG. Natural-gas fueling systems may therefore work best for centrally fueled fleets like buses and delivery trucks, rather than trying to cater to everywhere trucks may go. Both for safety and because natural gas (methane) is over 20 times as potent a greenhouse gas as CO_2, scaled-up natural-gas fueling would need careful engineering and procedures to avoid leakage.

All these fueling challenges we've identified for heavy trucks can be met, keeping the freight moving with far less liquid fuel, no oil, and ultimately no fossil fuel. The resulting oil savings—18% of all U.S. oil use today—would be the most important nonautomotive way to get the nation off oil by 2050. And the more truck fuel we save and diversify, the less the burden on the backstop technology—biofuels.

Pumping Biofuels

Leilani Münter is one of the world's top ten female race-car drivers. She made it to Daytona by 2006 and became a fixture in NASCAR racing—America's most-watched sport with 100 million viewers. But she's also been listed by *Newsweek* as "surprisingly green" and was named *Discovery*'s Planet Green #1 Eco Athlete. Why? For every race, she buys and protects an acre of rain forest to offset her carbon footprint. And now her mission is to convert NASCAR entirely to biofuels.

No matter how quickly we follow the paths to enormous fuel savings, as described in this chapter, the nation will still need liquid fuel—lots of it, falling over decades. As discussed above, planes and heavy trucks can't yet be cost-effectively electrified. Hydrogen-based designs face transitional barriers. But wherever electricity and hydrogen can't ultimately displace oil, biofuels can. With recent technological advances, there's nothing oil can do that ethanol, green diesel, and other biofuels can't—including powering race cars like Leilani's.

So how much biofuel do we need? What type should it be? And where will it come from?

If we speed down the road to Revolutionary+ automobiles and other dramatically improved vehicles and smarter uses, we can cut the total amount of liquid mobility fuel needed in 2050 to about 3.1 million barrels per day (fig. 2-23). Only 20% of this remaining demand is for automobiles. Forty percent would go to heavy trucks, 25% to planes. Buses, the military, medium-duty trucks, trains, ships, and pipelines would consume the remaining 15%. Can biofuels cost-competitively meet this demand without harming the world's food supply or environment?

From the Ground Up: Biofuels

In the giant vats of biofuel plants across the U.S., yeast transforms sugar from corn into ethanol in the age-old process of fermentation. This first-generation biofuel has become big business. U.S. ethanol producers made 13 billion gallons in 2010, the equivalent of 0.6 million barrels of oil a day.[157]

That's one-fifth of the total 2050 need for mobility fuel, so this first-generation technology wouldn't be enough for all mobility needs nor suitable for some (notably airplanes), and it might interfere with food production, despite co-produced feed.

Because they can be made from crop residues or dedicated energy crops grown on lands *not* taken out of food production, second-generation biofuels avoid the conflict between food and fuel that can arise when corn or soybeans are turned into fuels.[158] Decoupling fuel from food also reduces unwelcome linkages of food prices to oil prices.

Second-generation ethanol can be made from crop residue such as corn leaves, stalks, husks, and cobs (together called "stover") or from inedible crops like prairie grass. Such "cellulosic ethanol" would nevertheless face barriers. Pure ethanol doesn't work in cold climates, so its U.S. sale would need to be seasonal and regional. But that's no reason for any new U.S. auto to lack the capability to burn ethanol. An "open fuels standard" (a bipartisan proposal in Congress since 2009) would require new vehicles to have "flex" capability—the ability to burn varying blends of ethanol and gasoline.[159] Of course, ethanol standards alone do nothing to insure against volatility in the food market resulting from food-based ethanol production. Restructuring U.S. corn and soy ethanol subsidies would allow inherently cheaper cellulosic feedstocks to compete fairly, accelerating the shift away from food-based fuels. Ultimately desubsidizing agriculture would help even more.

Researchers and companies are progressing not only with cellulosic ethanol but also with "drop-in" fuels chemically and functionally indistinguishable from today's medium petroleum fuels used in trucks and airplanes. These advances promise big gains in the fight against climate change, emitting 60–120% fewer lifetime greenhouse gases than fossil fuels—far better than ethanol or biodiesel. (The 120% reflects the potential to take CO_2 out of the air, put it back in tilth where it belongs, and reward farmers who do so.)[160]

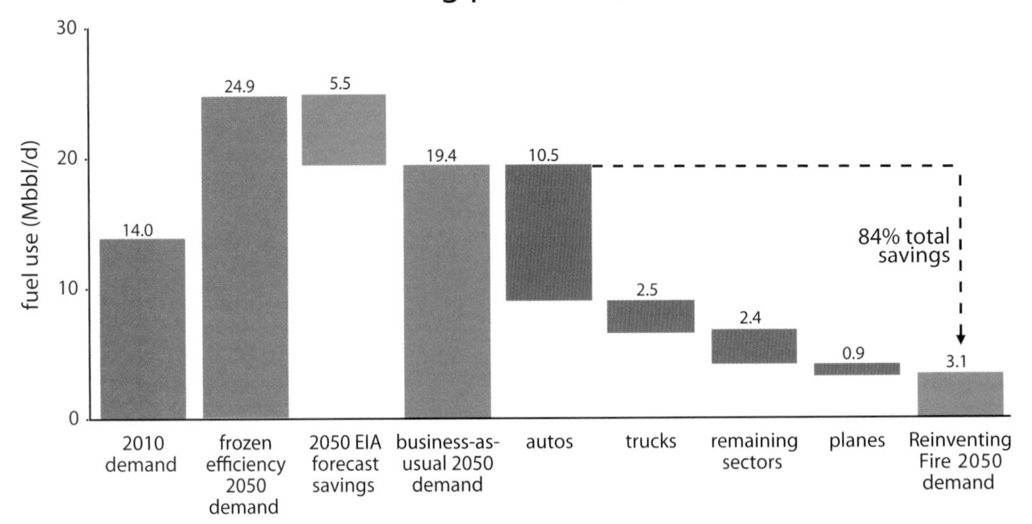

FIG. 2-23. By 2050, autos, trucks, planes, buses, and trains will still require about 3 Mbbl/d of fuel to operate.[161]

In total, second-generation biofuels distributed through existing infrastructure and run on existing engines could fuel all our land, sea, and air vehicles by 2050 if not outcompeted by natural gas, electricity, or hydrogen. The actual mix among these four competitors is impossible to predict and will be best determined in a fair marketplace. But do we have enough land and water to grow the crops needed to make three million barrels of biofuel per day? Can we do it without taking land away from the food production needed to meet growing world population and demand? And can we grow the feedstocks sustainably?

Recent studies say yes to all three questions. A 2005 USDA-DOE analysis concluded that U.S. farmland could sustainably provide each year more than one billion dry tons of collectable biomass *wastes*—enough to make three Mbbl/d of fuel—without taking food off the world's tables. About half of those billion tons would be agricultural crop residues such as corn stover. The rest would be mostly municipal waste and perennial non-food energy crops like switchgrass.[162]

And we needn't rely just on farms. The U.S. has 500 million acres of forests, covering one-fifth of the nation. Logging scraps, thinnings, and other types of wood can produce 400 million dry tons of feedstocks a year for another 1.3 Mbbl/d of biofuels. Figure 2-24 illustrates the volume of each of these feedstocks and their relative costs with projected 2050 technological improvements in three broad categories of conversion processes.

Producing advanced biofuels can also be more efficient than fermenting corn into ethanol. Some methods are 45–75% efficient in converting feedstock energy to fuel energy, compared to grain

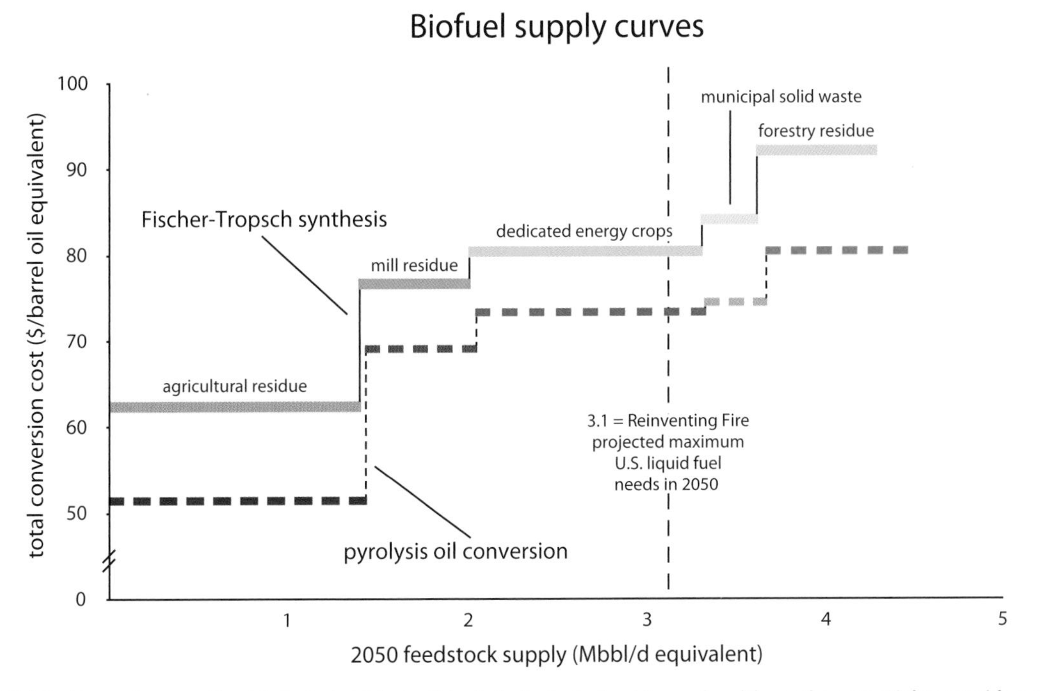

FIG. 2-24. Second-generation biofuels (which don't include fuels derived from algae) have the potential to provide a total of over 4 Mbbl/d by 2050 at unsubsidized costs (net of coproduct credits) lower than projected future oil prices. Cellulosic ethanol, not shown, runs between the two curves shown for thermochemical processes and has a slightly better yield, resulting in total supply of 4.7 Mbbl/d. See www.reinventingfire.com for details.[163]

ethanol's 38%. What's more, some of the processes create valuable by-products. Thermochemical gasification's electricity by-product lowers the effective cost per barrel of oil equivalent by about $10. Some biological conversion pathways that turn sugar into fuel can also produce a wide variety of specialty chemicals worth upwards of $5,000/ton, far more valuable than ethanol. Eventually it may be possible to apply integrative design to create biorefineries that combine biofuel conversion technologies, agricultural enterprises, fine-chemicals production, and algal production into highly profitable, zero-net-carbon facilities.

The Next Generation: Algal Fuels

Companies like DuPont Danisco Cellulosic Ethanol are building commercial plants to make ethanol from switchgrass.[164] Other major firms like Shell, BP, and Dow are in the cellulosic-ethanol race, as are private-equity-funded players like Amyris and LS9. But even as progress is being made on such second-generation fuels, venture capitalists from Bill Gates to Pierre Omidyar, entrepreneurs, oil majors[165] (Exxon, Chevron, BP, Valero), and governments have begun to invest in a more exotic feedstock—algae. Why? Algal fuels offer the promise of a major leap in biofuel productivity. Optimists claim that algae could turn 217 tons of CO_2 into nearly 10,000 gallons of oil, 18.5 tons of protein, and 18.5 tons of biomass per acre (versus about 400–500 gallons of oil for corn or sugar cane, and 1,100 for switchgrass). If such staggering yields are possible, an acre of algae could generate $50,000 per year—an order of magnitude larger than other terrestrial biomass crop production—and it needn't even use land.[166]

As with next generation Revolutionary+ fuel-cell autos, hydrogen-powered trucks, and advanced plane designs, the production and processing of algae face significant technology challenges at every step. Algae in open "raceway" ponds outyield conventional[167] fuel crops, but this approach needs flat land, water, sun, and, perhaps most critically, CO_2, and that may limit algae production to fewer suitable sites than land-grown feedstocks. Other approaches include growing the algae indoor in photobioreactors using solar-like artificial light, and cultivating "heterotrophic" algae that can grow in darkness. These indoor growing techniques have so far tended to be much more capital-intensive than growing algae in open ponds.

BIOFUEL INNOVATION

Dozens of firms are addressing the challenges presented by both algal and second-generation biofuels. Here are a few examples:

- Solazyme produces algae with genetically modified microbes that feed on sugar in large fermenting kettles. Mature algae are pressed to extract the oil. In 2010, the U.S. Navy ordered 150,000 gallons of algae-based jet fuel from Solazyme.
- Sapphire Energy recently broke ground on a 300-acre facility in New Mexico for growing genetically modified algae in ponds. Algal by-products like protein and nutrients will be kept in-house to feed more algae. By 2012, the test facility is expected to produce one million gallons of algal fuel a year.
- German firm Choren has built a large plant capable of converting 68,000 tons of biomass per year into 4.8 million gallons of diesel and 45 MW of electricity through a biomass-to-liquid approach that gasifies the feedstocks into synthetic gas and then converts the gas to second-generation fuel via Fischer-Tropsch synthesis.
- Colorado-based Rentech is building a beta pilot plant, funded by 2009 federal stimulus grants, that will produce nine million gallons per year of synthetic green diesel and 35 MW of clean power.

Despite the technological challenges of cost-effectively growing algae for biofuel at scale, a number of commercial ventures hope to scale to commercial production capacity within the next five years (see Biofuel Innovation sidebar). Continued progress with genetic modification should increase yields. DOE estimates that another 1.4 Mbbl/d of biofuel could be produced from algae by 2050.

In anticipation of second-generation and algal biofuels becoming available, airlines and military users have already begun testing engine compatibility, fuel consistency, performance, and logistics. Some airlines are skipping isolated testing altogether: Lufthansa burns a 50% biofuel blend in one of the two engines on its scheduled four-times-a-day flight between Frankfurt and Hamburg. With major test flights already completed worldwide on diverse engines and airframes, ASTM, a major standards organization, approved 50% biofuel blends for commercial airliners in December 2010. The U.S. Navy and Air Force have both flown advanced supersonic fighters on half aviation fuel, half biofuel derived from a mustard-like weed. The Air Force aims by 2016 to shift half its domestic aviation fuel off oil, and the Navy, to sail an oil-free Strike Group. By 2020, the whole Navy aims to be 50% oil-free.

New Business Models

Biofuels' rapid growth poses both challenge and opportunity to the petroleum industry—and presents some tough choices. Should oil companies and refiners stick to fossil fuels or move into biofuels themselves? If so, should they aim to produce ethanol, or should they move to more hydrocarbon-like fuels (like butanol) that are compatible with their existing massive infrastructure? Different companies are making different bets. Major refiner Valero Energy, for instance, snapped up corn ethanol plants at fire-sale prices after big ethanol producer VeraSun went bankrupt in 2008. And in 2010, BP acquired a leading second-generation biofuels company, Verenium.

From an agricultural perspective, moving into advanced biofuels will reduce U.S. biofuels' demand for edible crops but increase the demand for crop residues, perennial grasses, and trees. The potential emergence of algae could decrease the land needed to meet the remaining U.S. transportation fuel demand yet simultaneously increase the animal feed supply. Thus, whereas the challenge of first-generation biofuels was fuel or food, the promise of advanced biofuels is fuel and feed.

CONCLUSION: BETTER MOBILITY AT LOWER COST WITHOUT OIL

This chapter has explained how we can keep America's vast transportation system humming, growing, and improving—all without oil. By 2050, we'd drive superefficient vehicles fueled by a flexible mix of electricity, hydrogen, and sustainable biofuels (and, if desired, some natural gas for trucks), and we'd use those vehicles far more productively. To power our increasingly efficient heavy trucks and airplanes, we'd need, at most, biofuels equivalent to 3.1 million barrels of oil per day. That's less than five times the volume of today's U.S biofuels industry, which provided only 3% of 2010 mobility fuel.

This new transportation system would not only be cleaner and more efficient, reducing threats from both oil dependence and climate change; it also would cost trillions of dollars less to run than the business-as-usual alternative. All told, transitioning to more efficient autos, trucks, and planes in addition to changing how we use all of these vehicles invests $2 trillion to save $5.8 trillion (fig. 2-25).

Ending oil use for transportation by 2050 is possible (fig. 2-26) but will be a daunting task. It won't be instant or easy. Inertial drag will need to be overcome by the accelerating forces of automakers, real-estate developers, IT entrepreneurs, and others eager to make new fortunes from better ideas. We'll also need rapid innovation. It's not easy to create fit, safe, peppy, exciting autos that get the equivalent of 125–240 mpg with uncompromised or improved comfort, handling, and safety, all at attractive prices. Nor is it easy to overhaul other vehicles, tranportation systems, and human behaviors. But it's easier than coping with the consequences of not doing it. These ambitious goals are both possible and cost-effective, and first movers across the U.S. transportation system have already begun the journey.

There are three critical ways for business to lead this transformation:

Drive the transition to superefficient vehicles. A wealth of untapped efficiency remains. Airplane makers have made impressive strides but need to design radically different airplanes. Automakers need to exploit the virtuous spiral of ultralight-weighting, integrative design, and electrification to produce Revolutionary+ autos that are safe and affordable; the main obstacles are more cultural than technological or economic. Heavy trucks and other cargo carriers need to carry more weight using less fuel. First movers and fast followers will reap the rewards in fiercely competitive global markets. While risks must be intelligently managed, incrementalism is now the high-risk and transformation the lower-risk strategy.

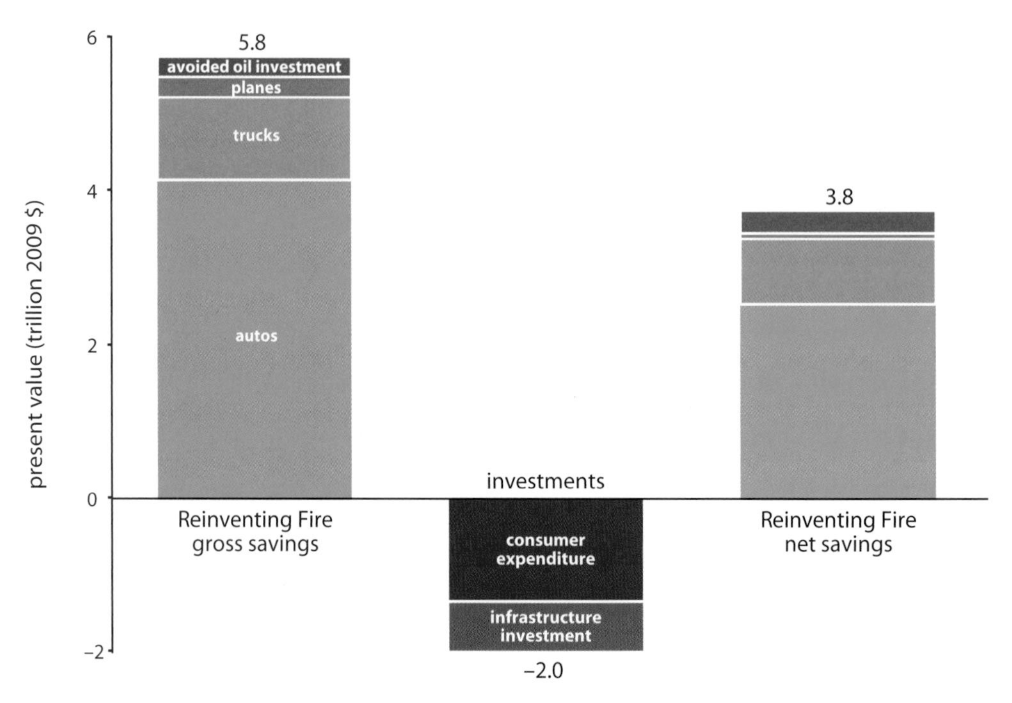

FIG. 2-25. The 2010 net present value (NPV) savings of $3.8 trillion includes the cost of building the distribution infrastructure needed to support a fleet of autos running on a mix of electricity and hydrogen by 2050 (the exact mix isn't important). Over the 40-year period, autos save $4 trillion in NPV ($400 billion from improved use), trucks about $0.8 trillion, and planes about $100 billion. About $270 billion of investment in domestic oil supply is also avoided.[168]

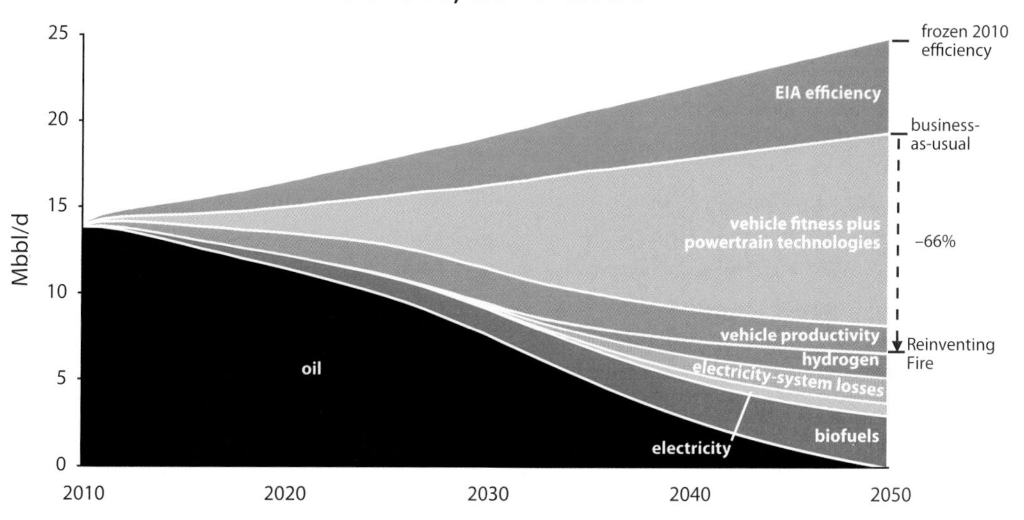

FIG. 2-26. A 2050 U.S. transportation system would need no oil if it productively used superefficient vehicles to provide the same services at lower cost and risk. The fuel mix of electricity, hydrogen, natural gas, and advanced biofuels—of which one illustrative possibility is shown here—cannot be determined in advance but is very flexible, supporting robust competition to 2050 and well beyond. If the hydrogen shown is all reformed from natural gas, it will come half from gas and half from steam.[169]

Decisive, inspiring leadership will be needed—and is starting to emerge, now that Detroit's recent near-death experience has concentrated minds wonderfully. Engendering a culture of innovation and establishing a long-term vision toward the large-scale production of revolutionary products starts at the top. Laggards will take the greater risk of having to catch up with competitors already well down the three synergistic learning curves. Leaders, though, can reap profits by locking up market share, supply chains, and a reputation for cutting-edge technology. Being any but the fastest kind of follower in such a fast-paced, multidimensional competition risks being left behind as a typewriter maker.

Invest across technologies and fuel types. We'll probably need, in some degree, all the vehicle technologies and alternative fuels presented in this chapter to move transportation off oil. Not all may ultimately be needed, but their diversity, especially in powertrains and fuels, provides valuable insurance against failures. That means investing now in everything from advanced-composites structural manufacturing processes and lightweight wheel motors to batteries, automotive fuel cells, and advanced biofuels. Casting the net wider lowers risk, widens opportunity, and bolsters competition.

For instance, investing in low-cost carbon-fiber precursors and production could position the U.S. as a major provider of this crucial raw material (vying with Japan, China, and Europe), as well as reducing cost to vehicle makers. Supporting the development of natural fibers and precursor alternatives like olefins could decouple carbon-fiber production from oil and its price.

Manufacturers need investment in retooling and in emerging manufacturing innovations to slash cycle time in producing breakthrough advanced-composite structures—an area underserved by venture capital. Start-ups, with additional capital support, could enter the market by designing and licensing new automotive technology.

Investing in service-oriented start-ups that provide mobility rather than vehicles and fuels, and in carsharing and ridesharing business models and IT enablers, could help companies expand their fleets. This would foster competition among manufacturers to provide fuel-efficient offerings for those large blocks of guaranteed market.

Biofuel technologies provide ripe investment opportunities, particularly advanced "drop-in" biofuels for heavy trucks and airplanes. With diverse feedstocks and conversion techniques already under development, many of these innovative approaches could be producing substantial amounts of biofuels as soon as 2020, even as rapid gains in vehicle efficiency increase those biofuels' share.

Support policies to speed the transition to radical vehicle efficiency and productivity. The right policies would provide a critical push. Size- and revenue-neutral feebates would offset advanced vehicles' initially higher prices, stimulating sales so manufacturers can rapidly scale up production and deepen cost cuts, accelerated by smart fleet purchases.

Harmonized trucking regulations could allow fewer trucks to carry more freight more quickly, with compounding benefits from reduced traffic, noise, congestion, highway wear, and fleet costs.

Land-use policies that now subsidize and mandate sprawl should be reversed to reward smart growth; otherwise the socialized costs of sprawl will continue to burden business by raising taxes, commuting times, and the salaries needed to offset commuting costs. Financiers and the whole business community have a strong stake in locationally efficient mortgages, which help reduce defaults, increase savings, and boost local economies. Taxing driving, not fuel, could restore depleted infrastructure funds, signal the societal costs of driving, let nondrivers invest in their own mobility rather than others', and level the playing field with other forms of or substitutes for mobility while reducing unwanted travel time for all.

So where could these actions lead?

At the start of World War II, Detroit switched *in six months* from making four million cars a year to making no cars—but instead churning out around the clock the tanks and planes, the Jeeps and munitions, that won the war. By 1945, one-fifth of the entire dollar value of U.S. war materiel came from the former auto industry—which then emerged from war with the scale and scope to create and dominate global vehicle markets.

That transformation was driven by coherent mobilization of an entire society to win a cataclysmic global conflict. In today's peaceful struggle for success in world markets, emerging transformation will be enabled and sped by innovative policies—but driven fundamentally by competitive forces and carried out by private enterprise. This is the sort of challenge for which a century of industrial development prepared us, wars steeled us, and the IT revolution inspired us. We need only rise to the occasion—or buy from those who do.

To help capture opportunities from the transition to energy efficiency and renewable energy, here are a few recommendations for the main stakeholders to focus on, other than the real-estate and community design suggestions above.

TABLE 2-1. Recommendations for key actors in the transportation sector

	NO-REGRETS	OPPORTUNISTIC	INNOVATIVE
VEHICLE MAKERS AND SUPPLIERS	Reduce rolling resistance, aerodynamic drag, and mass by conventional incremental improvements. Strengthen intellectual capital (design, analysis, and manufacturing) in advanced materials and electric powertrain. Fully count downsized powertrain when valuing improved vehicle fitness.	Master ultralighting and sharpen mass-decompounding analytics. Try a Hypercar-class concept vehicle to test new ways to organize small, fast design teams. Roll out non-fossil-fuel niche vehicles. Market the safety and performance of lightweight vehicles. Begin integrating composite parts into existing architectures. Develop service business models that sell mobility rather than vehicles.	Transform design process and culture to become bold and highly integrative. Develop and produce Revolutionary-fitness high-volume non-fossil-fuel vehicles. Launch high-volume manufacturing of advanced-composite structures; retrain repair shops to deal with them. Sell obsolete metal-stamping assets to competitors. Launch high-volume manufacturing of electric powertrains. Launch service businesses that offer mobility rather than vehicles.
FUEL PROVIDERS	Assess how core skills can best be leveraged in the post-fossil-fuel era. Invest and learn across non-fossil-fuel technologies. Develop strategy for peak oil on the demand side.	Invest in diverse non-fossil-fuel production and retailing. Invest in vehicle smart- and fast-charging infrastructure, allying with electricity providers. Launch non-fossil-fuel partnerships with test fleets.	Market non-fossil fuels on a large scale. Invest in vehicle efficiency technologies; if they succeed, make less money on oil, more on "negabarrels."
FLEET AND PRIVATE VEHICLE OWNERS AND OPERATORS	Consider public transit when making location decisions. Educate yourself on non-fossil-fuel vehicles; if you're a fleet, try some.	Test non-fossil-fuel vehicles. Test service business models that provide mobility rather than vehicles. Support integration, IT enhancement, and expansion of public transit. Implement employee parking cashout.	Switch to non-fossil-fuel vehicles on a large scale. Switch on a large scale to service business models that provide mobility rather than vehicles. Implement public-transit-only corporate mobility strategy, or close to it.

Continues

	NO-REGRETS	OPPORTUNISTIC	INNOVATIVE
GOVERNMENT AND NGOS	Increase fuel economy standards (CAFE). Provide affordable government financing to jump-start innovation, new industries, and retraining. Help low-income families buy very efficient new cars (while scrapping clunkers). Encourage innovation by procuring efficient and non-fossil-fuel vehicles, especially in fleets whose size can help speed automakers' innovation. Help make walking and biking safe, convenient, and popular.	Enact and scale well-designed feebates. Harmonize truck weight and size rules, ultimately nationwide. Speed the retirement of old, inefficient vehicles (cash for clunkers). Expand prize competitions for superefficient vehicles. Consider airplane landing feebates or graduated efficiency-based fees. Provide carbon-policy clarity to reduce investment uncertainty and risk. Facilitate P2P auto and parking rentals.	Implement long-term policies aimed at curbing or reversing sprawl in favor of smart-growth models. Offer multi-million-dollar prize for retrofitting a device that will improve existing auto fleet fuel economy by more than 20% for less than $500. Scrap vehicles older than 15 years (exempting bona fide collectors). Fully price, and perhaps tax, private and municipal urban parking spaces; use the tax income to modernize transit. Try surface rapid transit, CyberTran, and other novel transit modes.

FIG. 2-27. As our discussion shifts now from transportation to the built environment, this classic poster *Ruimte gebruik* ("Use of space") by the Dutch Cyclists' Union (www.fietsersbond.nl/english-info) reminds us that mobility choices aren't only about oil; they shape how we live. In bike-friendly Holland's morning rush hour, bikes outnumber cars and arrive faster. After introducing the ambitious Hypercar concept, RMI's 1995 *Atlantic* feature "Reinventing the Wheels" added: "Whether we also have the wisdom to build a society worth driving in—one built around people, not cars—remains a greater challenge. As T.S. Eliot warned, 'A thousand policemen directing the traffic/Cannot tell you why you come or where you go.'"

CHAPTER 3

BUILDINGS: DESIGNS FOR BETTER LIVING

FIG. 3-1.

CHAPTER HIGHLIGHTS

- **THE GOAL.** Make the 70% bigger stock of U.S. buildings in 2050 use 54–69% less energy than projected, by systematically applying well-known techniques (the first 38% savings) and capably executing integrative design (the other 16–31 percentage points of savings).
- **THE BUSINESS OPPORTUNITY.** Slash energy expenses and create important value for electricity companies by efficient and timely use, improve occupants' health and productivity, increase real-estate values, enhance corporate brands, and—especially if your company offers easy-to-access new efficiency products and services—exploit thriving new markets.
- **THE BOTTOM LINE.** A total of $1.9 trillion in saved energy costs in U.S. buildings by 2050, from an incremental investment of $0.5 trillion (both in 2010 present value), plus non-energy benefits that are often far more valuable, plus further potential savings from integrative design.
- **BUSINESS SECTORS THAT CAN PROFIT.** Real-estate owners, investors, and occupants; energy utilities' new and existing service providers (everyone from designers to contractors to financiers); and product suppliers.
- **POLICY ENABLERS.** Next-generation building codes and appliance standards with improved enforcement; rules that reward energy utilities for energy efficiency and landlords and tenants for cost-effective investments; access to energy use information and low-cost financing; reforms in how design is taught, practiced, and bought.

Drawing isn't the problem, *seeing* is the problem.

—BETTY EDWARDS (1926–)

On a raw day, with a cutting wind pushing low clouds just over his head, Anthony Malkin strides out onto a private balcony on the 103rd floor of the Empire State Building (fig. 3-1). At his feet is the great metropolis of New York City. Malkin points out buildings of every size and shape, from glitzy boutiques and humdrum shops to dazzling skyscrapers and dark tenements, with timeworn stone visages reflecting in modern glass facades and the old and seedy mixing with the new and glittering. In every building, he explains, there's a story.

Yet it is under Malkin's own feet that one of the city's more remarkable building stories is unfolding. A tall, strong-willed aristocrat of real estate, Malkin and his family own the Empire State Building, the world's most famous office building. In 2006, Malkin decided to do what many real-estate experts thought impossible—refurbish the iconic 1931 skyscraper into a shining example of modern amenity and efficiency. He's spending more than $550 million on a top-to-bottom renovation, from the glorious Art Deco lobby with its gold- and aluminum-leaf ceiling murals to its world-famous observatories.

Originally, Malkin planned to spend $93 million of that investment on refreshing the building's aging energy infrastructure—everything from the wires in the walls to the motors in the elevators. But after realizing that he could cut annual energy costs by $4.4 million by investing an extra $13 million (providing an incremental three-year payback), he increased the total energy-related investment to $106 million. The building got upgraded windows, new heating and cooling systems, new lighting, and other technologies that when combined with a few years' tenant improvements will shave 38% off its energy bills and CO_2 emissions, cut its peak electrical demand by 35%, and enhance its tenants' comfort and productivity.

Malkin, whose family's portfolio includes 10 million square feet (sf) of office buildings in the New York area, isn't slashing energy costs out of the goodness of his heart or to burnish his "green" credentials: he's doing it to make money. The bottom-line calculations for the building once scaled by King Kong show that Malkin will be able to achieve larger energy savings at a far lower cost than had been thought possible in such a difficult old structure. After the investments pay back in just three years, the increased profits will accumulate. "I make more money through energy efficiency," Malkin tells his rivals. "I'd like to show you how we did it—and how you can do it too."

If the nation were to follow Malkin's lead in upgrading its 120 million buildings,[170] the consequences would be dramatic and far-reaching. Buildings are energy hogs, consuming 42% of America's energy (more than any other sector) and 72% of its electricity.[171] Much of that energy is simply wasted. Standard forecasts suggest that by 2050, the growing U.S. building stock will annually use 53 quadrillion BTU of primary energy (see Building-Sector Terminology sidebar). But relatively straightforward changes, using existing and emerging technologies, can slash that projected energy use by, coincidentally, 38%. That adds up to a huge prize by 2050: *at least $1.4 trillion in net present value energy savings.* Those savings are worth four times the cost of capturing them. And juicy as that reward is, it doesn't even count often observed but rarely counted non-energy

BUILDING-SECTOR TERMINOLOGY

Primary energy has been extracted from the ground or from natural energy flows but has not yet been converted to other forms. For instance, it's the energy contained in coal before it's burned at the power plant, or natural gas before it's cleaned, processed, and put into a pipeline for delivery to customers.

Delivered, or site, energy has been converted to its commercial form and delivered to the facility where it is consumed. It's the energy remaining after the natural gas (or some other fuel) has been burned to generate electricity and the electricity has been transported through the power grid. Delivered energy plus the losses incurred in conversion and delivery equals primary energy.

End-use devices convert delivered energy into **energy services** like light, heat, torque, or motion. These in turn give **end users** the functions and experiences they want, like visibility, comfort, hot showers, or cold beer. Note that these functions and sensations are what the end user really wants—not lumps of coal, cubic feet of gas, or raw kilowatt-hours.

Commissioning is a technical process to make sure new buildings perform as designed. **Retrocommissioning** restores existing buildings to proper performance. Commissioning is best done by an expert independent of the original installer.

Passivhaus ("passive house") is a design standard (launched in Germany but now spreading rapidly in Europe and entering North America) for superefficient buildings that need no traditional heating and cooling equipment yet have similar construction cost to traditional buildings.

R-value is a measurement of resistance to heat flow through a piece of material (only 1⁄20th as much heat flows through an R-20 material as an R-1 material). Building scientists often use the reciprocal unit, $U = 1/R$, to measure the rate of heat flow through a given area, driven by a given temperature difference.

A **kBTU** (1,000 BTU) is the most common U.S. unit for overall energy use. One BTU is roughly the energy contained in one kitchen match. It takes about 2,000 BTU, the energy contained in about two cubic feet of natural gas, to make a pot of coffee. In 2010, the entire U.S. used 98 quadrillion (10^{15}) BTU or "**quads**" of primary energy or 65 quads of delivered energy.[172]

A **kWh (kilowatt-hour)** is the most common U.S. unit for electrical energy. One kWh is the amount of electricity used by a 100-watt lamp run for 10 hours. One kWh contains 3.4 kBTU of energy. A kWh should not be confused with a **kW** (kilowatt or 1,000 watts), which measures the *rate* of energy flow. Electricity flowing at the rate of one kW for one hour delivers one kWh of electric energy.

The **energy intensity** of a U.S. building is typically stated in delivered energy use per square foot per year (kBTU/sf-y). This metric helps us compare energy use for buildings of different sizes. The average U.S. commercial building used 90 kBTU/sf-y of delivered energy in 2003.[173] If your commercial building uses less than 50 kBTU/sf-y of delivered energy, chances are you're doing pretty well. If your building uses more than 100 kBTU/sf-y of delivered energy, you probably have significant opportunities for savings. The average U.S. home uses 44 kBTU/sf-y of delivered energy.[174] As we'll see, the most efficient buildings use a small fraction of these amounts of energy to deliver the same or better services.

benefits, ranging from improved worker productivity to higher retail sales (because customers prefer shopping in well-daylit spaces). Living and working in efficient buildings, says the modern evidence, makes us more comfortable and more productive—probably even healthier and happier. And that doesn't count wider societal values like security and environment.

The building energy efficiency revolution can create new business opportunities and strong new industries as companies gear up to install insulation in inner-city homes or to manufacture easy-to-install efficient office lighting systems. This expanded economic output means that reducing the energy used by our homes, offices, warehouses, theaters, shopping malls, and other structures could revitalize the real-estate sector and help rejuvenate the national economy. Consider the multiplicative effect of putting all this cash (and added value) back in the hands of businesses and consumers. In a sector that needs new ideas, energy efficiency in buildings creates exceptional opportunities for growth in jobs, new goods and services, and finance.

Such a transformation also touches off a reinforcing cycle that strengthens other key parts of the economy—and smooths the path beyond fossil fuels. If we succeed in cutting primary energy use in buildings 38% by 2050, they'll use 19% less energy than they did in 2010—despite 70%[175] more floorspace. More-efficient buildings free up electricity for transportation and natural gas for industry, electricity, and optionally transportation. Lower demand from buildings also reduces the pressure on the electricity sector, enabling a faster, cheaper transition from coal to more resilient and benign sources of electricity.

Let's be aware, though, that bringing America's buildings up to cost-effective efficiency levels is an enormously difficult task. Anthony Malkin is one of the idea's staunchest supporters. Yet even he sees it as a challenge as ambitious as the Marshall Plan that rebuilt Europe after World War II. Why? There are few cookie-cutter or blanket solutions. Upgrading existing buildings can only be done one building at time—which means 120 million times across the United States—and every building has a different story and often a different set of requirements. There are huge barriers, from regulatory rules that penalize utilities for making buildings more efficient to many building owners' and occupants' limited awareness of the many non-energy benefits of energy efficiency.

Perhaps the biggest problem is that the path to making money from the energy savings isn't always clear. Yes, efficiency improvements at least pay for themselves, usually several times over, and bring enormous savings to society. But can individuals and companies reap rewards big enough to justify incurring the hassle and making the investments? Building owners won't pay to boost energy efficiency, for instance, if their tenants are the ones to benefit from lower energy bills. Renters in leaky houses won't spend the money—even if they had it—to improve the efficiency of homes they'll never own. And for many individual companies and homeowners, the potential energy savings often look so small they're hardly worth the time and effort. It's no wonder that we're only scratching the surface of the cost-effective gains now possible in our buildings.

But we can overcome these challenges with both business-led innovation and policies to jump-start and speed investments in building efficiency. Such policies can be as simple as requiring energy checkups when houses are sold, or as complex as getting next-generation building codes widely adopted—and then kept up-to-date. They can be national in scope, like standards that require at least a certain level of efficiency in every TV (home electronics now use more electricity than refrigerators[176]), or they can be local, such as providing public or private capital to lenders to finance insulating homes at low interest rates. Such policies already have a proven record of success, from mandates that have raised the efficiency of automobiles or refrigerators to government research investments that created whole new industries like microchips and the Internet. Business leaders who aim to grab a share of the energy savings should support smart policies that help open doors to this opportunity. Transforming the building sector can't and won't be done by business alone; it will require a thoughtful marriage of business investment with business-supported public-sector involvement.

Yet in a society as complex and politically gridlocked as ours, business innovation can and must do much more than it has so far. Government can steer but shouldn't and often can't row. It's up to private enterprise to craft, test, break, fix, rebuild, and scale the hard work of earning that net $1.4 trillion: creating the value proposition, building the delivery mechanism, training the people, making the sale, corralling the finance, executing the deal, measuring the results, and continually improving the whole value chain. We'll offer some examples of the thousands of smart firms that are doing this every day, creating the next, efficient, superior American building fleet.

While the task is daunting, it is achievable. The Empire State Building is one beacon of hope in the enormous sea of U.S. real estate, and Anthony Malkin is rolling out what he's learned across his whole portfolio and to other big buildings. That's good for his profession, his city, his country, and our world. As this chapter will explain, there are great needs, vast opportunities, and practical solutions begging to be exploited.

BUILDING EFFICIENCY 101

What makes one building more energy-efficient than another? Energy use in an existing structure depends on the building shell, what's inside the shell, and how the stuff inside the shell gets used. All three are important.

Most people first address the stuff inside the shell without recognizing that the building shell itself strongly affects the type and amount of lighting, heating, and cooling equipment that's needed. If the building shell is as poorly insulated as a paper house or as leaky as a pasta colander, it will be very costly to heat and cool. If the building envelope is more like a freezer shell, it will be much cheaper—or even free—to heat and cool. Also, making a new building the right shape and pointing it in the right direction can strongly reduce its energy needs.

Inside the shell, buildings use energy in six main ways: space heating, water heating, space cooling, lighting, electronics, and appliances. The first four use the most delivered energy, but since lighting and electronics typically consume upwards of 3 kWh of fuel at the power station to deliver 1 kWh to your electric meter, and both this fuel multiplier and the electricity system's high capital costs make electricity the costliest form of energy, operating lighting and electronics often uses the most primary energy and money.

Beyond the shell and the stuff inside it, how we use our buildings is just as important as how we design them and what we put in them—maybe even more so. The most efficient lights, computers, and space-conditioning equipment are the ones that are turned off when they're not needed. This commonsense approach is surprisingly rare.

When all three factors are considered, the energy use in a typical commercial building adds up to about $2.16 per square foot each year, 77% of which buys electricity.[177] The average American household directly spends $2,200 a year on energy, 67% for electricity.[178] These energy costs vary not just because of big differences in how efficiently equipment is designed, built, maintained, and especially used, but also because of differences between building types—a data center, hospital, or refrigerated warehouse will use far more energy than an office building or school—and in energy prices and climates between locations.

UNDERSTANDING TODAY'S BUILDING QUAGMIRE

At the end of the 19th century, nearly all Americans worked outdoors, mostly in agriculture. Today, the average American is inside a building 87% of the time.[179] At first "we shape our buildings," said Winston Churchill, "and afterwards, our buildings shape us." Yet most of our buildings—however spiffy they may look—are profligate wastrels of energy, money, and business opportunities. They misshape our lives by making us less happy, healthy, wealthy, and wise than they should.

A Huge and Diverse Footprint

U.S. buildings consume a prodigious amount of energy—42% of the nation's primary energy, 72% of its electricity, and 34% of its directly used natural gas,[180] all the biggest uses by any sector. We spend more than $400 billion a year to heat and power them—even more than the government spends on Medicare.[181] In 2007, U.S. buildings used more primary energy than the total energy use of Japan or Russia, and twice that of India's 1.2 billion people. If American buildings were a country, they'd rank third, after China and the U.S., in primary energy use.[182] What's more, as shown in figure 3-2, their primary energy use is projected to keep on growing.

Sound like a problem? Absolutely. But that problem presents some of our most tangible opportunities to profit from new business initiatives, to drive innovation, and to save money—and to realize social benefits like creating jobs, reducing pollution, and strengthening national security. The biggest opportunities lie in designing new buildings right and retrofitting the existing buildings that waste the most energy. To see how to take advantage of these opportunities, though, you first need to understand a bit about how our buildings work, and what needs to change.

Any energy efficiency effort must swim against the tide of two powerful trends in both residential and commercial buildings: more are built and more uses are added.

Slightly more than half of U.S. buildings' energy is used by 115 million[183] homes, apartments, and other residences. Home energy use is climbing

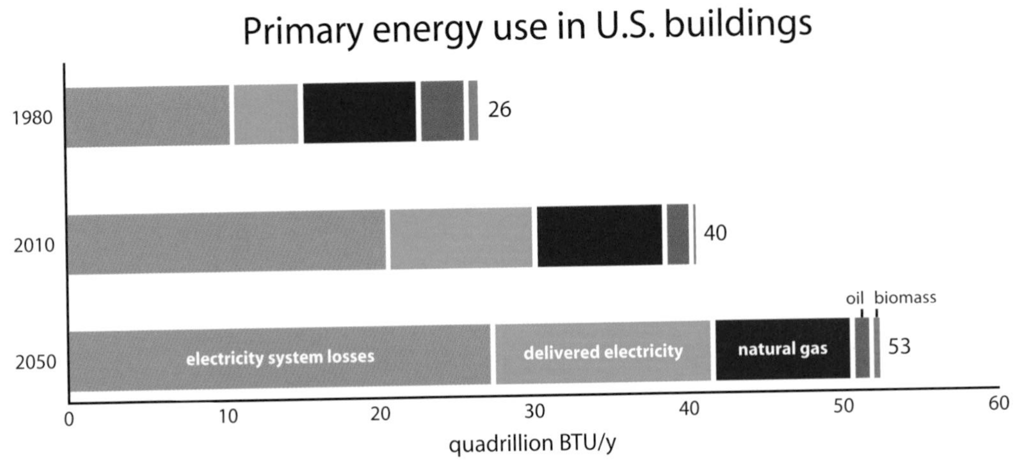

FIG. 3-2. Between 1980 and 2010, U.S. residential and commercial building energy use grew by 54%. Between 2010 and 2050, it is projected to grow by another 33%.[184]

partly because our houses have gotten bigger: the average new single-family home more than doubled in size since 1950.[185] We're also jamming more stuff—dryers and stoves, flat-screen TVs and phone chargers—into these homes. Even when not used, the clocks, controls, power supplies, and other standby power from these plug loads may cause up to 10% of residential electricity use.[186] Your microwave oven probably uses more electricity to run its illuminated clock than for cooking. Altogether, the U.S. runs at least eight giant power plants to power stuff that's turned off.[187] We're also simply demanding more comfort services. Over half of U.S. homes now have central air-conditioning.[188] Government forecasts assume that a continued rise in energy use is inevitable as floorspace and devices continue to outpace rising efficiency.

Our homes mirror our own diversity, ranging from New England saltboxes and Miami condos to Chicago public housing projects, Wyoming ranches, and Beverly Hills mansions. Not surprisingly, older homes in colder climates use more heating, while those in warm climates are the biggest users of air-conditioning. Apartments use the most energy per square foot because they tend to cram the same services into less space, but as shown in figure 3-3, detached single-family homes use five times more delivered energy because they're bigger and far more numerous.

We see the same trends of growth and diversity in commercial buildings. The most energy-intensive buildings are supermarkets and hospitals, where lights and equipment hum around the clock. However, the dominant users, simply because there are

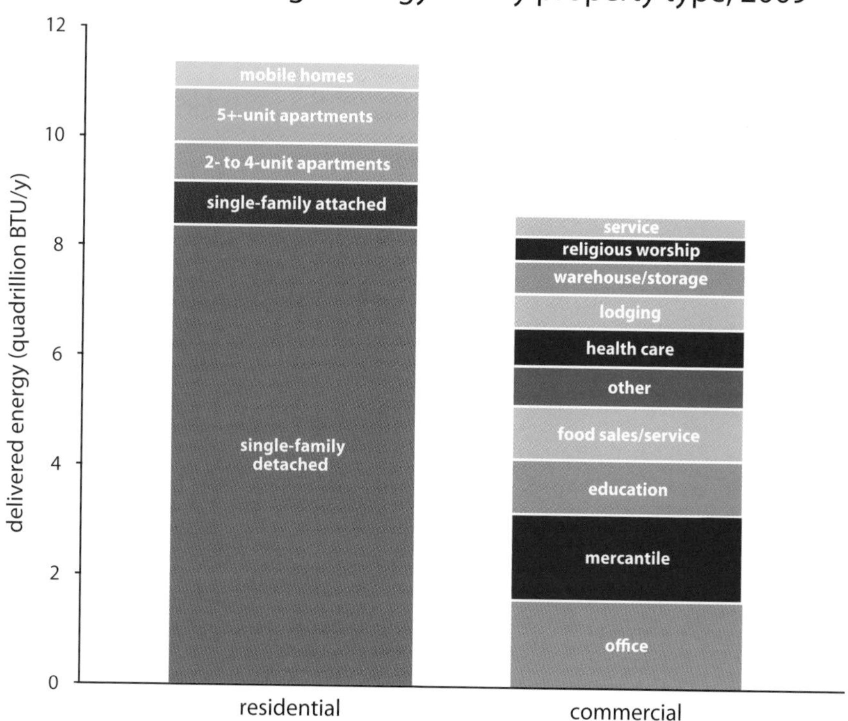

FIG. 3-3. In the residential sector, single-family homes account for over two-thirds of total delivered energy use. In the commercial sector, office, retail, and educational facilities use nearly half the delivered energy.[189]

so many of them, are offices, shops, and schools. Even though businesses may opt for work-from-home programs or hotelling (a reservation-based way to share unassigned desks among people who often travel or work elsewhere), total office space will continue to rise, though perhaps not as fast as once expected. As with houses, there's no "typical" commercial building. The needs and energy use of giant computer data centers or sprawling warehouses are worlds apart from those of dentists' offices or fast-food franchises.

Indeed, the colorful assortment of buildings that Anthony Malkin pointed out from the Empire State Building's observation deck only hints at the nation's enormously tangled building diversity and complexity. There is no simple solution. Achieving the gains that are possible and worthwhile will require myriad strategies and solutions, each tailored for particular types of buildings and different parts of the country.

Just a Few Key Energy End Uses

While the complexities of energy use in buildings can seem overwhelming, figure 3-4 reveals that nearly all their energy serves just a few main purposes—heating and cooling, lighting, water heating, refrigeration, and plug loads. Buildings need to keep us comfortable, illuminate our tasks, and power our gadgets while supporting our health and lifestyles. So we're not dealing with rocket science or brain surgery—just a few key energy end uses. Smart engineers and well-trained contractors implementing straightforward policies and business practices can wring far more work out of those uses' energy.

It's encouraging, too, that the nation has already made progress in improving the energy efficiency of its buildings. Delivered energy use per square foot per year fell nearly one-fifth in the average commercial building and one-third in the average residential building between 1980 and 2003.[190] Since

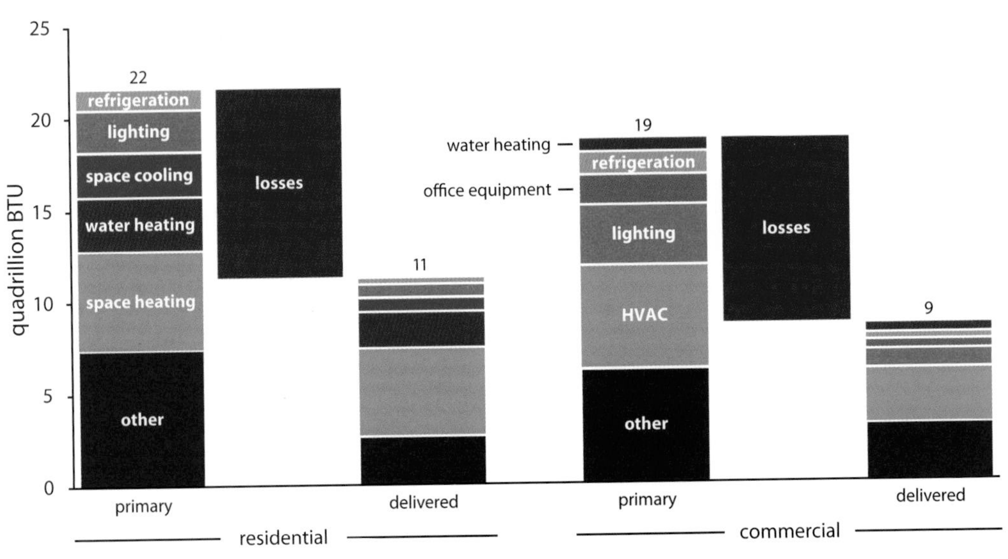

FIG. 3-4. The six key energy uses are different in residential and commercial buildings. Electrical uses entail far more primary than delivered energy because of the roughly threefold losses in converting power-plant fuel to electricity and the 7% average losses[191] in delivering electricity to your meter (plus a few percentage points more lost in the building's wiring).[192] Primary energy use drives your building's climate impact; delivered energy drives its operating cost.

then, additional modest gains have been achieved, mostly via building standards, appliance standards, and utility-funded programs. Further improvement has been spurred by the U.S. Green Building Council's voluntary but commercially valued Leadership in Energy and Environmental Design (LEED) rating system and the federal ENERGY STAR program, which saved $17 billion net in 2009 alone.[193] Though all these gains combined weren't enough to overcome the growth in floorspace or added services, that's a lot better than what was foreseen by most pre-1980 forecasts, which all but ignored efficiency.

Buildings use energy not just to operate but also to make and assemble their materials and install their infrastructure.[194] Buildings last so long that reducing their operating energy is more important than minimizing their embodied energy. That ratio will shift but generally remain large even as operating energy falls. Still, embodied energy—which, after all, represents cost—is well worth saving too by using less energy-intensive materials, using all materials more frugally, making them last longer, and reusing, recycling, or remanufacturing both the

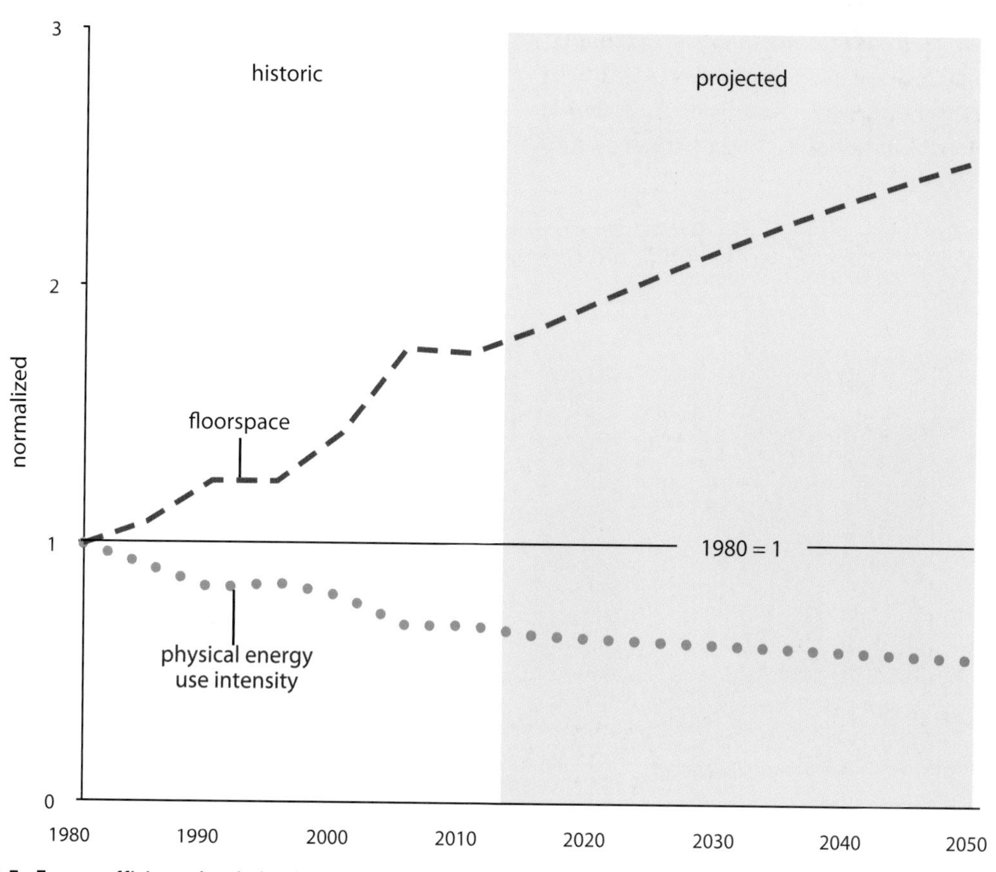

FIG. 3-5. Energy efficiency has helped cut the U.S. building stock's delivered physical energy intensity—delivered energy used per square foot per year—by 30% since 1980. Further reductions are expected to offset about four-ninths of the energy use from the expected 70% growth in floorspace between 2010 and 2050.[195]

materials and the buildings. Many clever architects now design buildings to evolve flexibly through many uses over their lives.

Many forecasts project that buildings' energy use will keep rising. But those forecasts aren't fated to come true. We can accelerate efficiency gains to outpace growth and service additions.

The challenges to implementation are daunting. But the prize is so big that its required rapid learning, retraining, financial and policy innovation, business formation, and scale-up have already begun. Businesses that help lead this effort will thrive; those that lag will do so at their loss. That net $1.4 trillion (plus its important non-energy benefits) has long sat on the table, with only a few sparrows darting in to grab a crumb. But now there's every sign of a concerted effort by serious players to grab big handfuls. We'll all win when they do—especially if we're in on the action.

THE EFFICIENCY REVOLUTION: WHAT'S PROFITABLE AND WHAT'S POSSIBLE

So what kinds of energy savings can we really capture in our homes, schools, office towers, and millions of other buildings? And how do we achieve them? The first step is to look at the numbers.

A Lot of Cost-Effective Energy Efficiency Is Available

The whole story of what's possible is shown in figure 3-6. The short of it is that we can, with significant investments and a transformation of the real-estate industry, save probably 38% and potentially 69% of a 70%-bigger building sector's projected use of primary energy in 2050. Further, this can all be done very cost-effectively: saving 38% requires an

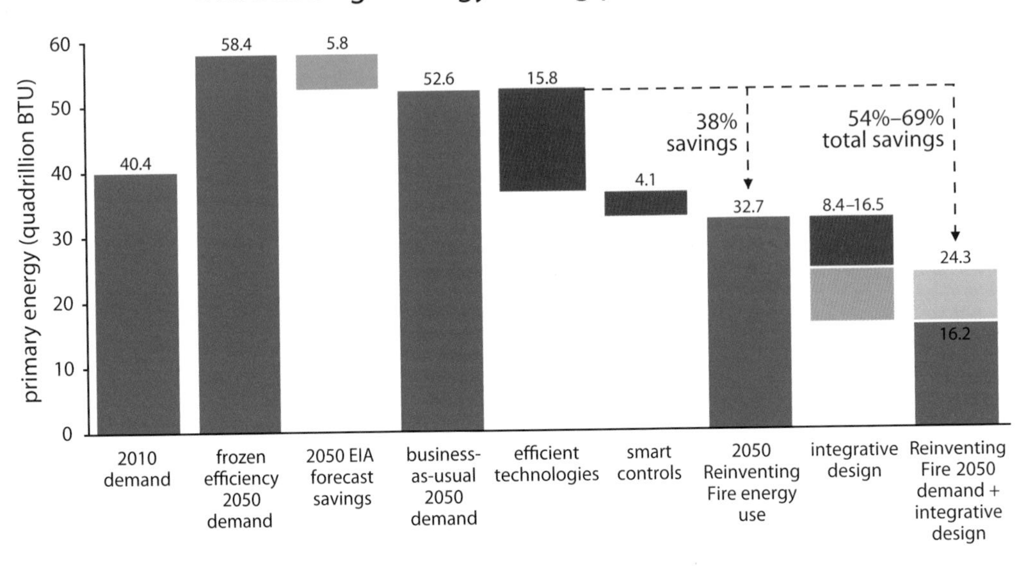

FIG. 3-6. The far-left bar shows that U.S. buildings used 40 quadrillion BTU of primary energy in 2010. More floorspace with the same energy-use intensity would use 58 quads in 2050, reduced to 53 by officially forecast savings (extrapolated to 2050) that don't count innovation or new policies. We next apply the National Academy of Sciences' findings on additional cost-effective efficiency and the American Council for an Energy-Efficient Economy's work on smart controls to reach 33 quads in 2050. Integrative design could then reduce this to 16–24 quads.[196]

incremental $0.5 trillion investment (in 2010 present value) spread over the next 40 years for a return of $1.9 trillion in saved energy costs. The average return on investment is roughly 24% a year.

These targets, if achieved, are very impressive. Never in our history has U.S. building energy use trended downward. Just leveling energy use from 2010 to 2050 would be quite a feat. But why should you care about this? Because a little extra spending now (just 2% of what real-estate developers invest anyhow in good times[197]) can create better and more valuable buildings, with healthier and more productive people in them, at lower long-run cost (fig. 3-7) while managing fossil fuel's risks and emissions. The dollars no longer wasted on energy will flow to the bottom lines of buildings' occupants, owners, or both. That big a prize, and what it implies for competitive advantage to leaders and disadvantage to laggards, merits vigorous effort.

These numbers are important because they also reiterate what many others have already said—energy efficiency is cost-effective. Which bears repeating. But we're not just saying savings of 10% or 20% are cost-effective. We're saying saving more than 30% *beyond* the official forecasts is nearly always cost-effective. If correctly designed and executed, considerably larger cost-effective savings are possible too.

One of the most powerful conclusions about the cost-effectiveness of energy efficiency (which, as we'll keep mentioning, excludes some of the more powerful benefits that make it even more compelling) is how it compares to the cost of

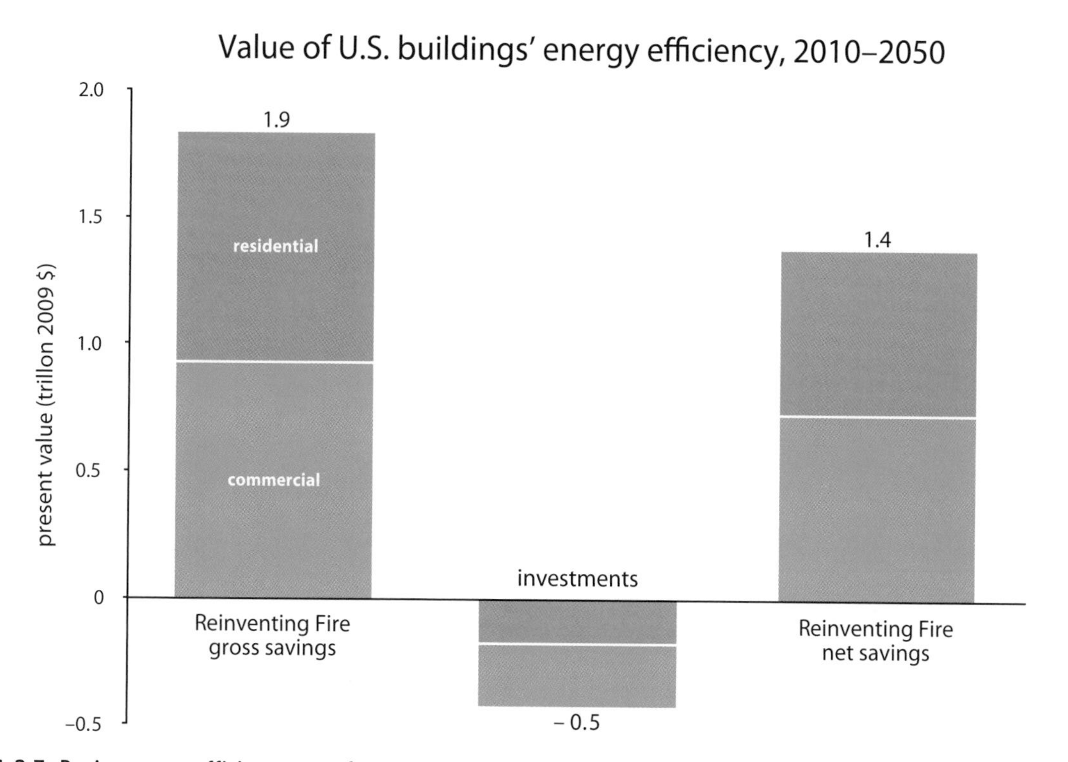

FIG. 3-7. Buying energy efficiency saves far more money than it costs. Spending an extra $0.5 trillion on energy efficiency in U.S. buildings over the next 40 years can save $1.9 trillion in present value (based on the Reinventing Fire 2050 building sector's energy savings of 38% and not yet counting integrative design, which we'll discuss later).[198]

the energy it displaces. As shown in figure 3-8, whether compared to electricity or natural gas, energy efficiency in buildings typically costs less than half as much as the energy it saves. And that's with today's technologies and prices.

It is important—though perhaps obvious—to note that it's not enough for these cost-effective technologies simply to be available; they already are. To get their benefits, businesses and homeowners will have to buy, install, operate, and maintain them. While the national track record in implementing building energy efficiency has been mixed, some states and regions have enjoyed sustained success. For our vision to come to fruition, the U.S. as a whole would just need to begin investing in energy efficiency as consistently and robustly as some of the top-performing regions already do. For example, the Pacific Northwest has been investing in energy efficiency for over 30 years, so its residential market share rate of compact fluorescent lamps is 75% higher than the national average.[199]

Across the country, companies are recognizing the energy savings opportunity by buying and selling efficient products. Impressive gains can be won just with today's conventional technologies, so let's start there to estimate what more can be done (see Assessing the Energy Efficiency Opportunity in Buildings sidebar).

Better Design and Equipment

Just drop in, for example, at Toyota Motor Corporation's office in Torrance, California, built in 2003. Toyota chose to design and construct this 624,000-square-foot (sf) office for 2,500 employees in Southern California to show its environmental commitment and demonstrate an energy-efficient building's shareholder value. Weather-normalized measurements show it uses 42% less delivered energy than allowed by California's Title 24

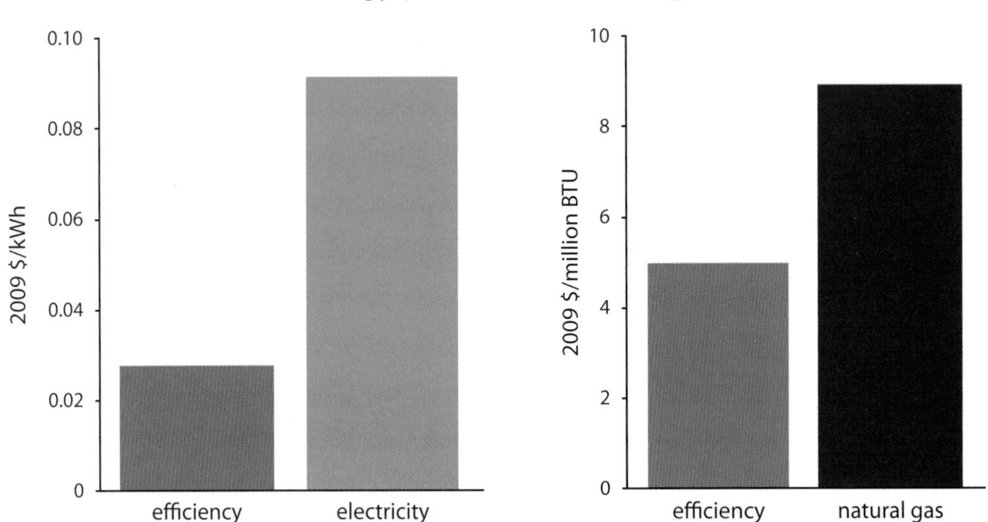

FIG. 3-8. The electricity and natural gas energy efficiency improvements analyzed cost much less than the energy they save. The "levelized" costs shown allow a comparison of energy efficiency with the energy supply it avoids, despite their unequal lifetimes and capacities.[200]

building code, one of the nation's strictest. How? The building was positioned to bring in the most light and the least heat, with double glazing, extra insulation, and a highly reflective "cool roof" on which Toyota mounted a large photovoltaic system, boosting its total saving to 60%. With utility incentives and tax credits, the whole package outperformed the firm's 10% rate-of-return target.[201]

In another case, the potential energy savings were simply too large to ignore any longer. In 2009, JCPenney spent $38,000 on energy efficiency investments in a 100,000-square-foot store in Orange County, California. The company replaced halogen lamps with light-emitting diodes (LEDs), saving money and better lighting the merchandise. Crews also repaired lighting and air conditioner controls, installed weather seals on all doors, and added occupancy sensors. The result: the store now uses 28% less energy, saving $41,000 a year—a pretty nice return for modest effort.[202] At thin retail margins, how much product would JCPenney need to sell to yield that bottom-line result—and how would its confidence in that return compare?

Right-Timing

There's more to saving energy in buildings than better design and newer equipment, however. It's also important to make upgrades and retrofits at the right time. Every building undergoes cyclic changes throughout its life. Major systems like facades and mechanical systems wear out, leases expire, sales and refinancings and market changes occur. As a result, astutely timed energy retrofits can piggyback on changes being made anyway, greatly reducing capital cost.

In Sacramento, foreclosed homes needed significant upgrades to sell. Seeing this as an opportunity to piggyback energy efficiency onto the upgrades, the Sacramento Municipal Utility District (SMUD), with help from the U.S. Department of Energy and the Sacramento Housing and Redevelopment Agency, set a goal: halve an old home's energy use with a simple upgrade package. Using utility funds to front the extra cost, the team's package of new windows, insulation, and mechanical systems got 53% savings at a cost of $12,800 over that of a standard rehab. The utility

ASSESSING THE ENERGY EFFICIENCY OPPORTUNITY IN BUILDINGS

After decades of successful programs and policies intended to drive energy efficiency, we know it's a clean, cheap, and reliable way to displace fossil fuels. To estimate potential savings over the next four decades, we analyzed three main variables:

Stock turnover. Every year, new buildings are constructed while others are demolished, and over time, the equipment in all buildings must be replaced. Each new building put up or piece of equipment sold is a new opportunity to save energy, because *new* generally means more efficient but less expensive. For instance, when an old window air conditioner breaks down, *any* new replacement unit will be more efficient. Our analysis compares the energy intensity of the efficient new building to what it would have been if the building were built only to current code.

The costs and savings of energy efficiency measures. Savings vary widely in new buildings and equipment depending on design and equipment choices. Many options with divergent cost and performance can far outperform the minimum standard. We assume that customers will choose efficiency only if it beats the average retail price of electricity or gas at the National Academy of Sciences' 7%/y real discount rate.

How many companies and individuals will choose to invest. Beyond how much energy can be saved at what cost for an individual building, we must estimate how many buildings will actually get such an investment. Past U.S. investments have been suboptimal. We project much higher participation, ramping up over the next 20 years to the rates now achieved by the top-performing states, as competitive pressure, internal buy-in, and policy support overcome traditional customer barriers to efficiency.

bill fell more than the mortgage payment rose to repay the rolled-in utility loan. "When you are working with houses with such extensive damage, we learned that it is actually fairly easy to upgrade to a higher level of energy efficiency," said Dennis Lanni, one of the project's leaders.[203]

In another example of right-timing, Eng Lock Lee was called to the Hyatt Singapore when one of its giant chillers was due for replacement in 1999. Most designers would simply have specified a new one the same size. But Lee's careful measurements showed that the plant was oversized and inefficient. The smaller, smarter system he designed slashed energy use by eight million kWh in the first year, saving more than $770,000 a year with a 45% return on investment.[204]

Similar stories of good design and right-timing can be found everywhere from New York City townhouses to homes in Florida. Some opportunities get missed because they seem too simple and obvious, though they can have far-reaching effects. Energy efficiency is not all about fancy equipment. Just light-colored roofs and paving, plus shade trees and revegetation to help bounce solar heat away, could cool Los Angeles by about 5 F°, cut the city's cooling loads by about 20%, and reduce by 12% its days of noncompliance with state ozone standards. The ozone reduction alone would save roughly $360 million a year in health-related costs.[205] And besides keeping buildings cool and roofs durable, each thousand square feet of light-colored roofing materials slightly cools the earth by bouncing away more solar heat—equivalent to keeping 10 tonnes of CO_2 out of the air, as a free bonus on top of the valuable energy savings.[206]

Admittedly, these examples don't yet reflect common practice. We'll explain why later in the chapter, and we'll offer solutions to the many challenges of implementing energy efficiency. But to begin, let's look at some emerging technologies that can help make today's standard practices first the norm, then obsolete.

Emerging Technologies

Fortunately, energy-saving technologies have decades, nay centuries, of history of replenishing the "efficiency resource" faster than it's exhausted, and at similar or lower cost. For example, the average new 2009 U.S. refrigerator used 72% less electricity than a comparable model did in 1972 yet cost 62% less, despite more and better features.[207] That ultimately saved the nationwide equivalent of more than 60 one-billion-watt power plants.[208] Some new refrigerators are more efficient still. Home air conditioners bought in California have also steadily improved their efficiency at a rate of 3–4.5% per year for decades and show no signs of abating.[209]

History demonstrates that efficiency improvements will continue until we approach the best that physics allows. That's why the savings we assume will accumulate between now and 2050 originate partly from emerging technologies[210]—devices that have yet to come to market and become cost-effective, but already exist and are entering or heading for market.

Emerging technologies are always a bit of a gamble. If the innovation doesn't offer significant additional benefits or help lower cost, it may never fully roll out to its marketplace. Even if it is a success, it can take years to reach that marketplace. Cellphones, one of the fastest-to-be-adopted technologies ever, took 15 years to approach market saturation as they got 20-fold cheaper. Yet just as cellphones were a disruptive communication technology—a gamechanger—plenty of disruptive technologies for buildings are rapidly emerging. Here are just a few.

Smart windows. One of the biggest bangs for the buck could come from smart windows. Already, windows that darken in response to a small electric current or heat, from firms like Pleotint and RavenBrick, are used in fancy conference rooms and in Boeing's 787 Dreamliner.[211] But a far more important technology is expected to reach the market. Serious Energy's "AdaptivE"[212] windows will use a printable liquid-crystal coating to vary the amount of incoming heat energy depending on the temperature of the outer pane of glass, while letting in the same amount of visible light. Such a selectively "thermochromic" window would automatically admit more than five times more solar heat on a cold winter day than on a hot summer afternoon. Just this one technology is expected to reduce a typical house's heating and cooling energy use by up to 30%.[213] The net cost could be negligible, since the house would need smaller, cheaper heating and cooling equipment.

Enhanced evaporative cooling. Thousands of years ago, the Persians figured out that spraying water into hot air caused the water to evaporate, absorbing heat and cooling the air. Today, that age-old technology is being updated to produce far more efficient air conditioners. Such evaporative coolers work well in dry climates—a version in Stanford's 2004 Carnegie Institute for Global Ecology lab gets 50 units of cooling per unit of electricity—but not in more humid areas like Miami or Houston. So Eric Kozubal, a senior engineer at the National Renewable Energy Laboratory, figured out a modern twist. He devised an innovative way to add a water-absorbing substance (a desiccant) to dry the incoming air so the evaporative cooler can make cool air in any climate. The process, which Kozubal calls desiccant enhanced evaporative (DEVap) air-conditioning, can save 50–90% of the energy used by a traditional air conditioner, depending on climate zone.[214] Such firms as Advantix Systems, Munters, Seibu Giken, and Trane have begun bringing advanced desiccant technologies to market.

Radical insulation. Major gains could come from a radical update of an old product—insulation. For decades, laboratories have been able to make extraordinarily light and fluffy silica-based gels that can insulate up to R-40—six times better than the best plastic foam—per inch. Now several manufacturers including Proctor Group Ltd., Aspen Aerogels, Cabot Corporation, and NanoPore[215] are scaling up production to make such aerogels and nanogels affordable. These exotic materials can turn even thin spaces like window frames, thin walls, and eventually paintable surfaces into potent thermal barriers.

Phase-change materials. Phase-change materials absorb heat by melting as the temperature rises. Used in walls or on roofs, they slow the buildup of heat in a house on a hot day. The stored heat is then gradually released at night. They're starting to emerge in building materials such as National Gypsum's ThermalCORE, which uses German chemical giant BASF's paraffin wax capsules.[216]

FIG. 3-9. A blowtorch doesn't damage a flower because of the impressive insulating properties of aerogels.

Light-emitting diodes (LEDs). The U.S. Department of Energy expects LEDs by 2030 to save almost as much electricity as all U.S. homes now use for lighting.[217] Analysts predict that the U.S. market for commercial and industrial general-illumination LED lighting will grow from $330 million in 2010 to $1 billion by 2014,[218] with giant companies like Philips, Osram Sylvania, Cree, and General Electric battling for share. Some of these semiconductor devices are already up to twice as efficient as compact fluorescent lamps (CFLs), and prices are dropping.[219] Starbucks is already using them to reduce lighting energy use by 80% in more than 7,000 company-owned stores.[220] A new generation of lighting fixtures is being developed to exploit LEDs' pinpoint light source, which permits exquisitely precise optical control, further boosting their energy savings. And new advances, from inventors like Tsutomu Shimomura, who has found a new way to wire and control LEDs with minimal wiring, could allow for fine-grained control of buildings' lighting and significantly reduce both energy and capital needs.[221] Meanwhile, next-generation LEDs that replace semiconductors with organic compounds, called OLEDs, are already appearing in cellphones and cameras and could make super-efficient, higher-quality televisions (fig. 3-10) widely available. Even better technologies are on the way.

FIG. 3-10. OLED technology is enabling high-efficiency and even bendable lighting and digital display solutions. Samsung alone expects to sell 600 million active-matrix OLED screens in 2015.[222] Sony's new XEL-1 OLED TV has a million-to-one contrast ratio, an extremely fast image response time, no backlight, 3 mm thickness (without the stand), and a 178° viewing-angle range.

Efficient rotors. Or what about a technology that can dramatically cut the biggest uses of electricity? That's the promise of pumps, fans, blowers, turbines, and propellers shaped like a nautilus shell. Nature uses this spiral shape over and over, and it actually follows a sequence of numbers called a Fibonacci series. Australian

FIG. 3-11. PAX Scientific's Fibonacci-spiral-rotor refrigerator fan (top) and pump impeller (bottom) are far more efficient than conventionally shaped components.

naturalist Jay Harman figured that nature must know something, so he started making rotors in these strange-looking biomimetic shapes (fig. 3-11). Now his technology is starting to improve fluid-moving rotors' efficiency by tens of percentage points—sometimes far more.[223] Such shapes can save 30% of the energy used by computer cooling fans, of which over a billion are made each year. Or they can stir a waterworks' seven-million-gallon tank with just 320 watts[224]—roughly five times as efficient as its nearest competitor.[225] Harman's northern California company, PAX Scientific, is now licensing the technology in many diverse sectors. Hundreds of other nature-inspired designs hold promise, too, for better building components and for overall building design. Even without biomimetic design, better aerodynamics and motors cut energy use by half in commercially available ceiling fans (Gossamer) and threefold in same-cost bathroom exhaust fans (Panasonic).[226]

Appliances. What about electricity losses from the mundane electric stove? Croatian-Swiss inventor Dusko Maravic has invented a new kind of pot whose bottom, rather than warping when heated, stays perfectly flat and in full contact with a computer-controlled resistance-heated ceramic element.[227] The result: better, faster cooking and large electricity savings (European measurements suggest about two-thirds). Further big savings are available from smarter controls and from several innovations in heat-trapping pots and their lids.[228]

Heat pumps. Further down the path to adoption and to significant energy savings are such innovations as miniature heat pumps that can both heat and cool buildings. The Swiss Federal Institute of Technology (ETH Zürich) has demonstrated a miniature high-speed heat pump that produces an astonishing 10–12 units of cooling from one unit of electricity at small temperature drops (around 23 F° or less). Run backward, the heat pump can also provide about eight units of hot water per unit of electricity, making it eight times as efficient as a standard electric-resistance water heater,[229] or almost three times as efficient as standard water-heating heat pumps.

These examples offer just a glimpse of where energy efficiency technologies are headed—and how clever businesses will find ways to make money from selling them. All these technologies, and scores of others like them, will deliver future energy savings. Practitioners who want to gain an advantage on stodgy competitors will wield emerging technologies as a strategic weapon, walloping anyone who doesn't keep up.

In our projections of the Reinventing Fire energy savings, we haven't posited *bigger* savings based on these types of technical leaps. Rather, we've assumed they'll continue to bring the *same annual percentage* of cost-effective savings between 2030 and 2040 that we think is possible between now and 2030. History suggests this is a very safe bet.

Smarter Behavior

We also expect our conservative projections can be exceeded because there's another route to significant energy savings that doesn't require new designs or major retrofits. The secret: changing how buildings' occupants use building services.

No, we don't mean pressuring people to choose uncomfortable thermostat settings or work in the dark. The behavioral changes we want to make are *not* about discomfort, privation, or curtailment. Rather, they're about intelligent choices. Most people are unaware of how much energy they are using—and how much is being wasted. When more than 1,000 guests at a 1997 Interface corporate conference at Maui's Grand Wailea hotel were given daily feedback on how they could save energy, water, and waste by following simple tips without sacrificing comfort, the property's energy use fell 22% in just six days.[230]

Providing just a little bit of information can change behavior. For instance, letting people know how their energy consumption compares to their neighbors' can stir the competitive juices and reduce use markedly. Opower, a start-up based in Arlington, Virginia, included reports in Sacramento households' bills comparing energy consumption among neighbors and offering recommendations on how to reduce it, such as turning down the heat when you leave the house. Participants cut their energy use by 3% on average.[231] Opower is now implementing this program with many utilities around the country. And new technologies, such as smart meters, are making this sort of information more widely available and easy to apply.

Even small, back-of-the-mind behavioral shifts can save large amounts of energy, because consumption varies dramatically between comparable buildings, depending on how their occupants behave. Many studies have already documented this. Based on the findings of 36 residential pilot programs, if 80% of customers got usage information by 2050, the resulting behavioral changes would shave more than 10% off U.S. buildings' energy use at a cost of only about $0.035/kWh, or one-third what we typically pay for electricity.[232] With better program design and use of social science research, the next generation of behavioral programs could save far more, perhaps over 25%.[233]

Having information on energy consumption also opens the door to cutting energy use without changing behavior. Smart controls, for example, can automatically turn down the thermostat when you're gone, or run your dishwasher or washing machine at a time that's convenient for you but cuts your cost. They'd also let an office building start precooling earlier on a hot day, when electricity costs less than in afternoon peak hours. (In concrete or masonry buildings this can often reduce total cooling energy, not just shift when it's used.[234]) Such unobtrusive technology can deliver the same or even better services at lower cost with no inconvenience or loss of amenity.

How to Move Beyond Conventional Design

So far, we've seen how some companies have begun to grab some of the energy savings that are now here for the taking, and how innovation will sustain the march toward ever more efficient technologies. We've also examined creative ways to nudge people to change their behaviors to save more energy, bringing big dividends without compromising comfort or other occupant values. Together these approaches can cut buildings' 2050 primary energy use by 38%. Later, we'll tackle the enormous practical challenge of spreading conventional good practice throughout the nation's huge fleet of buildings.

But first, let's aim higher. Indeed, we need to. Implementing the basic Reinventing Fire vision we've already outlined would still leave U.S. buildings using 33 quads of primary energy for electricity and natural gas in 2050. So how do we go further—preferably at comparable cost and ideally at even lower cost?

The answer: as long as we're going to the trouble of changing a building, we can take the extra trouble to change it more effectively by applying integrative design. Exploiting this opportunity could well raise the savings to over two-thirds of buildings' 2050 energy use. That's the last step—the two bars at the right side—in figure 3-6.

The costs of integrative design are quite variable. In a small but convincing set of new commercial buildings, integrative design yields greater savings than standard design with no (or occasionally negative) incremental capital cost—cost, that is, for the whole building, as some parts may cost more but may be offset by making other parts smaller and cheaper or even unnecessary. In other cases, the building may have a cost premium, yet it is often quickly recovered.[235] The wide scatter in reported savings and costs reflects differences in design and execution quality, because few designers are yet skilled in this approach. Those differences make the outcome highly uncertain—how quickly and deeply can integrative design be adopted across these complex markets? But the differences also imply a potentially important business opportunity in rapidly taking best-practice integrative design to scale.

So what exactly is integrative design? And how can we exploit it to transform the nation's buildings, designing better with artfully combined technologies rather than just with a grab-bag assortment?

Integrative Design

Instead of just upgrading, say, conventional lights or heating systems with more-efficient equipment, integrative design starts by asking whether there are smarter ways to design the whole building and all its interacting systems together. So far, we've sketched some important technological progress in specific technologies. Integrative design *combines* technologies (old and new) in novel ways.

Consider the many benefits Skanska USA reaped when it redesigned its 25,000-square-foot 32nd-floor offices in the refurbished Empire State Building in 2008. Rather than trying just to boost the efficiency of standard lighting and heating and cooling systems, the company's engineers rethought the whole design. They replaced conventional overhead ductwork with underfloor ventilation that lets each worker control airflow and temperature and reduces fanpower, noise, and the spread of infection. Since old-style ductwork typically covers the top one-fifth of the windows, eliminating it opened up the full window area. That plus glass-front offices brings daylight to 90% of the office, saving 35% of lighting energy. And workers like the natural light and new views: absenteeism dropped by 14%.

The design cost 4.6% more than a standard office fit-out to achieve the highest LEED rating. But it will cut electricity costs by 57% compared to Skanska's previous office, paying back in five years. Senior vice president Elizabeth Heider also points out that the lower energy use is a hedge against future higher electricity prices or carbon pricing.[236]

Switching to integrative design isn't easy. It requires architects, engineers, contractors, and owners to collaborate more effectively. It demands that designers rearrange their mental furniture and sometimes make end runs around entrenched practices. But it's hardly arcane. It rigorously applies orthodox engineering principles, but it often asks different design questions in a different order to get a better answer. And some of the insights it generates are obvious in hindsight.

Look, for example, at the typical big-box store surrounded by a sea of unshaded black asphalt. The heat absorbed and radiated by the acres of asphalt on sunny summer days taxes air conditioners in both the store and the thousands of parked cars. The blast of heat as you step out of your car is unwelcoming. At night, the dark surface requires powerful lights to provide safe illumination. Integrative design would quickly reveal that a light-colored parking surface would cut cooling and lighting costs. Then the store would keep customers more comfortable with less air-conditioning. The cooler pavement, too, would last much longer. If pervious, it can shrink or displace costly infrastructure to manage stormwater. And shading the parking lot with solar cells, as some merchants are starting to do, could further enhance comfort, while the electricity generated could either be sold to the grid or—even more valuably—used to recharge customers' electrified cars for free, tempting them to shop longer.

Right-Sizing

Whether outside the building or inside, integrative design has many powerful consequences. One of the most common and valuable is being able to shrink or even *eliminate* mechanical equipment—an idea called "right-sizing." That's what I was able to do with my own 4,000-square-foot

INTEGRATIVE DESIGN IN AN OFFICE BUILDING

In a new or major-retrofit office project, many advanced designers prefer underfloor displacement ventilation. Done properly, it's silent, individually controllable, healthier for workers, far more efficient, and comparable or lower in total capital cost than an overhead forced-air system. An underfloor plenum distributes the air, replacing costly ductwork and taking less vertical space. Lower pressure drop shrinks the supply fans and hence the chillers (because fan energy heats the circulating air). Ventilative mesh chairs, like Herman Miller's Aeron, keep people's buttocks cool (sitting on standard upholstered foam can make a person's buttocks up to 12 F° hotter in 30 minutes[237]). Other innovations expand the range of conditions in which people feel comfortable.

The building envelope is oriented, massed, shaded, and surfaced to capture useful natural energy flows and deflect adverse flows. "Tuned" superwindows—nearly perfect in letting in light without unwanted heat—optimize light and heat flows in each direction, improving comfort and shrinking cooling systems. Overhead ducts and the dropped ceiling to hide them go away; sprinklers become an unobtrusive architectural element. Smaller floor-to-floor height saves a strip of costly façade around each story and can even squeeze six rather than five stories into a standard U.S. 75-foot low-rise height limit, doing wonders for financial performance, while raising the ceilings for better daylight distribution.

Indirect lighting from directed daylighting and pendant fixtures provides glare-free light everywhere—light that's far more visually effective than if it were dumped downward from fixtures traditionally embedded in the dropped ceiling. Lighting surfaces, not volume, saves much of the lighting energy, prevents visual fatigue, and makes the whole ceiling plane bright and cheery rather than dark and oppressive. (Lighting *design* is even more important, yet more neglected, than efficient lighting *equipment*, but the two work together powerfully in all kinds of buildings.[238]) Less power for lights, fans, and pumps shrinks cooling systems further, saving more capital budget. Such capital savings help pay for the better windows, insulation, et cetera.

But now the ventilation and lighting combine to create a new benefit: The cost of reconfiguring the space when people move their offices ("churn"), which often costs several dollars per square foot per year, is practically eliminated. How? Since the whole ceiling plane is evenly lit, there's no need to move lighting fixtures or their wiring and controls when people move. Ventilation is available by replacing any floor tile with a diffuser. Power and IT wiring (if it's not wireless) can be moved by simply popping up a carpet tile and the floor tile beneath it—no need to tear into walls or ceilings.

The bottom line: such a building can deliver unprecedented thermal, visual, and acoustic comfort using only a small fraction of the energy previously needed; can cost about the same or less to build; and can be faster to construct.

house high in the Rocky Mountains. The house is superinsulated and about 99% passively heated despite temperatures down to –47°F, thanks partly to "superwindows" that look like they have two sheets of glass but insulate like 16 (or in a few cases 22). These measures *reduced* the house's construction cost, because eliminating the heating system saved more up front than the superwindows, superinsulation, and ventilation heat recovery cost. Reinvesting that saved capital then cut to less than a year the payback for saving roughly 90% of household electricity and 99% of space- and water-heating energy. A recent retrofit further boosted savings by updating the technologies from 1983 to 2010 state-of-the-art. Tellingly, the monitoring system to measure the savings appears to be using more electricity than all the lights and appliances combined.

The right-sizing approach, shrinking the size and cost of heating and cooling systems, is spreading. German engineer Dr. Wolfgang Feist launched the Passivhaus movement, which has created more than 25,000 certified passive structures in Europe[239] (including schools, commercial buildings, apartments, and houses) that need no traditional space-heating equipment, such as furnaces, even in cold, cloudy climates, and hence can cost no more to build than ordinary houses. A Thai architect, Professor Suntoorn Boonyatikarn, brought the same concept to hot, steamy Bangkok, saving 90%[240] of normal air-conditioning energy with normal construction costs and a tiny air conditioner. And at Anthony Malkin's Empire State Building, 6,514 double-glazed windows, remanufactured onsite into superwindows, now block at least two-thirds of winter heat loss and half of summer heat gain. That has cut the building's peak summer cooling load by one-third. So instead of excavating Fifth Avenue to dig up, replace, and enlarge the old chillers, engineers were able to renovate and reduce the existing

EXPENSIVE RETROFITS CAN COST *LESS* THAN ROUTINE RENOVATIONS

Sometimes right-timing plus right-sizing can yield even more surprising economics than either alone. For example, a 200,000-square-foot 1970s glass office tower near Chicago needed reglazing when its dark, light-blocking window units' seals failed, as they do about every 20 years. A normal reglazing would install identical units. RMI designers found that new superwindows could insulate fourfold better but let in six times more daylight and one-tenth less unwanted heat. Better daylighting, lights, and office equipment could then cut the peak cooling load by three-fourths. Replacing the aging air-conditioning system with a new one four times smaller and nearly four times more efficient would cost $200,000 less than renovating the big old system—enough to pay for all the improvements' extra cost with a bit left over. Thus a highly integrated, right-timed retrofit was simulated to save 76% of the building's energy, for slightly less cost than a normal renovation that would save nothing.[241]

This design was approved for construction but unexpectedly not executed: the owner belatedly discovered that the cash-short broker who controlled the building hadn't wanted to delay earning leasing commissions during the several extra months needed for the retrofit, so conventional quick fixes were done instead, losing the whole-building retrofit opportunity for another 20 years. The property was then so uncomfortable that it didn't lease anyway, so it had to be sold off. This illustrates the showstoppers that lurk at each of 20-odd steps in the commercial-building value chain.[242] The example does, however, suggest an important design opportunity from retrofitting with both right-timing *and* right-sizing. America has over 100,000 similar buildings awaiting similar treatment. Every year, thousands of them come due for reglazing. Alert owners will seize the opportunity for a deep whole-building retrofit.

ones. Electrical savings, cutting the peak load by a third, also avoided costly renewal of old interior cables. The savings: $17.4 million in avoided capital cost, recycled to help pay for the windows and other efficiency upgrades. That's why the building's 38% energy savings—actually over 40%—repaid the incremental cost in just three years, and why plans for a bigger utility hookup and an onsite cogeneration plant were cancelled.[243]

Right-sizing doesn't have to be done at the grand scale of a whole building (see Expensive Retrofits Can Cost *Less* than Routine Renovations sidebar). It can make a big difference when applied to small but ubiquitous devices. For example, a Danish study found that the pumps used to circulate hot water in normal European homes are 5–10 times bigger than needed and 4–8 times less efficient than they should be. Gradually replacing all 120 million household circulating pumps across Europe over a decade would eliminate the need for 8.5 one-billion-watt power plants and achieve one-sixth of Europe's Kyoto carbon-reduction obligation.[244] The smaller and far more efficient replacement pumps, now common in Europe, entered the U.S. market in 2007, offering 70–90% potential savings with a 1.5- to 3-year payback,[245] and are cheaper than replacing oversized pumps with efficient ones that are still too big.

More gains come from applying integrative design to systems that are usually copied-and-pasted—"infectious repetitis"—rather than properly optimized. In chapter 4, we'll learn how unconventional ways of laying out pipes and ducts, whether in a building or a factory, can save all but a tiny fraction of the pumps' and fans' size, energy use, noise, and investment.

Integrative design has many more steps and tricks than these few examples illustrate. But the result is simple: in certain situations, big energy savings can cost *less* than small or no savings. In other words, investments in energy efficiency can yield not diminishing returns—the more you save, the more and more it costs—but expanding returns. Some theorists think this is impossible, but good engineers think it's fun. And expanding returns, such as we see every day in electronics and information systems, change everything. Not only can we achieve astonishing reductions in buildings' energy use, but such reductions already are being made in select offices and homes all across the country and around the world. Telling their stories, in a quick tour, vividly demonstrates what's possible (see Integrative Design Standouts sidebar).

Additional Benefits of Energy Efficiency

We've made, as many others have over the years, a convincing case that based on avoided energy costs alone, America's 120 million buildings can be made dramatically more efficient, bringing cuts in energy cost that handsomely return the investment. But the overall story—and the economic equation—turn out to be more compelling than the energy savings alone would indicate. That's because the same upgrades that save energy also bring other valuable benefits as free by-products.

In fact, where measured, these extra benefits have almost always outweighed the value of the energy saved. Unfortunately, these benefits—often in the form of productivity, satisfaction, and comfort—are hard to quantify and thus not often monetized. Research to date confirms that the value of non-energy benefits is positive. But just how positive and how beneficial are questions ripe for researchers and entrepreneurs to answer more definitively. Further, different parties will weight the factors differently: Homeowners, for example, lack all the factors related to brand enhancement or attracting talent. How much more would building owners pay for efficiency if more of the non-energy cost benefits were quantified? Would more expensive investments be justified? Finding ways to structure agreements to collect on these additional benefits is a huge opportunity waiting to be unlocked.

INTEGRATIVE DESIGN STANDOUTS

The recent promotion of integrative design by architects and engineers offers many examples. Here we present several we find exciting or innovative in both technical content and ambition. Despite challenges, these projects found ways to move forward, proving that more and more designers and owners can achieve similar results.

Byron G. Rogers Federal Building (Denver, Colorado). This historic 1960s-era office complex of a half-million square feet, home to 11 agencies, is getting new superwindows, superefficient LED lighting, innovative mechanical and plumbing systems, energy-recapturing elevators, 100% solar water heating, and a host of other new features. The aggressive retrofit is expected to meet federal investment guidelines and cut the original energy use by 70%, meeting in 2013 the 2020 waypoint en route to the federal government's 2030 zero-net-energy-use goal.[246]

National Renewable Energy Laboratory Research Support Facility (Golden, Colorado). You'd expect the government's green energy technology incubator to be a model of efficiency—and it doesn't disappoint. Triple-glazed daylighting in every office, underfloor ventilation, digital controls, and efficient IT equipment are among the many technologies that cut delivered energy use to 35 kBTU/sf-y. That's especially impressive because the facility also contains a data center. To get to very low-energy buildings, "every watt has to count," explains IT project manager Craig Robben. Yet the 220,000-square-foot building, shown in figure 3-12, was cost-competitive with many big government offices nearby.[247]

FIG. 3-12. Despite its data center, this new NREL office uses half the normal energy, or one-third of the average for existing U.S. offices, and gets all its electricity from solar power on the roof. This image shows the building's south-facing windows, where outside ventilation air is passively preheated via transpired solar collectors (a technology developed by NREL).

Lewis and Clark State Office Building (Jefferson City, Missouri). Missouri is the Show-Me State, so when architect Bob Berkebile proposed to build an energy-efficient office for 400 employees of the state's Department of Natural Resources on a limestone bluff over the scenic Missouri River, there were plenty of skeptics. Yet at no increase in cost, his team delivered on time and on budget a comfortable, beautiful model project that saves 55–60%[248] of normal energy use.

Center for Health and Healing, Oregon Health & Science University (Portland, Oregon). Okay, you say, I get it: integrative design can save well over half the energy in a new office building at no extra cost. But what about a very energy-intensive building like a healthcare facility? Let's make it big—412,000 square feet, 16 stories—and put it in a cloudy place. Result: it was designed to save 61% of code-permitted energy use while costing millions of dollars *less* to build. How many millions? The original $30 million mechanical, electrical, and plumbing budget fell by 10% or $3 million. All other energy efficiency investments added back an incremental $0.98 million. Two solar systems added $0.89 million. Altogether, these improvements saved $1.6 million up front and garnered $1.6 million in financial and tax incentives, contributing a net $3.2 million to other costs. At a 10% cap rate, the energy and rentable-space savings will increase the building's value by about $8.6 million.[249] Total added value from integrative design: about $12 million, or $29 a square foot. Any gains in human health and performance are extra.

Deutsche Bank Twin Towers (Frankfurt, Germany). Deutsche Bank's iconic headquarters—twin 36- and 34-story towers—were nearly 20 years old in 2003 when tougher fire rules required a major retrofit. It would have been cheaper to sell the buildings and move, says green building manager Nils Noack. But the bank saw an opportunity to burnish its reputation for environmental leadership by staying put and turning the towers into two of the world's greenest buildings. The three-year project cost €200 million, much of it for ordinary renovation rather than for the green features. It will cut heating and cooling energy 67% and electricity by 55% while also saving space and accommodating 600 more employees. Every second window is operable. Smaller mechanical systems freed up a whole floor in one tower. LCD readouts in the elevators show each floor's progress in saving energy, driving informal competition. Efficiency plus hydropower will cut carbon emissions per employee by 91%—part of the bank's goal of becoming carbon-neutral worldwide by 2012.[250]

Fossil Ridge High School (Fort Collins, Colorado). In 2005, Poudre School District designed and built a 290,000-square-foot high school that's 60% more energy-efficient than comparable school buildings at no additional cost, saving $60,000 per year. The building uses lighting occupancy sensors, a sophisticated HVAC heat recovery system, and extra insulation. The design also emphasized better daylighting and indoor air quality, which both improve the indoor learning environment. The cost? Fossil Ridge's price tag of $179/sf, including design fees, furnishings, and equipment, compares favorably with that of other schools in the region.[251]

Passivhaus example (Urbana, Illinois). The U.S. Passivhaus movement is nascent in comparison to Europe but is starting to have its own dedicated following. Architect Katrin Klingenberg was one of the first to design to the stringent Passivhaus design standard in the U.S., building her own home in rather cold Urbana, Illinois, in 2003. With extremely high levels of insulation (average R-value of 56) and a super-airtight envelope, the building saves 76% of delivered energy compared to a new house built to code. You might expect one of the first Passivhaus examples to have an exorbitant cost, and it wasn't cheap at $18,000 more than normal, but Klingenberg notes, "It was a prototype. It took a lot of research, translation of certain materials, and experimenting. But the construction is actually standard balloon framing, and it is my belief that an experienced contractor could build such a house for about 10 percent more than a comparable home—an amount that could be easily recovered in energy savings over ten years." Others have begun to follow Klingenberg's lead, building certified Passivhaus homes in climates as diverse as coastal Louisiana, northern Minnesota, and coastal Oregon.[252] And German experience confirms Klingenberg's hypothesis, illustrating how the extra up-front cost of Passivhaus construction has fallen fourfold in the first 15 years and is approaching negligible.[253] Indeed, in the best Swedish passive houses it's already just on the negative side of zero.[254]

Brownstone rehab/retrofit (St. Louis, Missouri). The historic Hyde Park district of St. Louis, Missouri, is full of three-story antebellum row houses, solidly built of brick or stone but often derelict or abandoned. People in the neighborhood needed affordable housing, but conventional rehabilitation would yield unaffordable energy bills. Developer Ted Bakewell III devised and demonstrated in 1981 an alternative that could be carried out by local residents themselves, even if they lacked education and building skills.[255] At the bottom of each gutted shell, a new floor was poured using air-foamed concrete that would level itself. Building on that foundation, locally prefabricated airtight panels—sandwiches of foam superinsulation in fire-resistant drywall—were locked together and built up to line the masonry walls. Openings were cut in for the doors and windows, which each used two standard double-hung sashes. The only heating device was a pair of baseboard heaters costing about $100 a year to run. This simple renovation-plus-superinsulation retrofit cost only $4,700 more than an ordinarily inefficient rehabilitation. But it cut heating needs by more than 90% and enabled an entire 1,500-square-foot house to be cooled with one small window air conditioner.

As these diverse examples show, the extra energy savings available from integrative design can be very large. But these buildings still are relatively few. It might be unrealistic to suppose that today's best practice can be universally adopted, even over 40 years. On the other hand, the practices of 2011 are unrecognizably better than those of 1971; we see no reason why that progress won't continue and even speed up, and today's esoteric advanced practice has a way of turning into tomorrow's juicy competitive imperative. The major uncertainties in how quickly and widely integrative design might spread spell not doubt so much as opportunity.

TABLE 3-1. Looking beyond lower energy costs

INVESTMENT	RETURN
Capital cost	Lower operating and capital costs
Cost of capital	Reduced exposure to fuel price volatility and increase
Transaction costs	**Additional benefits**
Learning costs	Brand enhancement
Risk of execution and performance	Employee attraction and retention
	Staff productivity increases
	School test score increases
	Sales increases
	Reduced healthcare costs
	Faster absorption
	Higher tenant satisfaction
	More competitive rents
	Higher residual value

Energy efficiency creates many sources of value beyond just cutting energy costs. Though hard to quantify and monetize, these real benefits are often worth far more than the saved energy.

The most obvious non-energy benefit of energy efficiency is in saved installation and maintenance costs. For example, LED traffic lights' biggest saving comes from very seldom having to stop traffic and send a crew up a lift or ladder to change a lamp. They also save lives, because each LED array fails only a tiny piece at a time, so red lights keep on preventing crashes. In buildings, properly made and installed LEDs last three times longer than CFLs, and 25 times longer than incandescent bulbs. A National Academy of Sciences study concluded that where lights are hard to change (such as requiring a special lift crane to reach high lobby ceilings) or where paid staff change lamps, "the value of reduced maintenance greatly exceeds the value of the energy savings."[256] (For seniors standing on a chair to change a lamp, the avoided cost may be a broken hip.) Such savings are ubiquitous. Indirect lighting designs (where lighting is directed toward surfaces instead of spaces) often need fewer fixtures and less wiring, saving both capital cost and maintenance. Efficient motors and roofs run cooler and therefore longer. What happens to avoided maintenance labor? Some managers choose to cut staff; smart managers keep the people, redeploy their saved hours to ferret out more savings, and fire more of those idle kilowatt-hours.

There are myriad ways to boost office worker productivity, from offering incentive pay to providing outside perks. Few managers would think to add comfortable working spaces to this list. Yet a range of studies suggests that energy efficiency gains like improved temperature control and daylighting with even just a small window view can boost workers' output. Well-designed air distribution systems that prevent the recirculation of germs can reduce illness and absenteeism. When combined, energy-efficient lighting, indoor-air-quality, acoustic, and thermal strategies can drive productivity increases of 3–5%, theoretically repaying the initial efficiency investment very quickly and dwarfing the saved energy costs.[257] Indeed, increasing a typical U.S. office's labor productivity by just 0.7% would have the same bottom-line effect as *eliminating the entire energy bill.*[258]

The gains extend beyond just productivity. It's easier to recruit and retain employees who like their workplace or who see their companies as environmentally responsible. Lee Scott, the former CEO of Walmart, says the major effort he launched to make the retail giant greener and more energy-efficient ended up being worth doing just for the boost it gave to employee morale.

Another surprising non-energy benefit: increased sales. It turns out that natural light can make merchandise look more attractive to customers, lifting profits at retailers like Walmart and supermarkets such as Stop & Shop (fig. 3-13). An extensive independent study of West Coast franchise stores for another owner found 40%[259] higher retail sales pressure in well-daylit stores—and employees much preferred to work there.

The many non-energy benefits of energy efficiency extend to our children and our healthcare system. Consider whether you'd rather learn in a windowless fluorescent-lit classroom (as many of us experienced) or a mostly daylit school with abundant and perhaps even operable windows. Experiments in several states have tracked test scores and found student learning to be about 20–26% faster in well-daylit classrooms.[260] Or consider whether you'd prefer to recover from surgery in a windowless hospital room or one with

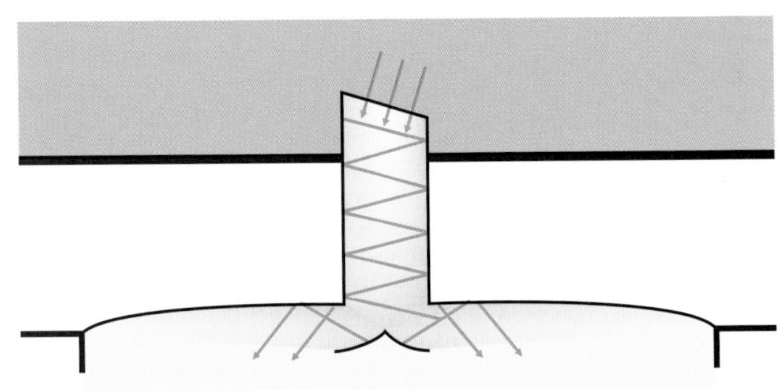

FIG. 3-13. This supermarket in Foxboro, Massachusetts, beams daylight (even on cloudy days) down from a rooftop skylight onto a perforated-metal reflector that bounces most of it up again off a curved ceiling. Both employees and customers prefer the natural light, lingering in the store, and contributing to higher-than-average sales.

streaming sunlight and views of trees or parks. Roger Ulrich, a well-known behavioral scientist, has compiled[261] more than 2,000 peer-reviewed papers showing that patients healed faster, felt better, and had shorter stays and fewer readmissions in green and efficient hospitals than in conventional ones.

Simply put, better working and living conditions provide greater value. An analysis[262] of the small yet growing ensemble of green buildings suggests that U.S. buildings labeled under the LEED or ENERGY STAR system charge 3% higher rent, have greater occupancy rates, and sell for 13% more than comparable properties.[263] These valuation trends reduce the risk of the buildings' loans going underwater when general market values decline, as they did in the 2008–2009 recession. And these properties face less risk from future fuel price spikes or interruptions.[264] LaSalle Investment Management is among the global asset-management readers that see growing evidence of cap rates differentiating to favor green, energy-efficient properties and portfolios.[265]

The timing for investment in energy efficiency could not be better. In a challenged construction industry, green and energy-efficient building has emerged as a safe haven with more upside, less downside, easier financing, and more eager and discriminating customers.

THE CONUNDRUM AND THE CHALLENGE

Now we must confront a huge, troubling question. If this vision of a far more energy-efficient building sector, with its rich prize of $1.4 trillion net (and a host of other, more powerful benefits) really is so compelling, then America should be rushing to embrace it. But we're not. The examples we've highlighted are but a few isolated islands in a vast sea of leaky, inefficient row houses, apartments, McMansions, shops, warehouses, factories, tenements, hospitals, schools, and skyscrapers. Yes, the Empire State Building is a beacon of hope, but the view from the top is a stark reminder of opportunities not yet seized. Why has the adoption of modern energy efficiency across the building sector been so distressingly slow?

Economist James Barrett explains that it's because "the connective tissue that links opportunities to execution is still being developed. It's not so much about leading a horse to water as it is about introducing the concept of drinking."[266]

A skeptic might even legitimately question whether the vision itself is overly optimistic or just plain wrong. After all, building owners and company executives are savvy business leaders. They constantly scour the landscape for any edge that will boost profits and the bottom line—or give them a jump on the competition. Equally smart and innovative entrepreneurs and companies are already aggressively marketing the technologies, services, and designs needed to make the nation's buildings more efficient. If they haven't yet succeeded in selling their wares to most building owners, then maybe the economic gains we envision really don't exist. Or as economists might argue, there can't be $20 bills lying around on the sidewalk, because if there were, someone would have picked them up already.[267]

There actually are good answers to this question. The energy savings and the huge non-energy benefits are really there for the taking. But standing in the way is a plethora of challenges that discourage most (but, as we've shown, not all) businesses, governments, and individuals from taking action.

To start with, one of the largest problems is that any single individual or company can reap only a tiny amount of these savings—so tiny, in fact, that in many cases the effort and hassle may not seem worth it compared to other choices. Why spend the time, money, and effort to retrofit an office to save a mere tens of thousands of dollars a year, when a shrewd business decision, such as a new marketing campaign or enforcing a patent, might win tens of millions of dollars?

Or consider the choices faced by homeowners. The average American house costs about $175,000 and will cost about $75,000 in energy bills over a 30-year mortgage life.[268] But that money trickles out slowly in monthly bills totaling nearly $2,200[269] a year. Let's imagine that adding insulation and weatherstripping, plus making a few equipment upgrades, could cut that by 30%. The savings would be about $700 a year. For most households, frankly, that amount looks trivial—less than the cost of buying a daily bagel with cream cheese, hold the lox. And to get those savings, you may have to shell out more than $5,000 up front for the upgrades (money many people won't have or won't want to part with), then wait several years until that outlay is paid back from energy savings. This doesn't even include the hassle factor of figuring out what to do, finding the right people to do it, arranging for contractors, and probably taking time off from work to let them in.

Frankly, buildings' energy costs and their reduction are low on the list of most folks' priorities. We'll almost always pick the office space surrounded by a sea of parking and conveniently located near the freeway off-ramp, or the home with the great kitchen and nearby park, over choices that eliminate the need for a car or optimize solar access. We'll choose big flat-panel TVs and gleaming refrigerators based on how they look and perform rather than on meticulous scrutiny of their energy consumption. In many cases, we don't even know how much energy is being used by a home, an office, a computer, or countless other buildings and products. Many aren't even specifically metered: some apartments, many commercial buildings, and whole military bases traditionally have just one meter.

This leads into another common problem—few firms or individuals track their energy use as a line item for which profit centers are accountable. Firms in rented space often have energy bills prorated on floorspace rather than based on measured energy use. If submetering is in place, it's usually at a whole-building level, giving no end-use information that lets customers link costs to opportunities. Many firms, especially chains and franchises, don't even see their energy bills, which go straight to a remote accounting department for payment. Amazingly, some large firms still think of energy costs as a fixed overhead, unchangeable and not worth examining.

The barriers are even higher when there's little chance for individuals or companies to reap directly the savings from efficiency investments they might make. J. Wayne Leonard, CEO of New Orleans–based utility Entergy, points out that many homes in New Orleans are among the leakiest in the nation, inflicting high air-conditioning bills on their residents.[270] But the people who live in those houses are often renters. Their landlords have no financial incentive to weatherize the homes, since the energy savings would go to their tenants' pockets, not theirs. The tenants, in turn, are unlikely to invest hundreds of dollars to improve a house they don't own and may not live in for long.

The same problem stands in the way of upgrades to many commercial buildings, since 45% of them are not owner-occupied.[271]

And if all these issues didn't deter you and you actually wanted to do an energy efficiency project, where would you get the funding? Capital for energy efficiency retrofits is costly, mainly because the dollar amounts are small (a relatively small fraction of total building value) while transaction costs to underwrite and service are relatively high.

Since many participants in the financing process, including property brokers, mortgage brokers, lenders, and even investors, are compensated based on the size of the deal, creating bigger deals by combining projects would bring economies of scale on underwriting and servicing, enabling better compensation for participants. But without easy-to-access capital, cost-effective projects ranging from insulating homes to deep retrofits of large commercial buildings won't get done.

Some financial reforms are available but little used. Few originators know about energy-efficient mortgages, which Fannie Mae introduced two decades ago. Analogously to a locationally efficient mortgage (see chapter 2), it counts efficient homes' lower energy bills when computing qualification ratios, qualifying costlier but more efficient homes if monthly energy-bill savings exceed monthly extra mortgage payments. But even if you demand that mortgage, the bank will probably count only a third of your actual energy savings.[272]

Further complicating the equation is that we're all human. Some people turn off lights, some don't. A study of military housing in Hawai'i revealed that energy use in homes with identical energy efficiency features varied by fourfold; a similar study of homes in Las Vegas revealed even greater divergence.[273] Same home, different people, widely different energy use profiles.

Better understanding the hidden triggers of these choices and behaviors is the first step to educating occupants. The challenges around occupant behavior are yet another huge opportunity for energy savings, now starting to be recognized and seized.

As you can see, despite seemingly reasonable economics, many structural and not very well-researched barriers hinder widespread energy efficiency upgrades. At the highest level, we're left with a vast chasm between the enormous societal benefits (including the large amounts of money to be made or saved) that accrue over many years from making the nation's building stock more efficient and the relatively small and diffuse rewards individuals or companies can reap. Or as economists would say, there is a market failure. The fact that energy efficiency is cost-effective just doesn't cut it. Most people won't act if the reward means saving only hundreds of dollars a year.

Fortunately, there are myriad strategies, involving clarifying and marketing the non-energy cost benefits of energy efficiency, offering new services and products that aren't just about energy efficiency, and adopting innovative policies, that can all help bridge this gap and correct the market failure. It's a complex story because of the enormous diversity of the building stock and the people who own and use it. But it is doable, and it is cheaper and less risky than continuing to let our buildings bleed cash.

SOLVING THE EFFICIENCY PUZZLE

Huge challenges are often paired with huge opportunities. Solving the puzzle of speeding buildings' energy efficiency will be difficult but rewarding. The resolution here is twofold—first, let businesses, homeowners, and government entities seek solutions to reduce their own energy use and develop services and products to help others do the same. Second, enable those solutions through smart policies that remove persistent barriers, open new opportunities, level the playing field, and align incentives.

Business-Led Approaches

For business leaders, there are two roles. First, save energy in your own organization. Second, find ways to expand your business and help others save energy. While the dollar signs loom even larger in the second step, they're big enough in the first step that many businesses will want to start in their own operations and learn by doing.

Starting at Home

That's what Gary Christensen, a developer, did in Boise, Idaho. He and his designers and constructors, none with special training or experience in efficiency, bet $23 million on designing it into an 11-story office tower. Their 2006 Banner Bank Building saved 38% of energy compared to code, at no discernible extra cost. Christensen also reported nearly 90% occupancy in year one, a $1.5 million boost in asset value, lower reconfiguration costs, and rents competitive with buildings 20–30 years older. His next goal: "net-zero" homes and small commercial buildings that produce as much energy as they use. Ambitious? Yes, but Christensen feels that it can be done sooner than people think. "We're striving for nothing . . . and when we get to nothing, we're going to do even less," he says. "We like to think of ourselves as the ultimate slackers."[274]

Large companies are finding basic energy efficiency more interesting than when energy was cheap and abundant. A 2010 survey found 52% of facility managers plan to buy more efficiency over the next year, and 84% said it's a priority.[275] Many large firms seek to cut carbon emissions and burnish reputations. Efficiency is also attracting investors (and mollifying activist investors), retaining tenants, and boosting real-estate value. Such progress is becoming highly visible. The Carbon Disclosure Project, for example, sent its 2011 request for greenhouse gas emission disclosure to nearly 5,000 corporate leaders on behalf of more than 500 institutional investors with over $70 trillion in assets.[276]

Making energy information available and transparent is important for buildings too. Simple energy use data from a sound building management system can uncover hidden treasure. A little sleuthing can determine whether high heating costs are caused by a poorly tuned boiler, controls that run heating and cooling equipment simultaneously, or wildlife that chewed holes in the ducts. An exorbitant electric bill may be caused by night cleaners who leave the lights on, employees who plug in electric heaters under their desks, or (as I once found) a long-forgotten snow-melting 40-kilowatt electric heater under the parking lot, not shown on the as-built drawings, but running 24/7 for decades. Constantly tracking and querying operational data is essential to optimizing how larger buildings work. Energy service companies, major property management firms, and specialist firms offer a growing array of end-to-end services to analyze data, spot opportunities, and devise, finance, and deliver efficiency upgrades.

But this isn't easy, even for a large building owner. Efficiency comes in many gradations, and ever-improving technology makes it a rapidly moving target, so if you overhauled your building's efficiency many (or sometimes even a few) years ago, it's time to revisit. Indeed, lighting technology has been changing so quickly that 20 years ago Nordstrom's financial managers cut its depreciation life from 10 to 5 years because that's about how often it was worth replacing.[277]

Efficiency is also not just about hardware. It depends critically on how well the equipment is kept up (like anything else, it deteriorates without maintenance) and how skillfully it's operated. Poor equipment well maintained and run will usually outperform good equipment poorly maintained and run. But good maintenance isn't just avoiding complaints. It invests in people, cultures, measuring tools, software, and procedures that keep buildings continuously tuned up. The payoff is impressive: a 643-building study found commissioning saved 13% of energy with a four-year payback in new buildings but saved 16% with a one-year payback in existing buildings.[278]

WHAT COMMERCIAL BUILDING OWNERS SHOULD ASK OF BUILDING PROFESSIONALS

With questions surrounding everything from building science to greenhouse gas reduction targets to green rating systems, it can be easy to lose perspective on what needs to be done to reduce buildings' energy use. To stay focused on what really matters, here are four questions commercial building owners should ask their engineers, consultants, utilities, and other providers:

How do we use energy, and how does it relate to how my building works? Do you have energy-intensive labs and data centers that dominate your company's carbon footprint, or do you have office buildings where lighting and air-conditioning dominate your bills? Understanding the basic drivers of your company's energy use is a critical first step to identifying where to focus your efforts. Once you know the basics, you can look at the costs and ease of implementing solutions. Unless something important has been overlooked (as is quite common), small opportunities may be quick wins, while the largest energy savings may be harder to implement. How and when you might bundle opportunities to do a bigger intervention at lower cost (e.g., by integrative design) can hugely affect what you decide to implement. Regardless of what actions you take to reduce energy use, clarifying your starting point is a key to success.

Can we avoid or combine large capital expenditures with energy efficiency efforts? Timing is the biggest driver of the capital cost of efficiency. If right-timed, efficiency projects often end up as a small incremental cost on efforts largely undertaken for other reasons, such as new construction, expansion, conversion, or major renovation. No building should be sold nor major tenant moved in or out without considering energy as a key opportunity to add value to the transaction. If not right-timed, as is often the case in incremental retrofits, you can end up trying to justify the entire cost of a new chiller plant or wall insulation through energy savings, and that seldom works out. Working in harmony and synchrony with, instead of fighting against, long-term building capital investment plans give you the chance to grab energy benefits much more cheaply, build them into plans, and boost projected building value.

How can I reward best-practice design? One way is through integrated project delivery (IPD). IPD organizes the relationships and functions of the parties who create or retrofit buildings so three potential adversaries—owner, designer, builder—become effective partners with fully aligned incentives.[279] Even without IPD, design professionals should be paid for what they save, not what they spend. This proven way of rewarding what you want ("performance-based design fees"[280]) can greatly improve design, both new and retrofit, and help distinguish good designers in a crowded market. Designers should offer bids that way (at least as an alternative) and educate their clients to demand performance-based bids; it's a good way to weed out inferior designers.

I want to cut my energy use by at least 30%. How can I achieve that? Often building owners are asking the wrong questions. Be clear on what your short-term and long-term goals are, and make sure you understand what they mean. For example, do not ask for a list of all measures that pay back within three years (it may well be that doing everything on a list of measures with 4- to 20-year paybacks will *collectively* pay back in one year as an integrated package). Do not ask for an energy model that will "predict" energy use (they can only inform comparisons). And don't ask for a wind turbine or rooftop PV system (efficiency comes first). *Do* ask for an *integrated package* of energy efficiency gains that will maximize net present value over a decade or more: it'll probably make your building more valuable. *Do* ask for a long-term schedule of major renovations so you can right-time and right-size them all. *Do* ask what other benefits the proposed bundle of energy savings may provide, like daylight, quiet, thermal stability and control, and air quality. *Do* ask for implementation advice and ways to engage your operating staff and employees to leverage the behavioral upside from energy consciousness. And *do* ask how you can clearly communicate your plan and track your progress to your senior management, make progress visible to employees, and enlist their knowledge, imagination, and enthusiasm.

To help you guide design by benchmarking based on RMI's experience, Table 3-2 shows some targets for key parameters, typical of a new midsize-to-large Class A office in an average U.S. climate like the Mid-Atlantic states. Adapting your specific targets to your actual climate zone and building type generally shouldn't do much worse and may do better. "Best Practice" is a moving target: a new library reportedly achieved 2,190 sf/ton.

TABLE 3-2. Benchmarking a new U.S. office building

DESIGN TARGET	UNITS	EXISTING (U.S.)	BETTER	BEST PRACTICE
Delivered energy intensity	kBTU/sf-y	90	40–60	<30
Lighting power density: connected load	W/sf	1.5	0.8	0.4–0.6
Lighting power density: as-used net of controls	W/sf	1.5	0.6	0.1–0.3
Installed computers/appliances/tasklighting	W/sf	4–6	1–2	<0.5
Glazing R-value (center of glass)	sf-F°-h/BTU	1–2	6–10	≥20
Window R-value (including frame)	sf-F°-h/BTU	1	3	7–8
Glazing spectral selectivity*	$k_e = T_{vis}/SC$	1.0	1.2	>2.0
Roof solar absorptance and infrared emittance	α, ε	0.8, 0.2	0.4, 0.4	0.08, 0.97
Whole-building airtightness	cfm/sf @ 0.3" w.g.	1	0.4	<0.25
Installed mechanical cooling	sf/ton	250–350	500–600	1,200–1,400+
Cooling design-hour efficiency**	kW/ton	1.9	1.2–1.5	<0.6
Level of installed perimeter heating	—	extensive	minimal	none

*A measure of how well the glazing lets in light without heat

**Whole system, including pumps, fans, and cooling towers as well as chillers

Some large companies have the resources and tools to devise energy efficiency projects in-house, without the help of expert firms or consultants. Texas Instruments improved the efficiency of its 33,000-ton central chiller plant by 5% after creating a single efficiency metric and displaying it prominently as a simple bar graph for the plant operators, who keep setting new records. This friendly competition between operators has saved 5 million kWh per year.[281] Many firms promote such rivalry between all their facilities, and their operators regularly share best practices. Many even swap with competitors because they learn more than they give up.

Another in-house strategy: translate energy savings into something not just engineers will relate to—bottom-line earnings. Several years ago, PeaceHealth, an eight-hospital system in the Northwest, standardized and improved energy management throughout its facilities. "We started out with the most cost-effective changes first—things like replacing outdated light fixtures, changing filters, cleaning coils, replacing gaskets," said resource conservation manager Scott Dorough. "The small changes really add up, especially if you do them system-wide." After three years, PeaceHealth accumulated savings around $800,000 a year. At the firm's operating margin of 3.5%, annual energy savings of $800,000 are equivalent to more than $22 million of new gross revenue. "These are dollars saved directly to the bottom line. It's like we're generating brand new revenue," said Gary Hall, vice president for facility services.[282] Translating energy savings into metrics—additional units sold, avoided employee salaries, or bottom-line profits—that resonate with different departments and audiences is the key to selling efficiency internally.

Communication and incentives are also important in-house tools. A clear signal from the top of the company, as Walmart's then CEO Lee Scott provided, does wonders to move energy efficiency from a minor concern to the front burner and to allocate resources accordingly. So does adding energy savings to the criteria by which managers are rated and compensated. At Microsoft, for example, yearly bonuses for data-center facilities managers are partly based on each year's efficiency improvements.[283] Adobe pays its property manager, Cushman & Wakefield, partly for how much energy it saves each year and can also claw back its management fee for underperformance.[284]

Early business leaders in self-imposed adoption of energy efficiency span the gamut from sustainable design firms eager to walk their talk to such global corporations as Cisco and Skanska to government agencies like the U.S. General Services Administration, Department of Defense, and FDIC. Yet they share something important in common—recognition of the value of energy efficiency beyond just cutting energy costs. Whether it's brand value, employee retention, shareholder demand, compliance, mission effectiveness, or just doing the right thing, these entities are motivated to look at energy efficiency based on far more than a simple payback period. If that is the case, will firms and industries that don't view energy efficiency as a driver of brand value or employee retention ever get on board?

Further, do the "start at home" strategies just described for large corporate buildings or portfolios—create strategy, design and analyze custom measures, dedicate staff, finance and implement projects, carefully monitor results—make sense for the typical 10,000-square-foot leased office or single-family home? No. Big and small buildings need different treatment. Many small-business owners are simply struggling to keep their companies afloat in a tough economy. They lack the expertise or time to develop energy strategies, hire the right contractors, and oversee the work. Size—building size, firm size, size of overall energy use—does matter, making big firms often more capable of implementing energy efficiency.

UNLEASHING FACILITIES MANAGERS AND THEIR STAFFS

Cutting facilities staffs' budgets and headcounts in hard times is an easy but false economy: that's exactly when you need them more enabled and empowered to wring out waste—treated not as an overhead to be minimized but as a profit center to be optimized. When the University of North Carolina at Ashville hired Paul Braese to manage its facilities, he brought with him team-building skills from the Navy. He found a staff of smart and resourceful people who lacked the equipment and authority to do their best. To start, money was invested to buy a lending library of measuring tools to provide each of his staff with the resources to have notional "ownership" of several buildings to diagnose and fix. Soon mechanical rooms were cleaned and painted and had their piping color-coded for easy identification. Pumps were repainted when repaired. The clean rooms gave the team a sense of pride and also allowed them to spot instantly such problems as water leaks or failing pump couplings. Switching from spot to group relamping leveraged time, so the same people could catch up on more recommissioning. Then Braese put together a deal to let his team keep and reinvest some of the money they saved, so they could move faster than legislative capital allocation and create a spirit of "intrapreneuring." After years of ups and downs, the savings helped make the campus one of the most efficient in the UNC system, without any major capital investment. Braese feels that most facilities teams can achieve such performance if properly led. All it takes is organizational and cultural change that starts at the top.[285]

Selling Efficiency at Scale

This is where the second and larger opportunity for business lies—selling energy efficiency services and products to mass markets that may not yet understand efficiency's methods and rewards. New and old players, from small home contractors and smart-meter makers to giants like Dow Corning's insulation business and Johnson Controls's comprehensive energy retrofit business, are doing just this. During 2015–2030, annual marginal investment in U.S. energy efficiency is expected to grow from $5 billion to $18 billion in residential buildings and from $3 billion to $13 billion in commercial buildings.[286] Even a modest part of a $31-billion-a-year market is worth a careful look.

The small commercial building market is particularly ripe for innovation, since over 70% of commercial buildings are smaller than 10,000 square feet.[287] Few have yet cracked the code to become profitable in this space. In this market, the challenge is simply providing energy efficiency efficiently. Business processes with the

WHERE THE PROFITS ARE

Today, improving buildings' energy efficiency is a rapidly growing industry blossoming with new business ideas and models. There are opportunities for all types of businesses and professionals—house doctors, contractors, product suppliers, and investors. How can you grab a piece of the pie?

Barrier-busting via products. What about products that not only deliver energy savings but also blast through barriers? Imagine a rapid whole-house insulation retrofit that at $200 would eradicate the cost barrier. After all, it used to be very costly to find leaks in air ducts—until Lawrence Berkeley National Laboratory developed, and an entrepreneur brought to market, aerosolized chewing gum[288] that goes wherever your ducts are leaking and plugs up the holes. Or what about cheap wireless energy stickers that, when placed on a device, a pipe, or a duct, will measure energy flow and report it back to a central hub where it joins other data and pops up instant diagnostics? Information technology as well as sensor, materials, software, and installation advances have not yet been applied to the energy efficiency industry with as much rigor as in other industries. That's a huge opportunity.

High-tech surveys and analyses. Energy efficiency is most challenging in existing buildings, especially homes. Cheap new information technologies are enabling companies like Recurve and Crowley Carbon to cut the time and cost for surveying a building. UK-based Crowley Carbon can send its entry-level professionals into a customer's building with a smartphone and come out with an energy blueprint. The blueprint is matched against an in-house database of energy efficiency products. Crowley Carbon then provides the financing, installs the products, and monitors the savings.[289] As one engineer says of previous practice: "We've made it such an art to just identify improvements, we've limited it to people who are interested in high art." Improving the speed of adoption and increasing the scale of energy checkups and modeling (while not losing too much accuracy or depth) will be critical to driving down costs. Why can't a wireless network of portable sensors and smart software similarly capture a snapshot of how your building is working and use artificial intelligence to suggest how it should be improved?

Energy monitoring and feedback systems. The old adage that you can't fix what you can't measure is true of energy too. To span the huge gap between energy use and energy awareness, companies are helping homeowners and businesses learn and cut their energy use. Such firms as Tendril, Powerhouse Dynamics, and Scientific Conservation are selling energy monitoring systems that show real-time energy use and suggest quick, easy fixes. For the average homeowner, these systems aren't yet cost-effective just for their bill reductions. But there's another party clearly interested in improving building intelligence—utilities. Integrating monitoring and feedback systems with communication systems is the next big step. This not only helps with cost (utilities will probably pay for a big chunk) but also helps ease the transition to a smarter grid.

supply-chain efficiency of Walmart, UPS, and McDonald's are needed to slash the transaction costs of designing, installing, and monitoring energy efficiency and solar power projects for the mass market—both small commercial and residential. Some owners of many similar small buildings—car-rental offices, banks, franchise food-and-beverage establishments—are setting the pace through Samba Energy. If successful, its model could spread rapidly.

Utilities are uniquely positioned to target smaller buildings and are increasingly relying on outside experts to help them. EnerPath in Southern California is using paperless mobile-computing, workflow-management, and transaction platforms that bring economies of scale to making large groups of small buildings efficient, from Los Angeles to Tennessee. In Los Angeles, representatives went door-to-door to small businesses and generated a custom energy savings report using smartphones. The results? They enrolled 85% of those customers and retrofitted about 25,000 small businesses in just 18 months.[290] Importantly, a large chunk of the savings will flow back into the local economy, supporting jobs by keeping previously exported energy dollars recirculating on Main Street.[291]

This illustrates the emerging lesson that cost-effectiveness is far from the only driver of energy efficiency adoption. Others—reducing hassle factor, providing measured results, comparing use to peers', and attracting capital—are just as critical to creating the right environment for energy efficiency. These non-cost factors are even more important in the existing-housing sector, where arguably efficiency progress is most sluggish. Recognizing this opportunity, several Denver energy efficiency entrepreneurs are creating a Facebook game with sweepstakes, comparisons to friends, and behavior modification suggestions integrated with your utility energy use information. By making energy efficiency upgrades or behavior modifications easy, competitive, and more rewarding, such efforts can not only cut energy waste but also launch energy education.[292]

Surprisingly, similarly streamlined delivery and marketing even works when superinsulating old stick-built houses en masse. In 1983–1985, the Hood River Conservation Project in the Columbia Gorge retrofitted 2,988 dwelling units for free. In 85% of the roughly 3,500 eligible homes assessed, residents agreed to implement savings projects, and ordinary local builders were trained and organized to complete the whole program in three years. Local opinion leaders helped galvanize interest in the program, showing that program design and community engagement has just as much to do with implementing projects as good economics.[293]

Two anecdotes suggest creative ways to market superefficient new homes. A Montana builder in the 1980s offered to pay all the utility bills for the first five years you owned one of his houses; he soon had a 60% market share in three counties, plus a waiting list of buyers from hundreds of miles away, but his bet was safe because his houses were superinsulated.[294] And Perry Bigelow's Bigelow Homes, a Chicago- and Austin-area homebuilder, has guaranteed the annual heating or cooling cost on every home it's built for the last 30 years and has had to pay homeowners only a handful of times. Bigelow performs a blower-door test on every home and submeters the air-conditioning condenser.[295]

Another across-the-board business opportunity is training and educating a growing energy efficiency workforce. How can we lure some of the best engineers and managers into energy efficiency? And how can we help them get the certifications and accreditations that reflect their skill level and help ensure quality? Some encouraging answers are emerging from existing trades (see Training from within the Tribe sidebar), unions, and entrepreneurs.

These signs of business-led progress in transforming America's buildings are encouraging. But will they be enough? Are there other levers we should consider if we're really serious about ending our reliance on fossil fuels? Yes. Those levers are held by policymakers, yet strongly informed, tuned, and exploited by business leaders. And to get where we want to be, we should be pushing, pulling, and turning every button, lever, and knob we can find.

Policy-Led Approaches

The slow adoption of building energy efficiency reflects poorly understood value propositions and the "hassle factor" of execution. Overcoming these obstacles in 120 million buildings requires five kinds of actions to reduce the generic barriers and thus scale the business opportunity.

Making Information Available

Smart policies to allow and accelerate energy efficiency create many benefits that can make business operations easier, not more burdensome. Rules can improve the efficiency of the market, helping unlock more opportunity. Requiring disclosure of buildings' energy use, as the EU began to in 2006 and some U.S. jurisdictions have initiated, gives everyone—from buyers and renters to brokers, and energy efficiency providers—the information they need to make intelligent choices. That's good for building owners seeking to improve, and good for service providers who want to help them. And it's good for the whole building stock and the country, because it rewards owners of efficient buildings and gives others an incentive to join them.

High-quality, timely, measured data—especially linking technical with economic performance—remain surprisingly scarce. Yet cheaper sensors and monitoring technologies, cheap and ubiquitous telecommunications, the dawn of the smart grid, and rising customer demand are creating momentum, capability, and markets for high-quality data collection and application. Businesses

TRAINING FROM WITHIN THE TRIBE

As the economy slowed in recent years, industrious folks in the construction business have used downtime to retrain themselves. Steve Jungerberg of Retro Green Homes, in Carmel Highlands, California, took a life-changing class on building science at the Building Performance Institute. Now he's teaching other contractors new skills, peer to peer, about how to build and retrofit better. As a former contractor, he has connections with residential contractors and large developers. He also works with owners and architects, focusing mainly on shell and heating: efficiency first, then renewables.

"This has to be local and done with enthusiasm," Steve says. "You have to love your neighbors and your community and want to make it better. You can learn these things fairly easily and apply them and make a real difference in your community. I look at every person from where they're at and want to get them one step closer."[296]

More specialized teachers are also emerging. Eric Walters at Pool Power in Modesto has been spreading the "big-pipes-small-pumps" approach (see chapter 4) to pool pumps. (Simple techniques like straighter pipes, better pumps and controls, and bigger—hence lower-friction and longer-lived—filters can save most of pool pumps' electricity.) Walters started his pool business 10 years ago. The first customer's pool equipment Eric looked at had elbow fittings everywhere. A huge, noisy pump was shoving water through a maze of piping but delivering little flow. The second pool he viewed was the same, and the third, and the fourth. One week confirmed that Eric's customers were wasting hundreds of dollars every month by ignoring basic flow physics in their pool systems. So he's grown his own retrofit business while working to train installers—and to improve California building codes to reduce pool pumping's hidden but large energy use.[297]

that collect, share, or sell such data, with due privacy protections, could thrive, because the latent demand is huge. Simple graphics help too. When the municipal utility in Osage, Iowa, posted in the town hall a winter aerial infrared photo vividly showing the heat leaking through every roof, it got the laggards' attention, and they quickly learned how to keep up with their better-insulated neighbors.[298]

Local directives can help speed this trend toward better information. Major appliances have ENERGY STAR labels comparing their energy use with national averages. Why not do the same for all devices and buildings? Such information is powerful. All owners of German commercial and residential buildings bigger than 540 square feet[299] must be able to produce, upon request, an "energy passport" that details the building's energy intensity for potential buyers or renters. The result: many inefficient buildings are being systematically retrofitted, some even to Passivhaus levels. Why? Because having the energy data in hand proves that upgrades are much cheaper than continuing to fuel and power the building's waste, let alone expanding energy supplies at even higher costs and risks.

Austin, Texas, figured that disclosing energy costs could have a similar effect in America. So in 2008, the city passed a law requiring that every house for sale must undergo an energy checkup, just as house sales now require real-estate appraisals. Some sellers complain that it's one more hoop to jump through, but the results are encouraging. In nearly all of the 5,000 checkups conducted in the first year, at least one improvement suggestion was made and 500 homeowners implemented them—a 10% uptake.[300] That's far less than Austin Energy's 25% goal, but it's a good start. As housing markets revive, efficient homes' selling advantage could encourage the others.

At the very least, sharing information can cause people to think about energy use and how to reduce it—an important step forward in itself. The sad truth is that U.S. awareness is so low today that a recent study found the average householder underestimated home energy use by a factor of 2.8 and was likely to believe that turning off lights was more effective than insulating the house (it's not).[301]

Sensible Incentives

Smart rules don't just inform choice and unlock opportunity but also level the playing field and help ensure quality (if enforced). If pursuing energy efficiency becomes more common, service providers won't have to spend so much time and money educating clients and struggling to get sensible ideas adopted. Well-placed rules can also increase confidence in energy investments.

Perhaps most importantly, smart standards can help realign incentives so we can better allocate capital, turn waste into profit, and create new business and job opportunities. Today, when equipment manufacturers promote inefficient but perhaps seemingly cheaper products, they're forcing society to bear the far larger costs and risks of more power plants and pipelines (see chapter 5). When those inefficient products are household appliances—most of which are bought not by the people who'll pay the electric bills, but rather by housing developers, landlords, and public housing authorities—that split incentive sets the stage for such enormous misallocations of societal capital that government has a duty to intervene to protect the public interest.

That was the rationale when, during the energy crisis in the 1970s, President Gerald Ford ordered voluntary appliance efficiency standards, which Congress made mandatory in 1978, initially without a single dissenting vote.[302] By 2010, those national standards had saved about 7% of U.S. electricity and created two out of every thousand jobs.[303] A 2011 study found equipment standards could save another 6–12% of 2025 electricity use beyond the federal forecast.[304] California has improved its performance much more than the U.S. as a whole. Implementing its own even tougher standards, continually updating them, and supporting them with incentives helped the state hold its per-capita electricity consumption essentially flat over the past three decades while the United States' average per-capita electricity consumption rose nearly 50%.[305]

Companies with leading-edge appliance technologies can gain a competitive advantage over less innovative rivals. That's why some corporations, such as Whirlpool, have supported more aggressive appliance standards, along with tax credits to help offset any higher costs.[306] We agree with them. The technologies exist to boost the efficiency of a vast array of products and devices—and more aggressive standards will speed adoption. The alternative, much as it was with car efficiency standards (see chapter 2), is to continue losing market share to overseas competitors whose higher energy prices and stronger policies often drive their efficiency innovations harder. Japan's appliance makers, for example, have become far more competitive under a "Top Runner" policy (now adopted also for cars) that makes all models' efficiency rise in step with the most efficient model on the market.

Energy Performance Standards

Another arrow in this quiver of policy solutions is building energy standards, which have long been proven to bring important economic benefits. The Northwest Power Planning Council estimates that state building codes achieved one-fifth of the four-state region's 20% electricity savings during 1979–2009. Updated codes and federal standards will join utility and nonprofit efforts to offset 85% of the Pacific Northwest's electricity load growth over the next 20 years.[307] If you're a manufacturer, supplier, or retailer of energy-efficient equipment, this long-term multi-utility commitment to energy efficiency helps build you a stable market.

The U.S. is far from tapping the full potential of efficiency codes to get buildings designed right so they don't waste energy for decades. Eleven U.S. states have pre-1999 commercial-building energy codes or none at all, while 10 states have no residential energy code.[308] Europe is far ahead of the U.S. Even the most aggressive U.S. commercial building code results in energy use more than twice what's allowed in Denmark.[309] The state of Vorarlberg in Austria even mandates Passivhaus standards for all new houses, and several major European cities plan to follow suit. Influenced by such leadership, a May 2010 EU directive requires all 28 member states to mandate "nearly-zero energy" buildings by 2020.[310] "Net-zero" buildings aren't some wild green fantasy; they're the official 2030 goal of (among others) the 2007 Energy Independence and Security Act, the American Institute of Architects, the American Society of Heating, Refrigeration, and Air-Conditioning Engineers, the U.S. Conference of Mayors, the National Governors Association, and the U.S. Army.[311]

U.S. codes can be better designed and implemented. Today's most effective and innovation-friendly codes set absolute numerical energy targets, ratcheting down predictably over time as cost-effective technologies and delivery methods improve (or targeted utility rebates and incentives rise). This flexible, performance-based approach reduces headaches from prescriptive requirements (mandating specific technologies that may soon become obsolete) or confusing baselines for calculating percentage savings (as in the LEED rating system). Codes have not typically accounted for plug loads such as computers, televisions, or manufacturing equipment in buildings—but they should (see fig. 3-4)—and have rarely addressed existing buildings. Absolute-target codes can address extreme plug-load scenarios (perhaps as is done in the ENERGY STAR Portfolio Manager) and can also more easily address existing buildings. And a priority should be to weed out old code provisions that inadvertently forbid better modern practices.

There are legitimate reasons for the low U.S. rate of code adoption, modernization, and enforcement. Building codes are regulated by states or even localities—in all, some 7,000 different enforced code districts.[312] It costs money to create energy code offices and train (then retrain) officials. This is hard enough for large cities to fund and manage, let alone for small communities. If there's code enforcement, it's more often for health and safety—making sure buildings have working fire sprinklers and won't collapse in an earthquake or blow away in a hurricane—than for energy efficiency. For governments facing budget crises and cuts in critical services in education and public health and safety, adding funding for energy code innovation, training, and enforcement is at the bottom of the to-do list. Herein lies the benefit of codes harmonized over larger areas. That doesn't mean one-size-fits-all: California distinguishes 16 climate zones and numerous building types and uses.

Setting targets is important too. About 20% of all U.S. commercial structures are occupied by government entities,[313] for instance, and all levels of government are under intense pressure to cut spending. Governments are already moving this way. Recently, the U.S. General Services Administration (GSA)—the nation's biggest landlord, managing all federal real estate except that of the Department of Defense and the Postal Service—released its plan to cut federal facility energy intensity by one-third by 2015. Targets for brand-new federal buildings are even more aggressive.[314] At a local level, Boulder, Colorado, recently required all rental housing to achieve certain efficiency ratings by 2019. The city has set up rental property owners for success by providing a long lead time for compliance, by offering significant supporting resources, and by articulating the need for the change (roughly 60% of the college town's housing is rented).[315] Perhaps most critical is that the Boulder market and citizenry are pushing for these changes almost as much as policymakers are pulling for them. Imagine the huge gains—and societal benefits—if legislatures decided that agencies at every level of government should make energy efficiency a higher priority.

Rewarding What We Want

Perhaps the most powerful policy push can come from a profound regulatory change in the business model of electric and natural-gas utilities. Utilities are uniquely positioned to become highly effective players in building efficiency. They're the only service provider with universal existing customer and billing relationships, knowledge of their clients' energy use, large scale, and access to major, low-cost financing. Given these enormous advantages, you'd expect utilities to be the dominant provider of energy efficiency in the United States. But in most states they aren't, in some they're bystanders, and in some they even see efficiency as a danger to their own financial success.

Historically, utility regulation has tied revenue to the amount of energy sold and the capital invested to provide it. Investing more money to build more energy-supplying infrastructure to sell more energy was the golden road to revenues and profits. Indeed, utilities and their investors tacitly preferred that customers be inefficient, since inefficiency could raise profits. But this traditional business model turned out to have a major flaw. When demand rises, the utility must eventually build costly new capacity. Utility executives—and the public utility commissions that oversee regulated utilities—began to realize that paying customers to boost their energy efficiency enough to eliminate the need for new capacity would actually cost less than building it, or even just running existing capacity. "When I tell big customers we would be happy if we sold them less electricity, they look at me like I've burned out a few brain cells," said

Peter Darbee, then CEO of PG&E, a major California utility, in 2009. But his logic is airtight. "Energy-efficiency programs cost electricity customers less than half what they pay to help fund a new power project," Darbee explained.[316]

While reduced energy bills benefit their consumers, the strongest motivation for a utility to conserve energy is financial and regulatory. Half the states have mandated that utilities meet aggressive energy efficiency goals.[317] To help utilities comply while staying financially healthy, many regulators have changed how utilities get paid for saving energy. Perhaps the most popular and effective mechanism has been decoupling. Decoupling breaks the link between earnings and total energy sold. To sweeten the pot, many regulators also reward utilities for cutting customers' bills or beating efficiency goals. Unfortunately, though, this new business model for utilities is taking hold slowly. In more than half[318] of U.S. states, electric and gas utilities are still penalized for cutting your bill and rewarded for selling you more energy. That's just as odd as it sounds: it rewards exactly the opposite of what we want, so that's what we get. Chapter 5 will explain how to fix that.

Not surprisingly, utility success stories of big, cheap savings are far more common where perverse incentives have been corrected. In the Northwest, despite some of the nation's cheapest electricity, more than 100 utilities have together met half the region's growth in electricity demand since 1980 with energy efficiency—at an average cost of less than $0.02/kWh, a penny below the wholesale price.[319] In Vermont, an innovative "efficiency utility" funded through a $0.009/kWh public-benefit charge on electric bills gains significant economies of scale, internal experience, and expertise that the 22 small utilities serving Vermont couldn't muster on their own. As a result, just in 2008, they were able to deliver 85 million kWh of electric efficiency (actually shrinking demand) at one-fourth the cost of comparable electric supply.[320]

Emerging Drivers

Another area where policies can open doors is creative financing. Traditional financing has been and will continue to be an important source of efficiency project funding. However, in several new financing options for customers, a utility or company funds the efficiency project on its own balance sheet. Another important new financing option created in 2008, the PACE ("Property Assessed Clean Energy") bond, was authorized in 25 diverse states by early 2011.[321] PACE bonds are funded solely by voluntary private investors and used solely by voluntary building-owner participants to save energy. Participants (but nobody else) repay the loan through property taxes over a time period typically much longer than the efficiency investment's payback period, so the borrower saves money from day one. Repaying via property taxes—as sidewalks, sewers, and many other kinds of infrastructure have been funded for about a century—automatically transfers the obligation (with its benefits) to the next owner if the property is sold. Residential PACE programs ran into unexpected regulatory roadblocks in 2010, but commercial PACE programs are growing. How the roadblocks are cleared in Congress or the courts remains to be seen, but PACE bonds are widely considered the most important innovation for financing energy efficiency in decades, and they deserve a chance to shine.

Other promising policies include point-of-sale energy efficiency inspections, universal building energy labeling, rewards for the top performers (not just penalties for the worst), and building feebates—sliding-scale hookup fees for new or extensively retrofitted buildings to connect to electricity and gas (and water and sewer) services. The size of your fee or rebate would depend on the building's efficiency. This would signal efficiency's value right up front when design choices are being made—and unlike static standards, which start getting obsolete as the ink dries, the dynamism of building feebates would drive continuous improvement and reward innovators who beat standards.

Thus many powerful incentives and levers do exist to help achieve the vision of a vastly more efficient building sector—if we enhance and exploit them.

CONCLUSION: MORE COMFORT, MORE PRODUCTIVITY, LESS ENERGY, STRONGER ECONOMICS

There's obviously no silver bullet—but there's a lot of silver buckshot and birdshot. Transforming America's building sector will demand national attention and action, an intensive ramp-up of investment and innovation, and broadly targeted policy enhancements and changes. Slight changes to business-as-usual will not suffice. The agenda for building owners and investors, building users, service providers (including utilities), and governments and policymakers is long, detailed, fine-grained, and highly diverse.

That said, experienced practitioners actually do know what to do. There are six main imperatives:

Attack buildings' inefficiencies with transdisciplinary insight and entrepreneurship. How can the operational efficiency of Toyota, the sales prowess of Avon, the scientific focus of Google, the product design of IDEO, the level of social engagement within Facebook, and the hassle-free transactions of Amazon be applied to energy efficiency to drive up demand and drive out cost? As discussed in this chapter, there is much room to address huge, pervasive challenges ranging from data accessibility to financing, leveraging integrative design, and training and education through high-tech, information-based strategies. We need the best problem-solvers, marketers, and operations minds to engage in energy efficiency and create the products and businesses that can boost it toward Internet speed.

Make energy use more transparent. We need to give customers measured (even real-time) energy use and pricing signals. Cheaper sensors and monitoring technologies, cheap and ubiquitous telecommunications, the dawn of the smart grid, and rising customer demand are creating momentum, capability, and markets for high-quality data collection and application. Information on energy use simplifies efficiency for businesses by speeding diagnoses, improving energy modeling and analysis, and enabling benchmarking against competitors and incorporation into building valuation. Simply put, we need a massive upgrade to how measured data on building energy use are collected, shared, and used.

Provide easy-to-access financing, priced commensurate with energy efficiency's exceptionally low risk. Energy efficiency requires up-front funding. This funding could come from added charges on utility bills (a mid-2011 law made such "on-bill financing" the New York State norm), from engineering-finance firms that understand and take on the risk, from property redevelopers that overhaul entire buildings, from energy service companies that provide or help find financing, or from innovations like PACE bonds.[322] Regardless of source, it must be easy to use.

Train and educate a high-quality workforce. A vast workforce needs to know how to find the savings opportunities in our buildings, design and install systems that capture the savings, and commission and properly finance the projects, all the while mitigating risk. This is no small task. Quality mustn't be sacrificed to quantity: as a wise old contractor said, "If you can't afford to do it right the first time, how come you can afford to do it twice?" Accreditations and certifications can help ensure qualified workers are on the job and can proudly stand behind their work so they earn customers' confidence. And trainers need strong hands-on experience, preferably taught by trusted peers.

Upgrade to next-generation building efficiency policies and align utility incentives. Unlike incentive- or market-based programs, most building codes have fairly strong compliance, on pain of red-tagging. But codes must be well supported, funded, enforced, and continuously reevaluated. Helping advance and reduce the costs of code compliance through improved supporting software and resources is a good business opportunity. Allowing utilities to make money on energy efficiency (rewarding independently measured savings, not expenditures or activities) is critical. States should also follow the lead of Washington, Massachusetts, Vermont, and California in continuing to set aggressive efficiency targets for their utilities, while encouraging flexible implementation and rapid learning. This will expand and mature the burgeoning markets for efficiency services and equipment and encourage entrepreneurs to take the plunge.

Begin overhauling how building design is done, taught, and built. To turn dis-integrated design into highly integrative design, architectural and engineering pedagogy needs reform, in-practice design professionals need mental retreads, clients need to value and require integrative design by experiencing its benefits, and incentive-aligning approaches like performance-based design fees and integrated project delivery need to become the norm.

We've seen how market forces and well-informed policies can work together to start us down the path to a radically more efficient and profitable U.S. building stock. But this chapter has also made it clear how daunting this task really is. It requires nearly every single American to make different choices. It requires money. It requires collaboration between business leaders and policymakers. It requires a long-term view of the built environment. Building a few million ultralight electric cars seems almost trivial compared to retrofitting or replacing everything from double-wides in a trailer park to high-rises in San Jose. Even if it were possible to physically retrofit nearly all of the nation's 120 million buildings at an average rate of 8,200 per day—and make the same proportion of new ones energy-efficient—will their owners really want to? Will you want to?

You should. Because we've also shown the enormous business opportunity behind these challenges. Anthony Malkin and his Empire State Building demonstrated that. And we've described the other benefits—everything from faster classroom learning and hospital healing to the ability to create more jobs and recruit better workers—which could be far more valuable than just the $1.4 trillion in net avoided energy costs that drops straight to the bottom line.

Superefficient buildings won't just use less energy, freeing up electricity for Revolutionary+ autos and gas for industry and flexible power production; they're also the key to supplying more, better, cheaper, safer energy to all sectors. Buildings are a future hub of energy storage, energy production, and energy markets. Intelligent buildings can make and perhaps store electricity. Their physical characteristics can shift based on weather and on wider energy needs and offers. Efficient, smart buildings can both learn and signal when to buy or sell energy and its services. Thus efficient buildings can become the foundation, and efficient vehicles and factories the pillars, of the vastly different U.S. energy system we're already starting to build—less risky, probably less costly, and far more robust and resilient.

Projected decline in U.S. buildings sector energy use, 2010–2050

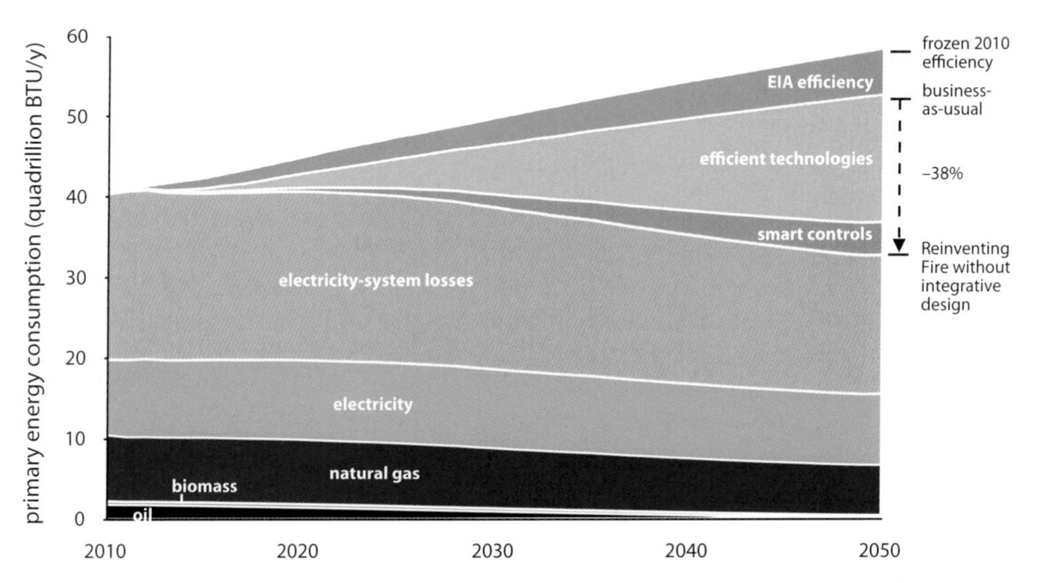

FIG. 3-14. Effectively implementing the basic Reinventing Fire efficiency opportunities described in this chapter could reduce U.S. buildings' 2050 use of electricity and natural gas by about 15% compared to 2010.[323] Adding integrative design could reduce buildings' 2050 energy use further—from 33 to 16–24 qBTU/y, or 40–60% below the actual 2010 level—as shown in Fig. 3-6, eliminating much or most of the remaining need for electricity and natural gas shown. With or without integrative design, the "other fuels" remaining in 2050 would include no oil—just firewood and occasional wastes, and a bit of bottled gas in rural areas without pipeline gas. As we'll see in chapter 6, these gaseous fuels could all be phased out over the longer term through fuller use of efficiency and renewables.

Embarking on this energy adventure is not just about enhancing the bottom lines of our nation's businesses and households; it's about the long future we and our children will compose and inhabit, and the kind of nation and world we want to live in.

If we do get motivated, pay attention, spark new businesses, and achieve our vision of American buildings that use energy in a way that saves money, what would that mean for fossil fuels? By 2050, those buildings, though their floorspace expands by 70%, can be using 13–55% less electricity and 24–68% less natural gas than they did use in 2010 (fig. 3-14). Their oil use, now 0.84 Mbbl/d, could readily shrink to zero.[324] That's all from efficiency—not from the additional option of onsite, especially renewable, power generation that efficiency makes far more profitable (see chapter 5).

For the first time in history, even from profitable efficiency alone, our economy's biggest single energy use can now head down, not up, while our prosperity, health, and security climb even more strongly and consistently. And where we live, work, shop, play, heal, learn, worship—the spaces where we spend nearly 90% of our lives—can become more magical: spaces that create delight when entered, health and serenity when occupied, regret when departed, and pleasure when remembered. Our buildings will at last become not just the shells we live in, but truly worthy of who we are and whom we aspire to become.

TABLE 3-3. Summary of recommendations for key actors in the buildings sector

	NO-REGRETS	OPPORTUNISTIC	INNOVATIVE
OWNERS, INVESTORS, AND USERS	Assess the energy performance of your buildings and benchmark it against competitors'. Optimize current performance. Seek out energy-efficient space. Insist *pro formas* use actual, not typical, energy use data. Train employees in energy management and make it part of everyone's normal responsibility. Be your own guinea pig: retrofit your own space first. Treat facilities staffs as profit centers and resource them to exploit opportunities.	Educate appraisers to value efficiency and agents to sell it. Pursue efficiency ratings and labels. Express energy savings in ways that are meaningful to specific stakeholders. Piggyback energy efficiency on other infrastructure investments. Include churn cost in total cost of ownership. Advise and finance your employees' energy efficiency gains at home—a great fringe benefit. Help organize bulk buys.	Make energy use prominent, with real-time feedback. Anticipate, support, and blow past regulation. Align your goals with your tenants' in retrofit improvements. Partner with utilities and municipalities to finance and capture full demand-side potential and, where possible, distributed generation. Adopt integrated project delivery and performance-based design fees. Help your suppliers and customers become efficient.
UTILITIES, SERVICE PROVIDERS, AND PRODUCT SUPPLIERS	Build and market your energy efficiency brand. Provide turnkey solutions that require little engagement from the customer but guarantee outcomes. Train your people effectively and regularly. Use effective indicators of energy use, including normative comparisons and suggestions of action. Scrap and recycle, rather than reselling, obsolete equipment.	Master integrative design and apply it whenever appropriate. Leverage the similarities in the building stock to gain scale through repetition. Adopt and promote project structures that align incentives around efficiency. Start measuring and counting non-energy benefits. Use financing mechanisms to offer no-up-front-cost solutions.	Get paid for performance, not effort: adopt, market, and require performance-based design fees. Bid into forward capacity and energy markets to get up-front funding for efficiency. Address customers' challenges that go beyond products to make energy efficiency hassle-free. Consider a "service economy" business model that sells comfort and light rather than chillers and electricity.

	NO-REGRETS	OPPORTUNISTIC	INNOVATIVE
GOVERNMENT AND NGOS	Implement modern codes and standards. Support voluntary investment through incentives and education. Create and maintain national, state, and local building databases and overlays to prospect for opportunities. Incentivize utilities to invest in efficiency. Fund R&D and design education.	Help identify preferred and prequalified resources. Enable transparent access to energy use data. Allow accounting and tax treatment of energy-saving investments as expenses rather than capitalized assets. Allow utilities to invest in and support efficiency and distributed generation and earn proper rewards.	Encourage point-of-sale surveys and retrofits. Make performance-based design fees, integrated project delivery, and similar process innovations the norm in government procurement. Reform design professionals' practice and pedagogy around integrative design.

CHAPTER 4

INDUSTRY: REMAKING HOW WE MAKE THINGS

FIG. 4-1.

CHAPTER HIGHLIGHTS

- **THE GOAL.** To save 27% of primary energy used in U.S. industry in 2050 (beyond savings already forecast) by realistically implementing known and cost-effective energy efficiency and waste heat technologies. Further potential savings are available from integrative design, and even more is possible with radical breakthroughs in how materials are made and used.
- **THE BUSINESS OPPORTUNITY.** Manufacturers that burn less fuel will benefit from lower operating costs, reduced waste, often better quality, and a hedge against volatile and increasing fuel prices. Small and large companies have the opportunity to develop, sell, and implement new technologies for advanced industrial efficiency.
- **THE BOTTOM LINE.** A total return of $0.7 trillion (2010 present value) on a $0.3 trillion investment.
- **BUSINESS SECTORS THAT CAN PROFIT.** Manufacturers of all sizes, researchers, entrepreneurs, and design and implementation firms.
- **POLICY ENABLERS.** Clear, complete, and stable fuel price signals and producer lifecycle responsibility will spur continued innovations and faster efficiency adoption, helping build competitive advantage.

As for the future, your task is not to foresee it, but to enable it.

—ANTOINE DE SAINT-EXUPÉRY (1900–1944)

Imagine that you're in one of the largest industrial complexes in the world. Your guides are two big hard-hatted guys who respectively run a petrochemical plant and an integrated steel mill. They inhabit a thrumming jungle of rolling mills, pipes, storage tanks, furnaces, cracking towers, mountains of ore and coking coal—a heavy-industrial landscape so vast and tangled that even a skilled industrial tourist is hard pressed to discern the basic flows of energy, products, and waste streams. Sprawling across a hundred square miles are refineries, chemical plants, paper mills, and other industrial complexes on a grand scale. Giant rail yards, airports, and gargantuan ships provide links to suppliers and customers all over the world.

You've been driving for hours along miles of pipes and conveyors, pausing to see red-hot metal pour from giant cauldrons to be cast into ingots or rolled paper-thin, or crude oil and natural gas being transformed into plastics and resins that ultimately end up in car seats and heart valves. Everywhere there are temperatures 500 F° hotter than the hottest days in the Sahara desert and pressures as great as those at the bottom of the sea. At night you can see the unceasing lights and operations from 50 miles away. They light up a satellite picture.

This landscape, where we heat, beat, and treat our way to prosperity, feels like the ultimate 24/7 hell on earth. And it's a steaming, insatiable maw for energy that, either directly or via electricity, comes mostly from coal, oil, and natural gas. Some of the plants and mills are constantly making valiant efforts to become more efficient, and a few factories even use excess heat and waste CO_2 to keep vast greenhouses producing food year-round. But overall, the whole complex, like the rest of the industrial world, is overwhelmingly dependent on a prodigious flow of fossil fuels.

A week later and far away, imagine that you're gliding silently through another kind of jungle with two new guides, biologists who have devoted their lives to understanding the ecological links and flows in this vast wilderness. Here the raucous cries of birds and monkeys intermingle with the gurgle of the river and a symphony of insect noises. This rainforest feels swelteringly hot and humid. Its virid canopy is so thick and intertwined that only at midday can enough light filter through for easy reading; the rest is captured overhead to drive a very different, far older, and more mysterious kind of production developed over billions of years.

Here vast quantities of raw materials—air, water, soil, sunlight, and organisms previously made of them—are converted into refined products. Spiders turn insects into silken webs tougher than Kevlar, plants emit perfumes more alluring than Chanel No. 5, and trees and their symbionts capture and convert CO_2 into precious hardwoods, fibers stronger than stainless steel, potent medicines, and innumerable other valuable products. The throughput rivals that of many industries. Biologists estimate that a single square mile of jungle has more than a hundred thousand tons of biomass[325] that is continually processed and refined into an immense variety of sophisticated products with no mines, no smokestacks, no fossil fuels, and no pollution. In addition, all these processes happen at normal temperature and pressure, with no human operators, no maintenance, no noise, and zero waste: Each step's products are food for the next.

It's a stark contrast to human industry, in which we mine ores, fashion them into products that we use briefly, and then throw it all away. Our processes use exotic elements from all over the periodic table, many rare and toxic, to drive unnatural chemistries under severe conditions inside costly furnaces and reaction vessels and then separate and discard most of the results as waste.

But nature shows us how we can meet our needs with a small fraction of the energy and material we consume today, without depleting scarce resources or choking on our own wastes. Nature makes myriad improbable products, even self-reproducing and self-repairing ones like human beings, from just air, water, soil, and sunlight. Its incredibly intricate chemistries use only a handful of common elements, their reactions catalyzed by enzymes at lightning speeds under ambient conditions.

Are these two worlds incompatible? Surprisingly, no. Industrial visionaries can learn key lessons from the rainforest: take nothing, waste nothing, do no harm. And many of the most successful companies today have discovered that Nature holds answers to design questions that we're just learning how to ask. Following Nature's lead is not only good for the earth; it can also bring crucial competitive advantages to business.

Meanwhile, industry can clean up its act and trim its appetite for energy—and it's made a great start. Over the past 40 years, U.S. industry has cut energy intensity—the amount of energy required per unit of output—in half, scrubbed its stacks to reduce acid rain, greatly reduced poisonous discharges into water, and clamped down on profligate flaring of "waste" gases. It has also made breakthroughs in catalysts and enzymes that drive low-temperature chemical reactions, learned to wrest valuable products from waste streams, and even begun to integrate differing industries into a nascent "industrial ecology." In Kalundborg, Denmark, for example, materials and energy flow in a symbiotic dance among a refinery, a power plant, a pharmaceutical factory, a drywall plant, and a fish farm, transforming waste from one operation into valuable fodder for another, and even supplying heat to the city of Kalundborg and fertilizer to surrounding farms.[326]

Yet despite all this ingenuity, the driving force of competition, and the demands of more discerning customers, we are still far off nature's benchmark. Why?

The answer comes in the complex interplay among Investment, Incentives, and Innovation. Solutions are slowed by one more "I"—Inertia. In these big industries, where pennies spent or saved can mean the difference between losses and profits, legions of clever engineers and executives are perpetually seeking to improve the balance between benefits and costs. They carefully total the capital and operating costs when choosing a product or process, sizing a heat exchanger, or selecting a reaction temperature. Then, to maximize shareholder value from investments that may total billions of dollars, groups of experts pore over computer simulations, balancing energy efficiency against throughput, reward against risk.

Yet the economics of investments don't arise in a vacuum. Society sets rules about taxes, emissions, and permissible behaviors. All these affect costs, prices, risks, returns—and hence choices. Changing these rules can get more of what society wants and less of what it doesn't want. Many of industry's advances over the past three decades arose directly from the Nixon-era Clean Air Act and Clean Water Act (and a host of other laws and rules) that forced industries to slash pollution and touched off a search for innovative ways to turn waste into profit. Now we can go further with new incentives, such as the removal of subsidies, new tax rules that create a level playing field between saving and using energy, requirements that industry take responsibility for the full lifecycle of its products (as it already does in virtually every industrial country except the United States[327]), or a floor on the price of energy. Changing the rules changes the game. Many other countries already do a lot of these things, and that helps them ensure long-term competitiveness.

Industry is continuously innovating along paths that make clear economic sense. When the game changes, innovation often shifts into high gear. Those '70s cleanup laws, far from wrecking the economy, gave Americans healthier lungs, more productive forests, and some of the lowest electricity prices in the world—and at the same time, they made industry more globally competitive, not less, because of the efficiency gains the laws stimulated. As usual, smart engineers competing to solve design problems found better and cheaper solutions than anyone had imagined.

A new wave of ingenuity and smart policy can transform industry in the next four decades as well. Let's consider what's possible (fig. 4-2). U.S. industry in 2010 used 24.4 quads of primary energy a year—roughly one-fourth of the nation's total

INDUSTRY-SECTOR TERMINOLOGY

For definitions of **primary energy, delivered (or site) energy, end-use devices,** and **kBTU** (1,000 BTU), see the terminology box in chapter 3 (page 79).

Feedstocks are the raw material inputs for processes that remain embedded in the final product. They do not provide heat or steam (the kinds of **process energy** needed to operate the production process), nor do they generate electricity. Many industrial processes use fossil-fuel feedstocks to make their final products. Natural gas, for instance, is often processed to create the materials for fertilizers and plastics. In addition, coal is often used to produce coal coke for steelmaking. Such feedstock uses are omitted from the Reinventing Fire analysis.[328]

The **energy intensity** of an industrial process is the amount of delivered energy used per unit of output. Energy intensity values are useful tools to monitor the energy use trends within an industrial subsector. However, they vary greatly between subsectors because of different process needs (e.g., the cement industry has a much higher energy intensity than the pulp-and-paper industry because of its inherently energy-hungry process).

Combined heat and power (CHP), also called **cogeneration**, produces electricity and useful heat in a single, integrated system. There are two major kinds: the topping cycle generates electricity and then uses the leftover high-temperature heat to run the process, while the bottoming cycle converts leftover low-temperature heat into electricity. The former opportunity is more efficient, the latter more abundant. CHP typically saves at least half the fuel, cost, and emissions of conventionally making electricity and process heat separately.

Fuel-switching substitutes one fuel for another to do the same task. The choice turns on costs, process needs, environmental regulations, constraints of existing equipment, and other factors. A common type of fuel-switching is from coal to natural gas—a cheap and convenient heat source for many processes, but sometimes replaced with costlier electricity for greater precision and cleanliness.

(counting, as usual, only energy used for fuels, not feedstocks). If we remained at status quo, then industrial energy use would swell to 44.4 quads in 2050. But we won't; the U.S. industrial mix will tilt further from heavy industries toward higher-value products. Such shifts alone will cut U.S. industrial energy needs by 21%. In addition, industry has a long track record, driven by relentless competition, of delivering year-on-year reductions in energy intensity. That durable trend, including more combined heat and power (CHP), is already baked into official forecasts and cuts 2050 energy needs by a further 4.4 quads or 10%. So how much further can this chapter's recommendations take us?

The 3.5 quads saved in the process energy needed to produce less and different transport fuel (see chapter 2) will leave 27 quads of industrial energy in 2050. That's only 10% more than in 2010 despite 84% higher industrial production—a laudable feat. But technically feasible, economically viable, and competitively necessary efficiency techniques can save even more. They come in two flavors: emerging technologies entering the market too recently to have been widely adopted yet or included in previous analyses, and more effective and widespread reuse of processes' waste heat. These save 2.3 and 2.4 quads respectively, bringing industry's total 2050 energy use nearly 9% below

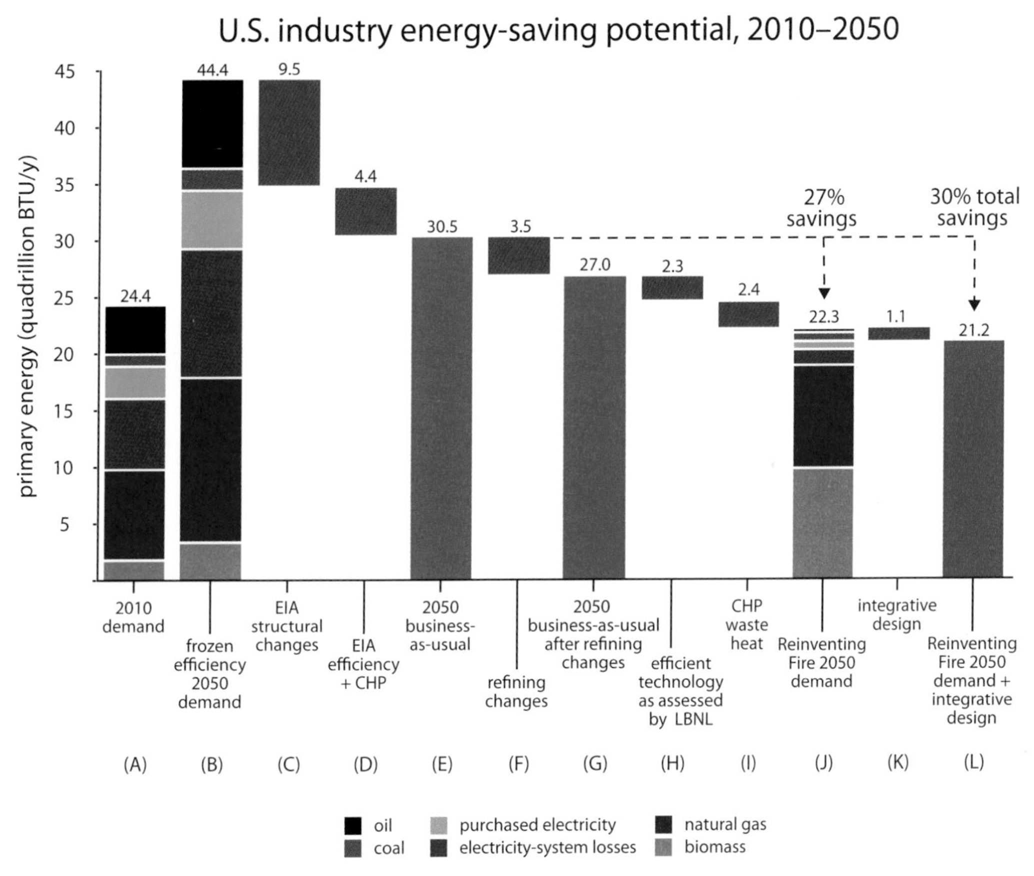

FIG. 4-2. By 2050, shifting industrial patterns plus energy efficiency, both predicted and additional, can roughly halve U.S. industry's primary energy use despite 84% higher output.[329]

the 2010 baseline. Finally, conservative estimates for integrative design (just in drivepower and fluid-handling systems) can save a further 1.1 quads.

So, by 2050, we assume the industrial sector will have grown by 84%, yet we have found ways to reduce its energy use 9–13% below the 2010 level, using known or emerging cost-effective technologies. An incremental investment of roughly $283 billion over business-as-usual would return more than twice as much—about $665 billion (both without integrative design).

Those results, while impressive, still leave us with a prodigious industrial energy appetite. Should we go on a further diet? We think so, for two reasons. First, in a world where demand and competition keep rising while natural resources become pricier, markets will reward those best able to manage costs, and energy is a critical and manageable component of industrial costs. Second, managing climate risks, as public policy now requires in an increasing number of U.S. states and in all but two major industrial countries, requires about an 80% reduction in fossil CO_2 emissions in the whole economy, so a 9% energy use reduction in industry is insufficient.

The next obvious question is: can we go on a further diet and perhaps emerge even leaner and fitter? This chapter provides a menu of tasty but slimming options ranging from smarter technologies to design breakthroughs to creative policies. Unlocking the vast additional potential for further energy reductions will take a coordinated effort to prime the right innovations, elicit ingenuity, and reward leadership.

HOW THE INDUSTRIAL JUNGLE DRIVES U.S. ENERGY DEMAND

Industry produces nearly all of the daily necessities and comforts we take for granted, from the paper and pens that we write with to the cars and trucks we drive, the fuels that power our travels and heat our homes, the materials those homes are made of, the equipment and appliances within them, and the food we eat. It's a staggeringly diverse sector of the economy—more so even than buildings. It's also a crucial part of the economy, employing almost 20 million people and generating more than 40% ($6 trillion) of the United States' GDP.[330]

Doing all this takes immense amounts of energy—three-fifths from directly burning fossil fuels (fig. 4-3). Industry also uses large amounts of oil, gas, and other hydrocarbons as feedstocks, the raw materials for everything from nylon to DVDs (see Using Fossil Fuel as Feedstock sidebar).

Diverse Energy Uses

Industrial energy use is unlike energy use in other sectors. Transportation moves people and goods

USING FOSSIL FUEL AS FEEDSTOCK

Industrial feedstocks—fuels used as raw materials rather than for energy—represent the equivalent of 15% of industry's energy use, or 5% of total U.S. energy use. Dependence on fossil fuel both for raw materials *and* for the energy to process them makes the industrial sector especially vulnerable to volatile fossil-fuel prices, helping explain why companies like Dow Chemical, DuPont, and many others have been working so hard to make their processes as energy- and molecule-efficient as possible. And although our analysis does not attempt to measure potential reductions in fossil-fuel feedstocks, some of the energy-saving steps we will discuss have the virtuous effect of using less, or even no, fossil fuels as raw materials.[331] Indeed, in 1999 the National Research Council forecast that biomaterials could ultimately meet over 90% of the nation's needs for carbon-based industrial feedstocks.[332] That transition is already under way.

in relatively standardized kinds of cars, trucks, planes, and other vehicles. Buildings, though dizzyingly diverse, provide shelter, comfort, light, cooking, entertainment, and other relatively standardized services via a limited palette of technologies: structure, insulation, glazing, vapor-compression cooling, and fuel-combustion heating. The electricity sector has essentially one product, mostly made by raising steam to turn turbines and generators.

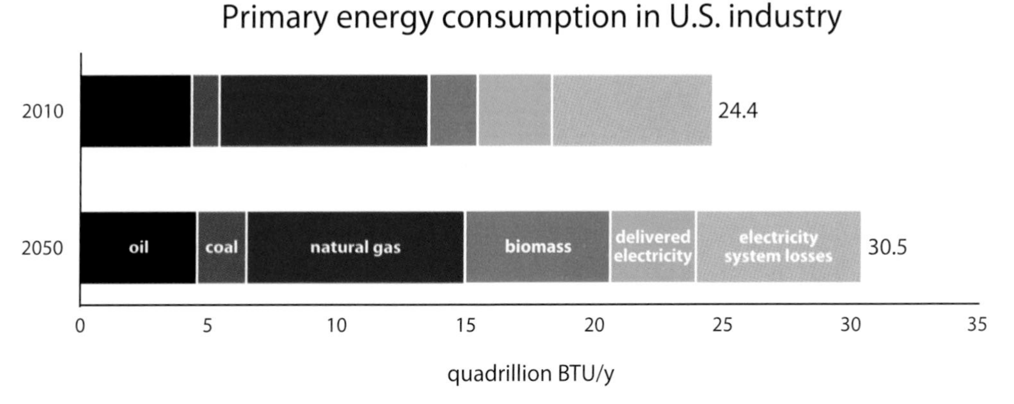

FIG. 4-3. In 2010, fossil fuels supplied 79% of the energy in U.S. industry. Two-fifths of industry's primary energy was used as electricity, three-fifths as directly burned fuels (52% natural gas, 28.5% oil, 7% coal, 12.5% biomass). Feedstocks are not included in this analysis but would add about 5 quads of noncombustion fossil fuel use per year.[333]

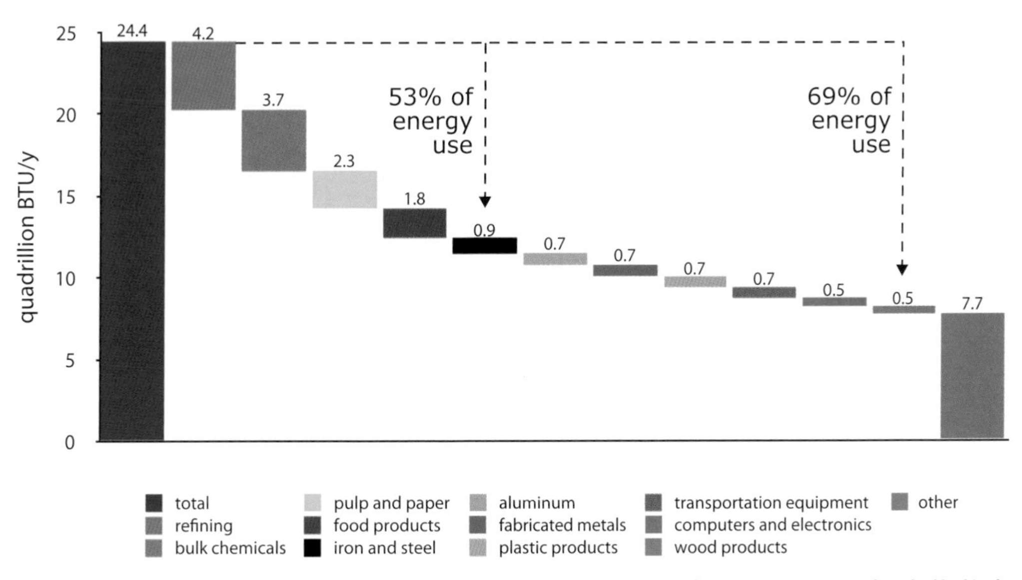

FIG. 4-4. Industry is complex and diverse, but the five most energy-intensive subsectors use more than half of industry's total primary energy.[334]

In contrast, the products of industry provide millions of services and are made in thousands of basic production processes, each with hundreds of variations. Hardly any two plants are identical. Yet, though industry has hundreds of subsectors, just 11 use more than two-thirds of all industrial energy (fig. 4-4), so the portfolio of solutions must focus on them while being applicable throughout industry. This is a huge hurdle, as each subsector has its own priorities on how to win on price in viciously competitive global markets. In the industrial jungle, sheltered market niches are few, change is constant, and risk is high, so managers are exquisitely sensitive to input costs (see How Industrial Attention to Efficiency Hinges on Fuel Prices sidebar).

HOW INDUSTRIAL ATTENTION TO EFFICIENCY HINGES ON FUEL PRICES

The U.S. chemical industry's energy intensity, mostly in natural gas, plummeted (fig. 4-5) until nationwide efficiency gains crashed natural gas prices in the mid-1980s and glutted energy markets. The chemical sector's energy intensity then stagnated or even rose for another 15 years before managers' attention refocused enough to rehire the energy-saving experts they'd laid off in 1986. For a two-thirds-of-a-trillion-dollar sector making more than 70,000 products, employing more than 800,000 people, and producing 19% of the world's chemicals,[335] such agility is impressive—but would be even more so if married with greater strategic consistency, such as policy can help reinforce. We'll return to that point at the end of this chapter.

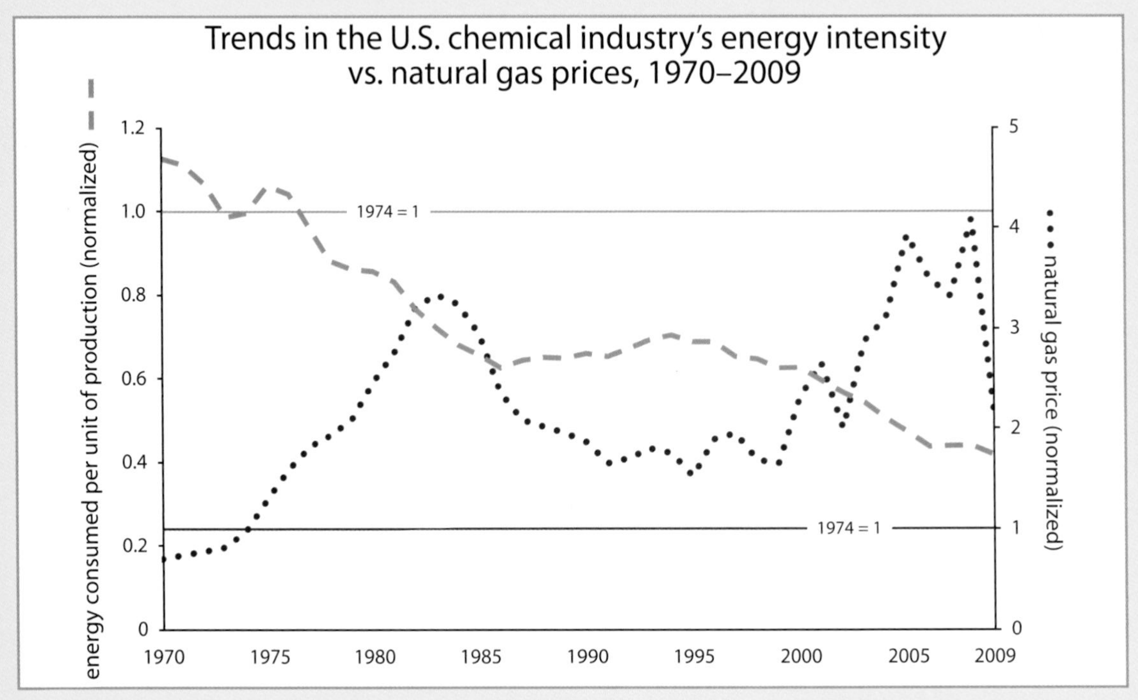

FIG. 4-5. Relative prices matter. The impressive drop in the U.S. chemical industry's energy intensity (the dashed green line) turned on a dime in 1986 when fuel prices fell (illustrated by natural-gas prices, the dotted blue line)—and then took about 15 years to start rebuilding momentum as prices rebounded.[336]

There's another way to think about types of energy consumption—what the energy is used for. Here, despite the incredible diversity of products, processes, and plants, just two purposes account for more than three-quarters of U.S. industry's primary energy use. Within manufacturing facilities, more than two-fifths of primary energy is used to heat things, whether giant vats of steel or tiny dabs of solder on circuit boards. Another two-fifths turns shafts to drive machines, from the clattering conveyor belts in soft-drink factories to the robotic arms on auto assembly lines. The rest goes into myriad processes and support functions, including smelting, reduction, lighting, and space conditioning.

These services can be powered in many different ways, so companies have picked whatever works best and costs least. America's historic roots in coal have long made iron-making coal-based; that's why the great Midwestern steel complexes are between Appalachian coal mines and Minnesotan iron mines. In contrast, Qatar has abundant cheap natural gas, so Qatar Steel's "direct-reduced" process for making iron, developed by the North Carolina–based company Midrex, burns gas. Sweden is rich in hydropower, so its companies have been developing iron reduction based on electricity or on hydrogen. These different choices heap complexity onto the industrial energy story. In

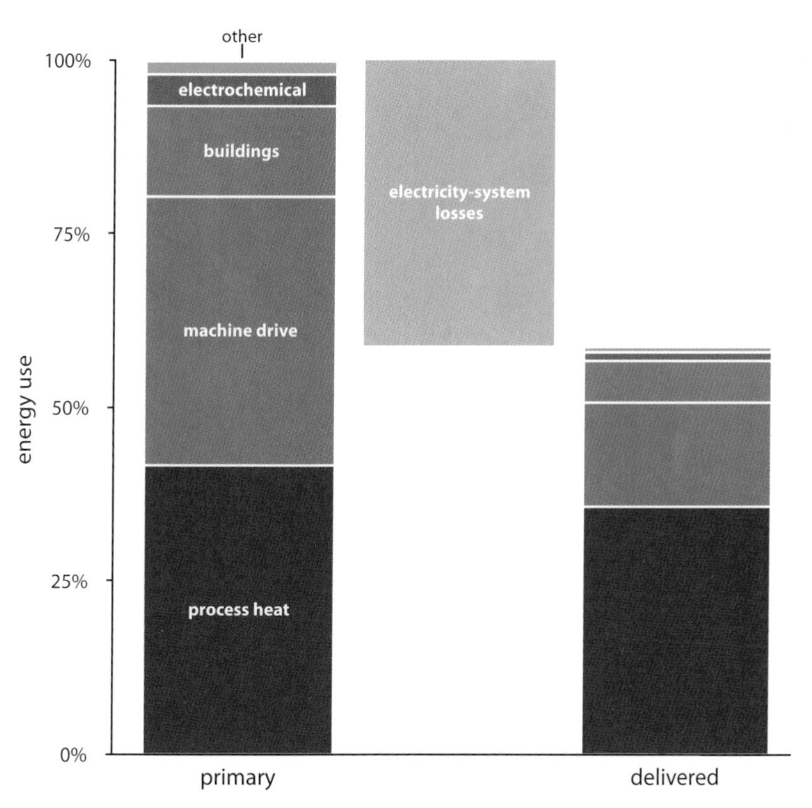

FIG. 4-6. Two main end uses—process heat (generally from directly burned fossil fuels) and machine drive—dominate industrial end uses. Electricity-system losses use nearly as much primary energy as process heat does.[337]

the industrial jungle, as in the rainforest, greater understanding only reveals greater diversity.

Driving Improvements in a Constantly Evolving Sector

The Reinventing Fire story becomes even more interesting when we consider that industry keeps changing—and improving—how it uses energy. Globalization, competition, and advancing technology have driven two encouraging trends: decreasing energy intensity (the amount of energy needed for a unit of production) and shifts to products that are more valuable but take less energy to make.

Why? As an economy matures, it needs a smaller share of energy-intensive products like steel and cement.[338] That's especially true in the U.S., where most buildings and infrastructure have already been constructed. That's why cement used per dollar of GDP peaked in the 1920s. The U.S. is also increasingly importing energy-intensive materials (petrochemicals from the Middle East, steel from China) whose world prices often undercut domestic producers'. If those foreign producers use energy less efficiently than domestic producers, this trade raises global energy use (see Energy Embodied in International Trade sidebar), and if they also get more of their energy from carbon-intensive fuels, global carbon emissions rise even more.

In addition to the benefits of ever-evolving energy efficiency, we'll also get a big reduction in industry's energy use simply as a by-product of weaning the transportation sector from oil. If cars and light trucks don't need gasoline or diesel fuel, then industrial refiners don't need to burn five quads per year of fossil fuels to make it. While energy derived from biomass will still be required to produce sustainable biomass-based transport fuels—process energy that accounts for most of the red biomass bar in figure 4-2 (column J)—the net effect is to save a total of 3.5 quads per year. So transforming the transportation sector affects other sectors' energy use, most of all industry's. With half of 2050 autos powered by hydrogen, half by electricity, and all trucking and aviation by biofuels, refinery operations can shift their fuel source from 76% fossil fuels today to 84% biomass in 2050. Alternatively, more autos could run on electricity, while trucks and planes could use hydrogen, cutting this processing energy further. Moreover, many advanced biofuels, like algal fuels that are harvested but need little or no processing, won't require such energy-intensive industrial steps at all.

ENERGY EMBODIED IN INTERNATIONAL TRADE

The U.S. in 2002 imported about five quadrillion BTU ("quads") net in the form of goods and services produced outside the country, equivalent to 5% of total U.S. primary energy use.[339] This in turn means that in 2004 (the most recent analysis available), the U.S. share of global fossil-fuel carbon emissions was not 22% from domestic fuel-burning, but actually 25–26%[340] due to the energy embodied in net trade to supply U.S. consumption.[341] During 1997–2004, as trade grew, especially with more carbon-intensive economies like China's, such imports of "gray energy" appear to have raised the U.S. share of global carbon release more than domestic efficiency gains lowered it.[342] Offshoring industrial production could become more worrisome not just for jobs but for climate as the U.S. continues its shift to a service economy. Analysis of such changes in the international division of labor in worldwide production of goods and services is complex and laden with uncertainty.[343] We have therefore limited our industrial analysis to U.S. domestic production. But industry is not limited by national borders. It is the only sector that can provide many of its products and services from abroad. Without conditions to encourage clean U.S. production, we risk exporting rather than eliminating fossil-fuel use.

The net effect from industry restructuring,[344] historical efficiency trends, and the virtual elimination of oil refining for transportation is to keep U.S. industry's fossil-fuel consumption almost static over the next 40 years while nearly doubling industrial output.

VIEWING INDUSTRY THROUGH THE EFFICIENCY LENS

The Dow Chemical Company uses a gargantuan amount of energy to make everything from millions of tons of plastics to solar photovoltaic roof shingles. It needs more than 3.7 GW of electricity, equivalent to the entire output of three big nuclear plants. It burns more oil than the Netherlands. As a result, even small boosts in efficiency bring a big bottom-line benefit. So Dow has been a relentless hunter of energy savings. The company has taken steps as simple as replacing the ethylene furnaces in its Freeport, Texas, refinery with more-efficient units, and as complex as developing a whole new process for making propylene oxide (the raw material for polyurethane plastics) in Antwerp using 35% less energy.[345]

Through thousands of projects like these, Dow slashed the energy intensity of its processes by more than 38% between 1990 and 2005, boosting both efficiency and profits. The company calculates that it saved $9.4 billion between 1994 and 2010 through energy efficiency costing $1 billion. That enhanced Dow's crucial cost advantage over less efficient competitors when energy prices spiked in 2008.

In fact, energy efficiency has proven to be such a good business strategy for Dow that the company is hungry for more. It now aims for a further 25% reduction in energy intensity by 2015. And in February 2011, executives announced an investment of $100 million in scores of new efficiency efforts. The projects that the company will fund "offer exceptional financial returns," explained Doug May, Dow's vice president for energy and climate change.

Dow is hardly alone in its focus on efficiency. Another pioneer is manufacturing conglomerate United Technologies: By focusing on energy use, it cut companywide energy intensity by 45% during 2003–2007 and greenhouse gas emissions 62% during 2006–2010 (while sales rose 13%, earnings per share rose 28%, and dividends per share rose 67%).[346] Or consider 3M, the innovator behind everything from protective films to Post-it notes. By developing a program to recognize teams that substantially improve energy efficiency, 3M boosted it by 22%, saving more than $100 million during 2005–2009.[347]

That said, in the same time period, 3M booked cumulative profits of over $20 billion, so its energy efficiency projects are still only a small fraction (about 0.1%) of total investment. This gap suggests that 3M has only scratched the surface of the opportunity. But 3M and many other entrepreneurial companies are rapidly going deeper and finding they can not only save more energy but also develop and market new technologies that save still more.

Understanding the Efficiency Levers

So what's behind the seeming magic of energy efficiency at Dow, 3M, and other industrial players? Energy efficiency across the myriad industrial processes and products can really be boiled down to a simple (though not often fully applied) recipe that should be the mantra of industrial executives, their design and operational captains, and the investors who hold their stocks. The recipe goes like this:

First, reduce the energy needed to run fundamental processes. Next, reduce the losses in the systems that distribute energy services within a facility. Third, improve the efficiency of devices like boilers and motors that turn energy into useful services. Finally, put energy that's now wasted into use.

Lever 1: Reduce the Energy Needed to Run Fundamental Processes

Consider one of the major challenges in making chemicals and drugs: many reactions are finicky. If reactants aren't mixed perfectly, or if the pressure, temperature, or timing is a little bit off, lots of other reactions will occur too, making undesired byproducts. That's why some fine-chemical processes yield only 2–20% desired product, and why the yields in pharmaceutical plants can be as low as 1%. Large amounts of energy and capital—often most of the plant—must then be spent to separate out the useful pure products from the dross. But modular, credit-card-size "microreactors,"[348] whose tiny channels and chambers let reaction conditions be exactly controlled, can make nearly pure product more quickly, cheaply, and safely with orders of magnitude less energy and waste. Stacked in hundreds or thousands,[349] their precision chemistry can make the plant's later separation stages largely disappear. Multistep chemical conversions without intermediate recovery also become possible just as in living cells, "drastically reducing operating time and costs as well as consumption of auxiliary chemicals and use of energy."[350] Even olefins—the highest-volume chemical product— may be more cheaply producible in microreactors.[351] Surprisingly, mass-produced "distributed" distillers and hydrogen reformers[352]—small versions of normally huge devices—may also be cheaper and more efficient than centralized ones.

In many industries, membranes that let only certain molecules through can replace brute-force distillation, which uses about 3% of all U.S. energy to purify oil and chemicals. Yet where distillation *is* needed, it can be improved. Product molecules shipped by a refinery may have been boiled and recondensed 15–20 times. Why so many? First, most operators monitor product quality not continuously but only periodically, so in between measurements, they're flying blind. They must redistill the product to unnecessarily high purity to be sure it'll meet specifications. Such "overrefluxing" often uses 30–50% more energy than a continuously monitored process. Second, process plants typically control flows and reaction conditions based on measurements of what *already* happened, rather than using algorithms that anticipate and optimize what's *about* to happen. This loosens quality control, wastes materials and capacity, and makes necessary extra processing to restore degraded output quality. The solution? Add sensors and controls to make sure that processes always run optimally: make it right, make it once, then stop.

That principle works only if operators can see and understand what they're doing. At a large Asian hard-disk factory, clean-room operators began saving millions once a pressure-drop gauge that showed when to change dirty filters was relabeled not just in green and red zones but in "cents per drive" and "thousand dollars' profit per year." In another plant, simply labeling the light switches saved $30,000 in the first year. Rapid advances in sensors, computing, and color-graphics display help apply operators' intelligence to their processes, with revolutionary effect.

High-tech equipment is often as wasteful as old-fashioned. A Caltech lab retrofit is saving 50–60% of the energy formerly used by equipment, like mass spectrometers, that is typical of much lab equipment in industry. Simple improvements like power-management controls, doubled-efficiency vacuum pumps, and using the building's chilled water instead of mini-chillers were welcomed by the manufacturers, who'd never before been asked for efficiency.[353] Or consider the three-quarters-of-a-million fume hoods humming away in chemical and biotech labs, which waste 50–75% of their $3 billion/y electric bill. Each uses as much energy as three houses.[354] Retrofit pays back in two to five years, or immediately in new labs when combined with "sniffer" controls that reduce flow when the air is clean.[355] Even inefficient hoods use far less energy when closed. Best of all: no hoods. Design out the toxicity.

Lever 2: Reduce the Losses in the Systems That Distribute Energy Services within a Facility

Big savings often come from simple technical improvements and maintenance, such as insulating pipes, maintaining steam traps and air filters, adjusting power factors, or properly installing pumps and fans so fluid can enter and leave them smoothly. In fact, some of the most lucrative but least visible savings are not in conspicuous devices but in the pipes, ducts, and wires that connect them.

Who, for example, ever thinks about saving the few percent of electricity lost in factories' and buildings' distribution wiring? The Copper Development Association (CDA) does, because it wants to sell more copper. In the 1990s, it published a new wire-size guideline. The old one that everyone used before, and most still do, was meant only to prevent fires. The CDA's new one is meant to save money. In a typical 1996 office lighting circuit, substituting the next fatter wire size yielded about a 169% annual after-tax return. Now Japan aims to adopt a similar standard and get it adopted globally. American industry and utilities still misallocate about $1 billion a year just by buying inefficient electric transformers with skimpy copper and lossy iron.[356]

Sometimes spotting waste in the energy distribution system requires only a keen ear. If a weekend visitor can hear the air compressor working when nobody else is, compressed air is probably leaking. Those leaks add up. Compressed air is ubiquitous, consuming 9% of U.S. industrial electricity. It runs machines like pneumatic screwdrivers, actuates old pneumatic controls from the pre-digital era, and dries products like circuit boards, among many other uses. But making and distributing compressed air is only about 10% efficient,[357] so it's extremely costly and best eliminated. Just fixing compressed-air systems commonly saves up to half their energy with six-month paybacks.[358] When Modern Forge of Tennessee cut its total electricity use 8% by optimizing its compressed-air equipment, the cost was less than zero because the plant avoided buying a new compressor.[359]

Lever 3: Improve the Efficiency of Devices Like Motors and Boilers That Turn Energy into Useful Services

U.S. industry uses more than 13 million[360] motors to run drill presses, chillers, grinders, mixers, blowers, and countless other types of equipment. Motors use three-fifths of all industrial electricity. Every few weeks, a constantly running motor uses an amount of electricity costing more than the motor. In fact, one 100 hp motor, costing about $10,000, can burn half a million dollars' worth of electricity in 20 years. A big factory can easily have hundreds of such motors. A gain of a single percentage point in one motor's efficiency is worth more than $60 present value per horsepower. So it's easy to see how to save vast amounts of energy and money by choosing very efficient motors and turning off unused ones. Japan's Mori Seiki Company's aggressive energy management system turns off machine tools' unused functions, reportedly slashing power use by about 80%, and saving 40% even during normal operation.[361]

Few managers notice, though, because motors lack a taxi-meter-like display recording real-time cash flow. So they may not use motors only when needed, or only as big as needed. Simply adding variable-speed motor controls can often cut their energy use by 15–30%, and sometimes much more. Further gains come from installing the most efficient motors. Curiously, for the most common kind of motor, the North American trade price in 2010 was completely uncorrelated with efficiency up to at least 100 hp, and only loosely correlated at 400 hp.[362] An astute buyer can get a very efficient motor cheaper than some inefficient ones—saving money up front, getting better motor life and quality, and saving nearly $20,000 worth of electricity in present value for one 250 hp motor. And motor efficiency advances keep raising the bar: The newest 2011 models are two to three percentage points better than the "premium efficiency" motors of

2010.[363] Some may even be cheaper too, because their compactness saves so much copper and iron.[364]

It's the same story with pumps and fans, the biggest uses of motor power. European[365] and North American pumps show an efficiency range of six percentage points for the biggest sizes and 15–20+ points for small ones. And even among the most efficient fans of a particularly efficient type, a 1996 U.S. survey found efficiency gaps of 4–10 percentage points.[366] Most buyers assume higher efficiency costs more and isn't worth it, but careful shopping may invert both assumptions.[367] "In God we trust"; all others bring data.

Similar efficiency gains are possible for all the other energy-converting devices, such as furnaces, compressors, chillers, blowers, and boilers. The Gas Technology Institute and Cleaver-Brooks's Super Boiler, for example, is half the size and weight of today's mostly '60s-vintage firetube boilers, and one-fifth more efficient. But whatever their efficiency in constant operation, most boilers still lack smart controls. Many smaller boilers use about 25–45% of their fuel just turning on, warming up to deliver lots of heat briefly, then turning off and repeating, because their simple controls know nothing about the load and deliver only temperature. Smart controls sense and match actual loads. A boiler in Butler University's Jordan Hall thus cut its cycles from six an hour to two a day. Such controls' typical 20–50% fuel saving pays back in months to a couple of years.[368]

Lever 4: Put Wasted Energy to Use

Industrial facilities produce titanic amounts of waste heat, excess pressure, and useless byproducts that took energy and money to make. Often they must expend additional energy to cool and treat those waste heat streams before being allowed to throw them away. Hot steel billets are often moved so far for rolling that they need reheating. In one metal refinery, material was being heated to 2,200°F, quenched with water, dried with electrically heated air, then often run through the same cycle once or twice more: the operators hadn't yet learned to keep hot things hot and dry things dry. They quickly did.

As mentioned above, one particularly compelling use of waste is harnessing the vast amount of heat produced by generating electricity. America's power plants turn fuel into one-third electricity and two-thirds heat—and that heat is typically thrown away, wasting more energy than Japan uses for everything, because there's no productive use for it nearby, and old U.S. practices (which don't burden most foreign competitors) encourage or mandate power-only plants. But industry needs enormous amounts of heat. Cogenerating power and heat together in factories—combined heat and power (CHP)—can more than double efficiency, saving energy, money, and emissions. It's standard practice in Europe. Tripling U.S. CHP capacity to 240 GW would cut America's total CO_2 emissions by 12%.[369] Moreover, many factories could sell low-temperature heat (typically not a regulated utility function) to other factories or buildings within affordable distances, again a common European practice, but not in our analysis—potentially saving about 30% of U.S. industrial energy or 11% of total energy.[370]

The beauty of these four steps is that they not only can achieve virtually every imaginable efficiency gain (and point the way to ones we haven't thought of yet), but they also feed on each other. Reduce the amount of steam heat needed for a chemical reaction, for instance, and you'll also reduce the losses bringing steam to heat the reaction and be able use a smaller, cheaper, more efficient boiler. And if you can recapture some of that heat, the efficiency gains compound even more.

Moreover, as in buildings, efficiency often triggers non-energy benefits. Utrecht University professor Ernst Worrell found that counting the non-energy values of 47 energy-saving retrofit steps in the iron and steel industry *doubles* the cost-effective savings by making 11 additional steps cost-effective. He found 76 more case studies of non-energy benefits, 52 of them monetized and more than halving the payback period.[371] If industry counted those extra benefits, it would buy far more efficiency—another saving conservatively left out of our analysis, except later when we consider a part of the integrative-design opportunity.

Thus, by taking advantage of these four fundamental principles, such companies as Dow and 3M have made great progress in using less energy to produce the same ton of product. However, there are additional gains to be had.

Emerging Efficiency Technologies

In fact, new technologies continue to appear throughout the industrial landscape. These technologies improve the efficiency potential beyond the U.S. Energy Information Administration (EIA) "baseline" efficiency, but implementing them will require attention and care. Scientists at Lawrence Berkeley National Laboratory (LBNL)[372] have assessed a wide portfolio of emerging technologies ranging from superefficient boilers to ultrasound-aided drying to new electrodes for electrochemical processes.

By 2050 these emerging technologies should reduce industry's annual energy use by 2.3 quads (fig. 4-2, column H). As these current measures are adopted, we assume that innovation will continue to improve, augment, and replace them with others, allowing adoption to continue. We saw in chapter 3 that energy-saving technologies have centuries of history of replenishing the efficiency resource faster than it's exhausted, and we find this true for industry as well.

California, for example, evaluated the potential electricity savings in its industries in 2006[373] and found it to be 5.5 TWh/y ("TW" here meaning terawatts, or trillion watts). A 2008 reevaluation[374] found the technical potential had risen to 5.6 TWh/y even though 1.9 TWh/y had already been saved.[375] Similarly, the U.S. Department of Energy's 26 Industrial Assessment Centers have offered no-cost industrial energy surveys to qualifying facilities since 1976.[376] A 2006 evaluation[377] of two separate five-year periods found that the energy savings per recommendation increased by 9% between 1985 and 2005, even though industry's energy intensity meanwhile fell 7%.[378] The average payback period of the recommendations rose too, but only from 3.2 months to 5.2 months (a 230% internal rate of return).

Beyond Cogeneration

Expanding combined heat and power (CHP) beyond official forecasts is another huge opportunity. EIA-based projections raise CHP from 78 GW in 2010 to 123 GW in 2050, but we believe industry could profitably achieve 187 GW—enough to displace nearly 60% of the nation's 2010 coal-fired electricity.[379] Most CHP burns natural gas or on-site wastes.

Beyond cogeneration, we look at landfills, smokestacks, and liquid effluents, as they will also reveal plentiful BTUs ripe for the picking. Wringing electrons from relatively low-temperature heat or modest pressure can make more electricity. For instance, most industrial boiler systems deliver steam at various pressures for different processes by producing steam at the highest pressure required and throwing extra energy away in pressure-reducing valves. A simple backpressure turbine can instead convert excess pressure into electricity and pay back in a few years.

Some of these opportunities have already been tapped. Pulp-and-paper mills, for instance, often meet much or all of their own power and steam requirements from their waste biomass. Furthermore, companies are emerging around the nation to develop technologies for capturing more of this prized energy. Alphabet Energy, a start-up based in Berkeley, California, for instance, is working on next-generation solid-state thermoelectric devices that produce electricity directly from waste heat.

In total, Pacific Northwest National Laboratory (PNNL) estimates[380] that more than 3.5 quads of waste energy is potentially recoverable, worth more than $5 billion in net present value. Analyzing CHP techniques from LBNL,[381] we found that about 580 TWh/y of electricity, equivalent to 5.9 quads of primary fuel per year, can be technically and economically recovered from waste heat and pressure. Aggressively capturing that potential and CHP by 2050 would save 2.4 quads a year more than EIA or LBNL assumed (fig. 4-2, column I).

These gains are huge, but the way is littered with many hurdles: technology limitations, regulatory roadblocks, and most industries' past ambivalence toward CHP (generating electricity is not their core competency). So the big untapped potential isn't surprising. Yet these opportunities are achievable, and many foreign competitors have already achieved them. The U.S. Department of Energy dedicates nearly a quarter of its Industrial Technologies Program (ITP) budget to distributed energy production and CHP,[382] with gratifying progress in both application and improved technologies, like engines and microturbines.

But the government is not the only stakeholder supporting these developments. In 2000, a group of CHP companies and stakeholders set an ambitious goal of doubling installed CHP capacity in the U.S. by 2010—and hit 95% of that goal. We believe the sector can nearly triple CHP capacity in the next four decades. Our model assumes the barriers are unlocked over time, as further discussed at the end of this chapter.

Combining the Levers: Integrative Design

However, as we've hinted already, energy efficiency has still more in store. The real magic—and the full potential of Reinventing Fire in industry—comes from pulling two or more of our four levers in the right order and linking them via integrative design.

Think about a data center powered by electricity from a traditional thermal power plant (fig. 4-7). Two-thirds of the power-plant fuel doesn't even make it to the data center's electric meter—it goes to waste heat. Half the electricity the power plant produces is used to run the data center's chillers and power-supply equipment. Of the remaining energy delivered to the equipment that actually matters to customers—the servers—half is lost in inefficient power supplies and in thousands of fans that move heat (mainly unnecessary heat) off the motherboards into the room. And most of the energy that reaches the chips doesn't run applications, because most of the servers sit idle most of the time and all computing resources are greatly underutilized. It gets worse: much of the computation runs threads and processes that aren't really needed, and then the results may support inefficient business processes. In the end, less than 1% of the initial power-plant fuel actually creates customer value.

But now, let's start pulling the levers, starting on the right of the green ribbon. First, we'll reduce the required computing by writing terse code, elegantly compiled, so every computing cycle is needed and wanted. Next, we'll choose servers that are about four times more efficient but cost the same as the existing ones. They'll need very little cooling and power supply, and there are far more efficient ways to do both. Finally, the utility losses can be roughly halved by replacing both the power station and the uninterruptible power supply with an onsite ultra-reliable fuel cell with heating and cooling as by-products. By starting on the right, and executing savings in the direction opposite to the energy flow, small savings multiply into ever greater savings of energy and capital as you go toward the left.

Such is the compounding power of integrative design. This technique can be applied to any kind of industrial design. Consider pumping systems, with their maze of pipes and motors, found everywhere from refineries to pulp mills to widget factories. Just putting in more-efficient motors is only one small step toward making these plants more efficient—and in fact, it's the wrong place to start. First, to shrink the motor system and use it more efficiently, let's start further downstream, with the system the motor is operating. Half of industrial motor power drives pumps and fans, both with the same physics and similar opportunities.

In 1997, Interface Nederland's chief engineer Jan Schilham—an intense, soccer-loving efficiency hound—redesigned a pumping system to circulate heating fluid through a new carpet factory in Shanghai. Europe's top engineering firm in that field had designed the pumping system to total 95 horsepower. Schilham redesigned it to use no more than 13 horsepower. And his system cost less

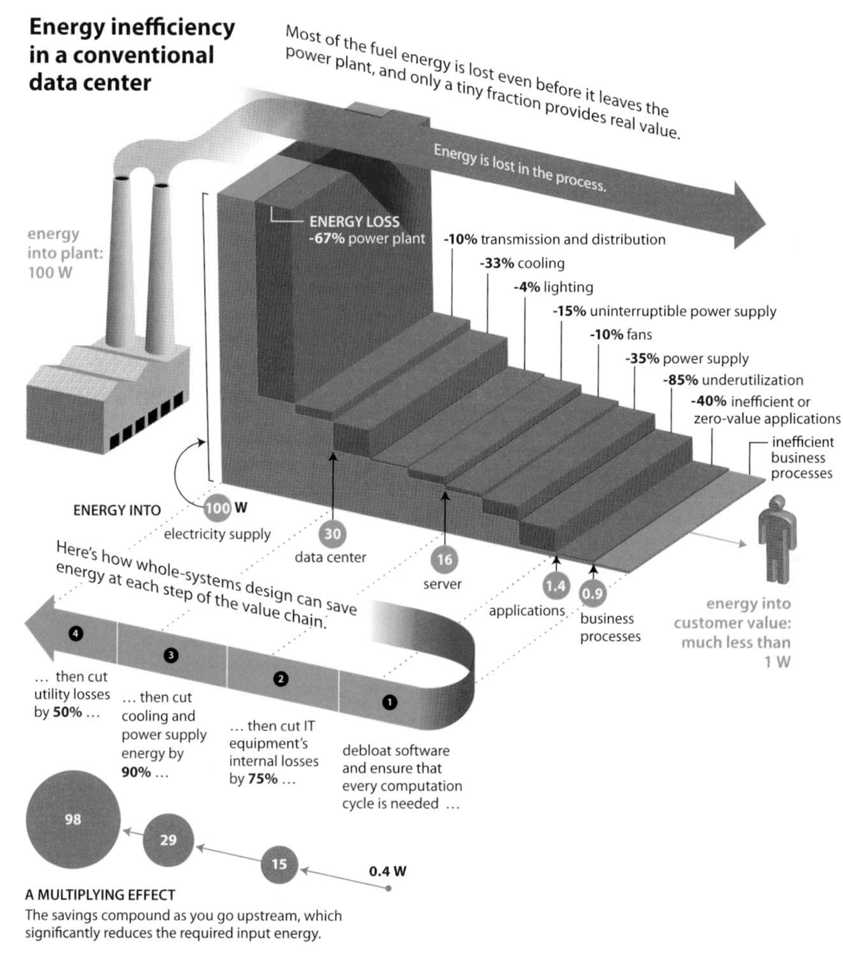

FIG. 4-7. Starting the savings downstream can achieve leverage of 10- or even about 100-fold in saving energy back at the power plant.

to build, fit in less space, was easier to build and maintain, and performed better.[383]

How did he do it? With two big rearrangements of his mental furniture.

First, he used big pipes and small pumps, not small pipes and big pumps. The smaller the pipe, the more energy it takes to push fluid through it. Fatter pipes' cost rises as about the second power of their diameter, but their friction falls as nearly the *fifth* power of their diameter—and so does the size, and hence the capital cost, of the expensive pumps, motors, inverters, and electrical supplies.

Second, Schilham designed the layout of the pipes first, *then* added the equipment they connect. Normal practice is the opposite, making the connected equipment typically far apart, obstructed by other objects, at the wrong height, and facing the wrong way, so the piping has about three to six times as much friction as it would with a straight shot. Fat, short, straight pipes have enormously less friction than thin, long, crooked pipes.

His design was inspired by lessons RMI had brought him from the restrained, avuncular, slyly hilarious Singaporean engineering genius Eng Lock Lee. Lee asked a simple but thought-provoking question: Why, when a critical pump must be connected to a spare backup pump, is the piping to connect them laid out so the fluid must pass through two right angles, multiple fittings, and typically two or three valves, all of which cause energy-robbing friction? Why not lay it out with no bends and no valves or one valve? (See fig. 4-8.)

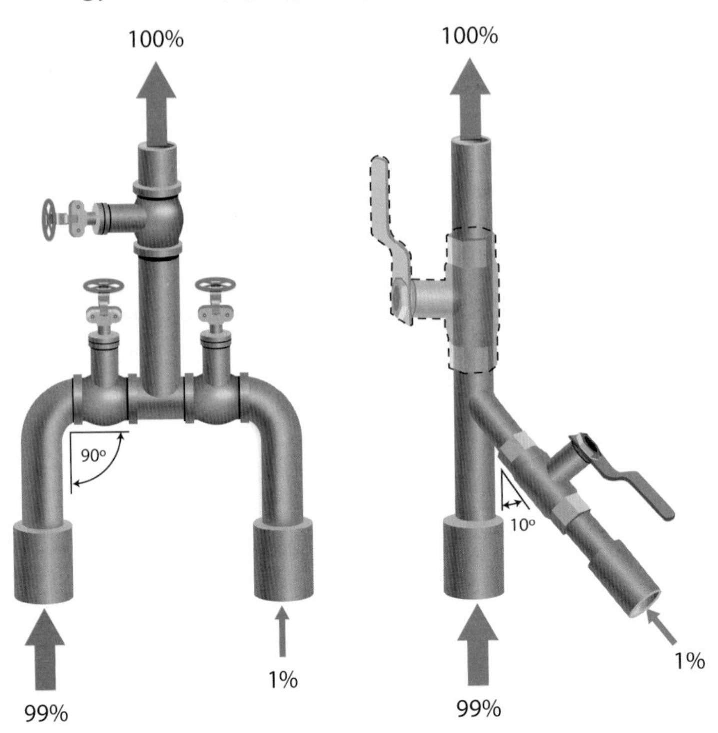

FIG. 4-8. Multiple feeds to a pipe are normally, but irrationally, plumbed at right angles (at left) rather than in a geometry that minimizes friction and cost (at right).

The reason: an old habit left over from what was easy to draw. Pipefitters like the traditional right-angled approach too, because they're paid by the hour, and they mark up a profit on the extra pipes and fittings—and they don't pay for your electric bill or your bigger pumping equipment. Yet all those sharp bends maximize friction and hence electricity costs—and they need bigger, costlier pumping equipment to fight the unnecessary friction. Nowadays the drawing is usually done by computers, so right angles are totally unnecessary. Big short pipes with only a few gentle bends brought Schilham those dramatic improvements in the carpet plant. And when Lee recently designed piping for a Singapore biotech plant, he was able to save 69% of the pumping energy at lower capital cost (fig. 4-9).

Such techniques can even be cost-effectively retrofitted. Lee's disciple engineer Peter Rumsey changed the pipe layout in the condenser-water pumping loop for his hometown Oakland Museum of California. He eliminated 15 pumps that will never again waste energy and maintenance. The system now uses 75% less pumping energy, with a one- to two-month payback for the redesign. And, of course, the more efficient the building, the less—if any—cooling it needs, so the less condenser water will need pumping.

Multiply all these savings from the end use and from the system, and almost no energy may be needed at the front end. This powerful way of systematically capturing radical energy efficiency throughout any complex system can be applied to almost anything that uses energy, water, or other resources. For example, each unit of friction or flow saved in a pipe or duct system saves about *10* units of fuel, emissions, and cost back at the

FIG. 4-9. A 69%-lower-energy, but cheaper, pumping system designed by Eng Lock Lee. Note the few and smooth ("sweet") bends, fat pipes, diagonal pipe runs, Y-junction, and—reflecting high Singapore land prices—vertically stacked pipes. The layout is utterly untraditional; that's why it saves energy and money.

power plant (fig. 4-10). That's because the 10-fold compounding losses from power-plant fuel to flow in the pipe get turned around backward into 10-fold compounding *savings*. Starting savings downstream also makes the upstream components progressively smaller, simpler, and cheaper, saving the most capital cost. Thus one less unit of flow or friction at the downstream end makes the motor about 2½ units smaller—and although most motor efficiency doesn't cost more, size definitely does. This is the same idea we saw in figure 4-7's data center. It's like chapter 2's sevenfold leverage from energy saved at the wheels of an auto back to fuel saved in the tank, shrinking the powertrain to help pay for the lightweighting. It's also like chapter 3's use of capital savings from smaller mechanical systems to help pay for the superwindows that help reduce those building loads. Whatever the system, *start saving downstream.*

Can we now apply the same logic to the design not just of a giant industrial plant's pipes and pumping systems, but also to its basic industrial processes? Yes, indeed—and that's the next big frontier of integrative design. It's a surprisingly common opportunity because many big process plants were designed when energy was cheap, so the economic incentive to be more efficient has grown along with energy costs. Unfortunately, many plant managers haven't had the

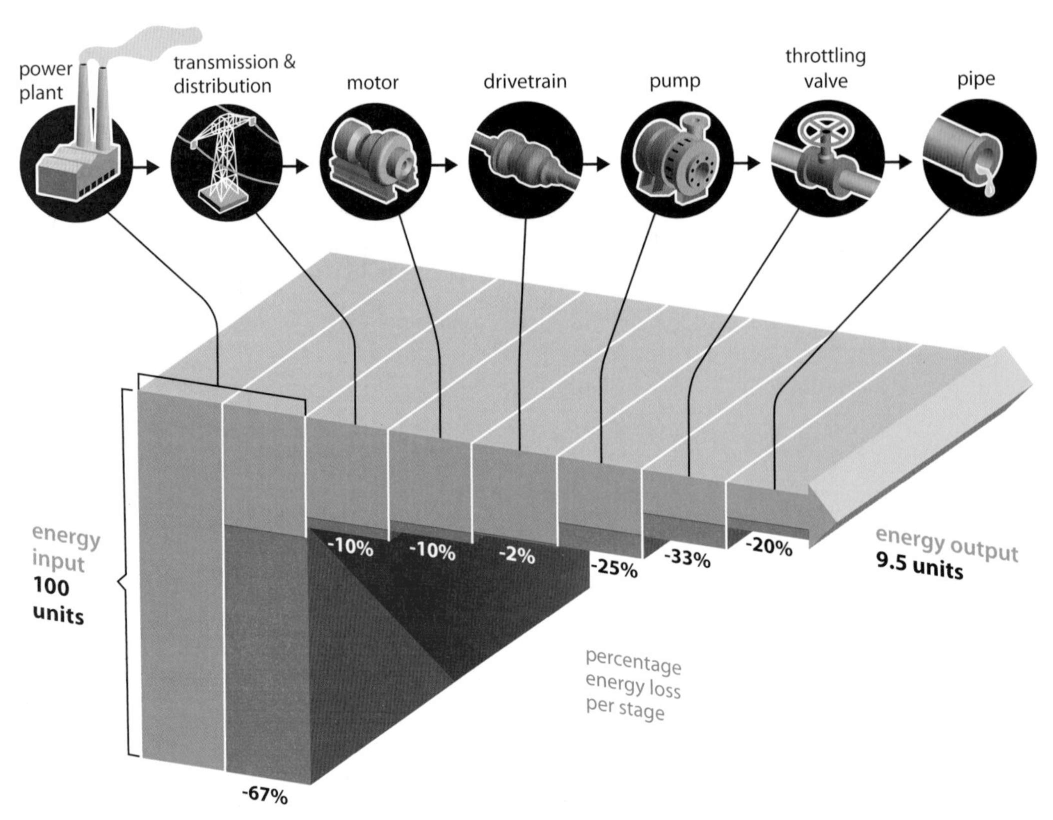

FIG. 4-10. Saving energy starting all the way downstream, at the end use—like flow from a pipe in this pumping system—turns compounding losses (left to right) into compounding savings (right to left) of both energy and capital.

luxury of time to contemplate improvements, so the opportunities remain untapped.

The Centre of Material and Process Synthesis, a South African group at the University of the Witwatersrand led by professors David Glasser and Diane Hildebrandt, noticed that individual *steps* in chemical-engineering processes produce CO_2, but that well-integrated *processes* need not. Applying that idea, they've designed processes to turn natural gas into liquid fuel that, instead of producing more CO_2 than fuel, produce none—and they built a modular coal-to-liquids pilot plant in Baoji, China, to prove it.[384]

Recent integrative designs and redesigns of diverse industrial plants by Rocky Mountain Institute and its partners have yielded retrofit energy savings of around 30–60% with paybacks of a few years, and new-facility energy savings of around 40–90% with generally lower capital cost. How broadly those findings apply is unclear; your actual mileage may differ. But these examples of improvement in what was considered already good practice suggest a big opportunity:

- A 2009 Electronic Data Systems (now HP) data center is using 73% less non-IT electricity and 98% less cooling and pumping energy than normal, with three times the computing per watt, at normal capital cost—but its full potential would have saved about 95% of the electricity and half the capital cost;[385]
- Texas Instruments' Richardson, Texas, microchip fabrication plant reduced the energy use in its facilities systems (as distinct from its tools—the chipmaking equipment) by 38%, total energy by 20%, water by over 35%, and capital cost by 30% or $230 million[386]—allowing it to be built in Texas, not Asia. Texas Instruments' next plant went one step further and designed out the need for the entire low-temperature chiller plant by using desiccants to dry the outside air.
- A retrofit under way at the world's number one platinum mining complex is expected to save about 43% of energy with a two- to three-year payback.
- Design workshops with Shell developed schematic retrofits for its most efficient refinery, the world's number two LNG liquefaction plant, and a North Sea oil platform that were respectively expected to save 42% and 40% or more, and about 100%, of their energy use, with paybacks of a few years—while a new $5 billion gas-to-liquids-plant design was expected to save more than 50% of energy and roughly 20% of capital cost.

These examples provide evidence that individual plants have much to gain from integrative design. However, what savings are possible throughout industry? Our analysis assumes, contrary to several examples just mentioned, that industry is so strong at optimizing thermal processes that only electric auxiliary systems can gain much from integrative design. To be conservative, we therefore apply integrative design principles only to electric drivepower (see Profitably Retrofitting an Industrial Motor *before* it Fails sidebar) and its largest use—moving fluids. Based on case studies and detailed published analyses of these drive- and fluid-handling systems,[387] we expect up to 1.1 quads of further 2050 efficiency opportunity, not to mention its uncounted net economic benefits (especially reduced capital costs). Our field experience suggests this is an underestimate.

Many benefits depend on others. For example, not only efficiency but also motor life depends on other energy-saving improvements: reducing voltage imbalance, improving shaft alignment and lubrication practice, reducing overhung loads (sideways pulls) on the shaft (e.g., by substituting toothed, nonstretch, low-tension "synchronous" belts for V-belts), and improving housekeeping, like not siting motors in the sun or next to steam

PROFITABLY RETROFITTING AN INDUSTRIAL MOTOR *BEFORE* IT FAILS

Immediately retrofitting an in-service standard induction motor to a premium-efficiency model, without waiting for it to burn out, is commonly assumed to pay back too slowly, in about 6–10 years. Yet many U.S. motors are so grossly oversized that probably half never exceed 60%, and a third never exceed 50%, of their rated load. This oversizing often makes them run less efficiently than their rating, further increases energy waste by spinning fans and pumps too fast,[388] and often enables the replacement motor to be smaller, and hence cheaper. Making the new motor the right size therefore reduces the average simple payback of immediate retrofit to about three years or less. Counting also the new motor's longer life (because it runs cooler, roughly halving the energy that makes it hot rather than making it turn, and it may also have higher-quality bearings) makes the immediate retrofit even more lucrative, because downtime tends to cause very costly lost production.

In addition, the new motor automatically eliminates any increased magnetic losses that may have been caused by improper repair of the old motor—a widespread practice that, according to GE measurements, used to cost the U.S. about $1–3 billion/y and probably still does. This plus proper motor sizing yields direct electrical savings roughly twice as great as would be expected from the new motor's greater rated efficiency alone. The high-efficiency motor also has better power factor and greater harmonic tolerance (hence better operation at variable speed). Thus it provides a half-dozen main (plus ten smaller) operational advantages—but it need be paid for only once.

pipes, nor smothering them beneath multiple coats of paint. Motor choice, life, sizing, controls, maintenance, and associated electrical and mechanical elements all interact intricately. A few interactions are unfavorable, but most make the savings of the whole drivepower package far greater and cheaper than would appear from considering just a few fragmented measures, as most analyses do.

HOW MUCH MORE PRODUCTIVE CAN INDUSTRY BECOME?

Summarizing so far: Adopting advantageous efficiency technologies not in EIA's projections[389] can trim 2.3 quads (7.5%) from 2050 industrial energy use. Adding cost-effective waste-heat recovery at the rate of equipment turnover saves another 2.4 quads (7.9%) of U.S. industrial energy in 2050. Overall, U.S. industrial energy demand can drop 8.6%, from 24.4 quads of primary energy in 2010 to 22.3 quads in 2050, while industrial output rises by 84%. And the remaining energy will come from quite different sources.

- The fossil-fuel share, whether as direct fuel or via purchased electricity, which is now 70% fossil-fueled, will drop from 81% of industrial energy in 2010 to 53% in 2050.
- Purchased electricity consumption will drop 75% from the 2010 level, both through greater efficiency and because industry is cogenerating much more.
- The share of biomass in industry's primary energy mix will grow from 8% in 2010 to 45% in 2050, mainly for the process energy (not the feedstock) to produce[390] the biofuels needed by trucks and airplanes (figs. 2-19 and 2-21) and in industrial mobility—from bulldozers to forklifts and container hostlers to farm combines.[391]
- EIA projections, extrapolated to 2050, suggest natural gas will cost about half as much as oil, so most of the remaining oil burned for process heat should switch to natural gas by then.[392]
- The remaining coal and natural gas also produce process heat and can be replaced by CHP heat recovery, biogas, biofuels, biomass, wastes,

solar heat, or electricity. This substitution is technically possible with the technologies we have today. Some are already cost-effective and will happen naturally. The rest may need an external driving force. We'll explore below how this fuel-switching could work.

These conclusions appear robust and unlikely to be disturbed by "rebound" effects (see Factoring in the Rebound Effect sidebar).

Addressing the Remaining Fossil Fuels

If most industrial firms become as aggressive over the next four decades as the pioneers are today in capturing energy efficiency and implementing CHP and waste-heat recovery, and if new technologies only sustain the historic adoption rate of old ones, then by 2050 U.S. industry will produce about 84% more output with 9% less energy and 41% less primary fossil fuel. Though this achievement will be cost-effective under existing rules and current and projected prices, needing no breakthroughs or mandates, adoption will be challenging, as we'll discuss below—but the prize seems likely to elicit the needed attention and effort.

But U.S. industry will still be using considerable fossil fuel, 77% of it natural gas. Have we the opportunity and the motivation to displace even more fossil fuel? We think so, for three reasons. First, in a world where competition keeps increasing while natural resources become pricier, markets will reward those best able to manage cost, and energy is a critical and manageable industrial cost. Second, reducing or avoiding volatile prices and potential supply interruptions reduces business risk. Third, actively managing climate risks

FACTORING IN THE REBOUND EFFECT

Much like the other sectors of the economy, industry has countless opportunities for using energy more productively—and the potential, at least in principle, for several forms of "rebound" to take back some of the energy savings through respending, price elasticities, or other effects. Rebound effects are particularly poorly understood and measured in industry and depend heavily on disputed macroeconomic theoretical models.

Consider a cellphone manufacturer who saves factory energy. Will she reinvest the savings to save even more, building on demonstrated success, or perhaps buy a very energy-intensive new furnace? Will she sell more phones (because they cost perhaps pennies less), expand the factory, and use more energy to make even more phones (but perhaps saving net energy because a less energy-efficient competitor then sells fewer phones)? Will these shifts slightly change energy prices? Will they perhaps make cellphones so cheap that they create and capture whole new markets?

Energy rebounds are likely to be held in check by several commonsense effects. First, because energy usually makes up only a few percent of total manufacturing costs, saving part of that few percent is unlikely to make products far cheaper and strongly stimulate sales. Respending saved energy dollars, whether they go to managers, shareholders, or customers, is unlikely to raise energy use much either, because only about 6–8% of the average dollar of GDP buys energy. Further, industry will produce only what it can sell. Innovation, novelty, marketing, et cetera, may boost sales significantly, but a more energy-efficient manufacturing process is a rather small and oblique driver.

However, measured rebound data are limited, especially in the industrial sector. Authoritative studies in 2000[393] and 2007[394] found the economic literature in such disarray that no meaningful conclusions could be drawn. So even though the rebound effect is not zero, it seems quite small and is unlikely to reduce industrial energy savings materially. Some commentators claim the opposite based on theoretical models, but they have not yet been able to explain or demonstrate their physical realism, and there's good reason to believe large industrial rebounds are a modeling artifact rather than a real and observable phenomenon.

HOW LOW CAN WE GO?

In the industrial energy limbo, we can get much closer to the floor, as a simple question reveals: What's the minimum amount of energy needed to make a billet of steel, a ream of paper, or the thousands of other products civilization requires? Only a modest fraction of what industry uses today. In fact, compared to theoretically perfect processes, the U.S. economy is only about 13% efficient;[395] the global economy, around 10%.[396] Even the most efficient industrial processes today use two or three times the energy that's theoretically necessary.[397] The potential energy savings (fig. 4-11), therefore, are huge.

Why does industry use so much more energy than is theoretically necessary? Often by operating at temperatures or pressures higher than required. Using high-quality energy when low-quality energy would serve—like heating a room with electric resistance heating—uses 100% of its quantity but only 6% of its quality, so as one architect put it, it's "like cutting butter with a chainsaw." Most factories do the same in many of their processes.

Of course, dedicated researchers have long been striving to approach these minima, with considerable success. In the 20th century, industrialists cut the energy intensity of making ammonia by nearly eightfold, and that of integrated iron and steel by more than threefold (fourfold for the best furnace).[398] We expect industry to continue this trend as it pulls ever harder on the four levers outlined above.

Yet even with the gains achieved so far, there's still a huge amount of energy overkill. To be sure, it's not practical or cost-effective to reach the absolute theoretical minimum amount of energy needed to make products. But figure 4-11 suggests that some of the gaps hold juicy potential—and a Cambridge team that showed a practical potential to save about 85% of the world's energy found that actually getting roughly halfway to theoretical perfection looks reasonable.[399] A veteran Dutch practitioner, too, concluded that for at least the next 10–20 years, energy efficiency gains upwards of 5% a year "are feasible and could cut industrialized countries' absolute total energy use by half in 50 years."[400]

It's hard to quantify the impact of future breakthroughs—as physicist Niels Bohr said, "Prediction is very difficult, especially about the future"—but we're confident that continuing technical progress will make our current projections of 2050 industrial energy use look conservative. And that's without counting some even more revolutionary opportunities we'll discuss later, transforming not industrial processes but what we do with them.

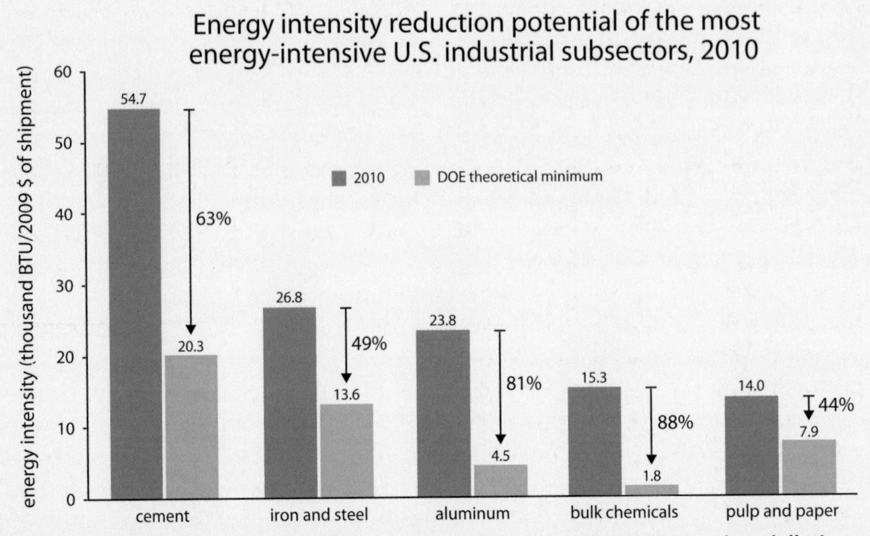

FIG. 4-11. The blue bars show projected delivered energy for various sectors for 2010 to make a dollar's worth of product. The green bars show the minimum energy theoretically required. The bulk chemicals sector shows an analysis for just 44 products whose manufacture uses about 60% of the chemical industries' energy.[401]

to a degree consistent with the Rio Treaty signed by President George H. W. Bush and ratified by the U.S. Senate implies roughly 80% decarbonization of the whole economy, so 41% fossil-fuel savings in industry falls well short.

The good news is that with the right incentives, innovation looks likely to get us there. Specifically, there are further opportunities at hand for lessening even more our reliance on fossil fuel. Basic chemistry and physics (see How Low Can We Go? sidebar) show that we're far from the bottom of the innovation barrel.

Ways to get close to the bottom of the barrel include fuel-switching, breakthrough approaches, and reductions in our use of virgin materials. However, the approaches we describe here all face technical, logistical, or economic barriers to large-scale adoption. Adopting them widely, rapidly, or both will therefore benefit from, and may require, policy initiatives.

Fuel-Switching

The savings from energy efficiency and integrative design by 2050 move industry off oil (except very small amounts not yet replaced due to incomplete stock turnover) and very largely off coal, and they also save 8% of industrial natural gas. Fuel-switching seems like an obvious way to defossilize industry further. Industry routinely switches fuels (see Recent History of Fuel-Switching sidebar) in response to relative prices or shortages and seeks to cut costs and hedge risks of all fuels by buying efficiency when that's cheaper.

Ending the Addiction to Cheap Coal

The remaining 0.7 quad of coal in 2050 provides process heat in steelmaking[402] and cement manufacture. Neither actually *requires* coal for heat, which can instead come from natural gas, biogas, biofuels, industrial wastes, solar heat, or electricity, but these processes still use coal because it's their cheapest heat source. Yet as steel complexes in China, India, Korea, and Brazil help drive up prices for prized low-sulfur metallurgical coal, natural gas is starting to replace coal as a heat source. Each ton of steel made with the gas-fired direct-reduced iron (DRI) process uses 30% less energy than traditional coal-fired processes.[403] DRI's economic edge isn't only in

RECENT HISTORY OF FUEL-SWITCHING

The energy crises of the 1970s made oil cost dramatically more than coal or natural gas. This nearly eliminated oil-based fuels from nontransport industrial use and electricity generation, except in the chemical and petrochemical plants and refineries that burn their own hydrocarbon waste products. The 1970 Clean Air Act also left natural gas as the only fossil fuel clean enough for many industrial applications, particularly at smaller plants. Big energy users like steel mills, cement plants, copper smelters, and pulp-and-paper mills clung to coal thanks to grandfathered facilities and remote locations, or they shut down and production went elsewhere. Industry was fortunate that, at the time of this transition, natural gas in North America was essentially a by-product of crude oil production and hence relatively cheap, so the only material cost of switching was building gas pipelines and retrofitting burners.

The electricity industry took longer to shift to natural gas, leaving more for industry, because the 1978 Powerplant and Industrial Fuel Use Act forbade new gas-fired power plants. That reinforced utilities' natural slowness, as regulated monopolies, in switching from coal and nuclear plants to gas-fired CHP or combined-cycle plants as they became superior. But as the legal bar to gas-fired generation was lifted in 1987 and utilities' transition gathered speed in the 1990s, natural gas prices began to rise, slowing heavy industries' shift to natural gas. Only the prospect of lower and more stable long-run gas prices, underpinned by major shale-gas production, began around 2010 to make gas again an attractive long-term utility and industrial fuel.

awash-in-cheap-gas Qatar: bellwether steelmaker Nucor[404] recently announced it'll build the first U.S. large-scale DRI project near New Orleans to burn bulk gas from the Gulf of Mexico. That this new gas-fired steel plant is expected to compete with old coal-fired ones augurs well for this fuel switch to spread. Meanwhile, long-term U.S. natural-gas prices are moderating, while long-delayed enforcement of 1970s environmental laws is pushing up coal prices.

Switching from coal to gas for process heat is no longer unusual: even in coal-rich South Africa, Mondi replaced coal boilers with gas-turbine cogenerators at a pulp-and-paper plant. Gas-fired boilers are cheaper to build and often more efficient and reliable than coal-fired boilers. The last significant holdout is old coal-fired cement plants, whose operating costs are almost one-third energy[405]—90% for heat[406] to keep the kiln around 2,500°F. Using EIA's fuel-price projections, switching U.S. cement plants from coal to gas would double fuel costs by the middle of the next decade, raising cement costs by an unthinkable 30%. But in the past decade, top cement makers like Holcim have displaced 12% of their global coal burn with alternative fuels discarded from other industries, from old tires to waste solvents, because those all beat coal's cost and manage its regulatory and reputational risks. The cement industry presents a unique industrial-ecology opportunity: the kiln's flexibility permits a range of fuels, including hard-to-dispose-of industrial wastes. In fact, one plant in Germany currently uses waste to supply 75% of its fuel, and it is applying for a permit to use 100%.[407] The gap between Holcim's average of 12% alternative fuels and getting to 100% is the availability of wastes. Regulations that support a waste-management loading order to prioritize the highest-value use of wastes can help divert waste from landfills to cement kilns. Combining municipal-solid-waste incinerators (now popping up all over China) with cement kilns solves the further problem of dioxin emissions, since the cement captures any chlorine from waste plastics. Such an integrated plant can also cogenerate electricity.[408]

Transitioning from Natural Gas

Natural gas too will be worth displacing as its prices rise, especially if carbon is ultimately priced. Two substitutes are obvious: substituting heat from increasingly renewable electricity, directly or via heat pumps, and solar process heat.

Direct Use of Electricity for Heating

In principle, electricity, being a uniquely high-quality form of energy, can displace heat at any temperature. In practice, this would normally cost several times more than fuel combustion because conversion losses and the high capital intensity of power plants and grids make electricity costlier per BTU than fossil fuels. Nonetheless, electricity is advantageously used for heat in an expanding range of niche applications because it can deliver heat precisely, intensely, and instantly without chemical contamination. The semiconductor industry, for instance, uses almost entirely electric heating to melt and cast silicon and for processes done in clean rooms. But electric heating is also moving into major uses long dominated by direct fuels.

The aluminum industry uses giant reverberatory furnaces to melt aluminum already made from alumina. These furnaces are often less than 30% efficient—as you'd expect, since the process is akin to trying to boil water by holding a blowtorch flame above the surface so some heat radiates downward. But now it's possible to immerse right in the liquid aluminum special electric heaters protected from corrosion by thin ceramic coatings. This process, called isothermal melting, boosts efficiency to 97%. It also yields higher-quality metal, sixfold less product loss to oxidation, 80% less air pollution, 80% less floorspace (making it retrofittable), and a potential energy saving around 50–75%, or 19 trillion BTU a year with 60% U.S. adoption.[409] The economic case is compelling. Even though electricity costs four to seven times as much as coal or gas per BTU (at 2010 prices), a typical foundry would save $1 million a year because of the efficiency gains, sweetened by the important non-energy benefits.

All- or mostly electric industrial facilities would spew far less or no air pollution from their own combustion stacks, dump less heat into sensitive ecosystems, and produce higher-quality products with fewer resources. Add these good-neighbor features to electricity's extra precision and control, and visionary industrialists with challenging process requirements may switch to electrified heat not despite electricity's cost but because of its value.

Heat Pumps

In more-common processes that simply need a lot of heat, electricity can often be made more competitive with burning fossil fuels by using electric heat pumps that recycle and "amplify" waste heat or ambient warmth already in the air, water, or soil. Small heat pumps already heat and cool houses and run refrigerators, moving heat from one place to another and concentrating it. Larger heat pumps in industry can also triple, perhaps more, the effectiveness of electric process heating. This saves fuel, can save capital cost and air pollution, and reduces the cost of removing unwanted heat. Today, industrial heat pumps are used in specific processes and plants where electricity is cheap relative to fuel and where waste heat is available from the production process (see Mechanical Vapor Recompression Slashes Evaporation Energy Use sidebar).

In most ordinary uses today, making an extra BTU from an electric heat pump costs about the same as recovering that BTU in a gas-fired industrial cogeneration plant. Making a BTU worth of steam by burning coal is about half the cost of either, but that margin will shrink as heat pumps improve or as coal becomes (or is made) costlier relative to gas. Coal's advantage over gas has already disappeared in U.S. electricity production (see chapter 5) and is shrinking in its few remaining uses in heavy industry. Pricing carbon, or other rules about carbon emissions, would shrink it further.

Solar Process Heat

Concentrating solar thermal systems have received a lot of press for their potential to generate electricity from solar steam. But another, less heralded application of this technology is nearing cost-competitiveness[410] today: solar thermal systems for industrial process heat. Replacing the 20- to 30%-efficient electricity conversion step with a 95%-efficient heat exchanger to provide direct

MECHANICAL VAPOR RECOMPRESSION SLASHES EVAPORATION ENERGY USE BY 89%

Making tomato paste means concentrating the tomatoes from 5% to 33% solids—normally by boiling off the water. In California's Central Valley, this single step burns over 10 trillion BTUs of natural gas a year to process 12 million tons of tomatoes (one-third of the world's supply).[411] Much ends up as low-quality steam, which a heat pump called a mechanical vapor recompression (MVR) system can upgrade for reuse. For example, paste producer Ingomar added MVR during a facility improvement. Recompressing the waste steam made it hot enough to heat the tomatoes. In effect, adding just 50 BTUs per pound of evaporated water could evaporate another pound. This saved 89% of the energy used by a more complex alternative that would have had to use even more energy to cool the waste steam before discharging it to the environment.

process heat rather than electricity delivers more than three times the energy for similar capital costs. The market is currently in its infancy, with only a handful of systems installed in the U.S. As capital costs fall and experience grows while natural-gas prices tend to drift up, solar heat's competitive range will expand—a migration opportunity for concentrating solar power (CSP) firms as ever-cheaper photovoltaics encroach on their market. In fact, accurately accounting for the volatility of natural-gas prices[412] adds $2/million BTU to the apparent price of the gas (just over the next five years) when comparing it fairly with solar heat, which has no fuel and hence stable prices once installed. That puts some emerging solar steam generators in the money right now.

Frito-Lay is already using the benefits of solar process heat. In 2008 the company installed a 2.4-thermal-MW system to provide 300 psi steam to the SunChips process line at its Modesto, California, facility. The system saves Frito-Lay the need to generate 15 billion BTU/y[413] of heat from natural gas and allows a new slogan: "SunChips are now made from the sun."

Despite its benefits for firms like Frito-Lay and for enhanced oil recovery, solar thermal will not be able to meet all industries' process heat needs. Round-the-clock operations would need to draw down stored solar heat—CSP plants normally invest in many hours of high-temperature heat storage—or, absent storage, to burn fuel at night, saving only daytime gas. Solar collectors also take space: Frito-Lay's collectors occupy over an acre. Some systems could go on factory roofs, mitigating space limitations but adding structural costs. And although some technologies don't require as much insolation as concentrating versions, all solar process heat approaches show better economics in sunny climates. Despite these limitations, solar thermal's avoided emissions and stabilization of heat prices will be attractive to industry. As a conservatism reflecting the high sensitivity of solar process heat adoption rates to the cost of capital and to future natural gas prices, our analysis assumes *no* solar process heat by 2050—not even its logical next step, solar cogeneration of process heat and electricity. But both look likely, and both will further reduce our natural-gas needs in the decades ahead.

Breakthrough Approaches

Over the medium term, fuel-switching can be accompanied or replaced by breakthrough approaches to reducing underlying demand for industrial production. Here are two examples.

Process Redesign

Roughly 12% of industrial energy is used to dry textiles, car paint, and many other materials, using huge gas-fired or infrared ovens.[414] But is it really necessary to heat up the entire autobody just to dry the paint? There's a far more efficient way, using electron beams targeted at and heating only the solvent in the paint. A standard automotive curing oven uses three to five million BTU/h. But the electron-beam curing units that can replace it need only 0.04–0.08 million BTU/h—a staggering 98% less. True, the electron-beam technology costs more than the oven, but the payback from energy savings can be quick.

Similarly, electron beams can sterilize food or drink containers or medical products, eliminating the standard heat, chemicals, and rinse water. And the technology is advancing rapidly. An electron-beam unit, loaded with big vacuum pumps and high-voltage power supplies, used to be the size of a school bus. Now its rack of power electronics and miniaturized emitter, designed to fit into existing process equipment to deliver heat exactly where required, would fit easily in the school bus's gas tank.[415]

Or consider mixing, which uses several percent of industrial electricity. Inside innumerable chemical reactor vessels, paddles spin to mix reactants together. Gooey reactants are mixed inside pipes full of deliberate obstructions to cause turbulent flow. Both methods use lots of energy. An Australian team invented a fat, unobstructed, slotted pipe that rotates within a smooth one. This "rotated arc mixer" causes a self-kneading motion that mixes as well or better with 80–96% less energy[416] and can homogenize temperature with 60–80% less heating.[417]

Since most industries are based on heat, heat exchangers that transfer heat from hotter to colder flows are ubiquitous. Simply dimpling the exchangers' metal tubes creates little vortices that in one refinery test boosted heat flow 50–60% while reducing pressure drop (and hence pumping energy) by 30–40%.[418]

Though lighting uses only 4% of industrial electricity, it's typically inefficient not just in energy but also in labor productivity, because lighting design isn't part of the traditional industrial vocabulary. Visualize a standard automotive assembly line lit by angled banks of eight-foot fluorescent tubes along both sides. Workers see mainly bright lights, shadows, or reflections—not what they're doing. Now hang a white metal or textile reflector above the line and point the lights upward, and the indirect lighting lets workers see properly, improving quality and reducing fatigue. Or in factories or warehouses with traditional high-bay lighting, replace high-intensity discharge lamps—lighting mostly where and when they're not needed—with precisely aimed, self-organizing networks of smart LEDs from firms like Digital Lumens, and you'll see better with up to 90% less lighting energy.

Ever smarter process designs are emerging to heat, beat, and treat materials using less energy (see Redesigned Process sidebar). But why do those medieval things at all? Isn't there a better way to make what we need?

Biomimicry: Glimpses of the Art of the Possible in the Natural World

The late Winnipeg renewables pioneer Ernie Robertson said there are three ways to make limestone into a structural material. One is to cut it into blocks—beautiful but unexciting. Another is to heat it to 2,500°F to make Portland cement—effective but inelegant. A third method is to grind it up and feed it to chickens. Hours later, it reemerges as eggshell stronger than Portland cement. If we were as smart as chickens, we'd have mastered this sophisticated low-temperature technology. And if we were as smart as clams and oysters, we might even perform the same task more slowly at about 40°F, or make that cold seawater into microstructures as impressive as the otter-resistant inner shell of abalone—tougher than the ceramics that form the nose cones of modern missiles.[419]

REDESIGNED PROCESS SAVES 30% OF A POLYETHYLENE PLANT'S ENERGY

Even with conventional equipment, redesigned processes can be big winners. Felipe Tavares, founder of Houston's Intratec Solutions LLC, found that a key reaction in a polyethylene plant used a tiny amount of "deactivator," just 0.02% of the process stream, to stop a key reaction just at the right moment. But the deactivator had to be dissolved in a carcinogenic and toxic solvent in order to work. That solvent later had to be removed by distillation, using 70% of the facility's total energy.[420]

So Tavares devised an elegant solution. He replaced the deactivator and its solvent with an equally efficient and safer chemical that didn't need its own solvent. Instead, it was soluble in the reaction solution. The switch enabled the purification process to be much simpler—and require 35% less steam heat. In fact, the change cut the whole plant's energy use by more than 30%. The savings? In one plant, they added up to three trillion BTU per year.[421]

Life's designs have been honed by 3.8 billion years of evolution and rigorous product testing. Those that failed the testing (probably 99% of all life's designs) got recalled by the Manufacturer. The 1% that survived can teach profound lessons about how things should be made, how they work, and how they fit.

Nature's design genius has already led to bat-inspired ultrasonic canes for the blind, tent flies that collect airborne dew as Namibian desert beetles do, and lotus-leaf-like self-cleaning paint. Palm-size plastic-film disks using adhesiveless gecko-foot technology allow Interface carpet tiles to be moved and laid in rolls, yet each tile can be unpeeled and plucked from the rest. And the tiles are gorgeously patterned like the fractal design of leaves on a forest floor so you can't see their edges. That makes installation easy (needing no color or pattern matching) and replacement less wasteful (changing as little as one tile, not the whole carpet).

Now imagine a time when we make solar cells like leaves, or underwater glue like mussels use. Or make bulletproof fibers as spiders make silk—under life-friendly conditions, in their bellies, from digested flies. Perhaps we could make cellulose fibers with bacterial enzymes, the way hummingbirds spin sugar and sunlight into nests. Maybe we'll get as smart as forests. More and more big companies are starting to understand this and to have a biologist at the design table.

Steering this life-imitating or "biomimetic" design revolution is forester and nature writer Janine Benyus.[422] She lives in the Montana Rockies, observes deeply, and lectures breathtakingly. By reorganizing the biological literature around function, not organism, she's revealing which organism knows how to solve your design problem, how, and who's seeking to emulate it. Janine and her colleagues at the Biomimicry Guild and Biomimicry Institute are starting to help the world of the made work like, and live harmoniously with, the world of the born. Biomimicry and integrative design are the two next big design revolutions. Each is important separately. Together they will transform industry and remake our world.

Elegant Frugality: Dematerialization and Reuse

We've seen so far a tremendous potential for energy efficiency and renewable efficiency innovation. But as with vehicles and buildings, it would be a serious error to think only about technologies. Why are we making so much stuff in the first place? Could lesser production plus smarter use do the same job more cheaply?

So let's go even further downstream, and before we figure out *how* to make something, ask *why* we're making it. Industry makes stuff—a lot of stuff. Global industry now makes four times the tonnage of major engineering materials—metals, plastics, cement—that it made 50 years ago. The amount is forecast to double again within 40 years.[423] That's despite a 26% drop in the amount of materials used per dollar of global GDP during 1980–2007. Many economies have matured, so expenditures have shifted toward low-materials-intensity items like electronics,[424] entertainment, and medical, financial, and recreational services.[425] Yet the take-make-waste juggernaut rumbles forward. What's it all for?

Mostly waste. The U.S. economy extracts, moves, processes, and uses more than 20 times your body weight per person per day (not counting water, unless it's returned too dirty to use).[426] Of that flow from the planet to industry, about 83% is mined, and the rest grown as food and fiber. Of the total, 93% is lost in extraction and manufacturing—in the form of overburden, tailings, scrap, or process losses. Then six-sevenths of the products actually made are discarded after one use or no uses: consumer ephemerals. Only the last 1% of the original extracted material sneaks through into durable goods—and of those, only one-fiftieth gets recycled. In all, therefore, only 0.02% of the originally extracted mass flow returns to nature as compost or to industry as a "technical nutrient" for recycling and remanufacturing. The other 99.98%, much of it toxic, is pure waste. It's hard to find a more wastefully designed system, or a greater business opportunity, on the face of the earth.

Some of the biggest materials savings can be free by-products of improvements we've already discussed. Living, working, shopping, and playing closer together can consume fourfold less lumber, fivefold less copper pipe, 15-fold less asphalt or concrete pavement, and 70-fold less water (pumped and treated with facilities that use industrial energy and are made of concrete and steel).[427] No wonder sprawl raises infrastructure costs and New Urbanist design saves them. Then think of the implications for vehicles: the 46–84% driving reduction we found in chapter 2 could imply (though we didn't assume this) fewer cars, hence fewer tons of materials to make them and their infrastructure.

Next, across the whole range of industrial materials and final products, dematerialization can further reduce materials flows. We can design out waste in mining and manufacturing, recapture resources now lost in extraction and manufacturing, and make products that last longer—then recover, reuse, repair, remanufacture, and recycle them. Systematically designing out waste and toxicity, and radically increasing the efficiency of using resources, can deliver more service per pound of material. In a world where producing stuff is a low-margin commodity business, a wealth of opportunities lies in the sophistication of designs and processes that provide more and better services from less for longer. And a "solutions economy" business model can reward both producers and consumers for that result—as when Dow leases you a dissolving service rather than selling you a solvent.[428]

Every aspect of our daily lives presents similar opportunities for multiplying materials savings by living lighter and smarter and closing materials loops. Consider these examples:

Recycling

Aluminum, glass, steel, plastic, and paper products have already gone through the energy-intensive processes of extraction, separation, and chemical transformation. They can be returned to service instead of being remade from scratch, thereby saving money, energy, and resources while reducing emissions. Aluminum, notably, takes 95% less energy to recycle than to make, yet America still throws away enough to rebuild its commercial aircraft fleet every few months. In Sweden, in contrast, aluminum cans and PET (polyethylene terephthalate) bottles must achieve 90% recycling or be banned.

Indeed, for most materials, recycling in the U.S. is alarmingly weak. It could be raised to the level of Japan—which cut its materials intensity 40% in just 11 years after the 1973 oil shock—or of Austria or Germany, which respectively recycle or compost 70% and 66% of their trash, perhaps the world records.[429] In many European countries and increasingly in Japan, manufacturers by law bear lifetime responsibility for their products and thus have a strong incentive to make them easy to repair, reuse, remanufacture, or recycle. Producer responsibility makes well-designed products not a disposal burden but a lucrative source of value and hence a competitive advantage.

Making steel from scrap is a powerful example of how the steel industry is minimizing its wastes. Recycling the steel in a given structure requires only one-third the energy and emits one-fourth the CO_2 of making it from primary ore. The steel industry already captures 65% of the available steel for recycling. Other industries could follow suit. Recycling electronics can be a veritable gold mine,[430] 60–70 times as rich as gold ore.

Recycling is a change not just in practice but in design mentality. In 1988, University of Zürich professors Hanns Fischer and C. H. Eugster decided to revisit their 1971-vintage elementary chemistry lab course. Every year, the students were turning $8,000 worth of pure, simple reagents into $16,000 of hazmat disposal costs. The course thereby taught once-through, linear, obsolete thinking. So the professors simply redesigned some lab exercises to teach how to turn the toxic wastes *back into* the original pure, simple reagents, saving costs at both ends. Students volunteered weekend and vacation time to recover reagents—far more fun than wasting them—and within three years the demand for wastes to reprocess outstripped the supply. Since then, waste has been just 1% of the original level. Net operating costs fell by $130 per student per year.[431] And those students are in strong demand, because the "cycle thinking" they learned can help save the chemical industry, by designing out the very concept of waste and creating only value.

Remanufacturing

When we recycle, the used material is typically melted down into an intermediate form before being remade into products. This saves lots of energy, but not as much as repair and refurbishment that reuses the product intact, like moving steel girders from a demolished building into a new one. Not melting and reshaping saves energy and money.[432] More broadly, many products, particularly those made from durable metals, are discarded when they're out of style, broken, or superseded. Refrigerators are rarely thrown away because their steel structures have become too worn. So why not remanufacture them with updated parts?

It's already happening. A special daylit factory owned by Herman Miller, the second-biggest U.S. furniture maker, returns to like-new condition every kind of furniture it's ever made. Its larger rival Steelcase vies with private remanufacturers over who gets that profitable business. IBM remakes its computers, Kodak its "disposable" cameras, Xerox its copiers and cartridges (the firm's green-design, zero-to-landfill photocopier is expected to save $1 billion over the long run). The U.S. Department of Defense is the world's biggest remanufacturer: giant Cold War–era B-52 bombers twice the age of their crews are still in service, their half-century-old airframes continually refurbished with new parts and technologies. DuPont's films division, once nearly broke, reestablished itself as a key player partly by collecting more than a billion dollars a year worth of used polyester film from its customers for remanufacturing—then remaking the film ever thinner and stronger, so it's cheaper to make but fetches a higher price. DuPont's clever chemists think they can keep up this less-is-more trick "indefinitely."

Reducing Process Waste

Remember how we need to start our savings downstream? Automakers and metal casters often buy twice the metal, and airframe makers ten to twenty times the metal, that ends up in their products. Designing out that waste saves all the energy- and capital-intensive process steps upstream, all the way back to the mine. Pratt & Whitney used to machine away 90% of its costly ingots shaping turbine blades—then designed out that waste by asking its suppliers to cast those exotic alloys into bladelike shapes. Carbon-fiber structures can now be made with less than 10% scrap (all reusable), and with holes molded in—zero waste—rather than cut out afterward. Smarter shapes and designs can save tens of percentage points of the steel and concrete in a beam; but why use a beam? German structural wizard Michael Schlaich's Stuttgart stadium, and many since, cut structural mass 8- to 10-fold at lower capital cost, with greater strength, just by substituting tension- for compression-based designs.[433]

The Advent of Additive Manufacturing

Another emerging advance, long a science-fiction staple, is now real: desktop manufacturing has arrived a quarter-century after desktop publishing. If you can imagine it, you can digitally draw it on a screen and hit "Print File." Instead of huge clattering factories that subtract most of a big chunk of stuff to leave the shape you want, far smaller devices can now quickly build up the same products layer by paper-thin layer, just as inkjet printing makes words appear on paper. Such additive manufacturing, using only what it needs, can create better products with up to 90% less material.[434]

A mail-order MakerBot the size of a microwave oven costs under a thousand dollars and can "print" in 3-D plastic practically any hand-size or smaller shape in a few minutes. Even that hobbyist version can make in one shot, with no assembly, things unmakeable by traditional methods: a ship in a bottle, or gears within gears, or superefficient nested heat exchangers become child's play. Digital control over every detail can open the door to mass customization. Customers can become co-creators, open-source design swappers, or even competitors.

Industrial "fabbers" starting around $5,000 (but falling fast) can print a large, complex object—a motorcycle, or a full-size airplane piston engine complete with propeller—so lifelike you must look very closely to realize it's all plastic. It can compete with injection molding for runs of about a thousand items and rising. Before long, an intricate plastic part to fix your car may be made on the spot, right at the dealer's service counter, rather than shipped from a faraway factory. The result: quicker and better service, just what you need when you need it, lower cost, often 50–80% less time, and much less energy.

Not stopping with silicone ears and prosthetic limbs, biological printing may soon print your own cells into organs like bladders, kidneys, even hearts. *The Economist* notes that a basic 3-D fabber "now costs less than a laser printer did in 1985" and, by undermining economies of production scale, "may have as profound an impact on the world as the coming of the factory did. . . . Just as nobody could have predicted the impact of the steam engine in 1750—or the printing press in 1450, or the transistor in 1950—it is impossible to foresee the long-term impact of 3D printing. But the technology is coming, and it is likely to disrupt every field it touches."[435]

Ultimately additive manufacturing may scale down to the atomic scale of "molecular assemblers."[436] We know that works, because it's how life turns light into leaves and mother's milk into babies. In time, our manufacturing will go there too, because whatever exists is possible. Markets strive to wring out waste. And waste is a terrible thing to waste.

TRANSFORMING THE INDUSTRIAL JUNGLE

McKinsey & Company's study of cost-effective energy efficiency opportunities shows that U.S. industries could cost-effectively save five quads per year right now.[437] So why haven't they? If the opportunities for efficiency and savings are so compelling, why do many profit-driven companies fail to use their energy in a way that saves money?

Barriers to Realizing Radical Efficiency

The first and most important barrier is that many companies have limited management capacity for, and expertise in, energy efficiency (which is often not seen as a core competency). This shouldn't be a surprise, because energy prices were so cheap for so long that energy efficiency was often the last thing CEOs needed to worry about. As one CEO of a Fortune 100 company said in the mid-1990s: "I can't really get excited about energy—it's only a few percent of my cost of doing business." That dismissive CEO had overlooked one simple calculation: that if an energy manager who'd just cut one of his factories' annual energy costs by $3.50 per square foot had duplicated that success companywide—90-odd million square feet—his whole corporation's net earnings that year would have been 56% higher. When this was pointed out, that engineer was quickly promoted and spread his best practices across the firm. But for many CEOs, that weighty penny hasn't yet dropped.

This lack of internal focus and expertise reduces the number of good ideas entering the project, design, and decision-making processes—and the likelihood that anyone will champion those ideas. And often it translates into inattentiveness. A famous company that hadn't needed steam for years still ran a big boiler plant, with round-the-clock licensed operators, simply to heat distribution pipes (many uninsulated and leaking) lest they fail from thermal cycling; nobody had gotten around to shutting down the old system that would never again be needed. And often energy efficiency simply signals neglect of continuous improvement. Many firms keep building plants the same old linear way (require, design, build, repeat) rather than in a learning loop—require, design, build, *measure, analyze, improve*, repeat—so they never improve.

Second, companies have limited capital. In the scramble for capital budgets, investments needed to maintain product quality and license to operate (for example, safety, environmental, and labor compliance) are made first. Then most firms divide the capital between departments, so the highest-return investments in the whole company can't compete across divisional boundaries; they'd overrun their vice president's capital budget while some other departments would grumble about getting none.[438] Only then are optional projects within each department executed in order of rate of return until the available capital runs out. The projects that usually win the race for investment dollars are those that boost productive output, for which efficiency is seldom properly credited at all, and pay back quickly—typically within months, at most a year or two. Few firms nowadays even exhaust that list, but if they did, they'd still leave many profitable but longer-payback energy efficiency projects undone, cheating the shareholders. Nor will most CFOs borrow more money on which they could earn a handsome return from efficiency, lest they violate financial ratios chosen by Wall Street sector analysts. And if energy managers can't even get their long-analyzed winning projects funded, why waste time thinking up more?

Also, most energy-intensive U.S. industries are mature and growing slowly, leaving few opportunities for efficient new facilities, so even when solutions and skills are available, there are few times when decisions to invest can be made. In boom-and-bust industries like semiconductors, there's almost no time to design the next plant right: designers are either too busy rushing to complete the next plant or so spooked by a downturn that they can't see it as a golden opportunity to design to win in the next upturn.

The third barrier is that the real rates of return required by companies to justify investments in aging plants are surprisingly high. The industrial world is volatile, and no one can really predict how a long-term investment will play out. A plant may be shut down by a new owner, or a new source of energy may be developed, such as shale gas, that drastically reduces prices, or there might be concern that future managers may not operate and maintain unfamiliar systems properly. It's not surprising that industrial managers may demand quick paybacks for investments in energy efficiency. Industrial firms are also under constant pressure from investors and shareholders to maximize returns on investment. Those investors want faster returns and lower risks than many of the emerging, hence still low-volume, energy efficiency techniques can initially provide.

That's why there's a crucial role for government and industrial consortia to help develop and demonstrate energy efficiency technologies, reducing their risks and lowering their costs. These groups can also spread the gospel of energy efficiency—and help identify cost-effective solutions—to firms that haven't time do it themselves. For example, the U.S. Department of Energy's Industrial Technologies Program has so far saved its participants a cumulative 9.3 quadrillion BTU of energy (about the annual use of all of California), worth $65 billion; through 2008 it had supported 118 established and 104 new efficiency technologies and was about to commercialize another 135.[439]

Getting It Done

So, clearly, the implementation of energy efficiency is not easy. But it's also not magic, and the prize is great. Industry would gain upwards of a half-trillion dollars in 2010 net present value, with savings 2.5 times the cost of being energy efficient, plus even more valuable direct and indirect gains in quality, throughput, other non-energy benefits, and of course global competitiveness. Furthermore, good energy management is key to good management, period. Energy management projects not only yield energy and cost savings, but they also lead to better understanding of processes and stimulate innovation on a broad front, pointing the way to higher yields at lower costs.

Big, cost-effective energy savings take not just leadership and skill but also unswerving persistence. But some managers have come up with innovative ways to drive progress at their companies. Energy efficiency expert Ken Nelson devised a contest to motivate employees to find savings opportunities, eventually savings hundreds of millions of dollars for his employer, Dow USA (see A Shop-Floor Competition with Big and Rising

A SHOP-FLOOR COMPETITION WITH BIG AND RISING PAYOFFS

Ken Nelson is an astute, congenial, low-key engineer who formerly led energy efficiency for Dow USA. For twelve years, Nelson organized a contest among the 2,400 workers of the company's Louisiana division, never above supervisor level. The contest sought to elicit suggested projects that save energy or reduce waste, that can pay for themselves within one year, and (initially) that cost under $200,000. After 936 approved and nearly 900 implemented projects, shown by audit to have matched actual average savings very accurately, the bottom-line benefit was stunning: $110 million a year saved, with 204% annual returns (triple-digit in all but one year), and both returns and savings heading *up*. It's as if each $100,000 bill they picked up exposed a couple of new ones underneath.

The contest's first year resulted in 27 projects costing a total of $1.7 million and an average return on investment of 173%. Many thought they'd exhausted the easy opportunities. They were wrong. The next year, 32 projects costing a total of $2.2 million averaged a 340% return on investment. Learning quickly, Nelson changed the rules to eliminate the initial $200,000 project limit—with such lucrative opportunities, why stick to the small ones?—and to include projects that would boost output. In 1989, 64 projects costing $7.5 million saved $37 million in the first year and every year thereafter, for a 470% ROI (the best so far). Even in 1992—the 10th year of the contest, nearly 700 projects later—the 109 winning projects averaged a 305% ROI, and in 1993, 140 projects averaged a 298% ROI. The engineers, as usual, learned new tricks faster than they used up the old ones.

This manna from heaven was being picked up by ordinary workers, for no special reward except their peers' recognition, and not because of the CEO's intervention but because the CEO didn't know about it and therefore couldn't get in the way. Though meticulously measured and documented, Nelson's additions to Dow's bottom line didn't come from fancy management theories, quality circles, empowerment processes, committees, or other managerial preoccupations. Rather, they came from a practical shop-floor process that translated volunteer ingenuity into saved money and continuous learning.[440]

Payoffs sidebar). But different cultures need different tactics.

Anita Burke, for instance, led a Texaco refinery from the nation's worst to its best environmental performer. One day her boss refused the capital for a vastly profitable project at filling stations because the benefits would accrue to a different account, so she lit a $20 bill on fire in front of him. "Put that out!" he yelled. "You're burning money!" She sweetly explained that his recalcitrance was burning far more money. She got her funding.

Or on a smaller scale, another efficiency pioneer is Paul Rak, president of Canadian steel fabricator VeriForm Inc. Rak spent $46,000 in 2006–2008 to improve his plant's lighting, automate its heating, and run its tools more efficiently. His 58% electricity and 90% natural-gas savings cut about $90,000 off the energy bills, nearly a 200% annual return on investment. Some paybacks were as short as a week and a half—and he's just getting started.[441]

What do pioneers like Ken Nelson, Anita Burke, and Paul Rak have in common? Guts, creativity, and perseverance. But making a real dent in industry's vast energy use will require building on these successes and making these stories commonplace. To do so, we recommend that leaders throughout industry take action on several fronts. First and foremost, educate and inspire your workforce. A broad grassroots recognition and attention to energy often has the most powerful leverage in cracking a diffuse problem like energy management. People on the shop floor usually have good ideas on how to save energy in their plant, and if they know they'll be recognized (perhaps even rewarded) for their initiative rather than punished for changing the status quo, the efficiency bug can go viral. This takes an energy management structure of appropriate size, rewarding what you want, and creating a culture of empowerment and discovery.

Second, business leaders must address energy efficiency hurdle rates that are often prohibitively high, let (indeed, make) projects compete across departments, and look beyond quarterly profits. Industrial firms that plan to be in business for the long term should adjust short-term goals and hurdle rates to get there. This means consistently providing project capital or financing in recognition of the long-term benefits, both energy and non-energy, so an efficiency culture like Dow's can become as durably embedded as safety and quality.

Last, critical is the need for business leaders to work in tandem with local utilities. Utilities must have their profits decoupled from energy sales and enhanced by a small share of what they help save to enable them to embrace and speed their customers' efficiency (see chapter 5). Utilities must not burden industry with prohibitive interconnection rules and costs for CHP or other techniques that generate onsite electricity more efficiently than the utility's electricity-only plants. Business leaders must look inward for energy efficiency improvements, but in the case of CHP, they will need to work with utilities to create a framework for success.

CONCLUSION: COMPETITIVENESS THROUGH RADICAL ENERGY PRODUCTIVITY

Between opportunities already technologically, logistically, and economically ripe today and others that can become so over the next 40 years, we have more than enough existing and emerging technology to move industry away from, and ultimately beyond, fossil fuels. Achieving this will require a different mindset, closer to the driving principles—and the long view—of a rainforest than of a chainsaw operator.

The key to survival for both rainforests and industrial firms is resilience, the ability to survive in hard years and flourish in good ones, learning from stress to become ever more adaptive. And a crucial part of resilience is hardy frugality. For industry, efficiency is key to surviving recessions, and the most efficient firms return to growth sooner, often with less competition for customers and for scarce resources. Trees don't produce byproducts that could damage their ecosystem but, on the contrary, support and are sustained by the rich biodiversity around them. These are lessons industry would do well to learn if we hope to create a system that will last—bypassing the several hundred million years of trial and error endured by the giant trees of eons past.

Some evolutionary pressures that would make industry more robust and resilient, in greater harmony with the ecosystem whose health undergirds its own survival, are obvious. They include desubsidizing fuel; mandating producer lifecycle responsibility; allowing and encouraging waste-heat recovery and reuse, including all forms of cogeneration; removing distortions that favor virgin over recycled materials; letting businesses expense energy-saving investments against taxable income (just as they now expense wasted energy) rather than having to capitalize them; and properly pricing the commons into which things get thrown "away," whether gunk in our water, junk in our landfills, soot in our lungs, or carbon in our air.

Such deliberate, across-the-board stimuli to innovation and adoption have long driven the competitive prowess of countries like Germany and Japan. Their high energy prices and strict environmental rules honed their efficiency, helped cut their healthcare costs and resource dependence, enhanced their economies' transparency and choice, and sped their broad adoption of energy efficiency. Yet efficiency's adoption has lagged in America, held back by deliberately false price signals, subsidies, and other lopsided policies that handicap the whole economy. An America hooked on artificially cheap fuel has trouble competing with countries whose more truthful fuel prices have driven bigger gains in wringing far more work from it.

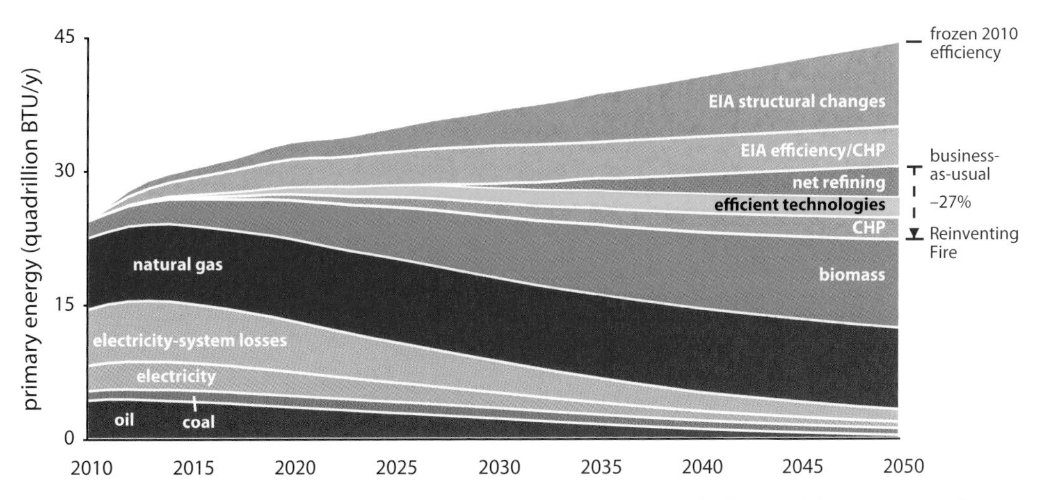

FIG. 4-12. Efficiency gains will halve industry's primary energy consumption, while the remaining energy requirements will shift away from fossil fuels. The limited integrative-design savings considered could raise the 27% savings to 30%.[442]

Let us be clear. Doubling industry's 2050 energy productivity will take strong commitment and sustained attention. It will take measurement, curiosity, intelligent risk-taking, aligned incentives, clear communications, persistence, courage, diplomacy, a little cunning, and perhaps a dash of theater and charisma. It will take pulling vigorously the four levers we've discussed, exploiting integrative design, and perhaps even starting to wring more work from less stuff.

And even after that, companies will most likely require a helping hand from new policies to move entirely away from fossil fuels. Coal, oil, and natural gas still look relatively cheap as long as they can impose social costs excluded from their price. Even more devastating, persistent uncertainty about future rules and prices is slowing all investment decisions by industry. And speeding up investments means changing the rules.

The overall goal is clear and achievable. We can radically reduce the amount of energy industry uses and get companies off of coal, oil, and ultimately natural gas.

The industrial efficiency improvements and fuel-switching we recommend are substantial. And they're not free. We calculate that achieving this potential will require industry to invest about $283 billion, present-valued to 2010. That's no easy feat for a sector that counts pennies and makes capital investments only when needed. But for those firms willing to bet on the future, the payoff is phenomenal. Just the technologies known to be cost-effective, earning at least a 12%/y real return, would yield more than $665 billion in present value (fig. 4-13). If fossil fuels become more

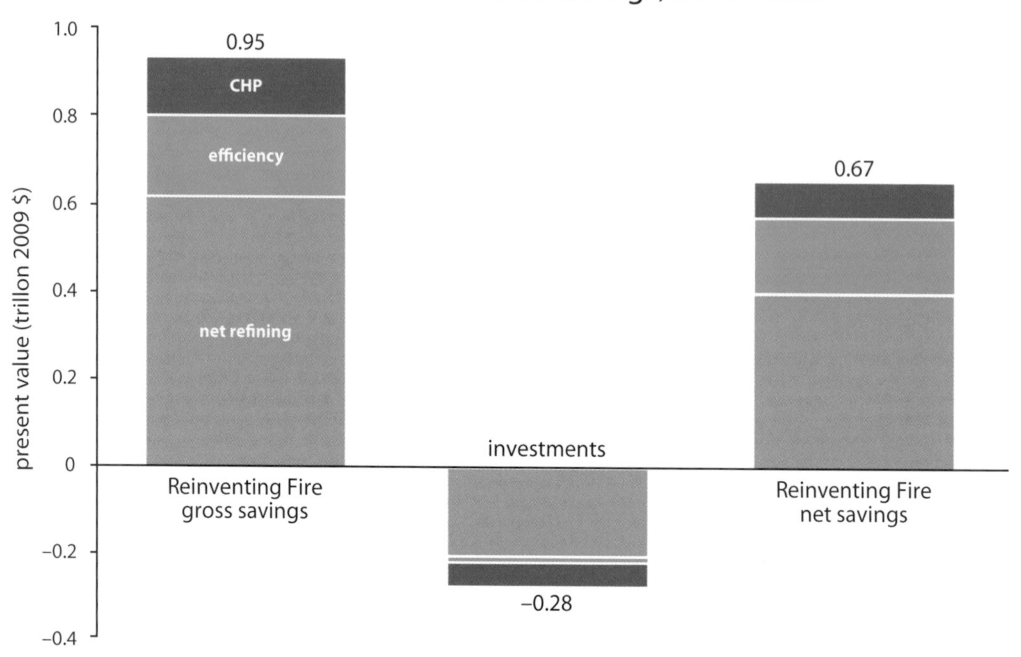

FIG. 4-13. Investment requirements and savings potential of energy efficiency in U.S. industry.[443] These savings come just from the four conventional levers and Lawrence Berkeley National Laboratory's technology assessments, not from integrative design.

expensive than the government forecasts, that return would rise.

This path to startling efficiency starts with the three concepts introduced at the beginning of this chapter: investment, innovation, and incentives.

Invest in factories and in the people on the factory floor. To exploit the four levers to the full, companies will need to upgrade facilities to highly efficient designs and processes—and hence to train and empower the legions of people who are ready to get to work saving energy and making money. From short-term plays (learning to quantify energy use in a plant or seeking out the latest efficiency technologies) to longer-term investments (funding R&D to commercialize new processes, or building and equipping an energy management structure to find ever more energy inefficiencies, as Ken Nelson did), there's a multitude of ways to invest money to save money. In fact, many of these opportunities may not cost much up front at all. Especially when harnessing integrative design, these steps in the journey may cost little more than one's time and the commitment to seeing waste as a nearly endless fount of profit.

Innovate to guarantee long-term competitiveness. Businesses are renowned for their ability to develop new markets. They are probably the most flexible institutions in the world, and their ability to innovate must be treasured, reinforced, and leveraged. Innovation arises when conditions make it necessary and nurture its development. With the coming of a new energy era, businesses will be forced to innovate in order to remain competitive. For instance, with the right spread in fuel costs, industry will be compelled to meet a portion of its process heat needs via fuel-switching, such as through electrification and solar process heat. As raw materials become scarcer or pricier, entrepreneurs will reduce waste sent to landfill with new business models that reduce waste and provide services instead of products. With sufficient support in R&D funding (from universities, governments, and firms themselves), businesses will find ways to scale breakthrough processes that will support U.S. industries' continued productivity and competitiveness. There's a lot at stake as business leaders make the choices that can permit long-term profitability and a robust future.

Incentivize the right behavior. Innovation doesn't occur in a vaccum. To decouple industry's successes from continued reliance on dirty fossil fuels, governments must allow and encourage full and fair competition between fossil fuels, competing supplies (such as renewables and electrification), and more productive energy use—all fully and honestly priced. Getting the rules right doesn't replace the need for managerial and cultural reforms, but it can help accelerate, reinforce, and sustain them. Recognizing the double-edged sword of excessive regulation and unintended consequences, we do not recommend any set of policies as the one best solution. However, uncertainty about future policies inhibits investments in tomorrow's industries, and this must stop. Industry-sector specific measures could include cradle-to-cradle product responsibility requirements, higher landfill costs, expensing of efficiency investments, R&D funds, demonstration projects or first-user funds for risk reduction, a fuel-price floor, or the sort of economy-wide carbon pricing that already hones rivals' competitive edge. Whatever the policies or rules, they need transparent, consistent, and long-term decisions that industry can plan for over decades. Only then will they unlock pent-up investment and renew the race for businesses to unearth hidden treasure.

One pioneer on this path, the late Ray C. Anderson, founder of Interface, had a "spear in the chest" epiphany in 1994. His firm had been rather wastefully turning oil into five billion pounds of carpet that, after brief use, went to landfills to rot slowly for millennia. Anderson—a brilliant

Georgia Tech–trained engineer, entrepreneur, ex-quarterback, and extraordinary business leader—set out to fix that, and to reconcile what he did at work with what he taught in Sunday school about stewarding Creation. Sixteen years later, more than 100,000 tons of carpet had been diverted from landfills for remanufacturing, recycling, and energy recovery. Eight of Interface's nine plants ran on 100% renewable electricity, and 40% of input raw materials was biobased or recycled. Energy use per unit of product was down 43%. Absolute greenhouse gas emissions were reduced 35%, waste to landfill slashed 82%, water use 82% dried up. Waste costs were down 42%, saving $438 million. The company's Mission Zero was well on its way to its visionary 2020 goals of "taking nothing, wasting nothing, and doing no harm—and doing very well by doing good, at the expense not of the Earth but of less alert competitors."[444]

Anderson figured that these steps have not only saved the company major energy costs but also helped gain and retain market share. And from its factories to its trucks to its installers, Interface enjoys the greatest independence from oil and its volatile prices of any company in its industry.[445]

As at Interface, the transformation of the U.S. industry sector begins with vision and continues with hard work, from redesigning and rethinking fundamental processes to getting the rules right, so we can all save energy, multiply wealth, and protect our habitat. We have nothing to lose but our waste.

TABLE 4-1. Summary of recommendations for key actors in the industry sector

	NO-REGRETS	OPPORTUNISTIC	INNOVATIVE
CUSTOMERS	Design out, reduce, reuse, remanufacture, recycle. Ask suppliers for lifecycle analyses.	Test service business models rather than buying products. Lean-engineer your use of products to make sure you're buying only what you really need.	Inform purchasing decisions by determining fossil-fuel impact. Switch from product purchases to service leases.
PLANT OWNERS AND MANAGERS	Measure energy use of plant/process. Optimize current plant/process performance and implement "lean thinking" throughout. Research energy-efficient technologies/practices. Test lifecycle analyses. Train employees in energy management and make it part of everyone's normal responsibility.	Put in place a continuously learning energy management structure. Reflect energy savings in reward systems. Develop efficiency and renewables intellectual capital. Cut hurdle rates for efficiency projects (to marginal cost of capital or less) and allow them to compete across the company, not just in each department, so the full profitable opportunity is actively financed.	Fund R&D to commercialize efficiency and renewable breakthroughs (in both processes and technology). Execute long-term energy strategy in existing and new facilities. Systematically apply integrative and biomimetic design companywide. Switch to service business models rather than product sales/purchases.

	NO-REGRETS	OPPORTUNISTIC	INNOVATIVE
PLANT OWNERS AND MANAGERS (CONT.)	Invest in and empower efficiency staffs. Market low energy use and fossil-fuel phaseout with customers. Educate customers on efficiency opportunities. Benchmark against competitors' energy efficiency and fossil-fuel independence.	Partner with product suppliers on efficiency innovations. Test service business models rather than buying or selling products. Experiment with integrative and biomimetic design. Anticipate carbon regulation/pricing. Beat competitors' benchmarks.	Partner with governments and utilities to optimize rules for efficiency, CHP, and renewables and fully capture demand response. Forget about competitors' benchmarks—forge your own trail that leaves them far behind.
GOVERNMENT AND NGOS	Educate and train for competitiveness. Remove utility barriers to efficiency, CHP, waste-heat recovery and resale, and on-site generation. Fund R&D to commercialize promising technology and enhance basic science.	Provide carbon-policy clarity to reduce investment uncertainty/risk. Require product life-cycle responsibility from manufacturers. Allow tax and accounting treatment of energy-saving investments as expenses rather than capitalized assets. Allow utilities to invest in and support efficiency and distributed generation and to earn proper rewards.	Internalize all costs in fuel prices. Eliminate fossil-fuel subsidies; ultimately desubsidize all energy. Extend renewable energy support to waste-heat recovery and solar process heat.

CHAPTER 5

ELECTRICITY: REPOWERING PROSPERITY

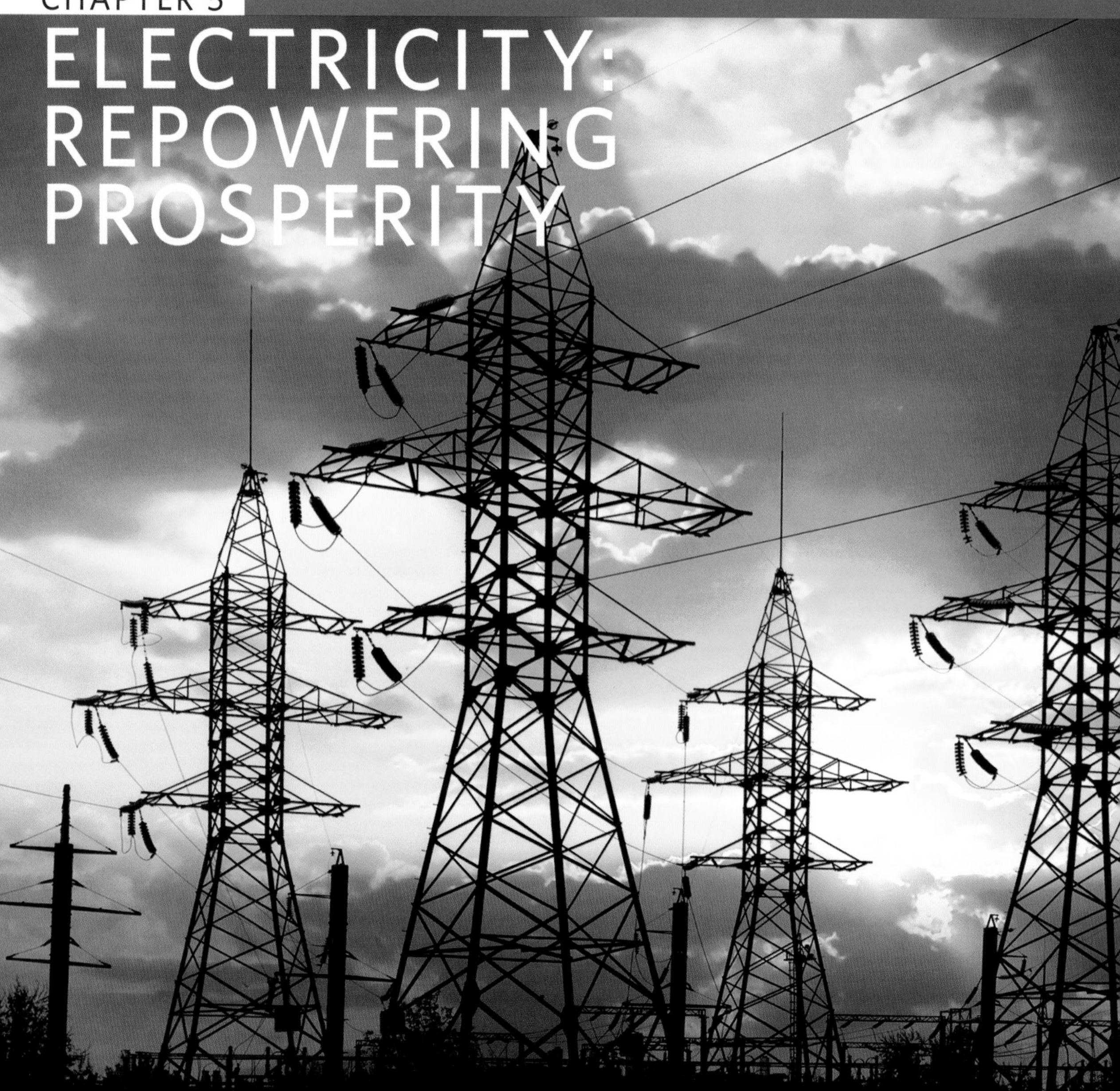

FIG. 5-1.

CHAPTER HIGHLIGHTS

→ **THE GOAL.** Eliminate fossil fuels cost-effectively from the U.S. electricity system. To do so, produce at least 80% of U.S. electricity from renewable resources by 2050, and more thereafter—perhaps ultimately all—while improving security, reliability, resilience, and public health.

→ **THE BUSINESS OPPORTUNITY.** Renewable energy is increasingly cost-competitive with conventional generation. It also manages financial risks, avoids fossil-fuel price volatility, and mitigates pollution. Renewables plus modular, local generation are a growing market open to many competitors, not just electricity providers.

→ **THE BOTTOM LINE.** Initial capital costs would be higher than business-as-usual, but fuel savings offset this (or more), and the investment attractively reduces or avoids important risks.

→ **BUSINESS SECTORS THAT CAN PROFIT.** Electric utilities, IT providers, resource aggregators, electricity generation and management technology suppliers, and other innovators can all find profitable business opportunities. Customers will also benefit from better-managed costs and risks and wider choices.

→ **POLICY ENABLERS.** Regulatory changes to the electric utility business model, aligning incentives to support energy efficiency and responsive demand, regional cooperation on system operations and planning, and incentives to invest in new technologies.

INTRODUCTION

You never change things by fighting existing reality. To change something, build a new model that makes the existing model obsolete.

—Buckminster Fuller (1895–1983)

Just south of the Wasatch Mountains in Utah stands a colony of aspen trees considered to be the oldest and largest living organism on the planet. It has thrived there for more than 80,000 years, since before humans walked out of Africa. The 47,000 trees in the colony, all sharing the same DNA, are sustained and united by a single, massive, and highly interconnected underground root system capable of transporting water and nutrients from one part of the colony to another.[446] Roots near abundant water supplies deliver water to trees in drier areas, while those with access to critical nutrients elsewhere return the favor. This underground network has helped the Utah colony to survive millennia of droughts and floods, heat waves and cold spells, and a changing landscape of resource availability.

Today, electricity—along with the digital information and communications systems it enables and requires—provides the vital root system that sustains our economy. Electricity has become the connective tissue of the Information Age. Virtually every transaction in our daily lives is now mediated, in real time, by electronic information. As recently as the year 2000, 75% of all the information stored by human societies worldwide was in analog formats such as paper documents, pictures, books, tapes, and X-ray films. By 2007, 94% of all stored information was in digital electronic form.[447]

Electricity not only animates the storehouses of human knowledge; it enables communications and control, and it delivers energy precisely to billions of devices throughout the economy. It is amazingly versatile: electricity is an energy carrier producible from virtually any primary source at virtually any distance from its ultimate use, and it can supply the energy for virtually any product or service. It's clean, efficient, precise, and flexible, ensuring that major infrastructure systems including communications, buildings, industry, and even transportation will continue to shift to electricity as an energy supply source of choice.

Underpinning the supply of electricity is a power grid that stretches across the U.S. like a colossal human-built version of that aspen grove, an intricate web of a quarter of a million miles of shimmering high-voltage wires linked to more than 5,000 electricity-generating facilities.[448] It reaches into nearly every dwelling and business in the nation, from remote cabins in the Ozarks and swank Manhattan condos to Tennessee auto factories and San Francisco skyscrapers. It charges our mobile phones, illuminates our offices and baseball stadiums, and powers our aluminum smelters and data centers. It is, as the National Academy of Engineering says, the greatest engineering achievement of the 20th century.[449]

Yet, as crucial and as ubiquitous as it has already become, electricity is poised for a profound leap in importance as the key enabler of the transitions in transportation, buildings, and industry described elsewhere in this book. To reinvent fire across the U.S. economy, our electricity system must accelerate the transition already under way to become renewable, diverse, distributed, resilient, and customer-oriented. This will enhance the physical, operational, and decision-making flexibility needed to thrive in a fast-changing world.

IMAGINING THE NEXT ELECTRICITY SYSTEM

Dramatically increased energy efficiency in the buildings and industry sectors, as discussed in chapters 3 and 4, will keep overall electricity demand flat or declining—even as we electrify vehicles. We can also continue to replace most of our aging fossil-fueled power plants with renewable energy sources throughout the U.S. Extensive modeling of the U.S. electricity system suggests that we can capture and integrate the renewable energy needed to meet 80% or more of our electricity needs by 2050. These renewable sources supply electrons not only from energy-rich areas like the windy Dakotas and the sunny Southwest, but also from electricity generation at diverse scales but closer to customers, reducing the need for new long-distance transmission.

Integrating information technology with electricity enables enhanced grid intelligence and price transparency, making every part of the system cheaper to run and better coordinated. An information-rich electricity system also enormously expands the range of offerings from which traditional and new service providers can assemble new value bundles for every taste and purse. And by enabling smaller, more granular, shorter-lead-time projects, this shift of electricity sources and scale can help utilities and capital markets manage their increasingly worrisome asset-related financial risks.

This more interactive, informed, rapidly evolving electricity system is not centrally planned from the top down. Rather, it can evolve at least equally from the bottom up through radically broadened, deepened, and accelerated innovation and competition. This competition is not confined to the United States: the costs of emerging and rapidly maturing power generation technologies are being driven steeply down by global markets, especially by aggressive production and investment in China—now the world leader in five renewable technologies and aiming to be in all of them.

The electricity sector's transformation will have far-reaching implications for businesses and customers, and for the competitiveness of the U.S. economy. First, the U.S. electric power sector could require investment as great as $3.5 trillion over the next 40 years to replace aging infrastructure.[450] The global market will be many times larger as emerging nations see rapid growth in both GDP and electrification. But *how* the system evolves isn't fated: it can be chosen from a very large and rapidly expanding menu. Countries that lead this transition most adaptively will be able to sell technologies and services around the world, gaining a competitive edge.

HOW WE MODELED OUR ELECTRICITY SCENARIOS

To assess the implications of possible future paths for the U.S. electricity sector, RMI developed and analyzed four scenarios or "cases" based on differing assumptions about how electricity might be generated, delivered, and used from 2010 to 2050 (fig. 5-2). We analyzed the performance of the U.S. electricity system using two models. First, RMI's own electricity dispatch model calculated the costs of meeting hourly electricity demand throughout any particular year, dispatching the lowest-cost reliable mix of resources within the assumed portfolios. Second, we determined

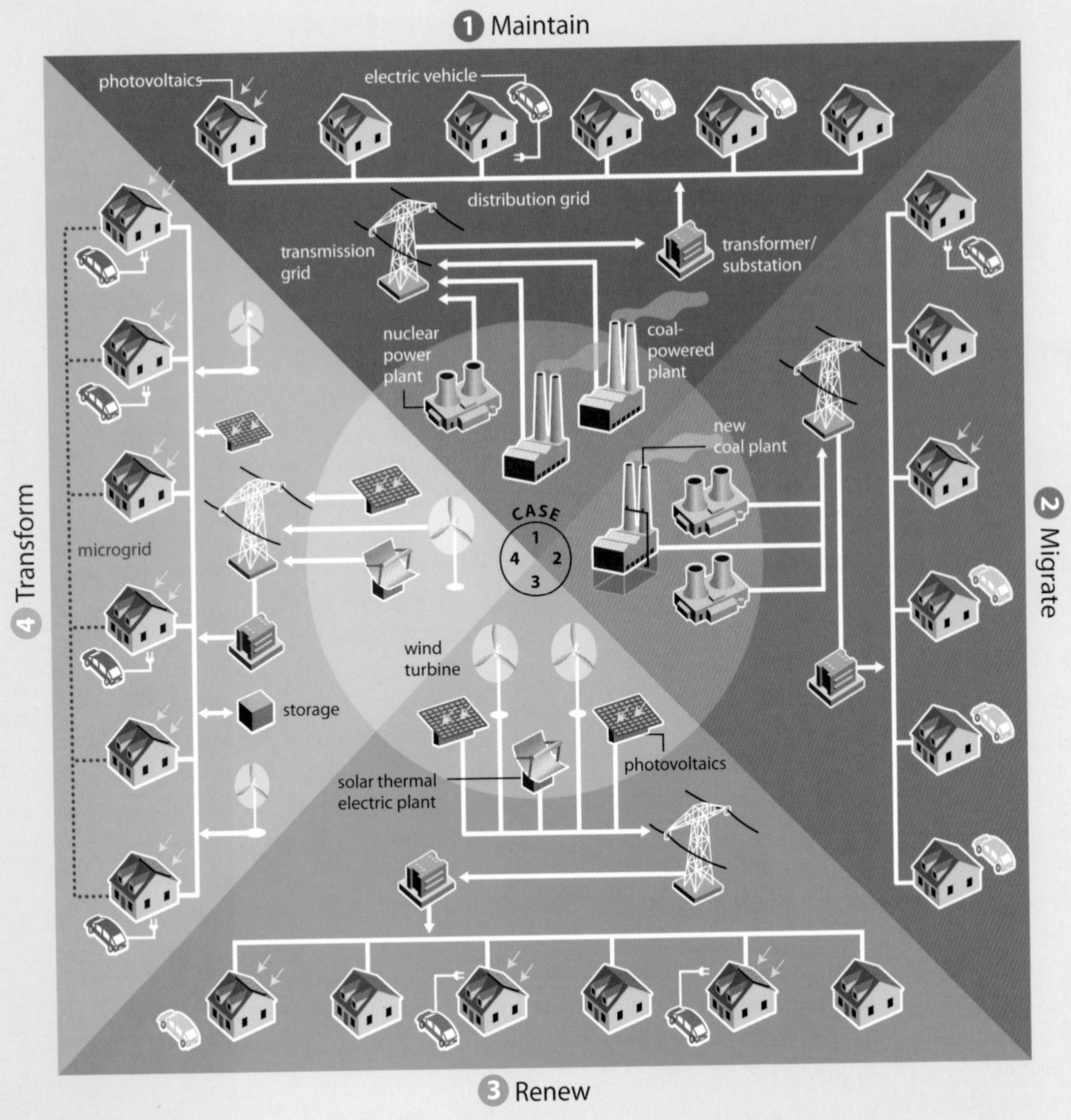

FIG. 5-2. Our four cases highlight key differences in how electricity can be made, delivered, and used in the future.

the total cost and performance of each case for each of seven U.S. regions using the National Renewable Energy Laboratory's Regional Energy Deployment System model (ReEDS), a sophisticated tool that simulates U.S. electrical capacity expansion and utilization for electricity generation and transmission, using simplifying assumptions to balance supply and demand at an hourly timescale (see www.reinventingfire.com for details).[451]

We evaluated each case on five criteria:

- ***Technical feasibility:*** Is there sufficient resource available, do the technologies exist commercially, and is the required scaling realistic?
- ***Affordability:*** How does cost compare with business-as-usual?
- ***Reliability:*** Can the system be operated reliably? How vulnerable is the system to natural or deliberate disruption, and can it bounce back quickly?
- ***Environmental responsibility:*** Does the case minimize health and environmental impacts?
- ***Public acceptability:*** Could this case actually be built under realistic political conditions?

The four cases we analyze and compare are:

- ***Maintain*** assumes that the future system looks largely like today's system, in both demand and supply mix.
- ***Migrate*** assumes that the anticipation of legislation to reduce greenhouse gas emissions drives a switch from conventional fossil-fueled generation to more nuclear power and to new coal plants equipped with carbon capture and sequestration (CCS).
- ***Renew*** examines how renewables like solar, wind, geothermal, biomass, and hydro can provide 80% of U.S. electricity.
- ***Transform*** picks up where *Renew* leaves off. It is powered by resources of varied scale but includes more distributed generators such as rooftop solar, CHP, fuel cells, and small-scale wind.

Second, the greater use of electricity and the greater integration of digital technology in many sectors means that an increasing share of the entire economy will depend on the reliability and quality of the electricity supply. Markets, transactions, supply chains, and operations will all require a steady supply of electrons. That means increased economic risks from electricity supply disruptions, small or large—from local outages in a building or on a college campus to regional blackouts. Just as the global financial crisis caught many off guard, businesses could be surprised by the vulnerability of the complex and highly interconnected electric power grid. A shift to more distributed sources of electricity, coupled with onboard control, storage, and power management in devices, could help mitigate these risks.

Third, electricity customers' choices will expand dramatically, bringing greater opportunities and risks for businesses just as it has in telecommunications. Not only will utilities and other traditional service providers offer a wider range of services and pricing structures, but the options for onsite generation, storage, and management of electricity will grow rapidly. An increasing number of electricity consumers will actually produce electricity themselves. These "prosumers" will be able to buy, sell, and store electricity according to fluctuations in their own needs and the economic signals coming from the grid.

In this chapter, we explore possible pathways for the transformation of the U.S. electricity system. First we will set the stage by describing a "business-as-usual" future and the major forces now driving change away from it. Next we'll explore three potential scenarios for change and evaluate their costs and benefits from economic, environmental, and societal perspectives. Based on these assessments, we'll recommend a path forward and identify the steps to harness emerging trends to reinvent fire in the electricity sector.

ELECTRICITY-SECTOR TERMINOLOGY

Power is the rate at which energy is transferred. Power is measured in watts (W), kilowatts (kW or 1,000 W), megawatts (MW or one million W), gigawatts (GW or one billion W), and terawatts (TW or one trillion W). The maximum amount of power that an electric generator is capable of sending out is its **capacity.** For example, the capacity of a nuclear reactor is typically around 1 GW, whereas the capacity of a standard solar photovoltaic (PV) panel is about 200 W.

Energy is the capacity for doing work. Electrical energy is measured in kilowatt-hours (kWh). Energy is calculated by multiplying the rate at which energy is transferred (the power) by how long it's transferred. For example, if a compact fluorescent lamp draws 13 W of power for 75 hours, it uses approximately one kWh of electrical energy. **Load** is the rate at which any or all customers demand energy from the electric system, either on average or at any specific moment.

Supply-side resources generate electricity. They can be **centralized**—large units, often of GW scale, connected to the **high-voltage** transmission system. (Voltage is analogous to pressure in a water hose, while current is analogous to the rate of flow.) Or they can be **distributed**—smaller units, or clusters of them, connected to the lower-voltage distribution system. Generators each have a **capacity factor**—how much electricity each actually sends out in a year, divided by how much it would send out if it ran at its full rated capacity for every hour of the year. Generators are sometimes shut down deliberately, usually for maintenance (**planned outages**), but sometimes unexpectedly (**forced outages**). The rest of the time they're **available** to be **dispatched** as needed.

Demand-side resources help customers use less electricity, or use it at more economical times, to deliver desired services. Demand-side resources broadly include energy efficiency and demand response. **Energy efficiency** means using less electricity more productively to produce the same or greater service with the same or better quality. Efficiency is about smarter technologies that do more with less—not privation, discomfort, or curtailment. **Demand response** voluntarily alters customers' consumption patterns in response to the changing price of electricity over time, incentive payments, or other signals of scarcity or abundance, saving money for both customers and provider.[452] In our modeling, we have assumed only demand response methods that are unobtrusive, so the customer doesn't mind or even notice.

The **grid** is the infrastructure that moves electricity from generator to customer. It includes the **transmission system**—large steel towers connected by high-voltage lines that move electricity over long distances, linked by special transformers and switches. As the electricity nears the customer, transformers, typically at **substations**, reduce its voltage to feed the low-voltage **distribution system** (poles and wires along the streets or in underground conduits) until it is finally transformed down to line voltage, metered, and delivered to customers. All these assets together constitute the grid.

"**Smart grid**" means many things to many people, but we use the term to describe the hundreds of technologies and applications that provide new levels of communication, information, and control to utilities and customers. For example, **smart meters** and related infrastructure can send price signals to customers, enabling them to adapt their usage to save money if they wish. The smart grid can also correct a longstanding limitation of classic grid design: Transmission lines routinely handle power flow both ways, but distribution lines don't because their protective equipment is typically designed for one-way flow only, from central generators to dispersed users. A smart distribution system will switch that "tree" structure to a "web" that can gracefully handle power flows in all directions.

A **utility** generates, transmits, or sells electricity (and perhaps natural gas). There are many types of utilities, distinguished by ownership and/or by the services they provide. **Investor-owned utilities (IOUs)**, **cooperatives (co-ops)**, and **publicly owned utilities (POUs)** are distinguished by their ownership structures: shareholders for IOUs (nowadays many use the term "**shareholder-owned utilities**"), customers for co-ops, and federal, state, or local government for POUs. Vertically integrated utilities perform all functions from generation to power transmission and distribution to retail sales. Some states separate these functions: a **generating company** only generates electricity for the wholesale market, and a **transmission company** owns transmission wires, while a **distribution company** owns distribution assets, contracts with providers for power and with transmission companies for wholesale delivery, and sells electricity to retail customers.

MAINTAIN: THE ELUSIVENESS OF "BUSINESS-AS-USUAL"

Perhaps the easiest future to imagine for the U.S. electricity sector is one that looks like today's—the past writ large, as expressed in our *Maintain* case (see Case 1: *Maintain* sidebar). Today's system is a 120-year-old story in the making, driven by the complex interplay between the laws of physics, the principles of prudent engineering, the evolution

CASE 1: *MAINTAIN*

Our first case, *Maintain*, expands a system much like today's in both demand- and supply-side technologies. There's little smart grid or demand response. Most utilities continue to be financially penalized for helping customers use electricity more efficiently, so they try not to. In such a scenario, coal and natural gas would generate 71% of U.S. electricity by 2050, up from 67% in 2010. Of the total present-valued cost of U.S. electricity supply in 2010–2050, nearly half would buy fuel.

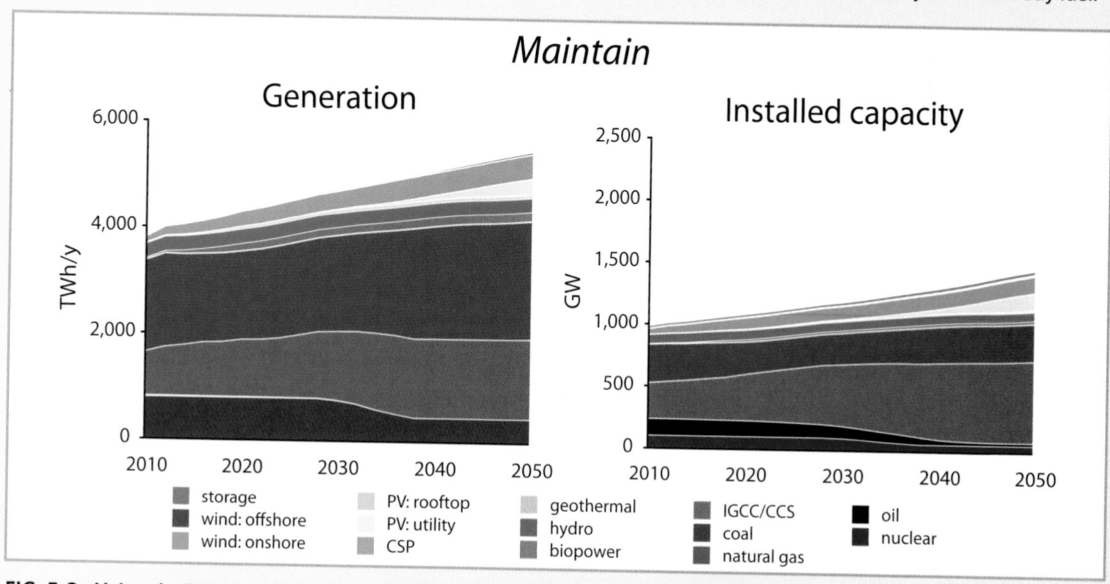

FIG. 5-3. Using the ReEDS model, *Maintain* attempts to reproduce the forecasted portfolio from the EIA *Annual Energy Outlook 2010* by largely using EIA's basic Reference Case assumptions, extrapolated to 2050.[453]

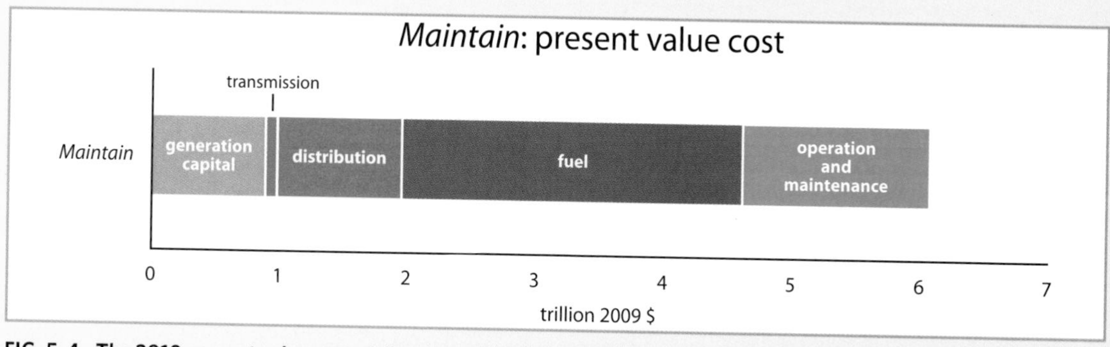

FIG. 5-4. The 2010 present value cost of the next 40 years' electricity is about $6 trillion, almost half of it going to fuel. The second-biggest cost is non-fuel operation and maintenance (O&M).[454]

of technology, and shifting economics and regulation. Working within these physical and institutional parameters, electric utilities have worked diligently to construct a complex system that provides least-cost, reliable electricity.

In 2010, nearly 86% of U.S. electricity was generated in large power plants from coal, natural gas, and nuclear fuel,[455] then sent often hundreds of miles via high-voltage transmission lines to lower-voltage distribution lines that finally connect to the customer. Powering this system consumed 976 million tons of coal and 7 trillion cubic feet of natural gas, producing 2,271 million tons of CO_2—40% of that year's U.S. fossil-fuel carbon emissions.[456] Two-thirds of this primary fuel on average—nearly half even in the most efficient power stations—is discarded as waste heat or used internally before the electricity leaves the power plant.

The electricity sector in the U.S., and much of the world, remains a largely regulated industry. State commissions generally regulate shareholder-owned utilities, which serve about three-fourths of national consumption.[457] The Federal Energy Regulatory Commission (FERC) regulates wholesale and interstate electricity commerce and interstate powerlines. Regulations largely dictate how the market evolves and therefore the behavior of private market actors. In the U.S., about 70% of customers are served by vertically integrated utilities regulated by state public utility commissions (PUCs), while some markets are served by publicly owned utilities or let retail customers choose from competing electricity suppliers.[458] The regulatory compact that governs most of the industry requires the electric utility to provide adequate service to all its customers and, in return, entitles it to recover reasonable costs and have an opportunity to earn a fair rate of return. It is up to the elected or appointed state regulators, who have considerable discretion within state and federal laws, to share benefits and costs equitably between customers and shareholders and to hold utilities to standards of prudent investments.

As a result, the electricity sector is considered a safe but unimaginative financial haven. From a regulator and customer perspective, consistency and reliability have been the guiding themes. Neither likes surprises in the form of price hikes (decreases are OK). In such an environment, incremental progress trumps innovation, avoiding risk is the watchword, and protecting the status quo becomes the norm. As Standard & Poor's notes, "Lower levels of risk [coupled with] the highly regulated environment has resulted in lower profitability and return on capital than in many other industrial sectors. In the regulated marketplace the level and margin of profitability has often primarily been a function of regulatory leeway, with the contribution of operating efficiency and revenue growth taking more of a back seat."[459]

This regulatory structure rewards building a large asset base and selling more electricity. Understandably, utilities shy away from strategies that place their shareholders' financial interests at risk. This institutional structure served well in times of steadily growing demand and increasing power-plant economies of scale. If those conditions didn't change and no major risks intruded, meeting future electricity supply needs from fossil fuels, as in our *Maintain* case, would have the virtue of relying on mature, proven technologies with seemingly abundant domestic fuels (coal and natural gas) to run the system, at least through 2050.

But the future is not the past, and current trends show there's no such thing as "business-as-usual." The electricity system is facing a convergence—some would say a perfect storm—of changes including technology development, reliability and national-security concerns (prolonged electricity blackouts are just as serious as oil interruptions), and environmental issues that together create some of the largest opportunities for innovation and investment seen since the industry got its start over a century ago.

So let's consider some of the challenges our business-as-usual case faces over the next 40 years.

HOW DID WE GET HERE?

The development of the electricity industry in the late 19th and early 20th centuries shaped its institutional structures to this day. The first evolutionary leap came out of a legendary debate between industry titans Thomas Edison and George Westinghouse. In the 1890s, they vociferously debated the best method for transmitting electricity—direct current (DC) or alternating current (AC). Edison favored DC, which at the time could be transported only locally without incurring large power losses. High-power semiconductors have since eliminated that limitation, so now high-voltage DC lines can compete with AC over very long distances, but in the 1890s, this constraint forced Edison to use a highly distributed system in which power plants were near customers. Westinghouse advocated AC, which could easily be transformed to high voltages to transmit over long distances with only minor losses, but at the cost of having to keep all the generators in exact synchrony, lest they get out of step and damage each other. Eventually Westinghouse prevailed, laying the foundation for the centralized architecture that dominates the AC electricity system today.[460]

The second evolutionary leap occurred just a few years later and solidified the industry's still-dominant business models. In the early 20th century, many electricity providers competed to build generation and distribution infrastructure in cities around the world. Many U.S. cities sought lower prices and a high quality of service by granting nonexclusive franchises to harness Adam Smith's "invisible hand" and enable competition among electricity suppliers. However, as companies invested and competed to serve the same customers, this competition sometimes led to duplication of plants and wires, which had high fixed costs. In fact, per unit of delivered energy, the electricity system has historically been about 10–100 times as capital-intensive as the traditional oil and gas systems on which modern economies have been largely built.[461] That capital intensity, coupled with investment uncertainty, created a significant challenge for market participants seeking to secure capital and operate efficiently.

Due to the nature of these investments and decreasing cost of production per kWh produced, electricity was declared a "natural monopoly"[462] where regulation should play the role that competition fills in a free market: "control of entry, price fixing, prescription of quality, and conditions of service."[463] As a result, the question became not *whether* but only *how* to regulate the electricity sector.

Since electricity crosses boundaries of cities, states, and regions and mixes freely in the grid regardless of its origin, the U.S. federal system of government presented several options for potential regulation. While the Commerce Clause of the U.S. Constitution gives the federal government the power to regulate commerce between the states, the 10th Amendment reserves to each state the right to regulate internally. Meanwhile, several city governments had already begun to take the reins and municipalize the electricity infrastructure within their boundaries so they could own and manage the system themselves. Today, municipal governments control 10% of electricity systems in the country.[464] Some city governments still threaten municipal takeovers as IOUs' monopoly franchises expire.

Recognizing imminent regulation and the threat of municipalization faced by incumbent electricity providers, Samuel Insull—Thomas Edison's successor—led the electric utility industry in lobbying for state public utility commissions (PUCs) as the industry's primary overseers. The first state PUCs were created in 1907 and rapidly proliferated. Every state except Nebraska (where all utilities are publicly owned) has one. The effect of this institutional structure cannot be overstated. As electricity historian Richard Hirsch notes, legitimizing electric utilities as natural monopolies without competition "allow[ed] them to pursue continued growth and consolidation, without public outrage or [mostly] calls for municipal takeovers of private firms." Legalized monopolies "eliminat[ed] investors' fears that utilities would lose market leadership" and reduced competition for capital and the cost of needed funds.[465]

Flat or Falling Demand

Unlike the case in earlier decades, demand growth cannot be counted on for increasing revenues. While U.S. demand for electricity has risen in all but four years since 1949, the rate of increase has been trending steadily down, so EIA predicts a meager +1% annual growth rate to 2035.[466] Successfully implementing the energy efficiency in buildings and industry discussed in this book could drop that growth rate to –1% and keep it there. While the electrification of vehicles represents a new source of demand, it will only partly offset projected efficiency gains (fig. 5-5). Some utilities have proven they can drive efficiency programs aggressively,[467] and they're not the only market actors who could.

Aging Infrastructure

Even though demand is likely to stabilize or decline, the need to build new generation sources will continue because U.S. power plants and infrastructure are becoming old and obsolete. More than 70% of U.S. coal plants—half of U.S. coal capacity—are more than 30 years old, and 33% are more than 40.[468] If they can all be affordably maintained and run until age 60, twice their normal accounting life, 94% of today's coal capacity will still have retired by 2050 through sheer old age (fig. 5-6).

Coal power plants, as with all steam-cycle power plants, have also long outrun their historical economies of scale. They stopped getting more efficient in the 1960s, bigger in the 1970s, cheaper in the 1980s, and bought in the 1990s, when U.S. ordering rates fell back to Victorian levels.[469] Highly efficient gas-fired combined-cycle plants derived from mass-produced aircraft engines took over most of their market in the 1990s, and today, as we'll see, renewables with even higher production volumes are exploiting similar advantages.

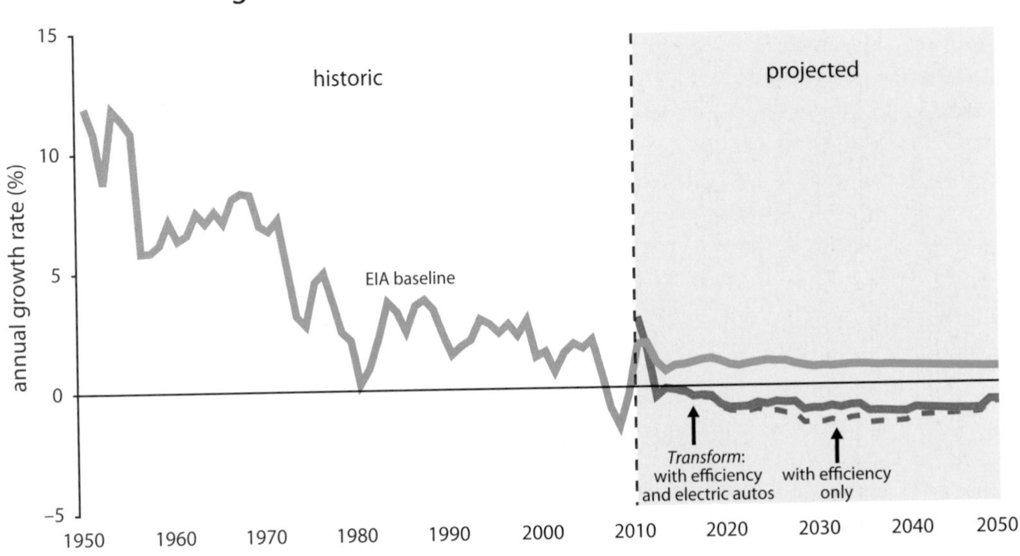

FIG. 5-5. The rate of growth in U.S. electricity demand has fallen rather consistently for 60 years and, with the efficiency gains described in chapters 3 and 4, will turn negative despite a modest increase from chapter 2's electrified autos.[470]

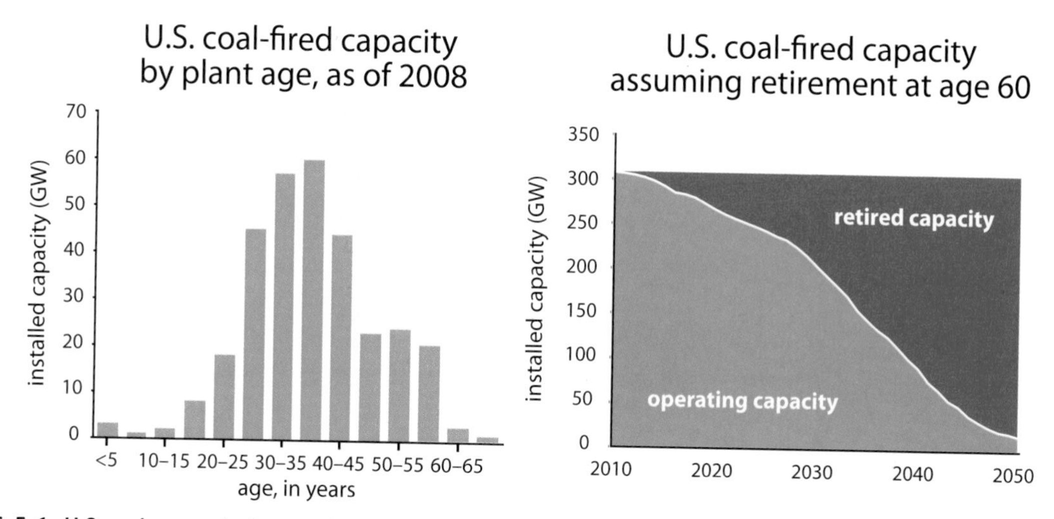

FIG. 5-6. U.S. coal power plants are aging and, unless lifetimes are significantly extended, will mostly be retired by 2050.[471] "Nameplate" capacity is under nominal conditions, but actual capacity can decrease when cooling water is unusually warm.

Environmental Constraints

Environmental and health impacts are increasingly likely to drive decisions in the electricity system. On *Maintain*'s conventional path, expected growth would still drive up the electricity sector's carbon emissions 38% by 2050, to levels nearly 600% above levels that would be required should the U.S. align with global reduction targets of 80% below 2000 levels by 2050.[472] While the future of U.S. climate-change legislation remains in flux, it's prudent to hedge the risk that the U.S. will eventually price carbon, as nearly all OECD countries already do. In 2007, over 20 Fortune 500 CEOs acknowledged their carbon dioxide (CO_2) emissions and resulting contributions to global climate change. They created the U.S. Climate Action Partnership (USCAP) and lobbied for federal carbon legislation.[473] In its 12th Five-Year Plan, released in 2011, even China capped its future carbon emissions. In the *Maintain* case, a $30–$100/ton price on CO_2 could raise the average price of U.S. electricity by $0.01–$0.05/kWh (roughly 10–30%) in 2050.

Beyond climate change, the likelihood of future environmental restrictions based on public-health needs poses significant, and potentially prohibitive, risks and costs for coal-fired generation. According to the National Academy of Sciences, the total annual health-related damages from sulfur dioxide, nitrogen oxides, and particulate matter created by burning coal in the U.S. were $68 billion in 2005[474]—*not* including ecosystem damage nor the health effects of some other air pollutants such as mercury.[475] Independent estimates by the Clean Air Task Force found in 2010 that coal-fired plants' air pollution led to more than 13,000 premature deaths per year in the U.S., raising health-care costs by $100 billion per year.[476]

Most coal plants predate and cannot comply with basic Nixon-era environmental laws designed to regulate emissions of these damaging air pollutants. In 2010, half of U.S. coal plants had no scrubbers for removing sulfur dioxide, 57% no nitrogen oxide reducers, 96% no modern controls for particulate and mercury emissions.[477] Just complying with impending U.S. Environmental Protection Agency (EPA) regulations on mercury, nitrogen oxides, and sulfur oxide could cost $120 billion,[478] falling disproportionally on the oldest and dirtiest plants.[479] Altogether a dozen rule-makings are under way to enforce long-deferred or long-ignored laws, and almost all will raise the cost of mining or burning coal. The Brattle Group estimates that more than 50 GW of coal plants could be retired rather than pay for the upgrades.[480] That's already starting to happen. In 2009, for example, Progress Energy chose to scrap 11 coal-fired generators and build $1.5 billion of new gas-fired capacity rather than pay nearly $2 billion to clean up its coal plants.[481]

A further issue likely to worsen, potentially constraining fossil-fueled generation in the decades ahead, is water scarcity. Each hour, a 1 GW coal-fired plant burns 500 tons of coal but also withdraws 24 million gallons and evaporates 1 million gallons of cooling water.[482] Experts expect water scarcity to constrain power production by 2025 in 10 water-hungry states, including California, Florida, and Texas,[483] and that may worsen with climate change. Already, 49% of all water withdrawn in the U.S. cools thermal power; of that, about 2.5% evaporates—nearly half of all domestic and commercial consumptive use.[484] *Maintain* would raise withdrawals by 30%, from just over two trillion gallons in 2010 to nearly three trillion gallons in 2050. Many jurisdictions are starting to charge market rates for that long free or cheap privilege. And in 2010, California limited once-through seawater cooling, so a fourth of the state's capacity must refit, repower, or retire by 2022.[485]

It's unlikely that these retiring plants will be replaced with more of the same: Wall Street is increasingly reluctant to finance them. Kevin Parker, global head of asset management for Deutsche Bank, says, "Coal[-fired generation] is a dead man walking. Banks won't finance them. Insurance companies won't insure them. The EPA is coming after them . . . and the economics to make it clean don't work."[486] In 2010, 6 GW of coal plants begun many years earlier came online, no new plants were started, and 38 planned coal plants were abandoned and 48 mothballed.[487] Even coal's biggest U.S. user, the giant American Electric Power, plans to burn gas for any new generation.[488] In 2010, EIA forecast only 10 GW of coal plants would be added by 2035; in 2011, that forecast dropped to zero.[489]

Global Technology Shifts

Unsurprisingly, concerns about climate change, pollution, and fuel price volatility have propelled governments around the world to spur demand for renewable technologies through R&D investments, tax incentives, subsidies, and mandates. These policies have elicited new technologies that can not only solve many of the old fossil-fueled

HOW MUCH COAL IS ECONOMICALLY MINEABLE?

Even America's domestic coal supply, traditionally taken for granted, is not guaranteed. There is enough underground coal in the U.S. to suffice for more than five centuries at the current rate of use, but only one-fourth of that is thought to be economically recoverable.[490] Even that estimate may prove optimistic. The Gillette Coalfield in Wyoming's Powder River Basin, which today produces almost half of the nation's coal, was once believed to contain over 200 billion tons of coal. But as the *Wall Street Journal* reported in 2009, the U.S. Geological Survey downgraded the economically recoverable resource by 94% to a little over 10 billion tons.[491] Caltech professor Dave Rutledge predicts a very substantial downgrade for all domestic coal fields. He estimates that the nation's economically recoverable coal amounts to just a 60-year supply at current rates of U.S. consumption.[492]

plants' problems but also present stiff economic competition.

Thirty-two U.S. states[493] have renewable portfolio standards (RPS) requiring 775 TWh/y of renewables by 2020 and 900 TWh/y—roughly 20% of projected electricity demand—by 2030.[494] In part due to these policies, U.S. windpower capacity has grown 1,300% in the past decade and solar capacity 8,800%.[495] And these shifts, some of which began in the U.S., have lately gained even more momentum abroad. Half the world's total 2008–2010 additions of generating capacity were renewable. By early 2011, more than 118 countries had some type of policy mandating or incentivizing renewable energy, versus 55 in early 2005.[496] Renewable markets are now immense, global, and dominated by developing countries, and will become increasingly so. Population and economic growth around the world, chiefly in developing countries, is spurring rapid growth of electricity infrastructure. Through 2035, official projections say China and India will together add nearly twice as much new capacity as the U.S. and Europe combined, continuing to drive renewable markets.[497]

As a result, technological development will be fueled by changes and innovations that happen *outside* the U.S. market at least as much as *inside* the U.S. As experience, scale, and technological innovation continue to drive down cost in fiercely competitive global markets (fig. 5-7), cost-competitiveness will speed adoption. This is already happening with astonishing speed, as we'll see in our *Renew* case.

Technological advances haven't been limited to the *supply* of electricity, nor to its end-use efficiency. Information technology has driven remarkable innovations in how electricity can be monitored, controlled, and delivered. These "smart grid" technologies not only can make the grid more reliable and secure, they can also make it

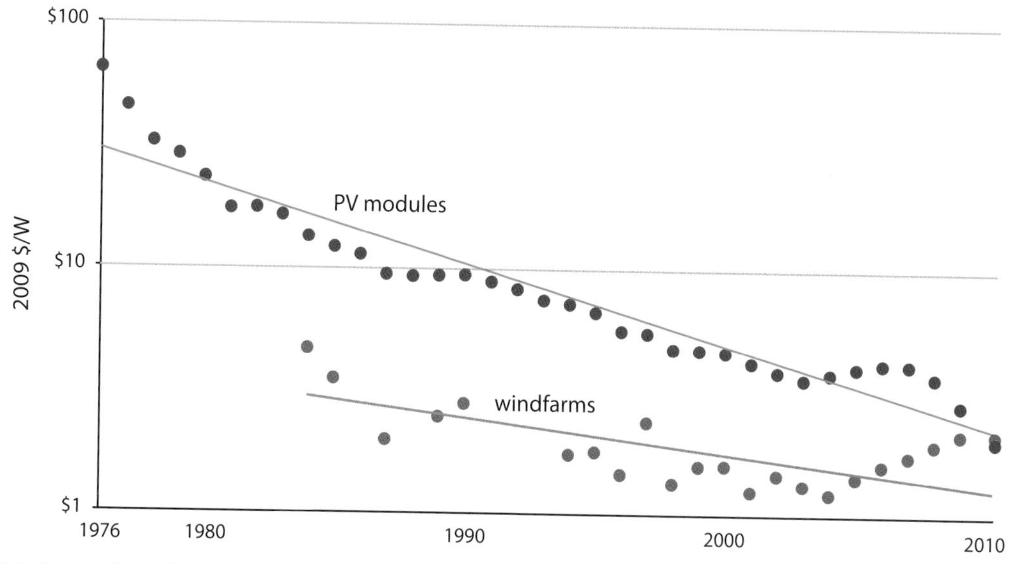

FIG. 5-7. Renewable technologies have largely followed cost-reduction learning curves, just like fossil-fueled power plants and hundreds of manufactured products. The recent bulges in both windpower and solar PV's costs were due to temporary bottlenecks as supply struggled to meet soaring demand; both bulges have since disappeared.[498]

more efficient, reducing the amount of power generation needed in the first place.

Meanwhile, competition introduced through the wave of electricity-market restructuring in the 1990s has allowed many new actors to enter the electricity business. Even more opportunity exists for companies and customers who can make unregulated investments in efficiency and locally sited renewables. Some companies are offering highly innovative products. Information technology providers are quickly infiltrating the electricity business with products that greatly enhance the level of information supplied to customers and utilities and enable advanced communications between customers, utilities, and even energy-using devices.

So even if for some reason we wanted to sustain today's familiar approach, innovation pushes us inexorably forward into a new world.

New Security Concerns

Finally, the reliability and resilience of the electricity system are becoming more vital and valuable. In August 2003, powerlines in Ohio, overloaded by a shutdown at a poorly maintained nuclear plant, sagged into the branches of insufficiently trimmed trees, setting off a cascade of failures that winked out the lights over 9,266 square miles from Toronto to New York City. Fifty million people lost power. The blackout closed 13 airports and snuffed out \$5–\$14 billion of economic activity in lost production and wages, spoiled food, emergency services, and other costs. Of the businesses surveyed, 24% lost more than \$58,000 an hour, and 4%—including autos, refineries, steel, and petrochemicals—lost more than \$1.2 million an hour.[499]

That blackout—and similar events in 1965, 1977, and 1996[500]—illustrates how businesses can lose millions of dollars[501] even when the power goes out for just a few minutes. Scientists at Lawrence Berkeley National Laboratory estimate these outages collectively cost U.S. businesses and residents up to \$160 billion annually.[502] Why? The interdependent technological systems that knit together across the economy—communications, finance, oil and gas (not even a filling-station pump works without electricity), water, sewage treatment, air and ground traffic control—all need a continuous supply of electricity.

It's a great tribute to the power industry's skill and dedication that such regional blackouts remain rare, about one per decade. But grid disturbances are on the rise,[503] and new threats of much wider and longer blackouts pose new challenges that cannot entirely be resolved within the design paradigm that created it. Any electricity scenario dependent on the frail aerial arteries of the transmission grid—without the ability to isolate demand centers from grid disturbances—carries a national-security risk.

In fact, physical vulnerabilities of the existing transmission system, first identified for the Pentagon in 1981,[504] persist today (fig. 5-8). In 2009, U.S. utilities and security agencies partnered with the Departments of Energy and Defense to simulate an attack on the electric grid. In this simulated wargame, attackers took out extra-high-voltage transformers at two substations and blacked out a city. They then threatened to black out 10 more cities in the next six hours if their demands weren't met. Forces scrambled to try to protect 2,000 extra-high-voltage transformers across the country. After six hours, the group attacked the control and communications systems instead, taking out 36 GW of a major utility's capacity. They threatened a repeat in five hours, but there was too little time to protect the control systems. Game over.[505] Threats needn't even come from attackers: the grid isn't yet protected against severe geomagnetic storms that, at their 1921 level, could black out 130 million people and cost \$1–\$2 trillion a year for 4–10 years.[506]

Cyberattacks were added to the list of grid security issues when many grid controls were shifted to the Internet with inadequate security precautions. The Idaho National Laboratory, the Defense Science Board, and industry experts have identified vulnerabilities that could let remote and anonymous cyberattackers instantly destroy very costly generators and turbines—possibly in large numbers simultaneously. Replacing them would take years and rely on a handful of foreign factories with limited spare capacity. Utility IT security professionals surveyed in 2009 were each handling some 150 "serious" cyberattacks per week.[507] Cyberattacks abroad had already blacked out cities, and attacks—said the Director of National Intelligence in his 2009 Annual Threat Assessment of the Intelligence Community—were becoming "more sophisticated, targeted and serious." Those threats are already being used in covert warfare, like the 2010 Stuxnet worm targeting Iran's uranium-enrichment plants.

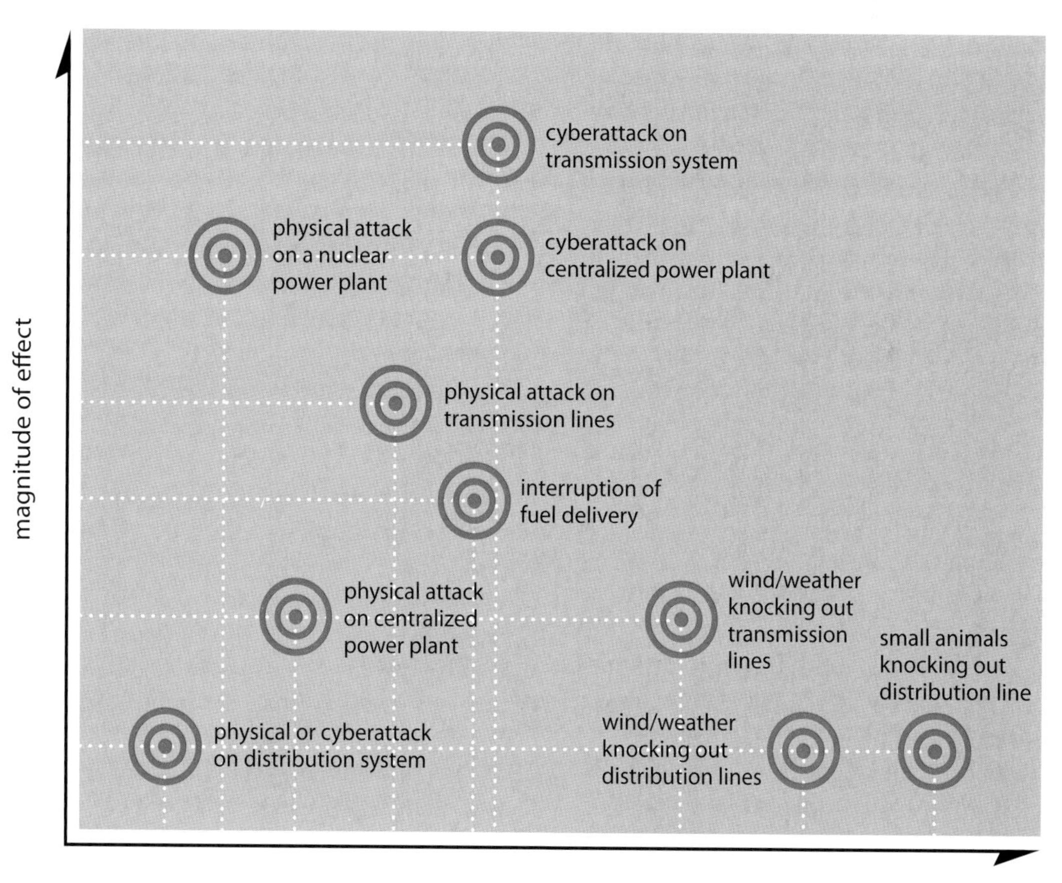

FIG. 5-8. The many threats to grid security and reliability vary widely in likelihood and severity.

Several characteristics inherent to the power system lend it both normal reliability and worrisome brittleness. To begin, a giant thermal power plant can lose a billion watts in the blink of an eye, often for weeks or months and without warning. The same or worse is true for a large transmission line. That's precisely why utilities are required to have enough capacity in reserve to compensate for the loss of the largest resource on the system. In fact, 98–99% of outages occur in the transmission and distribution grid—more than 90% of those due to weather, equipment failures, or small animals.[508] And since today's "dumb grid" provides no visibility beyond the substation, often your power company doesn't even know there's been an outage in the local distribution network until you call them and complain. While these seemingly simple and local disturbances mostly trigger only small outages, they can trigger cascading grid failures with far more damaging consequences.

Finally, if the grid goes down, it can take a long time to restore. The grid must be painstakingly reassembled to avoid further collapse, striking a delicate balance between supply and demand as each segment is powered up. Unfortunately, many large thermal generating units take up to a day to start, and most need grid power to do so. Abruptly shut-down nuclear plants are especially hard to restart, so when nine perfectly operating reactors were instantly shut down to protect them in the 2003 Northeast blackout, they took two weeks to return to full output.[509]

Inaction Is Risky

While predicting the future of the electricity sector is fraught with uncertainty, it seems exceedingly unlikely that *none* of these many diverse contingencies will happen. We're already beginning to see clouds on the horizon. To date, media and industry attention has been largely focused on the threat of climate change and carbon legislation, and debate about the science of climate change has overwhelmed constructive discussions about profitable ways to mitigate that risk. So why did the CEOs who created the U.S. Climate Action Partnership lobby for climate legislation? These savvy CEOs implicitly acknowledged that once the writing is on the wall, there are three main options: hope it disappears, wait for the ink to dry, or take up your pen and help write the story. If you're one of the nation's largest emitters, it doesn't matter whether you believe human activity is changing the earth's climate; it matters only whether you think your emissions might be restricted or taxed. If so, *Maintain*'s uncertainty on this score imposes risk and cost on your owners and customers. So while there are many risks to navigate in the electricity system, hedging the carbon risk seems a prudent first step—which the *Migrate* case takes by widely advocated means.

MIGRATE: THE CONVENTIONAL APPROACH TO "CARBON-FREE" ELECTRICITY

As climate and pollution solutions are sought, much attention has focused on nuclear and "clean coal" (capturing and sequestering carbon from coal plants). The approach seems straightforward—retire old carbon-emitting plants, build more nuclear plants, and invest in new technology that captures and sequesters (safely and permanently stores) the carbon emissions from new coal- and even gas-fueled plants. This vision, uniquely, can solve electricity's climate problem while bolstering rather than undercutting many of the power sector's century-old institutions. It supports utilities' existing business models, reinforces the traditional role of state regulators, and even preserves the coal-mining industry and coal-hauling railways. Not surprisingly, there has been a great deal of political and utility interest in moving the U.S. down this path. It's an alluring notion. We could become carbon-free without making dramatic changes in the system that's served us well for more than a century. So why not?

A future based on *Migrate*'s nuclear and CCS technologies would rely on familiar regulatory, institutional, and operational paradigms. But how feasible is a major buildout of nuclear and CCS to decarbonize the electric system? Nuclear is currently a major electricity supplier in the U.S.—one-fifth[510] of U.S. electricity in 2010 came from 104 nuclear units.[511] In contrast, there are no commercial-scale CCS plants—only four demonstration-scale projects (one in the U.S., three abroad) that trap coal plants' carbon emissions underground.[512] And while the total size of the investment is not unworkable, the degree of financial risk it focuses

CASE 2: *MIGRATE*

Migrate assumes that the anticipation of legislation to reduce greenhouse gas emissions drives a switch from conventional fossil-fueled generation to more nuclear power and new coal plants equipped with carbon capture and sequestration (CCS), all driven by the incentives of incumbent regulatory and business models. The buildout would require a large-scale ramp-up of nearly three new nuclear plants and nine CCS-equipped coal plants each year for the next 40 years at a total present value cost of $6.5 trillion. The electricity sector's carbon emissions would fall 64% by 2050. Increased nuclear capacity would require additional high-level waste storage about twice the size of the abandoned Yucca Mountain facility in Nevada.

Carbon-reduction goals also drive partial use of efficiency and renewable energy, but regulated utilities and their shareholders are conflicted because their investments in large, extremely capital-intensive, long-lead-time plants need assured capital recovery, enhancing existing incentives to sell more power. Electricity delivery, system architecture, and operations remain similar to today's, with limited smart-grid deployment.

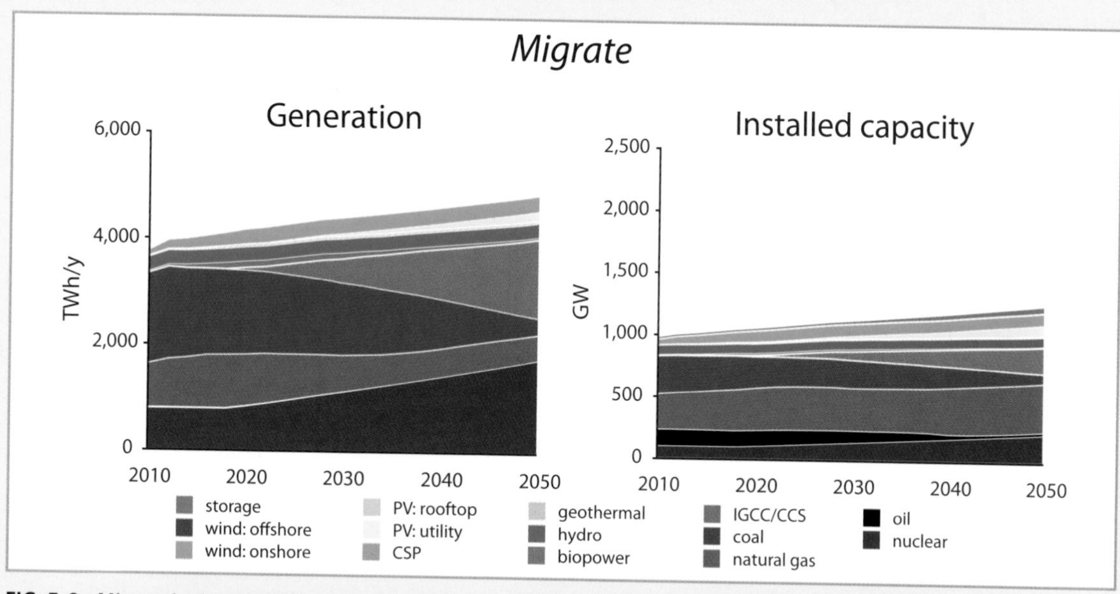

FIG. 5-9. ***Migrate*** **hedges carbon risks with new nuclear power plants and "clean coal"—integrated gasification combined-cycle coal-fired plants with carbon capture and sequestration (IGCC/CCS). Like** ***Maintain*****, the** ***Migrate*** **case improves end-use efficiency only modestly, so electricity consumption keeps growing, albeit slowly.[513]**

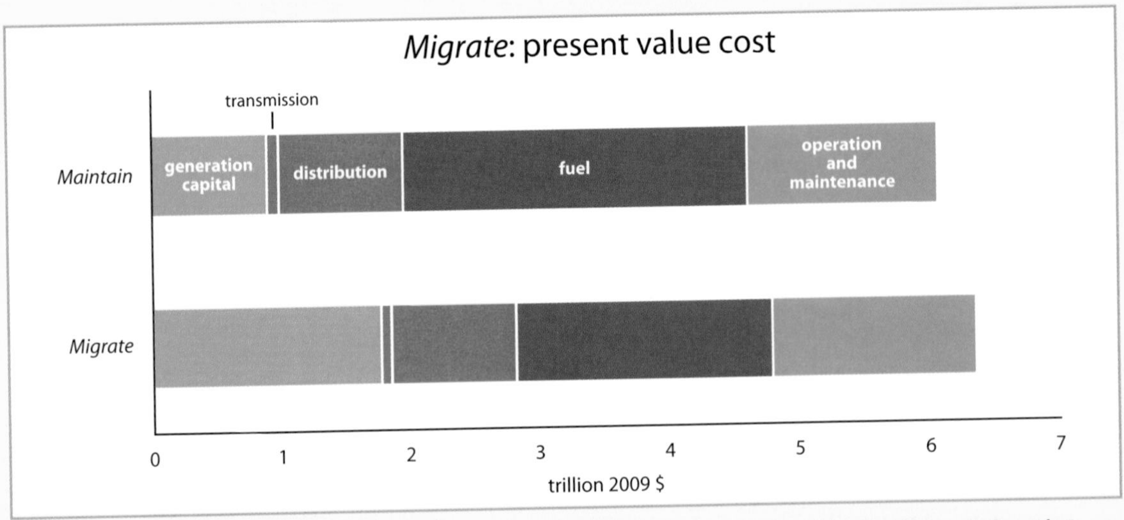

FIG. 5-10. Present-value electricity cost to 2050 is modestly higher than for *Maintain*, with twice the capital cost but lower fuel cost.[514]

into a modest number of very large and complex projects could make financing a showstopper. So far this has proven true for the purportedly more mature of *Migrate*'s two key components—expanding nuclear generation.

Nuclear Power's Financing Challenge

In the three years following August 2005, when nuclear power enjoyed the strongest political and policy support and the most robust capital markets in history, none of its 34 proposed U.S. projects was able to raise any private project financing despite federal subsidies rivaling or exceeding their construction cost.[515] The market verdict is similar abroad. Of the 64 nuclear power projects currently under construction globally, all are in centrally planned power systems, mainly run by authorities with a draw on the public purse.[516]

Why has U.S. nuclear power been unable to attract private capital? U.S. nuclear construction's track record doesn't give investors the confidence needed to wager billions. From the early 1960s to 1978, when momentum stalled, a year before the Three Mile Island accident, U.S. utilities pursued an aggressive nuclear building program. Of the 253 reactors ordered, three-fifths were abandoned as as result of threefold cost overruns (fig. 5-11) or were prematurely closed as lemons.[517] The massive overruns, totaling hundreds of billions of dollars, were caused by interrelated factors including rapidly evolving safety regulations, unstandardized and labile designs, challenges in managing very large and complex projects, and deteriorating finances as demand slackened and costs soared.[518] Participating utilities' debt downgrades, affecting 40 of 48 issuances, averaged four notches.[519]

Recent global evidence shows this isn't a uniquely American phenomenon. AREVA, the top global nuclear constructor, has seen its latest two nuclear ventures sour, its shares fall 71% in four years, and its CEO dismissed. Its Olkiluoto 3 plant in Finland is around if not over twice twice its original construction time and budget with no clear timeline for completion. Électricité de France, the world's most experienced nuclear utility, is experiencing similar delays with the Flamanville 3 project it began building in France in 2007: by mid-2011, it was four years behind schedule and $2.6 billion over budget.[520]

New nuclear plants' high and uncertain capital costs might be bearable in the short term if there were reason to believe a major nuclear construction program would reduce them. However, no country has yet demonstrated a significant or sustained learning curve for installed nuclear capital costs. For example, when Dr. Arnulf Grübler of the International Institute for Applied Systems Analysis carefully analyzed the official costs of France's uniquely ambitious nuclear program,[521] he found that capital costs did not fall with experience. In fact, the more plants were built, the more they cost. Real capital cost per kW in France's 1974–1990 construction effort rose 2.4-fold, and construction periods, though 30% below the U.S. average, nearly doubled.

Recognizing nuclear power's financial challenges, policymakers have tried to woo private capital with incentives, such as federal loan guarantees. These loan guarantees lower the cost of borrowing by guaranteeing debt-holders' principal, and they substitute direct Treasury financing for capital-market lenders. The U.S. Department of

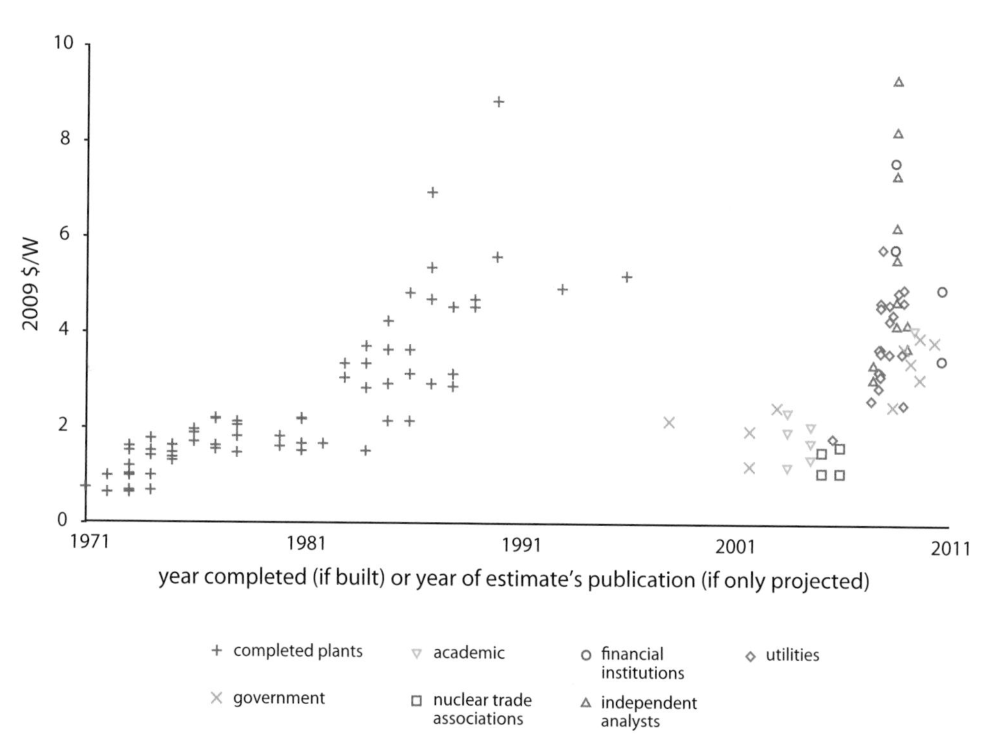

FIG. 5-11. The U.S. overnight capital cost of pressurized-water reactors (the main kind in the world market) has trended up over time, not down. (Overnight cost is what the plant would cost to build if it could be completed overnight. Actual project cost is nearly twice as great because it also includes financing and real escalation during construction.) The steep rise in 2007–2009 reflects the shift from vendors' initial promotional claims to actual bids putting the vendor at risk for part or all of any cost overruns.[522]

Energy's fiscal-year 2012 budget requested $34 billion, tripling its 2005 authority, in loan guarantees for new nuclear construction.[523] Yet the program has generated conditional commitments of just $7.9 billion to guarantee two Georgia reactors.[524] A South Carolina developer hopes for a better deal in the private market. Two other projects collapsed because the owners didn't want to pay the fees DOE assessed, as required by law, to compensate taxpayers for their risk. DOE hopes for other candidates, but those few who were willing to risk putting up the unguaranteed 20% of their projects' cost already have—sort of.[525]

To shed that remaining risk to shareholders, utilities with nuclear ambitions have lobbied state legislators and PUCs for authority to charge customers for "construction work in progress" (CWIP) before the plant enters service. A few have agreed. Georgia Power's twin nuclear units got federal loan guarantees, cheap financing (some via stimulus funds) by POU majority partners, *and* CWIP. But then the Georgia Public Service Commission, stung by customers' (especially big industries') reaction to price hikes and to the high return Georgia Power sought to restore its deteriorating ratings, wanted customers shielded from cost overruns, and the utility rejected that approach, clouding the project's future.[526] Similarly, Duke Energy's CEO said that if the North Carolina legislature rejected CWIP, they'd be "saying no to nuclear's future in this state."[527] As regulators recommended,[528] the legislature postponed consideration to at least 2012.[529] In other words, protecting investors at the expense of customers is proving difficult even for the utilities with the strongest political influence.

Fukushima Fallout

Finance already challenged a nuclear renaissance before the March 2011 magnitude-9.0 earthquake and ensuing tsunami devastated Japan's northeastern coast, triggering the worst nuclear disaster since Chernobyl. The natural disaster melted three and destroyed four of the six Fukushima 1 reactors, damaged the others, and caused the evacuation of nearly 100,000 people.[530] At this writing, six months later, the reactors remain far from stable, confining their radioactive releases remains daunting, and their shattered owner, Tokyo Electric Power Company (TEPCO), posted a $14 billion initial loss—with more to come as it faces years of cleanup and compensation costs—and fell under direct government control.[531]

Like seismic waves, Fukushima's effects circled the globe. Within weeks, the U.S. Nuclear Regulatory Commission launched a quick review and began framing a longer one to determine whether the 104 U.S. reactors—six identical to and 17 very similar to Fukushima's—need safety or procedural improvements.[532] Daily press features detailed troubling parallels between weaknesses in Japanese and U.S. safety regulation. A fifth of the world's reactors turned out to be in significant seismic zones. NRG Energy abandoned a $10 billion two-reactor nuclear project proposed in Texas (supposedly with finance from TEPCO) and wrote down a half-billion-dollar loss.[533] China suspended all nuclear construction and approvals pending review.[534] Germany suspended nuclear license extensions and shut down eight reactors. Then the conservative chancellor announced in May 2011 that Germany will exit nuclear power (which was providing 23% of its electricity) within a decade, switching as quickly as possible to efficiency and renewables.[535] The previous day, the prime minister of Japan (which was 30% nuclear-powered before Fukushima) had cancelled all 14 planned reactors in that country and announced a rethinking of energy policy, adding two new pillars: like the Germans, efficiency and renewables.[536] Soon he added, "Our country will put all of our resources into making renewable energy a mainstay of our energy supply" as part of an impetus for economy-wide structural reforms.[537]

Japan's wealthiest man proposed building enough solar farms to displace TEPCO, and announced support from at least 36 of the 47 prefectures.[538] Switzerland[539] accelerated its exit from nuclear power, and Siemens—now a leader in gas, wind, and solar power—announced its exit too.[540] Italy's voters rejected a nuclear revival by nearly 95–5.[541] Other dominos may fall.

The Japanese tragedy also revived the persistent issue of nuclear waste storage. Much of the radioactivity released at Fukushima came from overheated or burning spent fuel stored above the reactors. As in Japan, most U.S. spent fuel is stored onsite at reactors across the country because there's not yet a permanent storage facility nor a credible process for siting one. Some experts believe those spent fuel pools may pose a greater hazard than the reactors themselves; reprocessing the spent fuel would only add cost and risk.[542]

Fundamentally, though, nuclear power had been overtaken in the marketplace long before Fukushima, just as its U.S. orders had collapsed from poor economics a year before Three Mile Island. Its costs and risks are simply unattractive to investors, who've voted with their dollars.

"Clean Coal," Meet Markets

Like nuclear power, carbon capture and sequestration (CCS) could be used to eliminate carbon emissions but also faces challenges from its high costs and uncertain performance, which limit its access to capital. As the name implies, CCS is the technological capacity to extract carbon dioxide from fossil-fueled power-plant flue gas and store it in perpetuity to prevent its escape to the atmosphere.

Transporting captured CO_2 is relatively straightforward and well understood, but

CARBON CAPTURE AND SEQUESTRATION (CCS)

Carbon capture and sequestration (CCS) is a three-step process. It first separates and captures the greenhouse gases that normally escape from a power plant's smokestack, then transports the compressed or liquefied gas via pipelines, and then permanently sequesters it, probably deep underground. Developers hope to find profitable ways to create value (perhaps cement substitutes or construction materials) from the enormous amounts of CO_2 that could be captured, but so far CCS is simply an extra and unproductive cost that utilities would bear to meet future restrictions or taxes on carbon emissions.

CCS can be added to existing coal-fired power plants, although often, as in this chapter, CCS is discussed in conjunction with integrated gasification combined-cycle (IGCC)—a more efficient coal-fired power plant. An IGCC plant first cooks the coal to create a hydrogen-rich gas, which fuels a combined-cycle plant as if it were natural gas. On its own, IGCC doesn't dramatically reduce CO_2 emissions, but combining it with CCS can remove CO_2 before combustion, making capture more convenient and efficient. Even so, completely capturing CO_2 consumes up to 30% of the plant's output energy—a "toll" that engineers hope to cut to 15%.[543]

One of the world's first fully integrated CCS pilot projects began operation in 2009 at a 1,300 MW coal-fired power plant run by American Electric Power (AEP) in West Virginia. The project captures carbon from an experimental fraction—about 1.5%—of the flue gas and pumps the CO_2 1.5 miles below the earth's surface. At full-scale operation, this CCS process would make the plant's electricity about 80% costlier.[544]

AEP planned to take the next leap in 2014—building and operating a fully integrated CCS solution that captures 240 MW or 18% of the plant's flue gas at an estimated cost of $668 million. But in July 2011, AEP shelved the project despite a 50% DOE grant, because the two PUCs said their states' customers alone shouldn't pay the other $334 million. Absent federal climate policy, the investment had no business case.[545]

Such costly processes use chemical reactions to capture the CO_2—a daunting task because a 1 GW coal plant makes about a quarter-ton of CO_2 every second. But a simpler nonchemical process, originally funded by Shell, has been proven at the Eindhoven University of Technology in the Netherlands[546] and is heading for larger-scale demonstration and commercialization.

capturing and sequestering it are complex processes that have been little tested on the scale necessary for typical coal plants. CCS has been demonstrated to work in power plants at the pilot scale but has not been proven in a full-scale demonstration, which equates to scaling up the pilot demonstrations by 6–10-fold.[547] Failed storage could leak large amounts of carbon back into the air. Important questions remain around storage's governance and regulatory arrangements. How will the operator be able to access geological rights everywhere the underground CO_2 migrates? Who's responsible for stewardship of the site over the centuries it must be monitored to ensure CO_2 doesn't leak? Who'd be liable if it did?

These uncertainties make it hard to finance pilot projects or full-scale first-of-a-kind projects with private capital. The U.S. government has stepped in with nearly $1.5 billion in grants to help fund CCS at six new and existing 60–400 MW coal-fired power plants.[548] Similar projects are being planned around the world. If they succeed, many billions in private capital will be needed to add CCS to new coal-fired power plants. Making much difference to climate could require major investments well before some of the basic uncertainties are resolved.

Risks beyond Carbon

Despite the carbon benefits promised by this *Migrate* future, moving toward nuclear and CCS does nothing to address the industry's other critical issues about fuel, security, financial stability, and above all competition. Companies making multi-billion-dollar, multi-decade bets must be mindful that they're standing on an active seismic fault. What if the greatest threat is not carbon regulation but disruptive technologies like the $0.03 kWh windpower sold in mid-2011, or solar power priced below retail electricity, driving more customers to distributed generation? What if innovations in end-use efficiency's technologies, value, marketing, and delivery systems turn demand growth into durable demand destruction? What if the looming threat is not policymakers in Washington but a new generation of savvy consumers who don't read *Congressional Quarterly* or *Public Utilities Fortnightly* but who use automatic bill-pay, live on the Web, and are glued to Facebook and Twitter?

Further, putting large bet-the-farm assets in the ground creates strong incentives to ensure revenue is generated to recover the outsize capital costs. Faster deployment of customers' or third parties' smaller-scale demand-side and renewable resources could threaten the financial security of utilities that had made major commitments to coal or nuclear projects. In fact, the whole aura of the utility as a safe widows-and-orphans investment could lose its allure. Stagnant or falling demand, upward price pressures, and nonproductive investment burdens to modernize or clean up old assets are all shrinking utilities' domain of financial stability. Building big, slow, lumpy, costly plants could shrink them further—perhaps triggering a repeat of the "death spiral" of rising price and falling demand that many utilities suffered in the 1980s.

Nuclear power and "clean coal" lack not only the investment attractions but also the operational flexibility[549] needed to integrate the renewables already joining the grid. Since wind and solar generators' zero fuel cost makes them cheapest to run, their rising contribution makes fossil-fueled generators run fewer hours, reducing wholesale prices and fueled plants' profits. In Germany, increased solar generation has cut fossil-fuel plants' operating hours and hence their fuel costs enough to lower wholesale electricity prices by an estimated $4–$16/MWh. (On four days in 2010, surplus windpower even sent bulk power prices below zero.[550]) Studies in Denmark, Belgium, Ireland, and Texas confirm that windpower investments reduce average wholesale market prices.[551] This benefits customers but reduces revenues and profits for existing generators.

So while the benefit of reducing carbon risks with few changes to familiar institutional and business models has appeal, its costs and risks seem high. What alternatives do we have? The rapid growth of the renewable energy market in the U.S. and globally opens the door to considering the potential for these resources to meet our future electricity needs.

This is not a new idea. The Paley Commission in 1952 predicted oil shortages by the 1970s and recommended to President Harry S. Truman a massive shift to renewable energy.[552] For that matter, Thomas Edison exclaimed to Henry Ford in 1931:

> *We are like tenant farmers chopping down the fence around our house for fuel when we should be using Nature's inexhaustible sources of energy—sun, wind and tide. . . . I'd put my money on the sun and solar energy. What a source of power! I hope we don't have to wait until oil and coal run out before we tackle that.*[553]

He too was ignored. But now the message is coming not from wise men of yore but from today's marketplace, and it has become far too insistent to dismiss. As the International Energy Agency wrote in 2010,[554] "A profound change in the way we generate electricity is at hand . . . entering a period of transformation" from central, steam-driven, chiefly fossil-fueled power stations to distributed and renewable sources. Renewables, says the IEA, "will have to play a central role in moving the world onto a more secure, reliable and sustainable energy path." Our next two cases explore that path.

RENEW: TAPPING NATURE'S INEXHAUSTIBLE ENERGY SOURCES

Texas is known for many things in its long, storied history as the Lone Star State. Texans' fierce independence is legendary, prompting even its anti-litter campaign to proclaim, "Don't mess with Texas." The state's oil and gas fields, a fertile ground for wildcatters in the early 20th century, established Texas as the U.S. oil and gas capital for most of that century. And it's not just its acreage or attitude that lead people to quip, "Everything is bigger in Texas": if Texas were a country, it would have the 14th largest economy in the world.[555]

But Texas is also earning another reputation—as a profit-driven leader of U.S. renewable energy development. In 2010, Texas had more wind energy installed than any other state—by a factor of three—generating 8% of its electricity and surpassing its 2025 target.[556] The state's roughly 10,000 MW of wind makes Texas sixth in the world among *countries,* after China, the entire rest of the U.S., Germany, Spain, and India.[557] That's even more impressive because less than 20 years ago, Texas had no commercial wind projects and produced less renewable energy per person than any other state.

CASE 3: *RENEW*

Our *Renew* case examines a future in which renewables like solar, wind, geothermal, biomass, and small (plus existing big) hydro provide at least 80% of U.S. 2050 electricity. More generation by variable sources—wind and solar photovoltaics (PVs)—makes responsive demand from residences and businesses more valuable, so they spread faster. These customers use home area networks to respond automatically and unobtrusively to price signals from smart-grid-enabled two-way communication. Energy efficiency potential becomes easier to capture as customers gain more information about their energy consumption, can respond to price signals, and gain comfort with new technology. The increased energy efficiency flattens electricity demand growth, prompting many regulators and utilities to address the declining "build and grow" business model with regulatory reform and new value propositions.

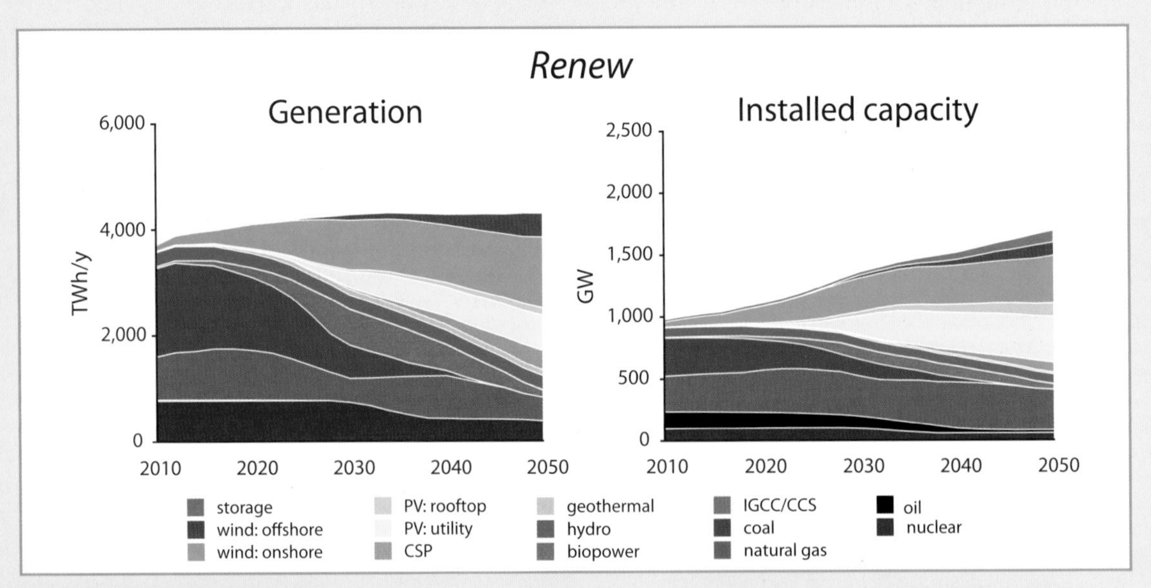

FIG. 5-12. The *Renew* case levels off electricity consumption through aggressive adoption of energy efficiency, but installed capacity grows substantially due to many renewables' lower capacity factors.[558]

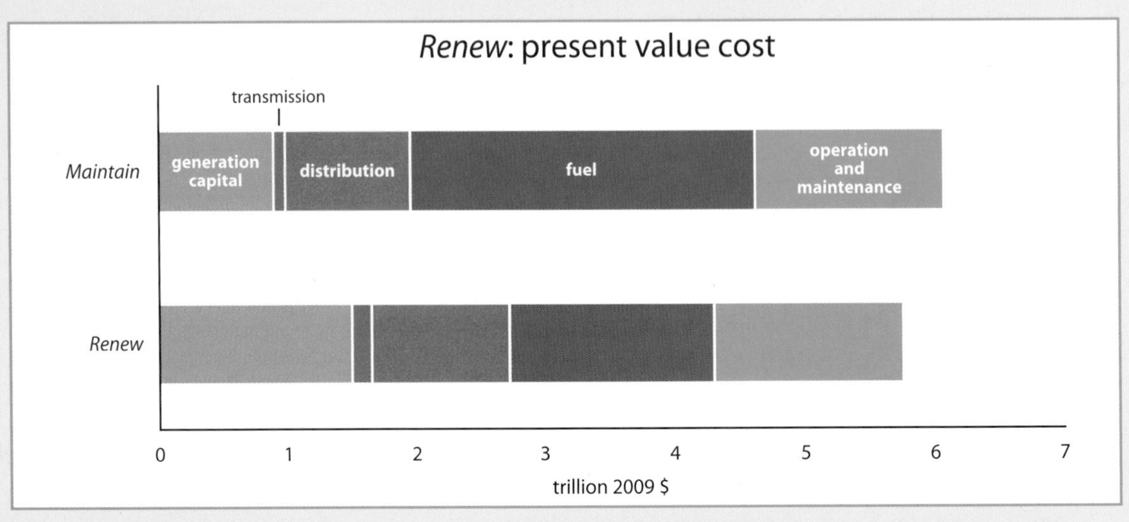

FIG. 5-13. The 2010 present value cost of *Renew*'s electricity is slightly below *Maintain*'s—in part due to greater efficiency adoption—with somewhat higher uncertainty due to reliance on continued technology cost reductions, but far less reliance on fuel with its uncertain prices.[559]

If costs for key technologies continue to make their way down the learning curve as projected, this transition to 80% renewable generation would need about twice the *capital* investment of *Maintain* (or if renewables' costs stick at 2010 levels, *Renew* could cost $7.3 trillion). The higher investment is offset by lower *fuel* costs, though, so the cost of the next 40 years' electricity could be less than *Maintain*'s.

What prompted this meteoric rise in renewable electricity generation? Did Texans suddenly get environmental religion? Hardly. As Jim Suydam, the press secretary for the Texas General Land Office, says, "Here it's all about making money."[560] That's a strong statement about an industry considered a niche player for most of the past century. If it's as simple as making money, why isn't every state sharing in this purported gold mine? What is Texas's magic ingredient?

The truth is Texas doesn't have one magic ingredient; it has several. The first is obvious—abundant renewable resources. From the Gulf Coast to the Panhandle, the state has more than twice the onshore wind potential of the second windiest state, Kansas,[561] and more solar potential than California.[562] But while access to resources is key, it's only part of the equation. The rapid expansion of the Texan wind industry owes as much to the state's policy, business, and culture, which have all enabled the state to harness its renewable resource effectively, as to the natural resource itself. In fact, the Texas experience provides a window into understanding the opportunities and challenges of accelerating the rapid growth of renewables in the rest of the nation—as our *Renew* case examines.

Abundant Renewables

Texas may have abundant renewable capacity, but what about the rest of the country? Is there enough renewable energy to meet the entire nation's electricity needs? An oft-quoted statistic notes that an area about 90 miles by 90 miles square, if covered with solar panels or solar concentrating plants, could produce all the annual electricity the U.S. now needs.[563] However, that tells only part of the story. Studies by the National Renewable Energy Laboratory (NREL) and the U.S. Department of Energy show that the U.S. is blessed with abundant and geographically dispersed wind, biomass, water, sun, and natural steam. Onshore wind alone, in suitably windy sites on available land, could generate 9.5 times as much electricity as the U.S. used in 2010.[564] All told, these renewable resources have the potential to generate 75,000 TWh/y of electricity[565] using today's commercially viable technologies—over 20 times the total national 2010 use.

Clearly, America's renewable energy resource base is ample and diverse. But can it be harnessed cost-effectively and delivered reliably, and can traditional electricity system operations be feasibly adapted to these fundamentally different resources? These challenges are real but not insurmountable. In fact, many are being tackled today in control rooms, boardrooms, and labs around the country and the world.

Costs Pass the Tipping Point

Renewable technologies generally have had higher capital costs than fossil-fueled power plants, but their fuel is free, their energy price is locked in for decades, and their capital costs are falling. Figure 5-7 showed historical cost reductions for wind and solar PV, which are driven by, and in turn drive, their rapid rise in capacity and output at home and abroad. Quick and substantial ongoing cost reductions will be key to continued growth (see fig. 5-14). Solar technologies—both PV and concentrating solar power (CSP)—and nearshore wind are poised to achieve big reductions, in line with their track records and their recognized cost-cutting opportunities (see Realizing Renewable Cost Reductions sidebar). Several promising but not-yet-commercial renewable technologies—advanced geothermal, deepwater offshore wind, wave, tidal currents—have much more uncertainty around future costs and were conservatively excluded from our analysis, as was new small hydropower for lack of reliable data. Meanwhile, many mature renewable technologies like biomass and waste combustion, conventional geothermal, and hydro will probably see only modest cost and performance improvements.

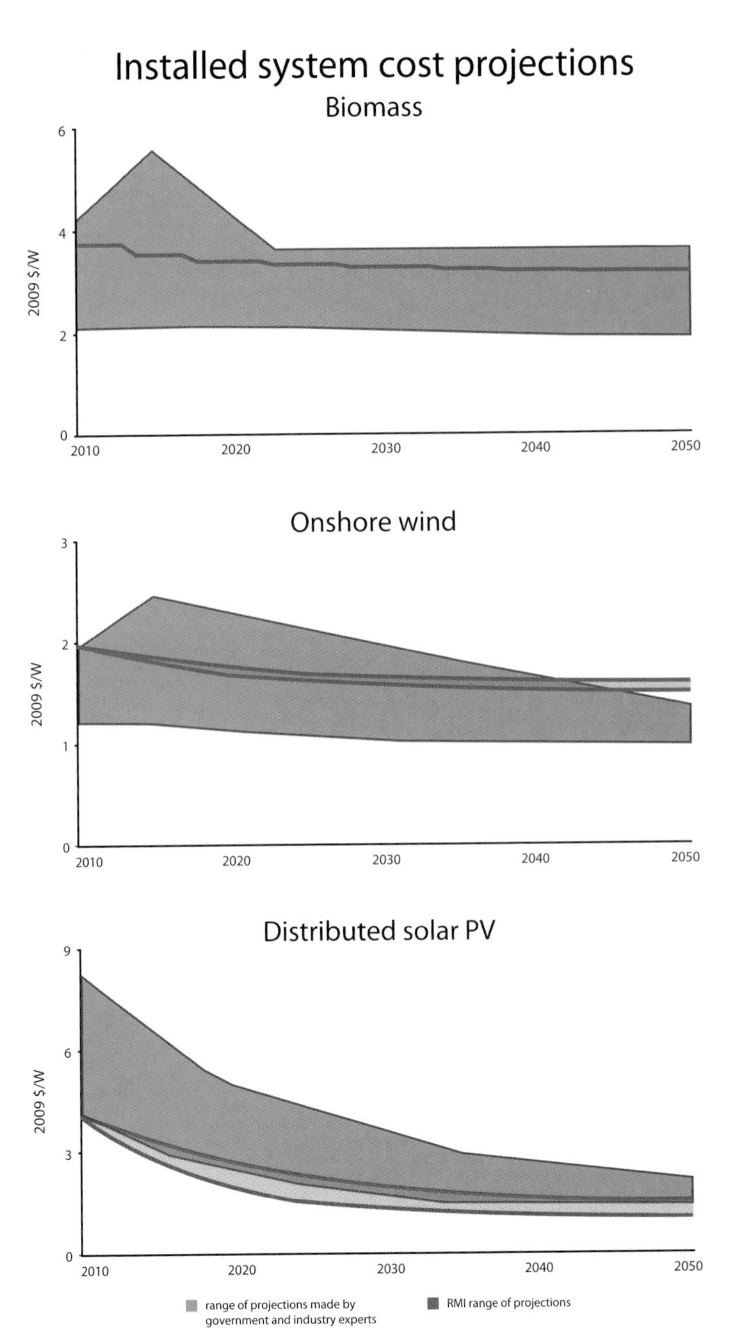

FIG. 5-14. Far from being theoretical, forward cost-reduction trajectories are rooted in observation of actual historical gains and in assessments of practical levers that can drive continuing reductions. By mid-2011, market prices for wind and PVs were already below the conservative projections shown here, which reflect fig. 5-7's price bulge already working its way through the supply chain; some 2011 wind projects cost less than $1.5/W. Our PV costs (blue) are at or slightly below the range of other projections (green) simply to reflect the latest market data (see Realizing Renewable Cost Reductions sidebar).[566]

REALIZING RENEWABLE COST REDUCTIONS

The solar photovoltaics industry offers one of the most compelling examples of ongoing cost reduction. Since the 1970s, every 10-fold increase in production of crystalline silicon modules has made module production 50% cheaper.[567] Today, such modules sell in bulk for below $1.4 per watt, down from best-in-class module prices in 2009 of over $2.5 per watt.[568] A major installer network[569] reported in April 2011 forward bulk prices around $1.0 for mid-2012. In May 2011, GE's global research director said that PVs "may be cheaper than electricity generated by fossil fuels and nuclear reactors within three to five years because of innovations."[570] Confirming this, by May 2011 California shareholder-owned utilities had contracted for 8.6 GW of new PV power, 4.4 GW of which is cheaper than the benchmark price (from a new combined-cycle gas plant) but is to be built in 2012–2017, before a gas plant could be. Another 45 GW or so of additional PV bids are expected in the state power auction, many at grid-competitive prices.[571]

Such dramatic cost reductions have been driven by a combination of improved manufacturing processes, economies of scale, and technological advances that have made it possible to create more efficient solar cells with less material inputs and outputs less sensitive to temperature and to the sun's angle and intensity. These trends are expected to continue. Most PV makers use technology and equipment borrowed from the highly refined and hypercompetitive microchip manufacturing business. Transistors became a hundred millionfold cheaper over the past 40 years;[572] PVs became about ten thousand times cheaper over the past 55 years. Nowadays a leading PV firm can make 1–3 GW of modules per year and build cookie-cutter plants in a year or two. The world added 17 GW of PVs in 2010[573] and by the end of 2011 will probably be able to make about another 50 GW *every year*.[574] There's no obvious reason why this industry can't scale like the chip industry, become even bigger, and provide a substantial portion of global electricity by 2050. Multi-GW-scale PV farms are already planned in the Chinese deserts and have been proposed in North Africa, and California, at the rate it's going, could be 20% solar-powered as early as 2020.

Meanwhile, a relentless stream of new solar technologies is keeping investors on their toes, using both silicon in various forms and a dizzying range of other materials and structures. First Solar, which produces modules at a cost (not a price) of $0.72/W,[575] has commercialized a rapid, thin-film, materials-frugal production technology. Other thin-film methods, high-efficiency materials, concentrating devices, and applications integrated with building materials (now common in Europe and Japan) promise step-change cost reductions. Entire new classes of devices are being invented.[576]

Even before those novel technologies reach the market, today's installed PV system prices have declined (by half since 1995, with one-third of that decrease since 2008)[577] to the point where the *non*module costs are the majority of best-practice total installed cost for utility-scale PV projects. Fortunately, these "balance of system" (BoS) costs are also ripe for major reduction.[578] Through smarter and smaller power electronics, streamlined installation technologies and processes, and project development approaches that leverage low-risk capital and better customer education, these project costs will decline as the market matures. In parallel with customer-facing and financing innovations, new preengineered system designs offer potential gains because they don't require site-specific electrical engineering. For example, some companies are using the module itself for the supporting structure while satisfying electrical grounding requirements without additional parts.[579] This eliminates additional racking components and cuts installation complexity.

Other renewables are also innovating rapidly. From 1999 to 2009, U.S. wind turbines increased rotor diameter 69% and hub height 39%, while better components, operations, and siting also boosted output.[580] In 2010, Siemens launched 3 MW gearless turbines with 25% more power but less weight and maintenance, half the parts, and a nacelle that fits in standard vehicles[581]; 18% of the 2010 world market was gearless. But the best may be yet to come. Clusters of smaller rotors cowled to capture wind from a larger area, as FloDesign is developing, can dramatically cut cost per kWh. Seamless carbon-composite airfoils whose shape morphs in a fraction of a second, optimizing aerodynamics with every gust, can cut blade stress, permitting longer blades that boost energy capture 10–15%, raise capacity 20–45%, and cut windpower cost 5–9%.[582] Broad-Star Wind Systems' AeroCam design packs more wing area into a smaller package for commercial roof edges, generating power with about two to three times less space at half the cost while shielding PVs from wind stress, which drives PV mounting costs.[583] It's one of several new designs well suited to urban settings and moderate windspeeds.

Geothermal projects exploiting deep hot dry rock could benefit from new field designs using novel downhole heat exchangers (a Geothermic Solutions innovation), plus cheaper drilling to reach a heat reservoir and superior well completions to contact and exploit it. Foro Energy is leveraging advances in high-power lasers to enable step-change gains in drilling ultra-hard crystalline rock and improving flow in innovative well completions.[584] Using shallower hot-water resources, Germany's geothermal sector is now booming and, says its trade association's CEO, "could supply Germany's electricity needs 600 times over"[585] despite its complete lack of volcanic geothermal resources.

Our *Renew* analysis doesn't count on such step-change innovations—but they do bolster confidence that our learning curves, based on the best industry and national labs data, are likely to be realistic for the technologies they include, and conservative in omitting the many others now rapidly emerging.

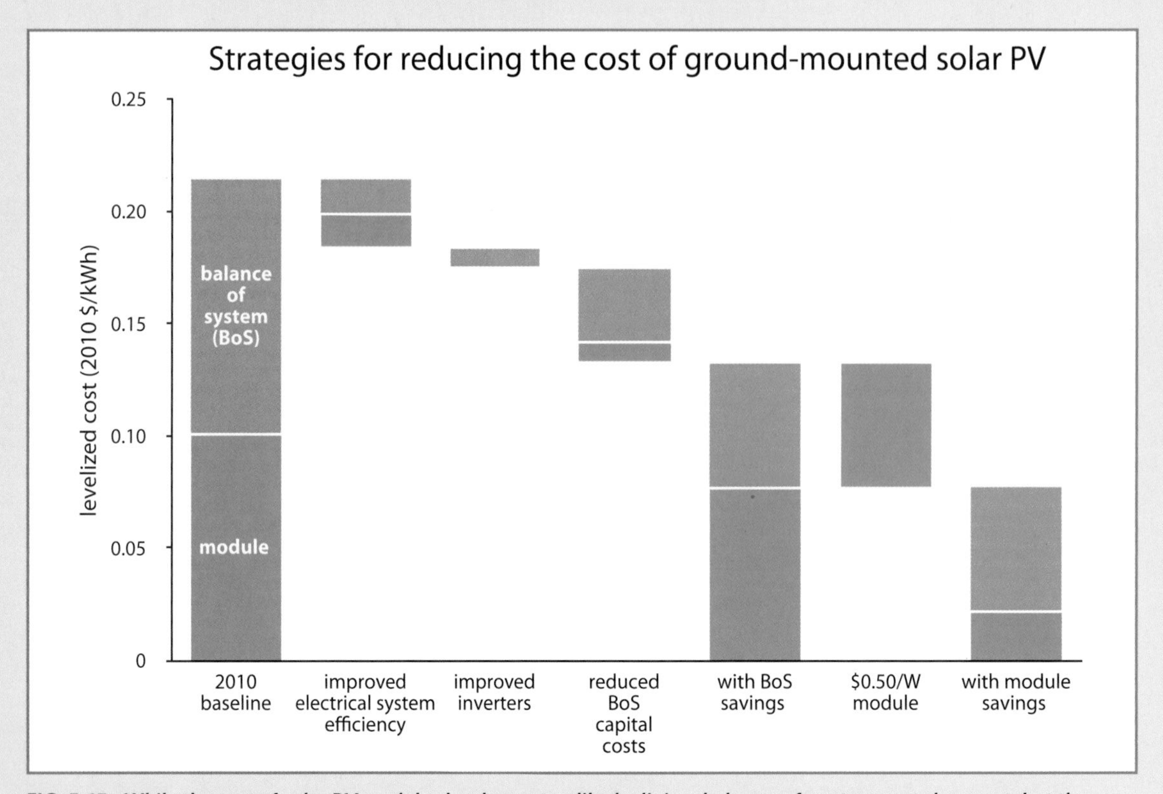

FIG. 5-15. While the cost of solar PV modules has been steadily declining, balance-of-system costs have not, but they can—by around half—if the many levers shown are pulled vigorously, as is now beginning to happen.[586]

Shifting to renewables must consider more than raw costs, too. Retiring old plants early can mean accounting write-offs as well as earlier replacement. Revenues can change or shift, too, if coal and nuclear plants retire or run less as renewables replace them. But this frees up fossil-fueled capacity already built, especially the more flexible gas-fired plants, to provide system backup as variable renewables produce more energy. In our analysis, by 2050 dispatchable capacity falls by much less than its own output as old plants shift from being traditional generators to flexible reserves—or even get shut down but maintained in readiness in case extreme events disrupt other resources. This shift may need new rules to recover capital in a way that fits resources' long-run system value.

Keeping the Lights On

Renewable technologies' historically higher cost has not been the only reason for utilities' skepticism about their large-scale practicality. A key concern has been whether renewable energy sources are too unreliable and unpredictable to keep the lights on and the factories humming in our modern economy. Windfarms and solar panels can't generate electricity when the wind dies down or the sun sets; their output fluctuates with the weather (fig. 5-16).

Having spent more than a half century relying mostly on large, controllable power plants, grid operators find wind and solar generation unnerving for two distinct reasons—its variability and its uncertainty. Wind and solar are variable because their output fluctuates throughout the day outside the control of grid operators, and they are uncertain because we cannot predict with complete accuracy what the output level will be at any given point in the future. Many believe that, as a result, renewables can never reliably supply more than a few percent of grid power. Getting to 50%, much less 80% or more, is considered fantasy.

But grid operators actually already have considerable experience dealing with variability and uncertainty. For a century, they've coped with constantly changing demand at every moment of every day. Furthermore, all power plants fail from time to time—fossil-fuel power plants are down about 14% of the time due to unexpected equipment failures (accounting for 6–8% of shutdowns) or planned maintenance.[587] To manage these daily sources of variability and uncertainty, grid operators rely on diversification, flexibility, and sometimes storage, each explored below.

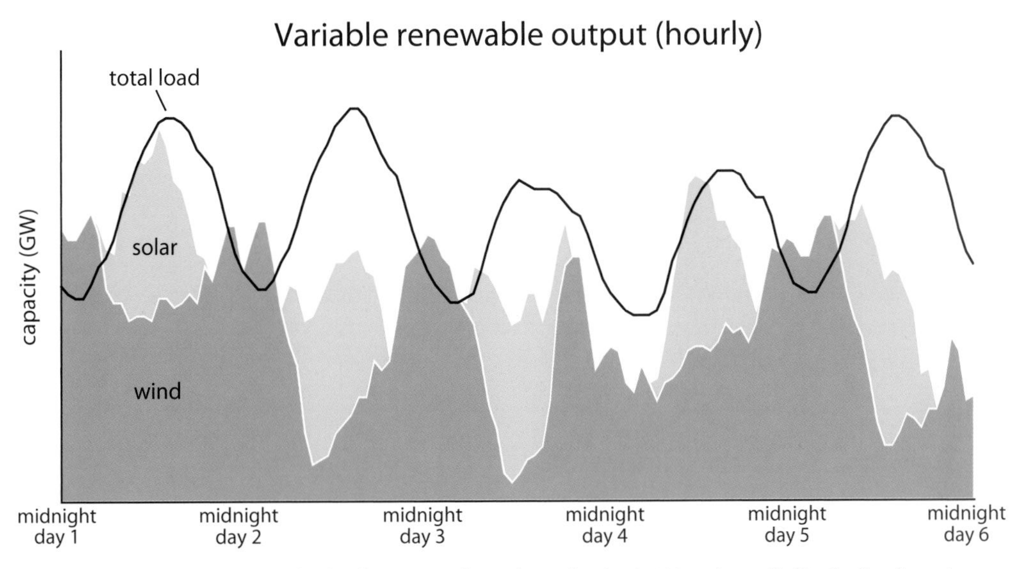

FIG. 5-16. The output from wind and solar fluctuates throughout the day in this schematic illustration based on actual data.[588]

ELECTRICITY SYSTEM OPERATIONS 101

Electricity is the only important "energy carrier" that cannot yet be easily and cheaply stockpiled or stored. Therefore, electricity must be produced and used at the same instant; it is the ultimate perishable commodity or just-in-time manufacturing system. Historically, this has meant expert operators must coordinate, schedule, and manage the grid in an intricate dance to speed up or slow down large generators' power output to match the magnitude and timing of electricity demanded instantaneously. Everything connected to the grid must work in harmonious rhythm in a symphony of frequency, voltage, and power quality. Too much or too little and the system crashes.

Every day, millions of customers flip off their light switches, dry their clothes, turn on steel furnaces, or ramp up production lines. The diverse timing of these individual demands produces broad patterns called "loadshapes" that change over hours, days, seasons, and years (fig. 5-17). System operators can usually plan rather accurately for these patterns, but unexpected events can and do happen, so operators must keep power in reserve that can be called upon within minutes or less.

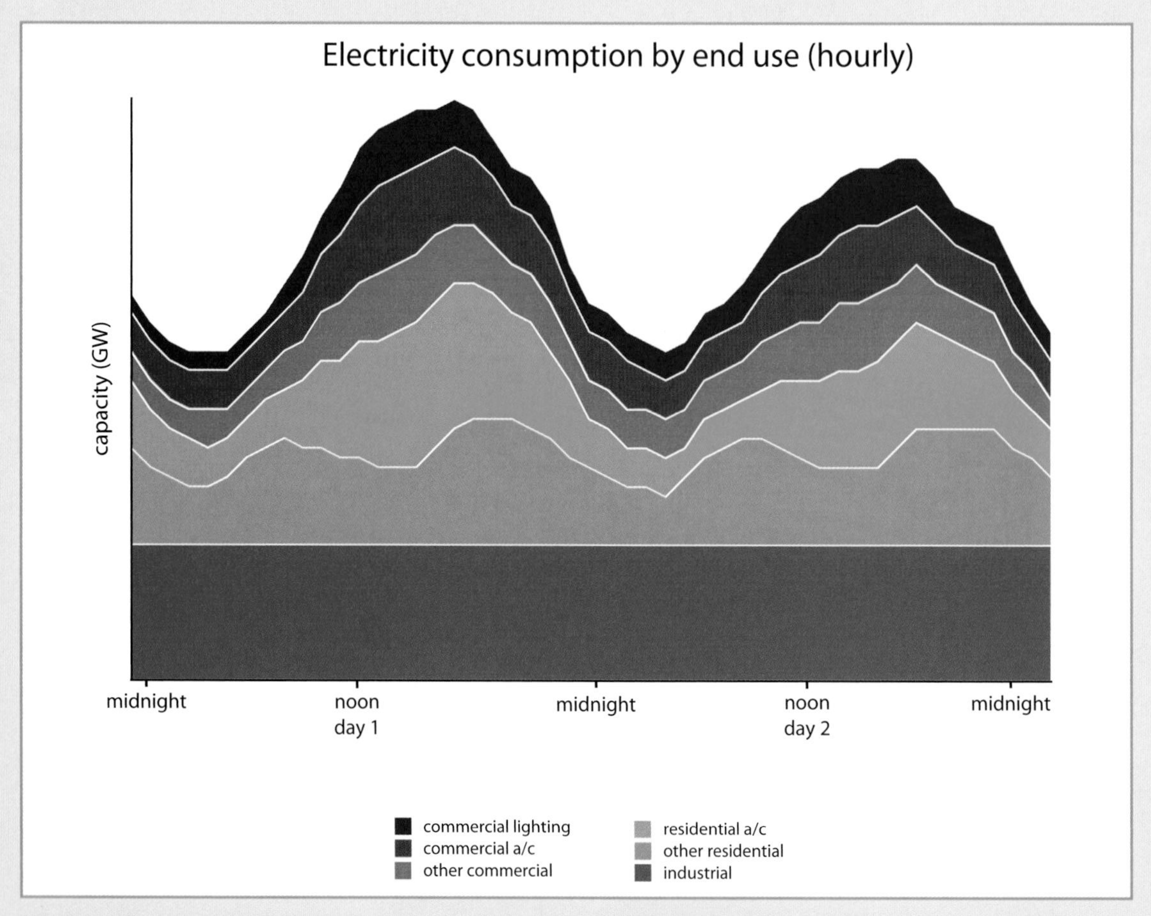

FIG. 5-17. The smooth loadshape reflects the diverse timing of individual loads across a large utility's entire service territory. The curves get more jagged and volatile if examined more closely in time or space.[589]

To meet this ever-changing demand, system operators use generators with the lowest operating costs first. These generators are commonly called "baseload" plants because they are used to meet the lowest expected level of continuous aggregate demand. Traditionally, baseload plants have been big steam plants, such as coal and nuclear, which are costly to build but cost only a few cents per kilowatt-hour to run, and which operate more efficiently at a constant high output. When demand rises moderately—so-called shoulder demand—operators tap higher-operating-cost generators. When demand peaks, operators may need to use the costliest-to-run generators, typically combustion turbines. These generators are flexible, able to start up quickly and to ramp electricity production up and down rapidly to meet fluctuating demands. In the hottest hours of the year, when air-conditioning loads strain utilities' capacity, the last, least efficient plants dispatched may cost as much as 80¢ per kWh or more to run[590] (or nine times the average retail price of electricity in the U.S. in 2008)[591]—and up to several dollars for a system-peak kWh *delivered* to customers in congested parts of the distribution system.[592]

A mix of baseload, shoulder, and peaking plants optimizes the cost of supply over the whole year (see fig. 5-18). Thus, the cost of producing the next kilowatt-hour of electricity changes throughout the day, rising with increasing hourly electricity demands. Traditionally, customers don't see these fluctuating costs, but instead are billed a flat price by their electric utilities based on average production costs plus capital recovery, overhead, and profit.

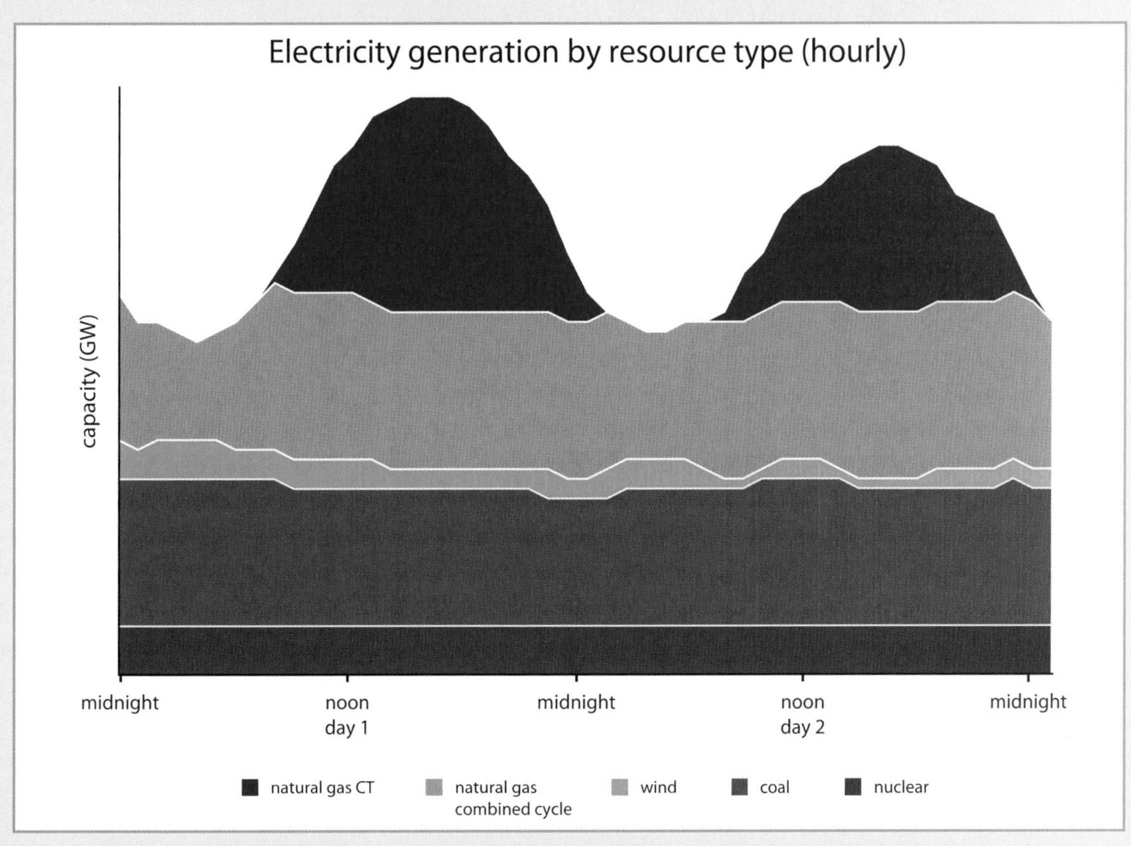

FIG. 5-18. Electricity supply is produced by a stack of various generating resources, in order of increasing operating cost, to match demand at every hour of the day.[593]

The electricity system takes advantage of diversity of both demand and supply. Demand diversity means that individual decisions—to turn on the lights or start the dishwasher—meld together into smoother loadshapes that utilities have learned to predict with reasonable accuracy. Similarly, supply diversity means that the grid relies on not just one but many power plants working in tandem. The output of any single *generator* is largely irrelevant; it's the reliability of the *system* that matters. Just as a diversified stock portfolio produces smoother, more predictable returns, a diversified portfolio of many power plants that fail at different times for different reasons produces a more reliable, predictable supply of electricity. It is this portfolio effect that led Federal Energy Regulatory Commission (FERC) chairman Jon Wellinghoff to tell the U.S. Energy Association, "People talk about, 'Oh, we need [firm output] baseload [power plants].' It's like people saying we need more computing power [so] we need mainframes. We don't need mainframes, we have distributed computing."[594]

The same principle of diversity can make integrating variable renewables much easier. The output from wind and PVs[595] can be made less variable by linking renewables from different areas—and of different types—via transmission lines to send or receive power over a wider region. The wind may be calm in one place but strong a few hundred miles away; if the wind is calm, the sun may be shining. Well-diversified sites can reduce wind's variability by up to half.[596] And other kinds of renewables—small and big hydro, biomass and waste combustion, geothermal, and solar thermal electric with heat storage—are dispatchable.

Operators also need flexibility. Flexible generators include fast-ramping power plants, such as peaking hydro or combustion turbines, which can quickly modulate output to match demand.[597] Those same versatile power plants that stand ready to substitute for suddenly failed big power plants or power lines can also, or instead, help meet the increased variations as wind and solar fluctuate. Utilities also routinely keep at least 15% extra power capacity in reserve.[598] That's a part of the cost of reliably integrating traditionally large power plants and lines. Variable renewables incur such costs too, but theirs can be roughly comparable and may even turn out to be lower, because a portfolio of many smaller units of differing kinds isn't likely to lose so much capacity unexpectedly all at once, and they may be nearer the users.

As many utility executives have realized, they already have a lot more flexibility on their existing systems than they may take credit for. But utilities' flexibility is not limited to turning their own flexible power plants on or off, which can be costly. They can cooperate with other utilities to share these flexible reserves. They can schedule power every five minutes instead of every hour, and over wider "balancing areas." And they can integrate forecasting of weather events into their grid operations to prepare better for cloudy or windless periods.

Flex Your Power

A new source of flexibility could come from demand itself. Up to now, customer demand, especially residential, has been considered to be relatively inflexible, only partially predictable, and not responsive. Customers can plug in their computers and refrigerators and know that they'll get electricity. But in most cases, deciding to plug in or not is the only choice they have. They can't choose to use electricity when it's cheapest, for instance, since most households and businesses pay the same price for electricity no matter when they use it, even though the utility's cost of providing it can vary by as much as tens of times over the course of a day.[599] For most people, electricity service is a bit like buying an early Model T: you can have any color you want as long as it's black.

But this passive consumption of electricity is on the brink of a radical shift. Advances in information technology (IT) and the smart-grid technologies that combine IT with the electric grid are enabling bidirectional control, distributed intelligence, two-way communication, ubiquitous real-time price information, and demand response. These new technologies are creating the opportunity to capture new value for customers and providers, old and new. For example, automated controls can imperceptibly respond to price signals by adjusting appliances' usage, allowing customers to minimize their costs with no hassle and without uncompromised experience, just as ATMs and the Internet have done in banking.

This new frontier of responsive demand builds on conventional utility programs that have paid large customers to curtail their demand to lower the system's peak demand in only a handful of hours per year. These programs have proved to be extremely valuable tools for cost-effectively meeting grid-balancing needs in times of system stress. The Texas experience shows how these demand response resources might prove valuable in the future. Texas's grid operator, the Electric Reliability Council of Texas (ERCOT), manages a voluntary demand response program, Loads Acting as Resources (LaaRs). On February 28, 2008, ERCOT detected a significant supply-and-demand imbalance driven by an unforecasted increase in demand coupled with a decrease in energy production from several conventional power plants and a drop in wind production (accounting for 5% of the imbalance, and accurately forecast). The event triggered an emergency response that activated the LaaRs system, adding approximately 1,100 MW of resources within 10 minutes. Other than the customers in the LaaRs system, no other customers lost power and system balance was restored.[600]

Building on this history, companies offering increasingly advanced demand-side resources are winning customers. Homes and commercial buildings are sprouting digital displays of the actual price of electricity at that moment, so people can decide whether they want to dry their clothes now or later for a few cents less. "Time-of-use" prices can shift several times a day, and "real-time prices" hourly, to signal the changing cost of utility service. In 2009, 169 utilities, ranging from PG&E to PPL Electric, were implementing residential time-of-use rates, and 19 utilities were implementing real-time rates designed to promote demand response.[601] In that same year, more than one million residential customers across the country were participating in time-of-use pricing programs.[602] Collectively, these demand-side resources can provide a valuable new source of flexibility that grid operators can use to integrate variable renewables and manage a continuously dynamic electric grid.

Calling on Storage

Another source of flexibility is electricity storage. Some bulk power storage systems already exist, such as 21 GW of hydroelectric pumped storage[603] or compressed air stored in caverns.[604] American Electric Power has demonstrated islanding (serving local needs from local sources, both isolated from the surrounding grid) with distributed storage, of which it's set a goal to install up to 1 GW—initially for conventional reasons like shaving substation peak loads to extend transformer life, but with an eye toward facilitating distributed renewables and automotive power exchange.[605] New storage from thousands of plugged-in electric autos could become cost-effective in just a decade. If researchers and entrepreneurs do succeed in making electricity storage cheap, that would make a renewable future easier. However, contrary to a widespread assumption, lack of cheap storage is not a showstopping obstacle to a renewable future—the needed flexibility, as we've just seen, can come from a variety of sources.

Conducting the Symphony

The combination of diversifying variable renewables by type and location, forecasting their variation, and integrating them with dispatchable renewables, flexible fueled generators, and demand response can together make a very powerful toolkit and create a power system that has the potential to meet our needs reliably. Figure 5-19 illustrates what these combined strategies might look like in practice. First, energy efficiency reduces the total load that must be met and disproportionally reduces the peak. In this example, wind that blows mainly at night and solar that peaks at midday provide the majority of electricity demand, but their sum does not perfectly match the loadshape. Sources of physical flexibility ranging from plug-in hybrid electric autos to centralized compressed air storage are used to fill in the gaps, drawing power when there is an excess

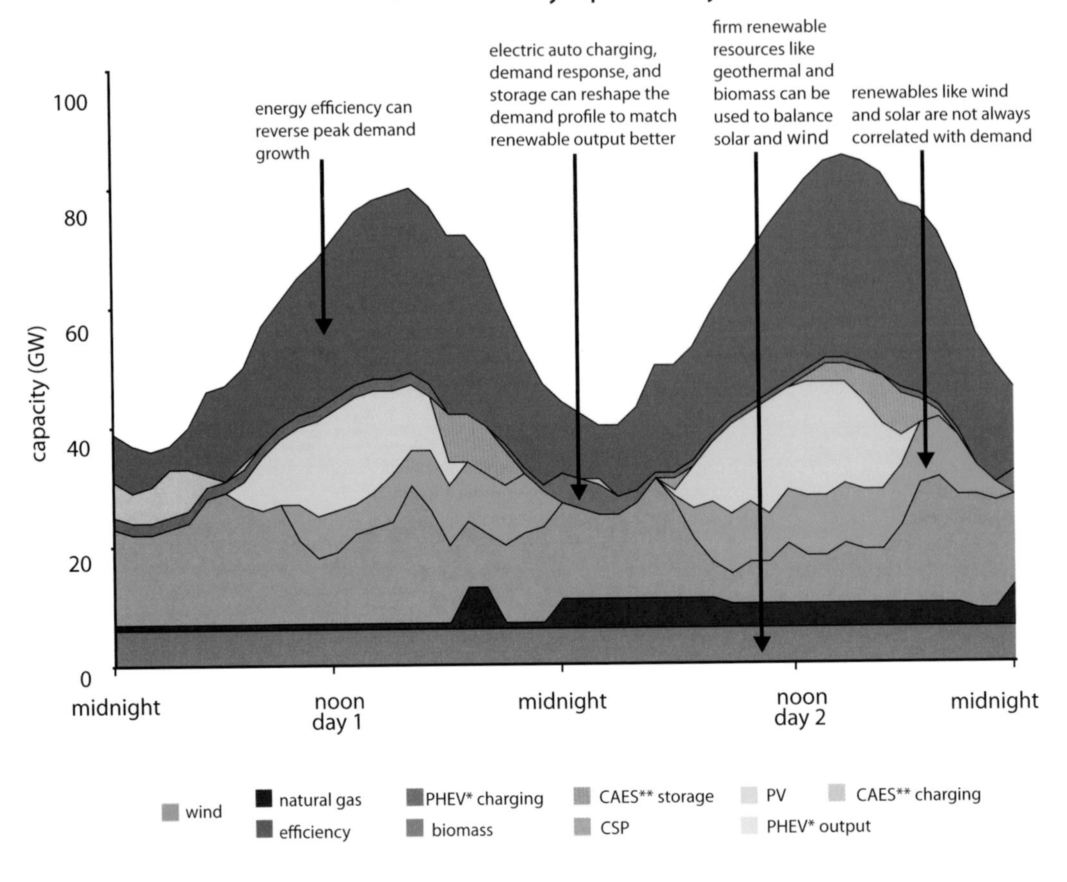

* Plug-in hybrid electric vehicle
** Compressed air energy storage

FIG. 5-19. This illustration uses actual renewables data and projected 2050 loadshapes. In this case, diversified renewables can reliably meet 80% of the annual electricity need by choreographing sources of flexibility ranging from bulk storage (where justified) to demand response to electric autos.[606]

and providing power when there is a shortage. Together, these resources can meet hourly loads and minimize the amount of renewable electricity that would be "spilled"—available but unneeded.

Integrating ever-higher levels of renewables is being successfully demonstrated in the real world. In 2009, eight American and three European authorities, writing in the leading electrical engineers' professional journal, didn't find "a credible and firm technical limit to the amount of wind energy that can be accommodated by electrical grids."[607] In fact, not one of more than 200 international studies, nor official studies for the eastern and western U.S. regions,[608] nor the International Energy Agency,[609] has found major costs or technical barriers to reliably integrating up to 30% variable renewable supplies into the grid, and in some studies much more.[610]

Meanwhile, renewable electricity supply in the 20–50+% range has already been implemented in several European systems, albeit in the context of an integrated European grid system. In 2010, four German states, totaling 10 million people, relied on windpower for 43–52% of their annual electricity needs.[611] Denmark isn't far behind, supplying 22% of its power from wind in 2010 (26% in an average wind year).[612] The Extremadura region of Spain is getting up to 25% of its electricity from solar, while the whole country meets 16% of its demand from wind.[613] Just during 2005–2010, Portugal vaulted from 17% to 45% renewable electricity.[614] Minnkota Power Cooperative, the leading U.S. wind utility in 2009, supplied 38% of its retail sales from wind.[615]

Reaching and exceeding these levels does require change—not in the physics of how grids work, but

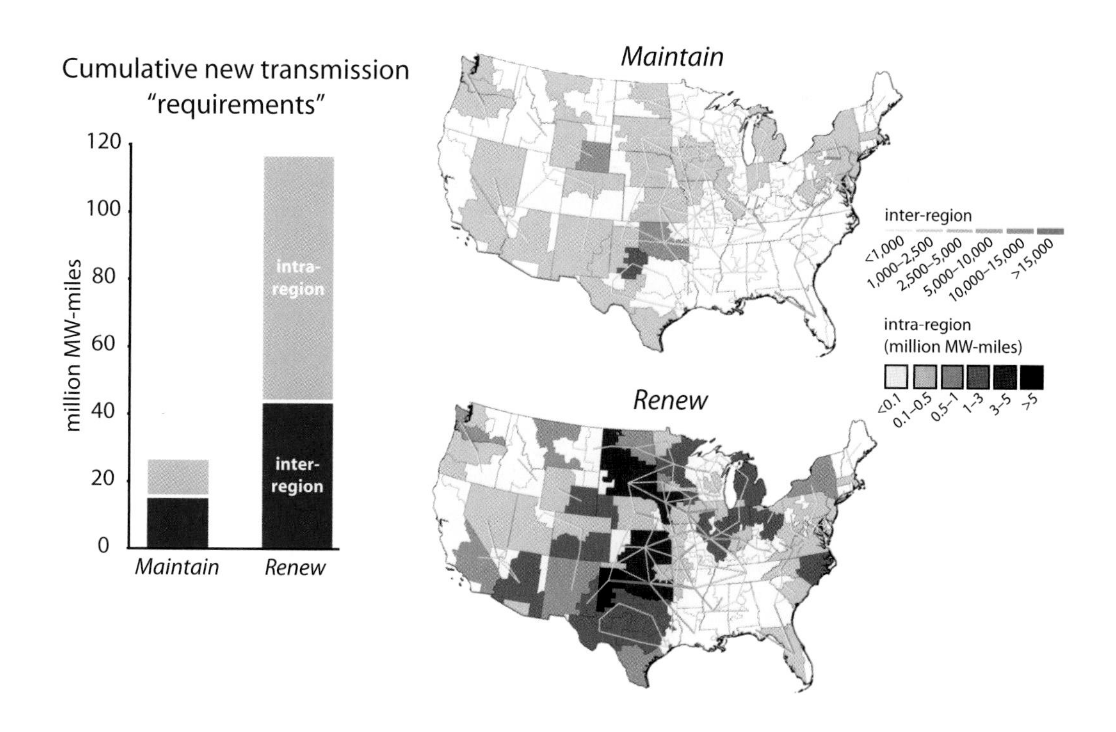

FIG. 5-20. Substantially more transmission is needed in *Renew* to connect the best renewable sites to load centers.[616]

in the strategies, rules, and procedures operators use to run them—a big institutional challenge.

Powerline Blues

An equal physical challenge lurks outside the control room. A major risk to a future that relies primarily on large-scale renewable energy is inherent in the infrastructure that connects these resources—large-scale transmission. Some of the best wind is in the sparsely populated Great Plains and offshore, while the greatest solar energy is found in the Southwest deserts. To tap these rich resources and bring them to coastal load centers, we'd need to build new transmission lines. As we've pointed out, transmission enables geographic diversification and eases balancing across the system because variability is smoothed over larger areas. A key question is just how much transmission could be needed. Our analysis shows that 116 million MW-miles of new high-voltage transmission lines, costing an estimated $166 billion, would be needed through 2050 in this *Renew* scenario (fig. 5-20). (*Maintain* and *Migrate* need more transmission too, but about 58% less than *Renew*.)

The challenge is significant. Transmission projects crossing utility or state boundaries are slow and hard to get sited and approved. The increased requirement for transmission to interconnect the best renewable regions with distant load centers poses a significant risk, because siting and permitting transmission projects is slow and difficult and is already posing challenges around the country.

One way to ameliorate these risks is to achieve new levels of regional and federal cooperation and partnership in planning, siting, financing, building, and managing new transmission. Another, not yet well analyzed, is the potential for efficiency, demand response, and local generation to free up capacity on existing lines, or to obviate the need for additional lines, more cheaply than building them. Those freed-up lines might then make nearer resources (like Great Lakes wind rather than Dakotas wind to serve the upper Midwest) cheaper than better resources farther away but needing new lines. Finally, creative transmission layouts could assuage some transmission challenges, like the offshore subsea New York–to–Virginia transmission networks that Google and Good Energies have announced,[617] or those that Europe plans to build beneath both the North Sea and the Mediterranean.[618]

Are We There Yet?

This *Renew* case, then, has much appeal. There is plenty of renewable resource available to generate 80% of U.S. electricity cost-effectively from centralized renewable sources, with an ample supply-and-demand balance from a diverse supply portfolio in each region (fig. 5-21).

Pursuing this future would essentially free us from the depletion, logistics, health, environmental, and climate risks of coal. The renewable technologies needed are modular, mostly mass-produced, and built faster than large coal and nuclear plants, tying up less capital for shorter periods and offering more option value for responding to changes. This lowers financial risk for utilities and investors. While continued study and experimentation is required, the evidence from ongoing experiments around the world and from detailed modeling convinces us that centralized renewables can be effectively and reliably integrated into the grid even at very high fractions, given new "rules of the road" and the comfort of experience.

Renew has a few issues, though, that give us pause. First, siting and building transmission will continue to be a challenge around the country. Second, a system dominated by central renewables has the same security and reliability risks as *Maintain* and *Migrate*: a heavily centralized grid and vulnerable transmission. Third, the operational and technological conditions that are necessary in a centralized renewable future lay the foundation for a new frontier in the electricity sector—one that puts the power of choice and control in the customer's hands.

Several trends—the growth of modular renewables, the rapid innovation of smart grid technologies, and the increased value of flexibility on the grid—create new possibilities that pervade both sides of the customer's electricity meter. Consider the renewable energy market. Rapidly growing global demand for renewable electricity technologies, chiefly in large-scale installations, will increase cumulative experience and adoption

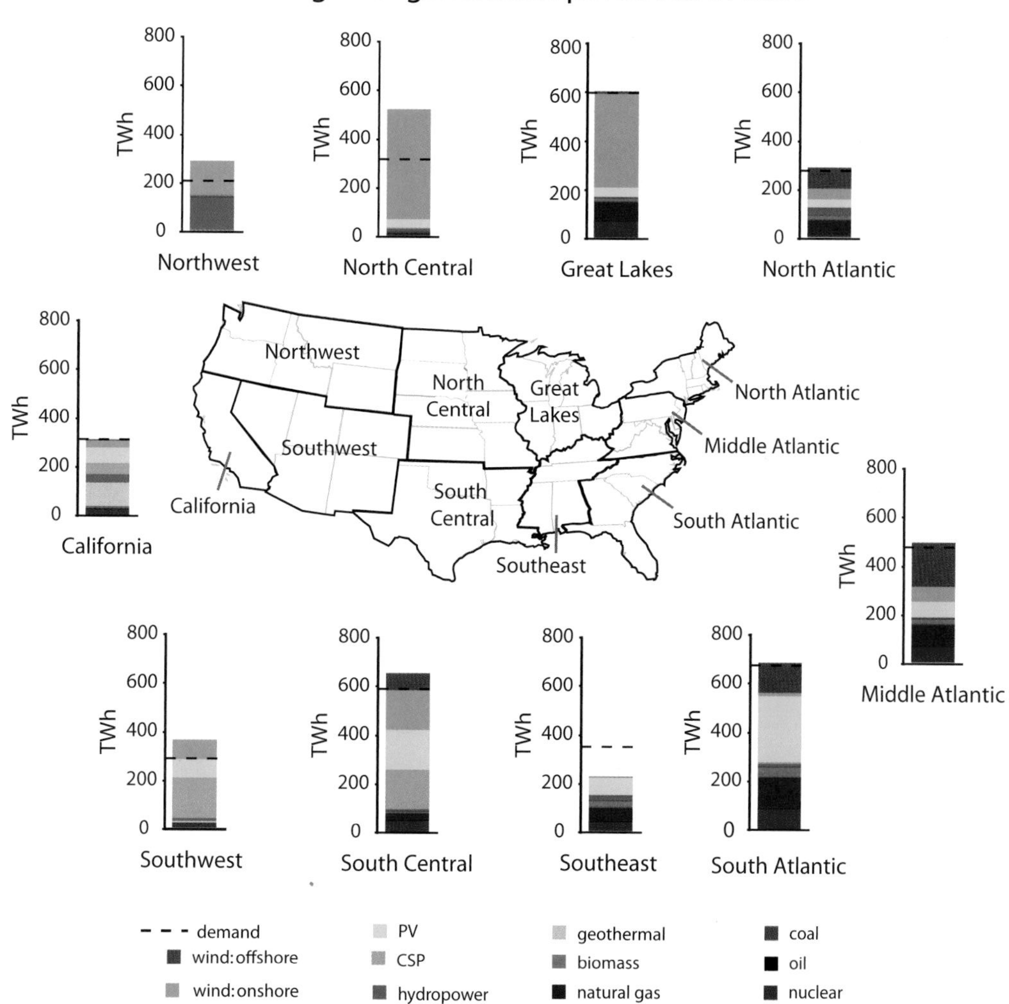

FIG. 5-21. Every U.S. region has ample renewable electricity potential. Its composition varies widely between regions, and some, like the Southwest and North Central states, have large surpluses compared to plausible loads, so the excess can be exported to less renewables-rich regions. These illustrative resource portfolios for an 80%-centralized-renewable scenario, used in our *Renew* case, show that the renewable resources used may vary widely by geography.[619]

that will benefit not only their centralized but also their smaller-scale applications. After all, a centralized solar PV array is built from the same modules you put on your roof. This means that *a growing market for centralized renewables will drive cost reductions for more locally sited generation, too.* And while a smart grid can ease the integration of centralized variable renewables, it simultaneously drives an increase in customer-sited generation and demand response. The *consumer* will become a *prosumer* who has many more ways of saving—and making—energy and money.

Today's one-size-fits-all, top-down approach could begin to shift to a fine-grained, highly diverse, more customer-centric future, accelerating the transition from regulated markets to more competitive and open power markets. The electricity sector's century-old linear *value chain* could morph into a complex *value web* where traditional roles blur, eroding the traditional cost-of-service regulation and "natural monopoly" business model.

In fact, the U.S. electricity system could be headed for a tipping point that will shake the traditional electric utility business model to its core. Will the result be functional, secure, safe, reliable, and affordable? While precise answers to these questions are at the edge of the best experts' understanding today, we can describe the emerging outlines of this revolutionary new world by fleshing out specific examples.

TRANSFORM: A SEISMIC SHIFT IN SCALE

An infectious idea is spreading rapidly through the world of electricity: the notion that power plants needn't necessarily be big and, for many uses, may better be made the right size for the job.[620] As Malcolm Gladwell remarked, "We like to use words like contagiousness and infectiousness just to apply to the medical realm. . . . Ideas can be contagious in exactly the same way that a virus is. . . . So what happens when you look at an infectious idea under a microscope?"[621] Let's find out by exploring our fourth case, *Transform*.

This kind of revolutionary change doesn't usually start from the top down—it starts from the bottom up, at the grassroots. Many seeds of change are emerging on military bases, data centers, college campuses, and housing developments as customers start to take greater control of their power quality and costs. Consider these:

- Finding space for a solar farm in the middle of Chicago's business district might seem like a challenge, but only if you're looking for vacant land. Chicago's 108-story Willis Tower is now exploring the possibility of becoming the nation's largest vertical solar farm, with up to 2 MW of solar capacity supplied by Pythagoras Solar. On the ground, 2 MW would cover 10 acres. Besides PV and solar water-heating panels on the roof, south-facing windows will be replaced with photovoltaic glass that preserves the view, shades out heat and glare, and also makes electricity. Energy efficiency retrofits plus solar generation will cut electricity use by as much as 80%.[622]
- State laws and utility regulations typically make it hard if not impossible for anyone but a regulated utility to sell power generated at one site to a buyer at another site. In New Jersey, however, state regulators have approved a pilot program that will allow developers of a solar farm being built on a reclaimed landfill site to sell power to residential and commercial customers within the redevelopment area. The 6.5 MW solar farm, with 24,600 solar modules, will occupy about 35 acres in Stafford Township. Its electricity will serve 216 apartments, nine local government buildings, an existing Target store, and 250,000 square feet of new retail space.[623]

- A superefficient affordable housing development in Sacramento, California, will use a first-of-a-kind private, commercial microgrid to manage and distribute intelligently the generation and storage of solar power among 34 single-family homes. The project, known as "2500 R Street," aims to achieve net-zero efficiency levels, with each home generating as much clean energy as it uses. Smart-grid technology from Sunverge Energy will let the utility manage generation and storage discretely or in aggregate—improving utility transmission and distribution, deferring capital investment, improving grid reliability, and reducing greenhouse gas emissions.[624]
- Across town, the corporate headquarters of the Sacramento Municipal Utility District (SMUD) is adding a different kind of microgrid. Three modular combined-heat-and-power (CHP) systems, each generating 100 kW, will supply a portion of its electrical, cooling, and hot-water needs year-round. The units can run normally in conjunction with the grid, but in a blackout, they can seamlessly disconnect and continue operating as an "island," providing power to a selected portion of the headquarters site. Reconnecting to the grid is equally seamless. Reusing previously wasted heat will achieve more than twice conventional power plants' overall efficiency.[625]

CASE 4: *TRANSFORM*

Renew's confluence of trends provides the basis and momentum to move beyond the centralized grid architecture developed in Edison's day to *Transform*—powered by resources of varied scale but with a greater portion of supply coming from distributed sources such as rooftop solar, CHP, fuel cells, and small-scale wind. The grid in this future exploits renewables' geographic and technological diversity as *Renew* does but needs half its transmission (subject to uncertainties about distribution, not yet fully modeled) because distributed generation is at or near customers. These mini power plants could enable the grid to be clustered in interlinked "microgrids" that could stand alone at need and hence could better withstand and recover from grid failures. We've analyzed this case using the following illustrative resource mix:

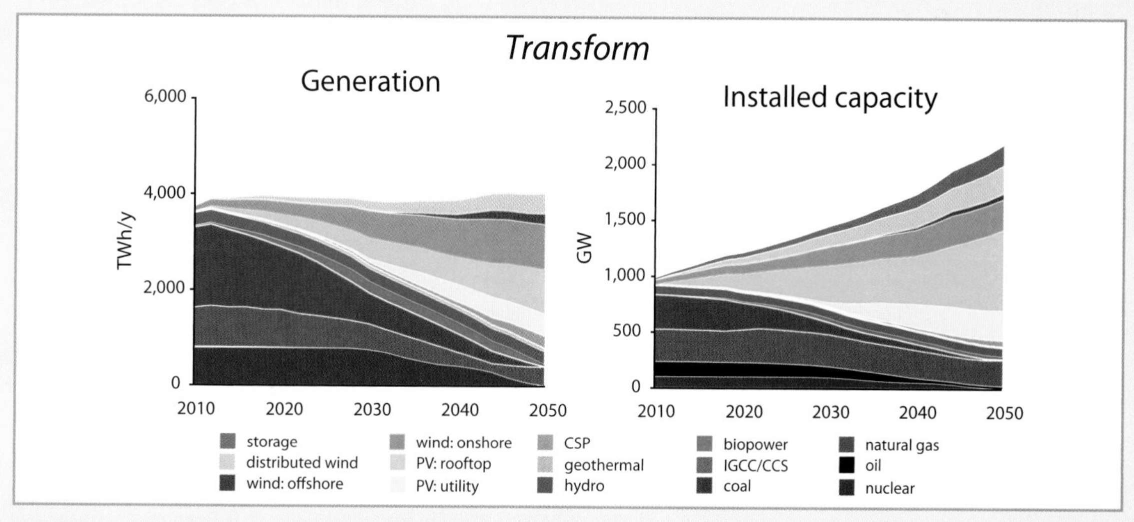

FIG. 5-22. Aggressive energy efficiency (fig. 5-23) flattens electric loads in the *Transform* case, but installed capacity grows substantially due to certain renewables' lower capacity factors. That growth (and natural gas use) could decrease if CHP in buildings were also modeled; as a substantial conservatism, CHP is modeled only in industry.[626]

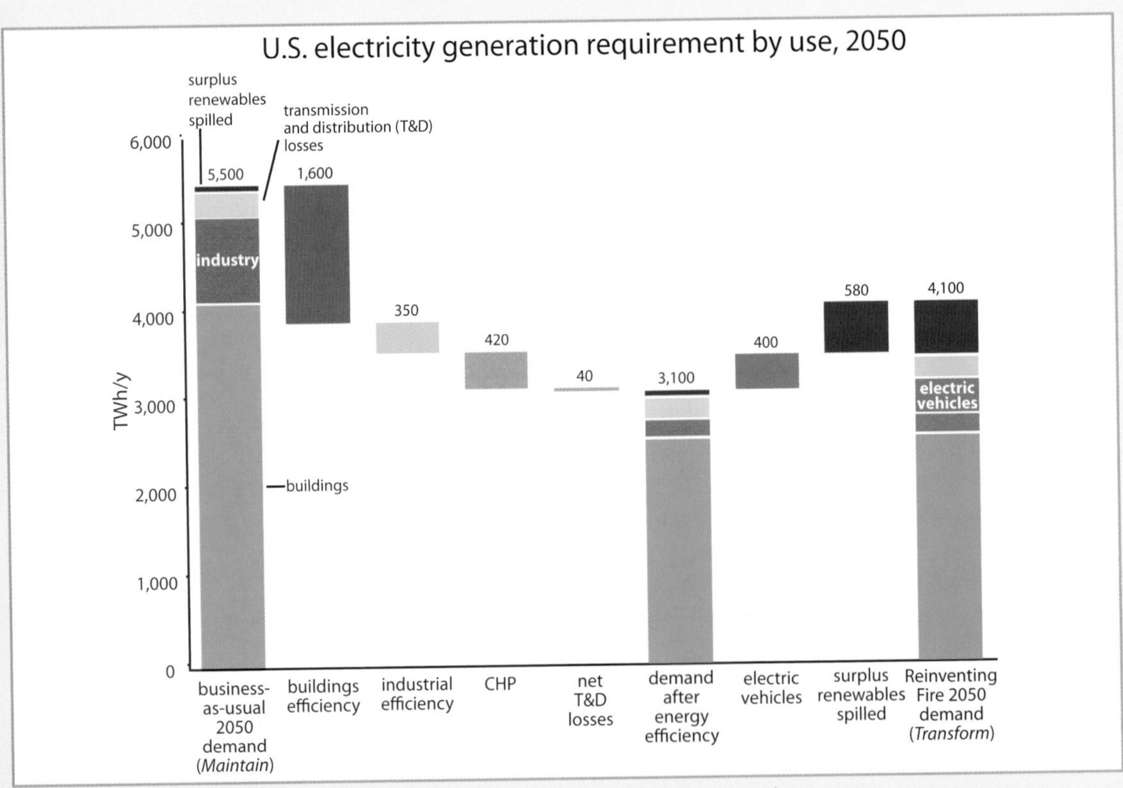

FIG. 5-23. Actions by customers can prepare the electricity sector for supply transformation. The end-use efficiency gains discussed for buildings and industry in chapters 3 and 4 can reduce traditional generation requirements. So can widespread adoption of cogeneration. Electric vehicles use more electricity, but less than savings elsewhere. Total use falls 25%. Conservatively, integrative design's savings are *not* shown.[627]

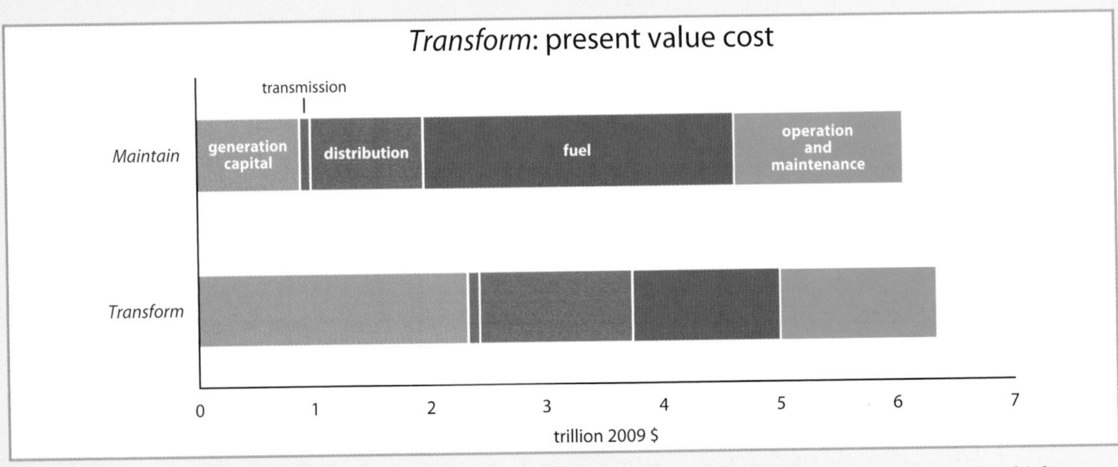

FIG. 5-24. The 2010 present value cost of *Transform* is slightly above that of *Maintain*, though with somewhat higher uncertainty in its capital cost (dependent on continued technology cost reductions), partly offset by avoided uncertainty in its much lower fuel costs.[628] **However, financial economics (sidebar, p. 207) changes the picture.**

What do these examples have in common? They reflect the potential for a fundamental shift from the top-down centralized approach to a localized approach that combines distributed generation, intelligent demand devices, and digitally empowered infrastructure. The foundation of this shift is the distributed generation market, which is capturing a steeply rising share of new generating capacity. Stimulated by policies ranging from investment tax credits in the U.S. to feed-in tariffs in Germany to even cheaper auctions in Brazil, the global distributed-generation market grew 91% in 2010 alone to $60 billion.[629] In the past decade, micropower—CHP plus renewables minus big hydro—has more than swapped with nuclear power their respective shares of global electricity production, and in 2008 micropower provided roughly 90% of the world's additions of electricity generation.[630] And the potential market is large. In fact, just installing flat-panel solar PV on well-suited U.S. roof space could generate enough total electricity to meet one-third of U.S. annual electricity demand.[631] Moreover, falling costs (fig. 5-25) will dramatically expand markets both for the centralized resources shown and for the distributed versions that use many of the same components.

Grid Operations Become More Distributed

The shift to a more distributed future is as much about the grid as about generation technologies. Wider use of variable renewables will create demand for more flexibility to match fluctuating supply and demand. In response, smart grids and advanced control systems will meet a larger share of the balancing responsibility as buildings, factories, and households automatically respond to system needs.

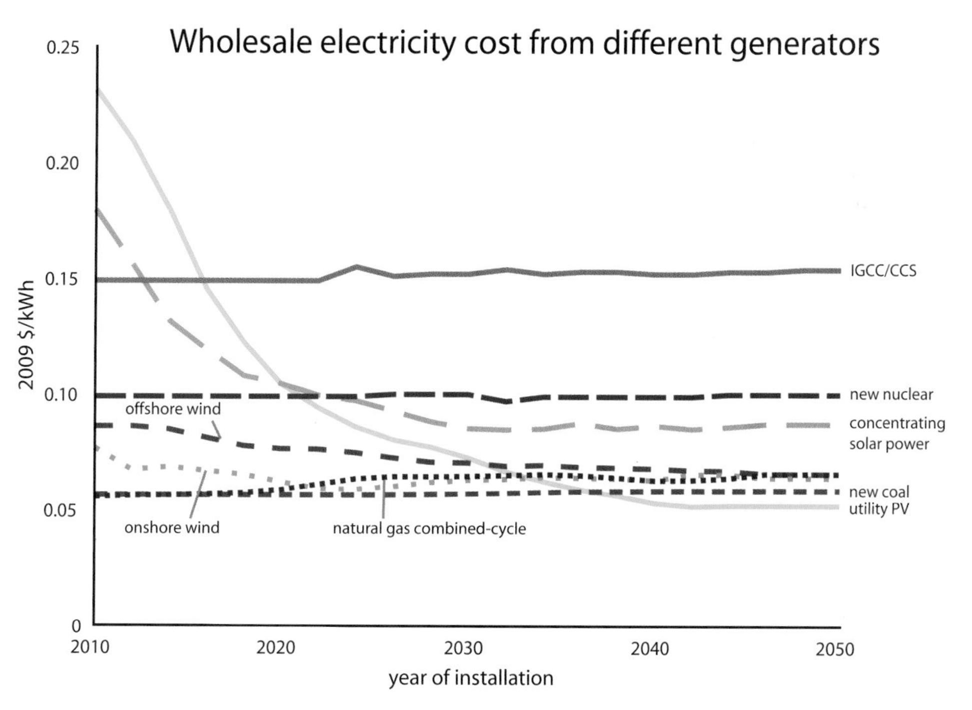

FIG. 5-25. Our levelized costs of electricity sent out from the renewable sources shown appear to exceed initially the costs for nonrenewables, but renewables win over time as their capital costs decline. The renewable costs shown *exclude* tax credits and similar subsidies, while the fuel and financing costs shown for nonrenewables implicitly *include* many complex subsidies. These comparisons also understate renewables' competitiveness by not counting their constant prices, distributed resources' avoided delivery costs and losses, or other "distributed benefits"[632] (see sidebar, p. 207).

Electric grids are currently operated by "dispatchers"—professionals sitting in rooms full of panels and switches, commanding the delivery of electrons from giant power plants to millions of customers. But with smart grid technology, distributed coordination could provide the system-level operability that used to require top-down control. This represents a transition from a centrally planned system to a market-based system that can govern and stabilize itself through millions of smart chips in digital conversation, just as ecosystems manage their intricate nutrient, energy, and signal flows with no manager, or as many consumer products are priced not by a czar but by the everyday dance of supply and demand.

Imagine an office building with an electricity budget. When appliances are turned on, they communicate with a computerized manager that looks at the priority of the new requests, compares them to other loads within the building, and considers the temperature and the setpoint of the air conditioner. Then the manager chooses how to juggle loads, or even onsite generators and storage devices, to meet the new request and remain under budget. Should it be necessary to exceed the budget, the manager requests additional grid supply. This complex decision-making would all happen seamlessly behind the scenes. Ambitious entrepreneurs are already creating and deploying such systems.

The Shift to a Granular Grid

Distributed intelligence can also pave the way to a more granular grid. Since distributed generation is sited locally, meeting basic electricity services needn't depend on the bulk power grid. And since an estimated 98–99% of outages originate in transmission and distribution (mainly distribution),[633] distributed generation presents a unique potential to improve reliability. To do so, distributed

WHAT ARE DISTRIBUTED RESOURCES?

When we say "distributed," we are actually referring to several distinct characteristics of an electrical resource, each important for different purposes. "Distributed" usually means dispersed geographically and connected to the distribution system rather than the transmission system, so the resources are nearer consumers, saving grid costs and reducing losses and failures. But "distributed" resources are also often modular—made in small, similar chunks that can be linked together. Modular technologies get their economies from mass production and rapid learning, not unit scale, and they improve reliability by replacing vulnerable big units with a diversified portfolio of many small units that are unlikely to fail all at once. Finally, in some contexts, "distributed" resources are small, in the kilowatt- to several-megawatt range, versus the hundreds or thousands of megawatts produced by a coal or nuclear plant. The smaller scale can save money by better matching most needs: in the mid-1990s, three-fourths of U.S. residential loads didn't exceed roughly 1.5 kW, nor three-fourths of commercial loads 12 kW.

Distributed resources in some or all of these senses can include a wide variety of technologies, including end-use efficiency, demand response, and supply-side resources that we've already defined. Relevant distributed resources include:

Storage—everything from distributed batteries (such as those in thousands of plugged-in electric cars, those used by data centers to boost power reliability, or even those in your laptop or cellphone) to more exotic devices like ultracapacitors and superconductive loops.

Distributed generation—while traditional forms include CHP powered by natural gas or waste heat, we focus here on renewable forms including rooftop solar photovoltaics; smaller-scale, higher-performance wind turbines; small hydro; cogeneration fueled by landfill gas, biogas, waste, or solar heat; and perhaps more that we can't even imagine yet.

Microgrids—subsets of the electric grid that have enough generation, storage, and intelligence to operate independently of the larger grid if it fails, but normally in collaborative interchange with it.

generators must be allowed to "island"—that is, to work with or without the grid. Such islandability should be the default choice so distributed resources make the grid resilient. An in-progress revision to IEEE Standard 1547 specifically allows such planned islanding.[634]

As more and more distributed generation and demand-side resources are adopted and smart communications enhanced, clustering those resources together to take advantage of technology diversity may become the logical next step. Individual distributed generators—variable renewables or others—could suffer from rapid supply swings, but clusters are less likely to. One potential model is a microgrid—a small subset of the electric grid that has enough generation, storage, and intelligence to operate reliably, independently of the larger grid. Imagine a campus of office buildings or a neighborhood where each building is not only connected directly to the grid, but also connected together (fig. 5-26). Just as you can diversify a financial portfolio with only a half-dozen stocks, you can diversify loads (and onsite supplies of different kinds) by linking a surprisingly small number of nearby customers.

THE HIDDEN VALUE OF DISTRIBUTED RESOURCES

While renewable distributed generation technologies, such as distributed solar PV and small-scale wind, have historically been more expensive than their centralized alternatives, economies of mass-production scale and the cost reductions driven by centralized renewables will reduce distributed resources' costs from markedly to only slightly higher than centralized resources'. But this is far from the whole story, because they compete in very different circumstances. Because distributed resources are smaller in unit size, can be constructed in shorter lead times, and can be installed closer to demand, they can produce benefits and costs not typically accounted for and not reflected in simple bus-bar costs.[635]

First, distributed resources can avoid the losses of delivering power. Total grid losses nationwide average 6–7%[636] but can soar as high as 14%[637] on hot system-peak afternoons. Avoiding distribution *costs* is even more important because these are the biggest capital spend for most utilities.

Second, onsite generators don't compete with high-voltage wholesale power in the bulk grid as big power plants do; they compete with far pricier low-voltage retail power on the customer's side of the meter. Onsite thermal generators like mini-CHP plants or fuel cells (or heat-capturing features added to PVs) can also often reuse waste heat nearby, displacing fuel and equipment.

Third, distributed resources' combination of short lead time and small unit size lets utilities reduce financial risk by building capacity in increments more closely matched to changing customer demand, easily ramping investment up or down as new demand information unfolds. In contrast, coal and nuclear units require many years to build, have no convenient off-ramps that preserve value, and are justified using forecasts that must try to peer through the fog for decades. The small, fast units' lower financial risk can increase their value by as much as severalfold..

Fourth, distributed resources can be strategically deployed throughout the distribution system, pinpointing congested, high-cost areas and delaying distribution investment. As the system evolves, these resources could even be relocated to sustain maximum value. Fifth, renewable and demand-side resources provide a free hedge against volatile natural-gas prices. Sixth and finally, as we'll explore below, some distributed resources can provide ancillary services like spinning reserves (generators kept rotating at their operating speed but not under load, ready to connect instantly to meet unexpected needs).

Of course, some of these resources can also add costs. For instance, high fractions of distributed resources may add more distribution costs than they avoid or defer. Aside from avoided delivery costs, though, these costs and benefits are not represented in our analysis because they're site- and technology-specific. Overall, we believe a proper accounting of these benefits, using the orthodox tools of financial economics and electrical engineering, could greatly enhance *Transform*'s cost and risk profile.

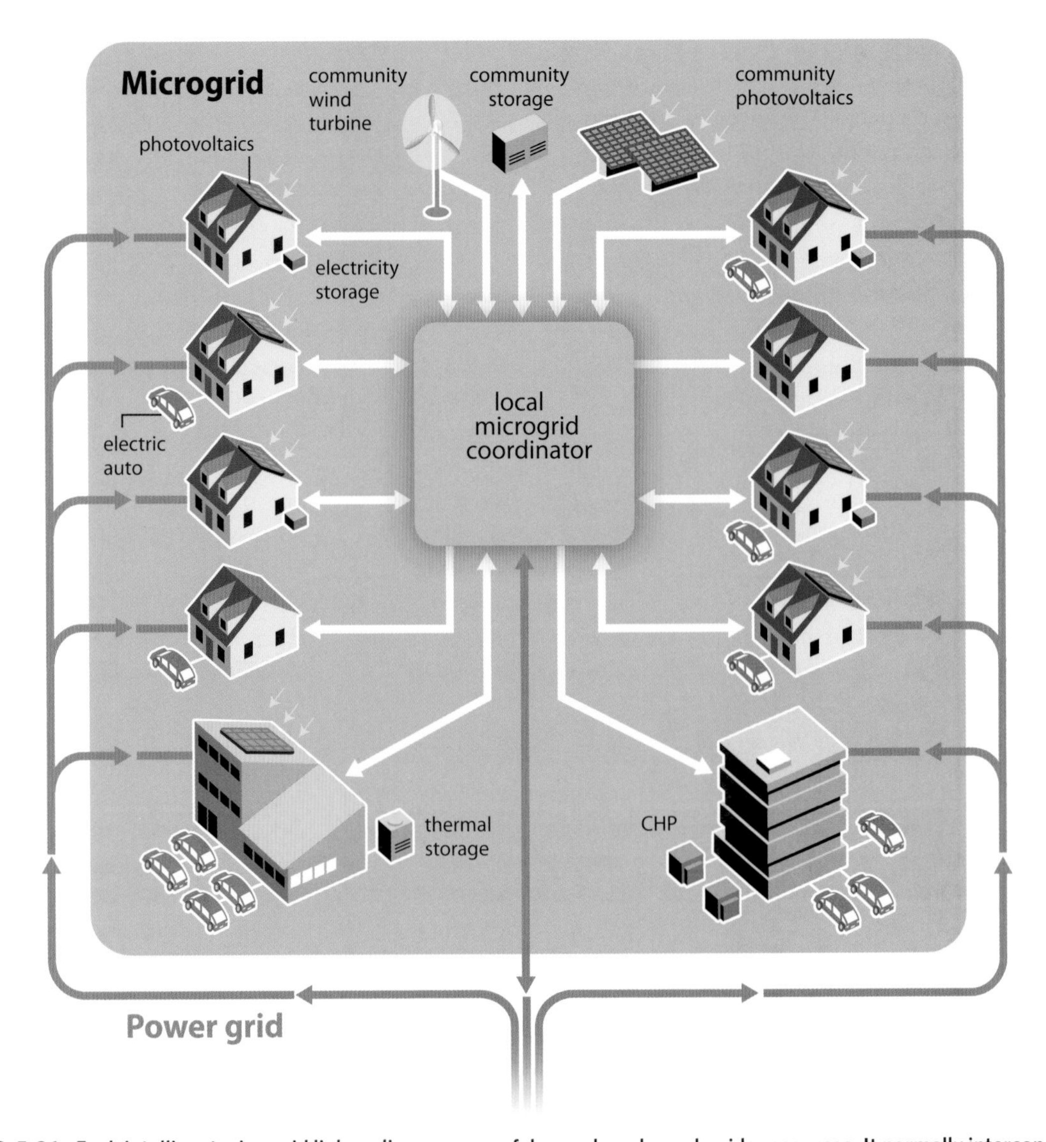

FIG. 5-26. Each intelligent microgrid links a diverse array of demand- and supply-side resources. It normally interconnects with the main power grid but can island from it if necessary. Distributed community storage can be even more valuable than household storage. Some storage looks like a pad-mounted neighborhood transformer box; other kinds are truck-mounted for easy relocation for greatest value.

In many ways, a microgrid is just what it says—a microcosm of the larger grid. And while a microgrid by itself loses some of the scale and diversity advantages of the larger grid, it improves reliability in several other ways. Because microgrids can be isolated from the larger system, power for critical loads will stay on for local customers as long as local resources are available, even when the regional grid fails. And since smaller microgrids can be nested within larger microgrids, the system can take the most advantage of large-scale diversity while still islanding at a variety of levels.

A microgrid by itself is typically less reliable due to its scale—its resources might be unavailable 10% of the time. But coupled together, the central grid and microgrid could reduce outages to 30 minutes a year, on average.[638] That means a downed transmission line from an ice storm need no longer cause widespread, cascading blackouts. And while outages totaling two hours may not seem significant to many homeowners, they can be quite significant to businesses that can lose millions of dollars an hour without power.[639]

There are at least 20 important microgrid experiments around the world,[640] including several in the U.S., with encouraging results. Take, for example, the University of California at San Diego. While many universities still rely on old coal-fired power stations, UCSD's campus is 82% powered by onsite generation from a gas-fired combined cooling, heating, and power unit (CCHP), which covers 95% of cooling and heating needs with a five-year payback, and from more than a megawatt of PVs, whose output is accurately forecast by a state-of-the-art sky-monitoring system that tracks cloud patterns across 25 square miles. A 3 MW waste-gas fuel cell is being added. The campus's 42 MW smart electric microgrid is also equipped with a 3.8-million-gallon tank that stores chilled water until it's needed, so the CCHP plant can run continuously (and most efficiently—66% overall average) rather than having to ramp up and down to meet demand. Power quality and reliability have improved. When wildfires knocked out the local utility's peaking turbines, the campus was able to shift in a half hour from taking 4 MW from the utility to sending 2 MW back. And with an integrated generation- and demand-management system built with smart-grid firms Power Analytics and Viridity Energy, the university can control demand as well as supply, cutting costs and boosting reliability.[641]

The Challenges of Breaking New Ground

Analyzing the economics of transformative grid infrastructure changes is a somewhat novel and daunting task. Smart-grid pilot programs have ranged in cost from $0.8 million to $610 million,[642] depending in part on the level of technology

OPERATIONS IN A DECENTRALIZED WORLD: DENMARK

In just two decades, the Danish electricity system has evolved from a handful of large-scale generating plants to an integrated two-part system, more than half of which is powered by thousands of distributed sources, largely wind turbines and combined-heat-and-power (CHP) plants.[643] Traditionally, distributed generation sources, especially renewable sources, create "unplanned power flows" that are anathema to conventional, top-down methods of dispatching generating resources. For a traditionally minded utility engineer, this may look like an unworkable system destined for a power outage. But for engineers developing the new control systems to manage the Danish grid, the increasing dominance of distributed resources is the impetus to create a new grid operating paradigm that divides the existing grid into islandable "cells" of control.

Instead of throttling central power plants to support the grid, the Danish system—equipped in 2010 for 36% renewable supply in an average wind year—uses real-time measurement and control systems, linked via existing telecommunications infrastructure, to monitor and control distributed resources. This so-called cellular control aggregates these distributed resources into blocks of supply that behave like virtual large power plants, allowing them to provide grid support services. By design, the islands of generation that support the larger grid can isolate from it, withstanding major system disturbances. Meanwhile, market mechanisms—aided by digital communications and real-time feedback—determine the least-cost ways to generate power and provide grid support services.[644]

deployed, but also on vendor and location. Extending beyond smart grid, the Galvin Electricity Initiative estimates that the annual cost per customer to implement and modernize microgrids would be about $200—buying integrated communications equipment, automatic sensing and measurement equipment, smart meters and smart switches, but not the generation technologies—while the annual benefits would total about $870.[645] But more data from real-world examples are needed.

Another challenge is the potential for public resistance. PG&E, one of the country's leaders in smart-grid deployment, has experienced considerable customer backlash, largely based on concerns about the health effects of electromagnetic radiation from smart meters. Several local governments, including those of Santa Cruz and Marin Counties, have banned further smart-grid deployment.[646] Across the country, Baltimore Gas and Electric's initial smart-grid proposal was rejected due to concern that families and elderly people might not be able to respond to dynamic price signals.[647] Such episodes usefully remind us that even simple technological innovations need careful design and broad stakeholder engagement to earn public acceptance.

Additionally, more choice also means some additional risks for customers. Paying real-time prices could mean that electricity bills rise—if using peak power is unavoidable. There are also serious questions about whether customers actually want more choices. Some households and small-business operators simply may not be interested in a more active, technology-intensive system, or they may not want what some might see as "big brother" technology in their homes. And the financial incentives utilities offer to reduce or shift demand may not be great enough to matter to many customers—electricity is a small part of most users' expenses. There are existing barriers to the transition: many customers today aren't yet allowed to sell electricity back to the utility (fairly or at all), nor to operate their own microgrids. And while distributed generation, demand response, and microgrids all have the potential to increase grid reliability, especially together, *Transform*'s greater dependence on smart grid and IT has the potential to increase cybersecurity threats to the grid. Tight security must be built in from scratch.

Diverse, Adaptive Futures

Despite these challenges, the *Transform* case offers the potential to mitigate many of the risks inherent in the previous three cases. Further, its more

THE MILITARY PURSUES MICROGRIDS

Recognizing the grid's vulnerabilities, a Defense Science Board task force co-led by former energy and defense secretary Dr. James R. Schlesinger recommended that the Pentagon cease its reliance on the commercial power grid for military facilities, turning instead to onsite or nearby power generation, preferably renewable, in islandable microgrids.[648] It turned out that about 90% of the 584 continental-U.S. military bases could do that, often to economic advantage. The Department of Defense (DoD) has begun its transition to resilience. Nine military bases are designing and implementing microgrids, and another six are scheduled to begin as part of the DoD's Smart Power Infrastructure Demonstration for Energy Reliability and Security (SPIDERS) project.[649] DoD is already the world's largest buyer of renewable energy, and it is expanding its purchases for forward operating bases where 10%-efficient power systems burn fuel whose delivery costs untold lives and up to hundreds of dollars per gallon. Now DoD can also set the pace for national adoption of resilient microgrids—and is starting to do so.[650] This suggests both a question and a potential answer for nonmilitary users: If the military is worried about reliable grid power for its installations, what does that say about the security of a national economy that relies on the same system? And shouldn't we consider a similar structural solution?

modular resources could fit a variety of customer needs and adapt quickly and nimbly to change. While onsite generation and microgrids may be the right solution for some customers or communities, others may choose to put solar PV on their roofs but remain connected only to the centralized grid. Yet others may not choose distributed generation at all but could use automated controls to respond effortlessly to real-time price signals from the utility.

Some of the technologies that could enable a more distributed future need continued technological development and testing. The many competing technologies and infrastructure designs currently being explored could drive the evolution of the system in widely varying directions. That's not a bad thing; competition drives innovation and improvement. But it does mean that society and businesses will have to place some well-informed bets, learn quickly, and adapt gracefully. Given this huge variety of costs, risks, and options, how can we choose?

FOUR CASES, ONE BROAD DIRECTION

To help illuminate and frame America's electricity choices, we've presented and analyzed four illustrative futures (fig. 5-27). In *Maintain*, the mix of supply sources continues to resemble today's. In *Migrate*, the system's structure remains largely the same but gradually shifts toward carbon-free sources such as nuclear and "clean coal." In *Renew*,

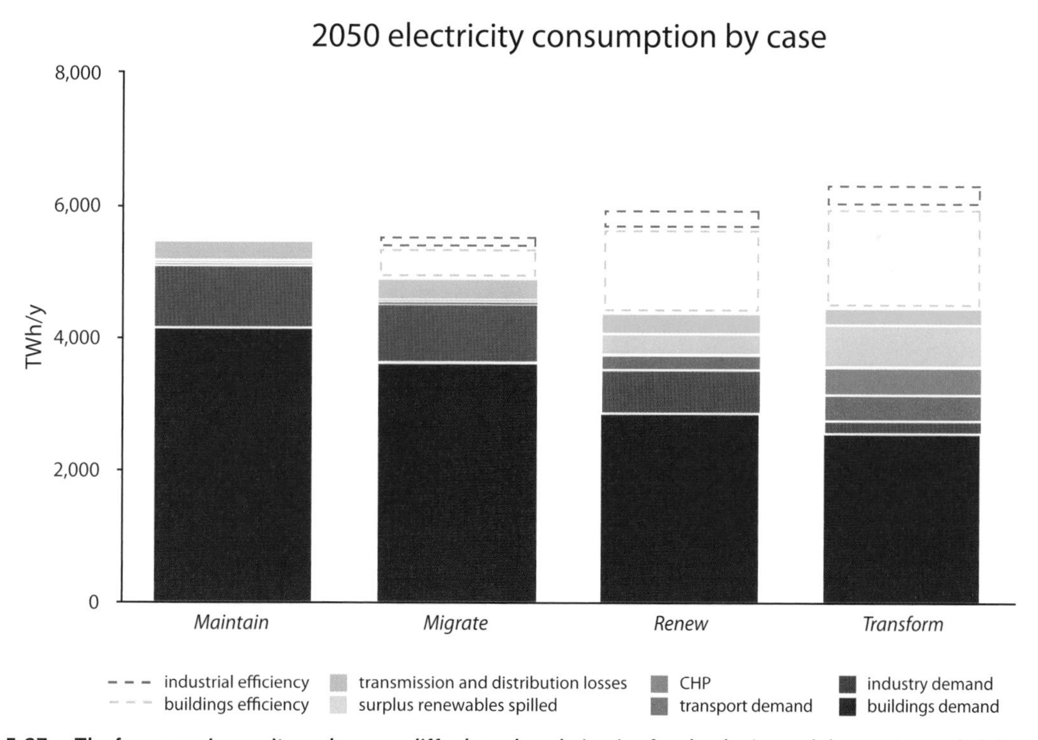

FIG. 5-27a. The four cases' capacity and energy differ based on their mix of technologies and those mixes' reliability impacts. The underlying total demand for electrical services doesn't change, but differing efficiencies change the amount of electricity needed to deliver those services.[651]

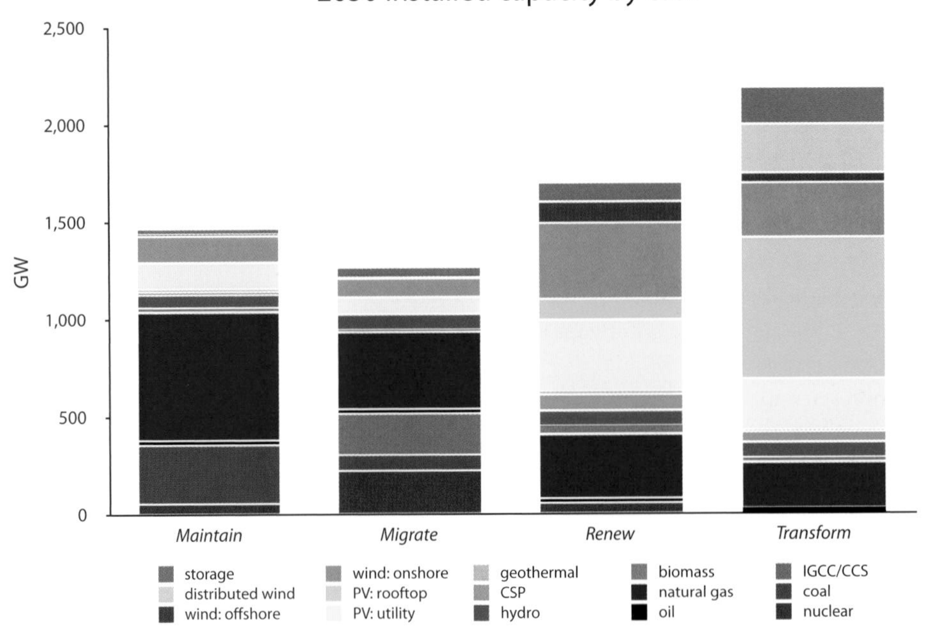

FIG. 5-27b.

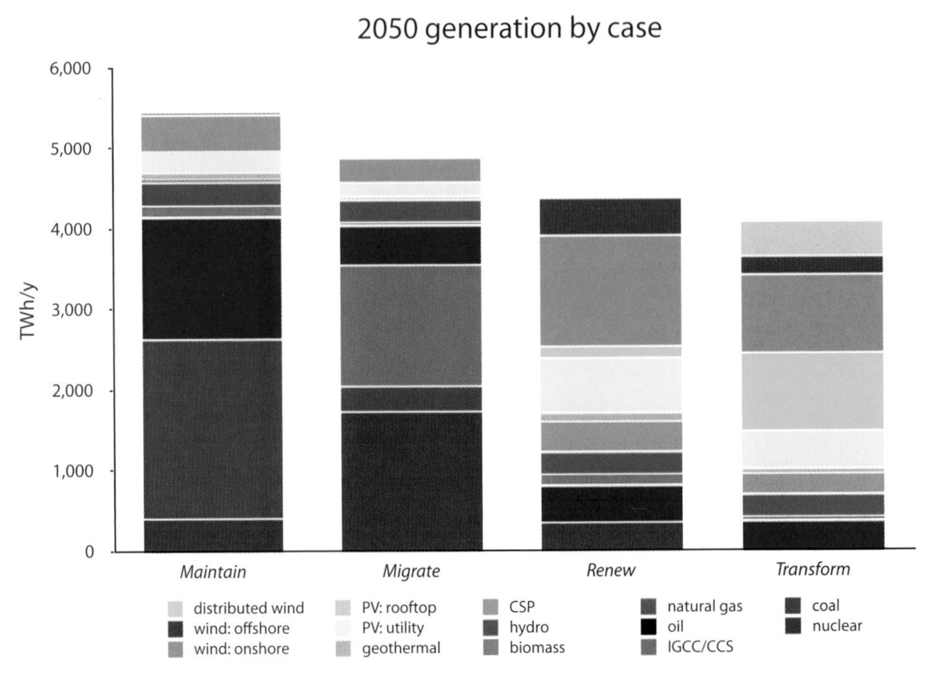

FIG. 5-27c.

the transition is instead to 80% renewable supply, largely from utility-scale wind, solar, geothermal, and biomass generation, and needs supply- and demand-side flexibility resources to integrate variable renewables. In *Transform*, the system is built from the load back (p. 222), reduces the bulk load via energy efficiency and distributed resources, and lets resources of all scales compete fairly.

While we have modeled four illustrative cases, there is actually an infinite variety of options, so how do we decide what action to take? We applied six main criteria to gauge success: affordability, technical feasibility, security, reliability, environmental responsibility and public health, and public acceptability. Our assessment of these criteria leads us to believe that the best path forward is to *Transform* the electric system into a combination of centralized and distributed renewables with a heavy emphasis on efficiency and responsive demand, enabled by enhanced communications and smart grid. Let's explore why.

Affordability. Despite highly varying sources of generation, present-valued cost varies only 12% across the four cases (fig. 5-28). The high capital costs of nuclear and "clean coal" drive up the cost of *Migrate*; additional capacity requirements and higher capital costs of distributed renewables drive up the cost of *Transform*; and to differing degrees, these extra capital costs are offset by saved fuel. While details are unforeseeable, *Transform*'s technologies have greater potential both for steady cost reduction

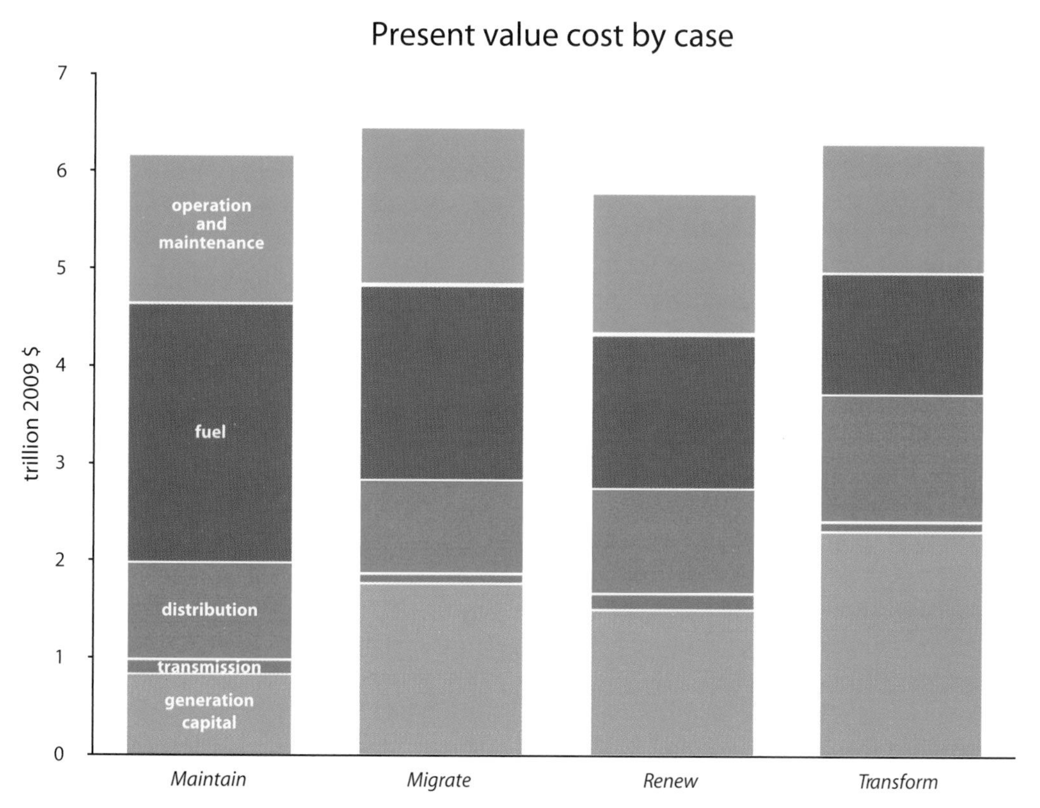

FIG. 5-28. The total 2010 present value of system costs varies by up to 12% across cases, but all forecasts are by nature wrong, and these estimates have substantial uncertainties.[652]

and for radical breakthroughs. Further, taking into account the reduced financial risk of distributed resources could considerably increase its value.

Total system cost is, of course, only part of the picture. Whereas the *Maintain* and *Migrate* cases rely on large, chunky investments, *Renew* and *Transform* rely on small, modular technologies that can be built quickly to match load more exactly, thereby reducing financial risk to utilities and ultimately to customers. Related is the shift across cases from fuel to capital as the dominant driver of cost, reducing fuel-price risk. A broad and fundamental conclusion thus emerges: *choices between such futures should rest much less on* costs, *which are roughly similar and all uncertain, than on* risks, *whose nature, gravity, and management differ profoundly.*

Technical feasibility. No showstopping issues make any case absolutely infeasible. Technology development is important in all cases—especially *Migrate* and *Transform*, both of which rely on some commercially unproven technologies. That said, real-world experiments in the U.S. and abroad have demonstrated that both demand- and supply-side technologies for the *Transform* case do work. Indeed, they're already a big global business. In 2010, renewables (excluding big hydro) got $151 billion of investment, added about 60 GW of capacity, and reached 388 GW of total installed capacity, slightly exceeding global nuclear capacity.[653]

Security. No scenario depends materially on oil, which in today's U.S. economy is less than 1% related to electricity (and less than 5% worldwide).[654] *Maintain* and *Migrate* do depend heavily on coal and uranium, and all scenarios on natural gas (though for different purposes). All these fuels raise different kinds of security concerns that in general are much smaller for *Renew* and *Transform*. Fossil-fueled, nuclear, and renewable technologies are all subject to extreme events, but in general, renewables can fail more gracefully, especially if distributed. All four cases depend on the transmission grid (*Transform* less so) with its inherent physical vulnerabilities even after key nodes are hardened and cybersecurity ensured. Only *Transform*, with its option to work around grid failure, offers a far more resilient grid architecture. The more distributed the generators and the more granular and islandable the resources, the more the large-scale cascading grid failures that now are nearly inevitable could be made nearly impossible by design, and the more the grid that undergirds our nation's economic and military might could stop undercutting it.[655]

Reliability. The current electric system is usually reliable—but we see little evidence that *Renew* and *Transform* could not also be (fig. 5-29). Both would require significantly more sophisticated control, forecasting, and use of demand-side flexibility resources. Even so, each month brings new breakthroughs in managing variable renewables, and each year new records are set for the fraction of electricity generated by these technologies—already from one-fifth to over half in some places—so the future is bright. *Transform*'s distributed resources, too, can in principle reduce outages by partly or wholly bypassing the grid where they nearly all start, and providing local backup supply to meet critical needs during outages.

Environmental responsibility and public health. Only *Renew* and *Transform* address the full array of environmental and public-health issues. While the *Maintain* case may feel comfortably familiar, it fundamentally fails to address major risks like climate change, environmental degradation, and harm to public health that are increasingly recognized as imposing important private and social costs. While *Migrate* mitigates carbon risk, it also drives up demand for water and the production of nuclear waste—impacts whose near-elimination is among the attractions of renewables and distributed generation.

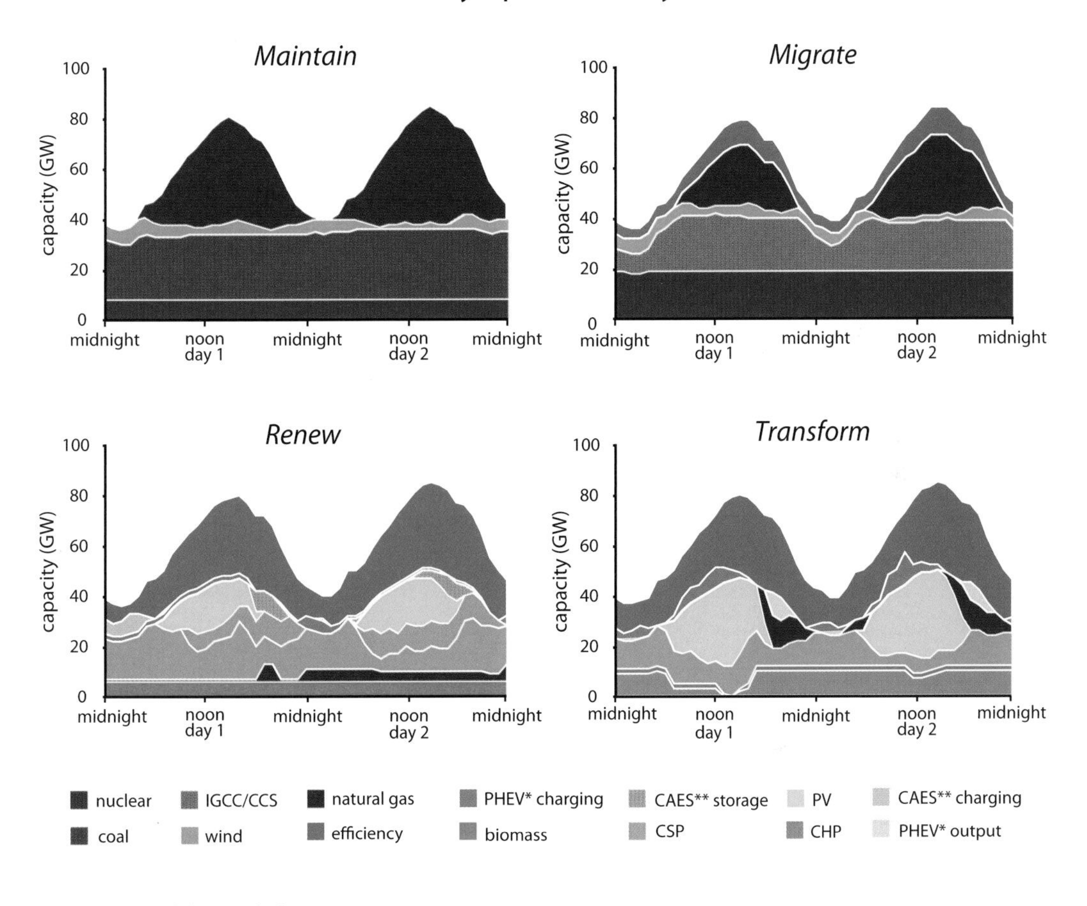

FIG. 5-29. Hourly simulations indicate that supply-and-demand balance is operationally achievable in all cases.[656]

Public acceptability. Energy efficiency and renewables have for decades enjoyed strong U.S. and global public support (though not in every local instance). Distributed architectures that tend to allocate energy benefits and its side effects to the same people at the same time—not to other people at the other end of the powerline or in future generations—minimize potential siting inequities and conflicts and tend to bring good jobs back to the community. However, finer-grained, more local energy choices turn remote, abstract, out-of-sight and out-of-mind choices into real ones right in front of us, so they place greater responsibility on informed citizenship and vibrant community discussion. Those of us with a more Jeffersonian bias will feel that this probably yields better results, but the process is not simple, easy, clear-cut, or predictable: it has all the messy vitality of real democracy. For choices as fundamental as energy, that's as it should be.

Conversely, smart meters and dynamic pricing, if not designed with equity in mind, could disadvantage those ill-equipped to use them or might create privacy concerns.

*Maintain*ing a business-as-usual extrapolation of the past does not guarantee business-as-usual results. It would expose the U.S. to financial risks, accelerated climate change and other health, security, and environmental problems, and increasing risks of blackouts. Addressing climate risk by *Migrat*ing to nuclear and "clean coal" may abate carbon risk but does nothing to address the rapidly changing needs of customers or the changing value proposition of electricity service providers, and it has even greater technical and financial risks than *Maintain*.

Instead, pursuing a future dominated by large-scale *Renew*ables tracks current technology development and cost trends and reduces carbon emissions, without the added technical and cost uncertainties of nuclear power and carbon capture. While this case is unfolding in front of us all, and it should be encouraged to flourish, we do see several drawbacks, including difficulty in siting sufficient transmission, and its continued vulnerability to disruption by as little as a rifle bullet. Rather, the *Transform* future can deliver the advantages of a central-renewables case while maximizing customer choice, entrepreneurial opportunity, and innovation. As this diverse system evolves, the market could steer the mix of different technologies and scales toward the most cost-effective and least risky options.

"All of the Above": A Recipe for Failure

Some will argue that we needn't choose, that instead we should pursue all options equally. In fact, today's conventional wisdom—within the electric utility industry and in Washington—embraces the idea that the best path to a clean and affordable future for U.S. electricity simultaneously pursues *all* available low-carbon supply options: nuclear, "clean coal," natural gas, and renewables. We disagree. "All-of-the-above" scenarios are easy to construct, but none has yet been shown to be necessary, let alone preferable. Indeed, they're undesirable for several reasons.

First, central thermal plants are too inflexible to play well with variable renewables, and their market prices and profits drop as renewables gain market share. Second, if resources can compete fairly at all scales, some, and perhaps much, of the transmission built for a centralized vision of the future grid could quickly become superfluous. Third, big, slow, lumpy, costly investments can erode utilities' and other providers' financial stability, while small, fast, granular investments can enhance it. Competition between those two kinds of investments can turn people trying to recover the former investments into foes of the latter—and threaten big-plant owners' financial stability. Fourth, renewable, and especially distributed renewable, futures require very different regulatory structures and business models. Finally, supply costs aren't independent of the scale of deployment, so PV systems installed in Germany in 2010 cost about 56–67% less than comparable U.S. systems, despite access to the same modules and other technologies at the same global prices.[657]

Utility and political systems tend to tip to efficiency-and-renewables or to coal-and-nuclear futures, but not both at once, because each future creates its own reality of institutions, cultural habits, perceptions, policies, and politics. Across Europe and the U.S., central-station dependence is not compatible with efficiency-and-renewables performance, because the former crowds out the latter in finite budgets and policy spaces.[658] For the U.S. electricity sector to take full advantage of distributed resources and enable the transitions in transportation, buildings, and industry described in chapters 2 through 4, a focused strategy is required. Combining fundamentally divergent futures, through hesitancy to choose a consistent and effective strategy, tends to create an incoherent mishmash, confused people, wasted resources, and lost time—the most precious resource.

HOW DO WE GET THERE FROM HERE?

Transforming the U.S. electricity sector to take full advantage of distributed and renewable resources along the lines we've described will entail a far-reaching set of changes in everything from the technologies used to generate electricity to how the grid is managed. Some of the changes ahead, such as ongoing improvements in the cost and performance of wind and solar generation, will be inexorably driven by the forces of global economic competition. But how far the U.S. can efficiently and reliably develop and integrate these resources will depend critically on the "rules of the road" that determine how the highly regulated electricity system operates.

The pace of progress can be greatly sped or slowed by regulations and statutes that create either a level playing field or a highly skewed one. Since many of the most important rules and policies are determined at the state level, achieving major change for the U.S. as whole will require a reasonable degree of alignment and coordination in state and federal policies, together with the development of industry technical standards for communications and power electronics that facilitate low-cost "plug and play" options and standard design principles like islandability, wherever possible.

In the years ahead, public- and private-sector stakeholders will need to change the rules of the game for the U.S. electricity sector in order to:

- Open the market to new actors
- Realign utility business models
- Coordinate power markets and system operations

In each of these areas, significant changes already under way in the U.S. and abroad could speed the development of distributed and renewable resources, as discussed below.

Opening the Market to New Actors

Nearly three-fourths[659] of U.S. utility-scale solar facilities and 85% of wind developments have been built by independents, largely because they've been better able to capture economies of mass production and develop core competencies around renewable technologies that differ markedly from fossil-fueled competitors.[660] This movement's members range from giants like NextEra Energy, Inc.'s subsidiary, NextEra Energy Resources, the largest renewable generator in North America,[661] to hungry start-ups like Cogenra Solar, whose PV panels produce both electricity and hot water.[662] Seattle, Washington–based start-up Clarian Power even offers a novel solar panel system that completely bypasses the normal connections to the utility's grid. You simply plug a cord from the PV system and its accompanying "SmartBox" into any wall outlet in your house, and its microinverters let electricity from the solar panels flow to all household lights and appliances, using only existing wiring.[663] Some of these firms are remarkably innovative. SolarCity, SunRun, Sungevity, SunPower, and a growing collection of other competitors offer rooftop PV panels for zero dollars down—eliminating the sticker shock that frequently deters customers.[664] Today, customers can go to these or any of a growing number of websites of service providers and independent information providers to get a bid to have solar power installed on their rooftop, and in some states they can even get a system for little or no net capital, just like signing up for a no-money-down cellphone plan. Using satellite images of houses, databases of utility tariffs, and information on incentive programs and tax credits, companies can provide, in minutes to days, detailed estimates or even fully developed bids for ownership or lease of solar PV systems. In some cases, if their rooftop is not suitable for a solar installation, customers may be able to participate in a nearby "community solar" project with all the same benefits but added economies of installation scale and professional upkeep. One way or another, customers can increasingly become prosumers. And renewable power's business case greatly strengthens when it's bundled with demand-side investments,[665] as entrepreneurs are starting to do at all scales.

The importance of creating the right rules and incentives that align profit-making incentives with technological abilities and societal desires is evident in Texas. In 1999, the Lone Star State adopted one of the nation's first renewable portfolio standards (RPS) as part of the law that restructured the state's electricity industry. The RPS mandated that the state's electricity providers must collectively install 2,880 MW of renewable energy by 2009. The law enabled providers to buy and sell renewable electricity credits (RECs) to reach the goal most cost-effectively. According to Jim Marston, regional director of Environmental Defense Fund's Texas office, "The utilities said it wouldn't work, that it would cost a zillion dollars." However, not only did the utilities of Texas meet the goal in 2005, well in advance of the RPS goal, but the law spurred the creation of the renewable energy industry in the state. When Texas legislators wanted to up the ante in 2005 by increasing the RPS to 5,880 MW by 2015 and 10,000 MW by 2025, there was little resistance. Marston says that even though "utilities kicked and screamed the first time we passed a mandate . . . but in 2005, they didn't even bother to oppose the stricter requirements, because they saw they could actually make money off it."[666]

A nascent industry is also growing around the management and use of electricity, providing customers with demand-side resources and smart grid solutions. Entrepreneurial companies like EnerNOC and Comverge are happy to install and manage equipment that enables customers to reduce their peak demand—and their electricity bills. The companies then bundle the savings with the reductions from other customers and sell the aggregated large-scale demand response to a utility or a market. It's good business. EnerNOC is now a public company with $280 million in 2010 revenues and 5.3 GW of demand that it can manage from 3,600 customers in 8,600 sites.[667]

The PJM Interconnection, a regional transmission organization in the Mid-Atlantic region, uses a market with clear price signals to meet its electricity needs at least cost. This PJM market allows demand response, in the form of individual customers' load flexibility aggregated by firms like EnerNOC, to bid in just like a large power plant would. Of the winning bids in PJM's 2014/2015 auction, over 9% were demand-side resources, and 68% of the new capacity available came from renewable and demand-side resources.[668]

New controls and power systems technologies can then provide different levels of power quality and reliability for different end uses. Your web server can get more reliable power than your water heater (which can be off for 15–30 minutes without you ever realizing it, thanks to its stored heat). Fire stations can get higher priority than pool pumps. Automated building energy management services offered by companies like Viridity Energy become even more valuable since they can seamlessly integrate demand, onsite generation and storage, and interconnections.

The success of these emerging businesses will depend in part on rules set by regulators, policymakers, and utilities that determine how distributed resources and renewable generation can be connected to the grid and on what terms. Achieving a transition to a largely renewable and distributed future will require transparent, fair, and nondiscriminatory rules that ensure the safety and reliability of the grid while minimizing barriers to entry. These rules of the game critically affect project economics and, in turn, the scope of competition.

For small-scale renewable energy generation, the most important rules are those set by state and local governments and by utilities. Under the Public Utility Regulatory Policies Act of 1978 (PURPA), the states set policies in two important areas affecting distributed generation: interconnection procedures and net metering. Policies to streamline interconnection procedures, while still preserving the stability and safety of the grid, help remove barriers to the development of customer-sited renewable energy and other forms of distributed generation. Many states have already implemented streamlined, fast-track interconnection procedures for certain types of distributed generation. Further evolution and deployment of best practices in design and implementation of these policies will ease the path of distributed and renewable resources.

Meanwhile, net metering rules determine the terms on which distributed generators can "spin the meter backward" by selling power back to the grid. Again, the devil is in the details, and the specific terms and utility regulations can make the difference between a thriving and a lifeless market for distributed generation. While 43 states already had statewide net metering programs as of the end of 2010, the quality of these programs varies widely.[669]

Even such seemingly simple steps as strengthening technical standards and communications protocols will help. "Plug and play" standards allow certified equipment to be installed with expedited reviews and approvals—or even no approval at all. Such standards remove the traditional burden imposed on distributed technology installers to prove, often through prohibitively costly engineering studies, that their equipment is safe, properly configured, and not harmful to the grid. On the utility side, distribution must become secure, omnidirectional, and friendly to islandable microgrids. Indeed, customer-side generators should be made islandable by default unless there's a strong reason not to.

Local governments also play a key role in determining the rules affecting small-scale distributed resources, for example in determining the permitting and inspection procedures for rooftop solar systems. These rules and procedures significantly affect the cost of solar installations.[670]

Finally, for large-scale renewables, rules made by federal and state regulators, and typically interpreted and implemented by electric utilities, are the most critical. Especially important are interconnection procedures and rules governing access to the grid. As the renewable share of generation increases, new questions will no doubt emerge about grid access and priority, pricing of "ancillary services" from generators, and how the costs of transmission, storage, and other grid assets are allocated among generators.

Realigning Utility Business Models

Achieving the transition to an electricity system largely dependent on renewable and distributed resources, and with far greater end-use efficiency, could be sped by changes in state utility regulation that let electric utilities embrace new ways of doing business. These changes could open new avenues for utilities to create value for their customers and to participate fully in a range of rapidly growing new business sectors ranging from energy efficiency services to developing distributed resources for customers.

A first important step would be the more widespread adoption of policies that eliminate the disincentive for utilities to provide enegy efficiency and distributed resources on the customer's side of the meter. In the past decade, a dozen states and the District of Columbia have changed the old regulatory compact by "decoupling" electric utilities' revenues from their sales, so utilities are no longer rewarded for selling more electricity nor penalized for selling less.[671] This is achieved by periodically adjusting prices up or down based on actual sales so that the utility's revenues are neither more nor less than what the regulators have authorized. Dozens of utilities decoupled over the past decade have found typical price adjustments amounted to less than $2 per month in higher or lower charges for residential customers and canceled out over time. But the impact of decoupling on utilities' incentives, behaviors, and cultures is powerful.

Decoupling doesn't go far enough on its own, though, because it doesn't reward the utility for buying cheap energy efficiency instead of building costlier new generation, so some states are experimenting with a second vital reform—shared savings. This lets utilities keep as extra profit a small part, typically a tenth, of the savings achieved through efficiency they've made possible.[672]

Rewarding utilities for cutting bills, not selling more energy, aligns their interests with customers' interests and society's larger goals. It can profoundly change utilities' choices and performance. California is a leader in these new regulatory models. Regulators allow shared savings to rise (up to 12%) with success, and utilities face penalties if they don't achieve at least 65% of their efficiency goals.[673] Now, a national campaign led by Natural Resources Defense Council, endorsed by Edison Electric Institute and American Gas Association, had by July 2011 brought decoupling and shared savings, in varying degrees, to 14 states for electricity[674] and 21 for gas, and such rules were pending in a further 8 states.[675]

Second, instead of treating distributed generation as a competitive threat, utilities can view it as an asset that they finance, own, and operate themselves on a customer's premises. PG&E, Idaho Power, and some other utilities have done exactly that for remote homesites, where it's cheaper to build generators onsite than to extend powerlines.[676] By mid-2010, about a dozen big utilities had filed plans to build a total of 1.1 GW of distributed PVs in ordinary, nonremote sites,[677] financing the projects by adding their cost to electricity rates in the normal way. This opportunity is immense for America's rural electric cooperatives, soon facing multi-trillion-dollar powerline reinvestments.

Utilities and other players able to get out in front, rethink their relationships with customers, and lead disruptive change could make more money at less risk, and deliver better service at lower cost, than they'd ever dreamed possible. There may even be a whole new business model: acting much like an Internet service provider, the utility could be the open source for myriad power generators and other companies, allowing these providers to get their electricity and services, like demand response, to customers on the utility's grid. It will take innovative regulatory reforms and new pricing policies to make such a model work, but such approaches are already being developed and applied in parts of Europe and in New Zealand.

Utilities now face competition from entrepreneurs they can't control. Existing franchise monopolies don't guarantee continued revenue, because customers aren't obliged to buy grid electricity. They're gradually figuring out that negawatts (saved watts) are often cheaper than megawatts, so they'll want to buy fewer electrons used more productively. The question is who will sell them that mix.

The same potential for bypass exists on the supply side too, albeit at considerably higher cost for most customers. A "virtual utility service" sellable without utility or regulatory permission can be assembled from distributed supply- and demand-side resources, IT and telecommunications, trusted installation and maintenance with reliability guarantees and insurance, green certificates or credits to resell, and new value propositions around security and price stability. Such free-market, unregulated services, if well executed and buyable with confidence, could be an attractive option for a small but growing number of customers in the years ahead.

Grid connection and utility relationship is likely to remain the best answer for most customers for the foreseeable future. It has the benefit of creating a system that can take advantage of supply- and demand-side diversity and risk management through a diverse portfolio of resources. But utilities will have to keep an eye out for non-utility service providers, whose portfolio of technologies, and solutions, will continue to expand.

Coordinating Markets and System Operations

A future electricity system with much larger shares of renewable and distributed resources would operate quite differently from today's system. With higher levels of distributed control and intelligence throughout the system, operational management will depend less on hierarchical command and control and more on distributed responses to signals indicating the state of the system. In the future, rooftop solar systems and windfarms may help automatically correct for voltage fluctuations on transmission and distribution lines, electric vehicles will intelligently charge and discharge their batteries at optimal times, and microgrid controllers will manage local resources in response to dynamic price signals from the grid.

The keys to making all this happen are:

- Efficient, competitive, and transparent power markets
- Dissemination of time-varying price signals to retail customers
- Secure and reliable communications to allow interactions across different levels of the system, from end-use devices to microgrids and virtual power plants to the high-voltage grid and wholesale power markets

As variable renewables' share of generation rises, transparent and competitive wholesale power markets will play a critical role both in the short-term management of resources on the system and in providing visibility of economic opportunities for new investment. Close integration of these markets with operations will allow market forces to address many of the most challenging issues about how to integrate high levels of renewables while keeping the system reliable. Scheduling power every few minutes instead of once per hour, and over a bigger balancing area—faster and wider markets—makes the whole system more flexible. Such "subhourly" scheduling is now used in the East (excluding the Southeast), in Texas, and in California and is scheduled to be implemented in the Southwest in 2013 and 2014.[678] In addition, these markets enable demand-side resources, including demand response, to compete head-on in meeting the flexibility needs of the system. That competition is starting to get fairer. In 2011, FERC's Order 745[679] required demand response to get paid fairly for its value, and a proposed rule values fast ancillary services above slow ones. FERC may next give end-use efficiency full market parity, letting it bid fairly against supply in many of the 37 states where it can't yet.

With higher levels of distributed resources in the system, time-varying prices at the retail level will facilitate optimal use of these assets. While many utilities already have time-varying prices, the widespread implementation of dynamic pricing, supported and enabled by advanced communications and controls, will be a key enabler in allowing distributed resources to play a major role in providing the flexibility necessary to manage a system supplied largely by renewable electricity resources.

One more powerful reform can come from utility CEOs: turn the utility inside out by planning from distribution toward generation, not the reverse. Three North American experiments[680] found that when utilities focused their demand-side investments like a rifle, not a shotgun, on the end uses and in the neighborhoods where they could defer or avoid imminent distribution investments (their biggest capital burden), they could sustain the same service with about one-tenth the investment. For jittery investors and CFOs, that's immediately welcome news.

CONCLUSION: THE PATH FORWARD

We've now pored over maps describing the possible future landscape of the U.S. electricity sector. While some parts of the terrain are well understood, others are labeled *terra incognita* or "Here be dragons." Taking a wide view of what's known and unknown, the best solution we found is a hybrid of centralized and distributed renewables, integrated by advanced communications and controls that securely choreograph supply- and demand-side resources nearly in real time, using islandable microgrids as necessary to ensure resilience. This solution will allow the U.S. to lead new waves of technology advancement and deployment in global markets.

The ensuing transformation of electricity will be complex and perhaps messy. The convergence of electricity and information, with rapid innovations in both, makes 21st-century technologies and business models collide with 20th- and even 19th-century cultures and institutions, often encrusted with regulatory structures and rules that no longer fit today's evolving needs.

While we can't predict unprecedented events and causations beyond our immediate view, we can bet that those already roiling the horizon will multiply rapidly. Intelligent choice today requires a steady long-term vision and an adaptable choice of route as we move ahead.

Just as the decisions made in the early days of electricity shaped today's system, so the investments we make now in facilities and system architectures will affect our ability to capture opportunity and manage risk later. Luckily, the technological and scale factors that molded today's electric system are largely history: technology has given us immensely wider and better choices just in time to avert serious problems. Rapid innovation, combined with society's urgent need to reduce fossil-fuel use, has created a golden opportunity to reinvent the electricity system—to the great advantage of clever and agile businesses.

Several lessons and core actions have emerged from our exploration. First, achieving the electricity transformation described here, thereby enabling earlier chapters' transformations, is just not a matter of choosing the right generating technology from among coal, nuclear, renewables, and other options. Rather, it's about choosing a *system* that can best exploit the full range of supply- and demand-side options in an integrated, least-cost fashion. Integration is the skill of a master chef, not a grocer's buyer. And, in the electricity sector, as in cooking, the best solution comes not from combining all the possible ingredients, but from an artful and harmonious combination of just the ones you need.

Multi-billion-dollar coal and nuclear plants don't coexist well with large amounts of variable and often distributed renewables. The best complements to renewables, the ones whose flexibility can help the system run most economically, are found on the demand side. Integrating electric vehicle charging (and, in time, discharging) into the electricity system is a key part of the solution. Advanced efficiency, demand response technologies, and the enabling power of microgrid-based communications and control systems are also vital partners, especially together. Some options once thought to be necessary, such as widespread bulk electricity storage, may not be. Some once thought to be unimportant, such as distributed PVs, may prove valuable, even vital.

Second, the economic and technical challenges are much smaller than the needed institutional shifts. Electric utilities' business models and regulatory structures will need to be reformed to level the playing field between investments in supply- and demand-side solutions, and between nonrenewable and renewable, and centralized and distributed, options. Many private and public stakeholders in diverse, often overlapping, regulatory jurisdictions will need to hammer out new rules and attitudes. Achieving these institutional changes will require persistence, patience, and process. Perseverance wins.

Finally, much of the power to change the system rests in the hands of electricity customers. *They* will choose whether and when to adopt new technologies, inform their choices, change their behaviors. Conventional economic analysis can't fully describe or predict these shifts. Insight into the behavioral economics of technology adoption will help inform marketing, but customers are complex and surprising, and like social media, sudden shifts that meld new ideas, behaviors, and technologies can abruptly transform the competitive landscape.

Leonard S. Hyman, longtime leader of Merrill Lynch's Utility Research Group, foresaw this turbulence in 1993. He wrote in *The Electricity Journal*: "The future will bring metamorphosis, convergence, disintegration. It will bring ulcers, losses, riches, confusion. All of this is certain. Nothing else is." Now that tumultuous change is going into fast forward. Rigidity and adherence to custom will lose to suppleness and rapid adaptation.

Humans' mastery of fire arose at a micro scale some millions of years ago, in the hands of individuals, families, and tribes. Today's highly interconnected global economic and social networks have accelerated to warp speed the diffusion of new ideas, technologies, designs, and systems. The seeds of change for the next revolution, the new fire, are still local, but they are growing and spreading. New solutions to many of our most challenging problems are already working around the world. Our job is to find and amplify these solutions, and to bring them into harmony with one another. The future of electricity is ours.

TABLE 5-1. Summary of recommendations for key actors in the electricity sector

	NO-REGRETS	OPPORTUNISTIC	INNOVATIVE
INDIVIDUAL CUSTOMERS	Take personal responsibility for usage. Participate in existing programs.	Learn more about distributed resources that might work for you. Assess your potential to provide demand response. Ready buildings for smart grid and demand response. Segregate critical circuits when rewiring/retrofitting/building, and build to be ready for distributed generation.	Advocate for the ability to net-meter fairly, own and operate islandable distributed generation and microgrids, and buy differentiated levels of reliability. Install your own efficiency and renewables, ideally with no money down.

Continues

	NO-REGRETS	OPPORTUNISTIC	INNOVATIVE
NON-UTILITY BUSINESSES	Understand the market for distributed resources and other technologies. Continue to innovate, both developing new technologies and reducing the costs of existing ones.	Partner to explore new integrative business models across old value chains. Consider islandability as an important resiliency feature. Segregate critical electrical circuits, physically or virtually. Seek and prepare for resilient supplies.	Advocate for regulatory reform (starting with decoupling and shared savings) and greater access to markets. Make islandability the default design. Install efficiency improvements and onsite generation.
UTILITIES	Thoroughly and immediately safeguard against cyber-attack, physical attack, and geomagnetic storms. Explore the value and sources of operational flexibility on your system. Master demand-side resources and build strong capabilities. Learn continuously about renewable and distributed technologies. Stress-test financial planning for stagnant or declining demand. Preserve and build optionality.	Maximize grid flexibility to ease renewables integration. Invest now to gain experience with renewables and distributed resources. Invest in smart grid—with security and privacy. Offer time-of-use pricing. Build alliances and seek approval for prompt regulatory reforms to align investors' incentives with customers' incentives.	Rethink relationship with customers and explore new business models. Consider acting as an open source for new entrants. Start resource planning at the distribution end. Purge obstacles to islanding and microgrids and start deploying both. Offer real-time pricing. Push for full competition between demand- and supply-side options.
REGULATORS	Institute decoupling with shared savings. Allow fair net metering of distributed resources. Support investment in R&D. Seek complete policy symmetry between supply- and demand-side resources, big and small, regardless of technology, type, or ownership.	Shift to subhourly scheduling and larger balancing areas. Plan at a regional level. Provide fair access to transmission and make new transmission compete with efficiency, demand response, and distributed generation; reward those resources for freeing up transmission capacity.	Create price and information transparency. Explore market transformation, including implementing markets more widely and shifting energy to capacity markets. Allow utilities to provide services on the customers' side of the meter.

	NO-REGRETS	OPPORTUNISTIC	INNOVATIVE
REGULATORS (CONT.)	Insist that utilities rapidly and thoroughly protect key assets from cyberattack, physical attack, and geomagnetic storms.	Strengthen standards and secure, flexible, interoperable communications protocols. Allow nonutility actors to island loads and operate microgrids. Phase out subsidies, or remove their effect on decisions by shadow pricing. Allow utilities to invest prudently in more-risky technologies. Allow utilities to invest in, support, and earn on efficiency and distributed generation.	Make plug-and-play interconnection and islandability the default design choices. Experiment with netted microgrids and explore how best to encourage them.

CHAPTER 6

MANY CHOICES, ONE FUTURE

FIG. 6-1.

The arguments for marginal, incremental change are not convincing—not in this day and age. The future, after all, is not linear. History is full of sparks that set the status quo ablaze.

—Peter Bijur (chairman and CEO, Texaco), World Energy Council keynote, Houston, September 14, 1998

INTRODUCTION

From the coal that powered the mighty steam engines of the Industrial Revolution to the oil from Edwin Drake's well and its many successors that became the lifeblood of transportation and commerce to the natural gas that has surpassed coal among America's fuels, fossil fuels have transformed human civilization. They made possible massive cities and endless leafy suburbs, giant industrialized farms and vast water transfers, jet airplanes and teeming highways, and an everyday profusion of inexpensive goods casually shipped from all over the planet. Human ingenuity and ambition being what they are, industrialization and economic progress could still have occurred without fossil fuels,[681] but the evolution of industrialization and of human history would have been very different, perhaps focused in different cultures,[682] and probably slower and harder.

Yet our economy's quick hit from mainlining fossil fuels has come with rising costs to our vitality and safety. As we've seen, our energy system is exacerbating, more than shielding us from, the turbulence of what military planners call VUCA—volatility, uncertainty, complexity, and ambiguity. That's true not just of our nation but of the world. As Thomas Friedman explains one example, "The world is caught in a dangerous feedback loop—higher oil prices and climate disruptions lead to higher food prices, higher food prices lead to more instability, more instability leads to higher oil prices. That loop is shaking the foundations of politics everywhere."[683]

Fortunately, we are not doomed to eternal punishment, as Prometheus was for stealing fire for humankind. Nor does the dwindling of the old fire of fossil fuels mean a return to the Dark Ages. Instead, we can create a safer, stronger, fossil-free world by tapping into a far greater resource than fossil hydrocarbons. The real underlying fuel of America and of modern civilization is innovation and ingenuity.

This book has taken us on a journey through many of the possibilities created by the flowering of innovation. We've seen how automakers and their suppliers are starting to build ultralightweight electric cars, how superwindows and integrative design can transform buildings from energy hogs to energy misers, how factories are making products with drastically less energy, and how giant companies and lone entrepreneurs alike are concocting new electricity grid designs and clever ways to learn when the next bus is coming.

Even though we're only on the cusp of this transformation to a fossil-free world—and even though the transformation will happen only if we put our might and will behind it—it is already creating vast new business opportunities and threatening entrenched industries. Who will become the Microsoft of making electricity demand nimble, for instance? Will micropower continue to push century-old central-plant designs off the market? Will insurgents continue to harry incumbent utilities? Will oil companies fade into history like buggy-whip makers and whalers, or will they become dominant players in clean energy and biofuels? As the greatest transition in industrial history unfolds, will America lead this transformation or trail behind others, condemned by old thinking and bad politics to lose the opportunity?

We don't know all the answers to questions like these. Nor can we predict innovations as yet undreamed of but whose distant rumble grows louder. As is always the case with technological and social transformation, the future may be far different from what we can conceive now. We are convinced, however, that transforming the energy system by Reinventing Fire will strengthen the economy and make the world safer, cooler, and better.

Though our crystal ball may be cloudy, let us try to visit the future that would be created by following the paths charted in this book. Will that future meet our lofty goals? Let's jump forward to a morning in America nearly 40 years from now.

LOOKING BACK FROM 2050

The coffee smells the same and the view out the window of the house onto a quiet neighborhood looks fairly similar. But the house is so well designed and insulated that it needs no central source of heat. Its appliances and gadgets sip a tiny fraction of the power their predecessors used in the first decade of the 21st century. The house no longer suffers chills and fevers, nor does it need yesteryear's noisy, costly mechanical equipment. Lacking those antiquated complexities, the house cost slightly less to build than the inefficient old one. Even though people now pay more for housing nearer their other destinations, they pay far less for energy and commuting, so, on the whole, housing takes up takes up a smaller portion of people's budgets as it did in 2011.

Not only that, but our house has become a modest income-earner. Instead of paying over $100 a month for household energy bills, we get a monthly check for the surplus electricity produced by the solar shingles on the roof, for the electrons that our Intergrid service provider buys back from the battery of our electric car to meet rare peaks in demand, and for the intelligence that invisibly coordinates that car's charging and some of the home's electrical services to use, buy, or sell power at the most profitable times, signaled to smart controls by varying prices based on real-time value. And when last year's record ice storm knocked down the interstate powerlines, the lights stayed on because the community microgrid instantly unhooked, ran autonomously, then seamlessly reconnected when links were restored. Large-scale blackouts are a remnant of a bygone age.

Though our electric auto is three times lighter, slightly cheaper to buy, and far cheaper to run than autos were 40 years ago, it's safer, peppier, just as spacious, and even more luxurious—the fruits of fierce competition in new materials, manufacturing methods, propulsion, and design. But now that we organize our communities around people, not autos, we drive much less because the places where we live, work, play, and shop are nearly all in easy walking distance. Some of the old sprawl suburbs survived as electric autos made commuting more affordable, but many were redeveloped into clustered, mixed-use communities surrounded by new parks and farms. Many of us get the majority of our food from farmers we know within a 10-mile radius—and it tastes better.

I'm one of the minority that still own personal cars. Most of my neighbors and both our teenagers get better access at lower cost from one of the guaranteed mobility services whose subscription packages integrate a rich menu of ways to get around—or not need to. Today, handheld telepresence is so good that going to see someone is truly optional, and if you choose to, your menu of ways to get there grows ever more diverse and attractive.

Besides our family's personal electric vehicle, my physical mobility options include jumping in an electric roadster from my carsharing program, catching a taxi or jitney or the ultralight rail two minutes' walk away, or using a social network to hook up with other drivers headed my way. Right now, though, I notice on my smartphone that the silent fuel-cell bus will soon arrive at a nearby corner. So, time to leave the house. The bus ride will give me a chance to do some work over the free network, then grab a hybrid-electric bikeshare to complete my short but hilly journey to work.

Once on the bus, other differences jump out. The roads are far less crowded than they were in 2011. The old zoning rules that ended up segregating housing by income level, causing isolation and dispersion, and requiring that you have a costly private car to get anywhere, are long repealed. Sprawl is no longer subsidized either: developers pay all the costs they impose on public infrastructure and services. Fewer trucks ply the highway. Fueling stations dispense biofuels and hydrogen, complementing the ubiquitous smart-charging points for electric autos. The air is clear and crisp. Engine growls have given way to birdsong. Traffic deaths, once a public-health menace as big as breast cancer, have become rare and injuries generally mild.

Today's carbon-fiber autos no longer rust. They still suffer occasional fender-bouncers, but real damage to their ultrastrong bodies is rare and maintenance almost nil. I dimly recall the days when autos needed to be fed over 20 kinds of fluids and consumables; now they need only wiper fluid. Wireless diagnostics and tune-ups in background make breakdowns almost unheard of. A mobile service van does the rare physical repairs at your home. Both the automaker and the certified app store offer wireless downloads to tweak everything from suspension to display styles, because the auto's functionality is all in software.

Reproduced all across America, these changes mean that liquid fuels for autos are down to a percent or two of their 2011 level, and nearly all those relics that require them run on advanced biofuels. The shiny old metal dino-juice-burning car-club relics that still use gasoline or diesel fuel refuel at their rallies from tank trucks sent by specialty chemical companies. In the 2040s, America even exported a little surplus domestic oil, but by now so many other countries have saved and displaced their own oil that it's rapidly becoming like whale oil after electric lights—a curiosity hardly worth selling.

American troops no longer guard oil, don't need it, and rejoice that it's no longer worth fighting over. While the world is still a dangerous place, it seems to be getting safer: military missions now are mostly humanitarian, disaster relief, and above all conflict prevention, which has become as routine as fire prevention. Most oil exporters are adapting to more diversified economies and often more open societies. The old energy quarrels that once showered sparks on the Mideast tinderbox are either resolved or at least becoming irrelevant to the rest of the world, and what we learned about energy solutions is helping with water. Petrodollars stay home, improving education and healthcare and funding R&D that reinforces American resurgence in innovation.

In 2050, workers still go to the office, typically three or four days a week (and otherwise work from home), but the buildings have changed. My workspace is bathed in the warm glow of natural light, filtered through trees and nourishing the verdant gardens inside. There's no noisy ventilation system but plenty of naturally flowing fresh air, and every room has operable windows. Our building, like most, exports net electricity from its solar roof and walls. The community hospital, with heavier energy needs despite its efficiency, gets all its space-conditioning, hot water, and electricity from cogeneration powered by biogas from the old landfill. Over a hundred million better buildings like these have cut the building sector's energy use by nearly three-fourths from 2011 levels, and it's still heading down.

New industries have sprung up, while some old industries have waned. In my office's neighborhood alone, the oil drilling rig designer has vanished, replaced by a wind-turbine maintenance firm. The geophysics company has switched from scouting oil deposits to finding geothermal hot spots. The car-rental office has become a mobility agency. A host of new companies manage energy demand, install insulation in inner-city homes, and monitor the health of the electricity grid. And some offer onsite manufacturing, not just customized made-to-order toasters and gadgets at mass-production prices—now that it's as cheap to make one item as a million—but also perfectly fitting shoes and garments, manufactured to your digitally captured body measurements in a quarter of an hour.

When the transformation to a fossil-free world began in earnest in the second decade of the century, the leading manufacturers of some key devices, like solar-electric modules, were Chinese. But once America got serious, supporting existing market forces to speed the transition, its mighty innovation engine kicked in. Researchers devised ever more efficient solar panels and ever cheaper ways to make them. Quantum dots, plasmons, nanoantennas, solar paintables, and other marvels first tripled, then quadrupled the now amusing 20% efficiencies of 2011's market-leading products. Power electronics, smart controls, and even storage are all built into not only solar modules but also roll-up fuel cells—or both together. As a result, the U.S. has become once again a leading supplier to a world rapidly following in its footsteps.

And we do business differently. We still fly to meet with clients and vendors, but via the new superefficient flying-wing jetliners fueled with the algal oils that took over in the 2020s. As soon as the relationship is established and sealed over a great dinner—those human interactions still matter as much as ever—future deals will need only telepresence, saving everyone time, money, and family life.

Across the U.S., waste heat from industrial processes is captured and reused repeatedly (and often milked for electricity too) until only uninterestingly gentle warmth is left. Fatter, shorter pipes soar in graceful curves between the sparse pieces of microreactors and other compact process equipment. Chemical processes mimic those of nature, based on intricate reactions between just a handful of common elements, catalyzed by enzymes at room temperature. Manufacturing also has become more agile, flexible, and local. Manufacturers whose product and process designs don't imitate nature are generally out of business. So are those who didn't realize that designing products for long-term value, not planned obsolescence, would create competitive advantage. And our economy flourishes with one-third the virgin materials it used to need.

Those industrial cogeneration plants all run on waste or natural gas, whose use continues to shrink as efficiency and solar process heat undercut its gradually rising price. Renewable electricity is also transmitted from afar where that's cheaper. But more and more electricity comes from nearby, especially from the buildings that use most of it. With even more efficient solar shells and building components, the net-zero buildings that took over new construction in the 2020s and are gradually replacing older buildings are now exporting much more electricity than they use. Guided by the invisible wizardry of myriad chips and software agents, silently exchanging their digital messages of value and substitution, the largely distributed generators and the demands they serve choreograph a continuous dance of power swaps that level loads and maximize value across the Intergrid. Dumb old buildings that can't do this aren't worth much.

So in my 2050, America no longer uses oil, nor can you expect to buy it where you recharge or refuel your car. Some trucks burn natural gas, but biofuels are now cheaper. The last shreds of coal for industrial process heat are being phased out. The last coal-fired power plant became an industrial museum in 2036. The last nuclear power plant will retire this year. Natural gas use has dropped by a full third since 2011. It first displaced coal for electricity generation, then competed as a flexibility resource. But this valuable transitional fuel, having overcome controversy in the 2010s through the adoption of strict operational standards of care and transparency, is finally in slow decline as efficiency and renewables squeeze out its remaining markets. Though it's now believed ample to at least 2100, its remaining "tail" is expected to dwindle nearly to zero well before then, so it can be saved for special uses and as a chemical feedstock—though some gas-fired generating capacity will be kept connected as a normally unused "active reserve" in case of rare emergencies like sky-darkening volcanic eruptions.

In short, one step at a time, we have quietly reinvented fire. Once this was a rare curiosity.[684] Now it's the new normal, a support structure invisible yet omnipresent—like water to a fish.

HOW CAN REINVENTING FIRE EVOLVE SMOOTHLY?

This vision of 2050 is a dramatically different future than business-as-usual projections. But the transition will seem much slower as it unfolds. And even more profound are all the other things that will vanish as we reinvent fire. Finally, we will have fuel without fear. Our big part of the global climate experiment will stop. Oil rigs won't blow up nor oil tankers break and leak. No more grieving families of lost coal miners, nor mountaintops inverted, nor mercury and soot in the air. No more wars over oil. No longer will petrodollars hemorrhage from our economy, some to fund foes. Oil diplomacy can be just plain diplomacy. Power blackouts can become local, brief, and rare. Energy can do our work without working our undoing.

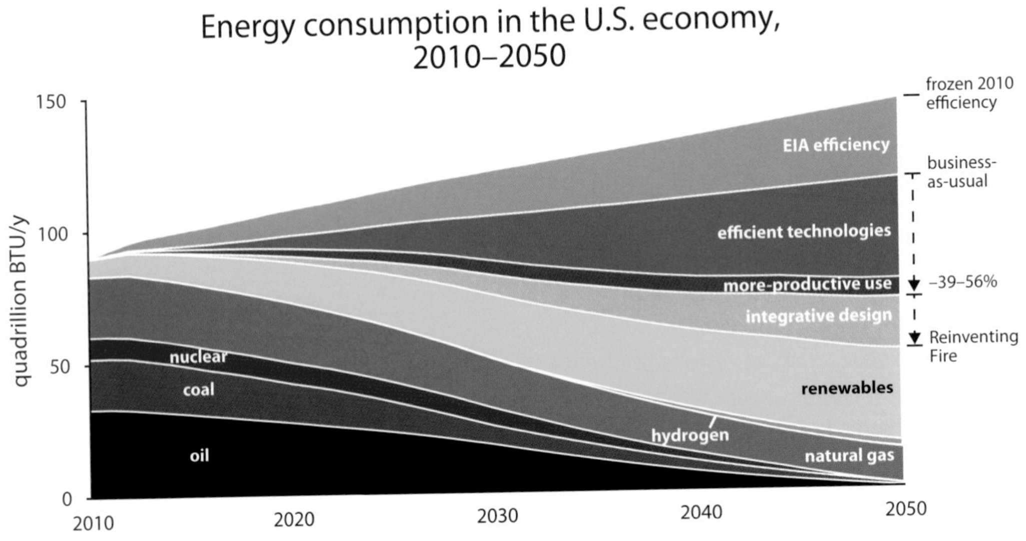

FIG. 6-2. The path charted in this book could advantageously phase out oil, coal, and nuclear energy by 2050. Natural gas use in 2050 would be about 14–36% below the 2010 level, depending on the level of integrative design from none to high (or 9–30% if all heavy trucks were fueled with natural gas). If the hydrogen shown is all reformed from natural gas, it will come half from gas and half from steam.[685]

NATURAL GAS: FUTURE OUTLOOK FOR A VERSATILE TRANSITION FUEL

Natural gas is an important domestic energy source (only 11% of 2010 use was imported) and is becoming more so. It's generally more benign than oil or coal. It's essentially free of sulfur and metal emissions, its combustion can be controlled to release little nitrogen oxide (NO_x), and per unit of contained energy, it emits about 29% less carbon than oil and 43% less than coal. As a transport fuel, natural gas emits about 20–30% less lifecycle carbon than oil.

In the United States, natural gas travels mainly in ubiquitous underground pipelines that are safe when properly maintained. Gas's infrastructure needs protection but is less vulnerable than coal's or especially oil's.[686]

Drilling for conventional natural gas has local impacts roughly comparable to those of drilling for oil (some fields yield both). However, gas wells deplete rather quickly,[687] so new drilling is constantly required. The biggest drilling issues today come from a rapidly emerging new resource called shale gas. U.S. shale gas production has risen 14-fold in the past decade and is now about one-fourth of total gas output. It accounts for one-third of U.S. gas resources and for most of the recent growth in U.S. technically recoverable gas reserves. EIA projects that shale gas will offset declines in other domestic supplies and also be able to meet slight demand growth, nearly eliminating U.S. gas imports by 2035.[688] Foreign shale gas resources are also vast. Their successful extraction could durably decouple gas from oil prices worldwide (they're still contractually linked in Europe and Japan) and help free Europe from dependence on Russian gas.

Extracting shale gas requires drilling horizontally into enormous basins of impermeable shale, then injecting high-pressure water many times to fracture the rock and liberate tiny gas bubbles.[689] (Most conventional gas wells also use such "fracking"—a technique developed in the 1940s—but less intensely.) If badly done, returned water can contaminate groundwaters or surface waters, but that's not necessary, and old toxic additives have benign substitutes, like glycerin for diesel oil. Wells must also be carefully completed so methane doesn't leak. Some operators have conspicuously misbehaved and some regulators have fallen short, making fracking controversial even in normally drilling-friendly places like Texas and western Colorado. (France, with nuclear interests to protect, is banning it altogether.) Besides cleaning up its act, the U.S. industry must also confirm that adequate gas output can be economically sustained[690] after the often-sharp initial falloff, and it must show that fracking won't trigger small earthquakes that might damage some older and Eastern buildings.

It will probably take a decade to resolve fracking controversies, reform bad operators, and build a stable regulatory regime that earns public confidence. A 2010 MIT study[691] found the environmental impacts "manageable but challenging. . . . It is essential that both large and small companies follow industry best practices, that water supply and disposal are coordinated on a regional basis, and that improved methods are developed for recycling of returned fracture fluids." Uncertainties persist about how well this will be done, how well initial claims of producibility will hold up, and what the long-term price will be. Some shales being fracked are also the caprock meant to contain captured carbon proposed to be sequestered below it.

We can take comfort in the fact that efficiency and renewables can gradually abate natural-gas needs, and that biomass-based alternatives exist that need no drilling.[692] The energy transition described here can thus take advantage of successful shale-gas development without depending on it.

So what will natural gas cost? After a bulge and crash, natural-gas prices in 2010 settled around $4 per million BTU (or, roughly, per thousand cubic feet or gigajoule), equivalent to $23/bbl. That's just one-fourth of the world oil price, an all-time record ratio. By spring 2011, prices were back to 2000 levels, and price volatility had greatly decreased. Gas is now a separate, independently priced competitor in North America.[693] Independent analysts expect $4–$7 gas to at least 2020. EIA projects $7 U.S. natural gas in 2035, $3 below its forecast two years earlier.[694] At these moderate prices, natural gas sets an important national benchmark for low-carbon fuels and will remain a strong competitor for carbon-free renewables, though its price volatility closes any gap between it and renewables by at least $2. Nonetheless, the uncertainties around shale gas's durability and impacts make it prudent to use gas frugally. That's worthwhile regardless, because efficiency generally costs less than any kind of gas, as well as avoiding its externalities.

Figure 6-2 shows how fossil fuels can be phased out—oil and coal by 2050, the rest of the natural-gas "tail" (see Natural Gas: Future Outlook sidebar) thereafter. And with abundant, benign, and affordable energy assured, there will be no worries that the energy party—or the civilization that depends upon it—will end. Fuel without fear, fuel without end, energy at last in harmony.

Is this actually a realistic future? Let's test how our vision stacks up against the commonsense requirements for any durable energy system.

Is the Reinventing Fire Vision Economically and Technologically Viable?

In the long and checkered history of energy technologies, one phrase keeps cropping up: a new energy technology could save the world—but only with a technological breakthrough. But our analysis, consistent with many other official and independent studies,[695] suggests that the Reinventing Fire vision is already viable, based not on miracles or magic but on purposeful application of what's already proven. Our analysis did not rely on breakthrough technologies or new inventions. Although Reinventing Fire includes some technological advances based on aggressive learning curves, especially for carbon fibers, batteries, and renewable technologies, the accelerating advances in material and biological sciences suggests that our assumptions might be conservative.

To be economically conservative, we picked only efficiency and renewables technologies that could attract investments using current business criteria and practices. We didn't invoke special economic dispensations such as lower discount rates, guaranteed government loans, tax breaks, or depreciation allowances. Most efficiency opportunities pay off today, and if renewables and Revolutionary+ electric vehicles aren't already cost-competitive with fossil-fuel alternatives, they will become so as economies of scale and the normal pace of innovation kick in. The one pump-priming policy we suggested to overcome an initial discount-rate barrier—feebates for new autos—should be revenue-neutral and temporary. Similar policies could be applied in other sectors, but conservatively, our analysis hasn't done that. Instead, as with heavy trucks, we suggested some new financial products that the financial industry could offer. We did mention the possibility of secured federal loan guarantees for accelerated purchase of superefficient new aircraft, but we didn't assume they'd happen.

However, because of all the noneconomic and nontechnology barriers mentioned throughout the book and summarized below, efficiency and renewables technologies aren't yet being adopted as quickly as the Reinventing Fire path shown in fig. 6-2 would need. Getting there will require tremendous improvements in their application and adoption. We've laid out the policies that can ensure the new technologies will be developed and adopted, and how to adopt those policies despite federal gridlock—especially by using the diversity of our states and regions as a fast-learning laboratory.

Such policy "boosters" will be necessary to unlock and speed each sector's potential for transformation. For example, the investments needed in new supply chains mean that without a feebate it's unlikely that Revolutionary+ cars will dominate the market by 2050—at least those from U.S. automakers. High information, diagnosis, and transaction costs also make our building retrofit rates doubtful without more aggressive and widespread energy standards and training. Industrial asset turnover and upgrading would go much faster with the producer lifecycle responsibility that's already driving competitors to design for closed loops. Decoupling and shared savings could supercharge utilities' efficiency investments.

Will Reinventing Fire Hurt the Economy?

The answer is a clear no. According to our analysis, getting the U.S. off fossil fuels by investing systematically in energy efficiency and renewables will actually stimulate the economy and create more wealth for far more players.

To begin with, the U.S. economy wastes over a half-trillion dollars' worth of energy every year. Cost-effective energy efficiency can put much of that lost money back to work. That plus switching from fossil fuels to renewable energy would save a gross $9.5 trillion (fig. 6-3). Those savings don't come for free, of course. RMI's analysis shows that Reinventing Fire will require an additional $4.5 trillion in capital investments compared to the business-as-usual path. That leaves net savings of $5 trillion.

Some $5 trillion in savings over 40 years may seem small in a $15-trillion-per-year economy. But it's the opposite of the economic collapse that some pundits predict.

So who pays and who wins in this multi-trillion-dollar sweepstakes? Reinventing Fire relies on businesses in each sector to invest at market conditions once the rules become helpful or at least neutral.

Descaling and recapitalization might also be needed in sectors where investments are truly disruptive, like carbon-fiber autobodies, smart grid, or some novel real-estate developments. More traditional financing could suffice for incremental changes like building controls, CHP equipment, and biofuels infrastructure. But overall, with the right policy and regulatory framework in place, we

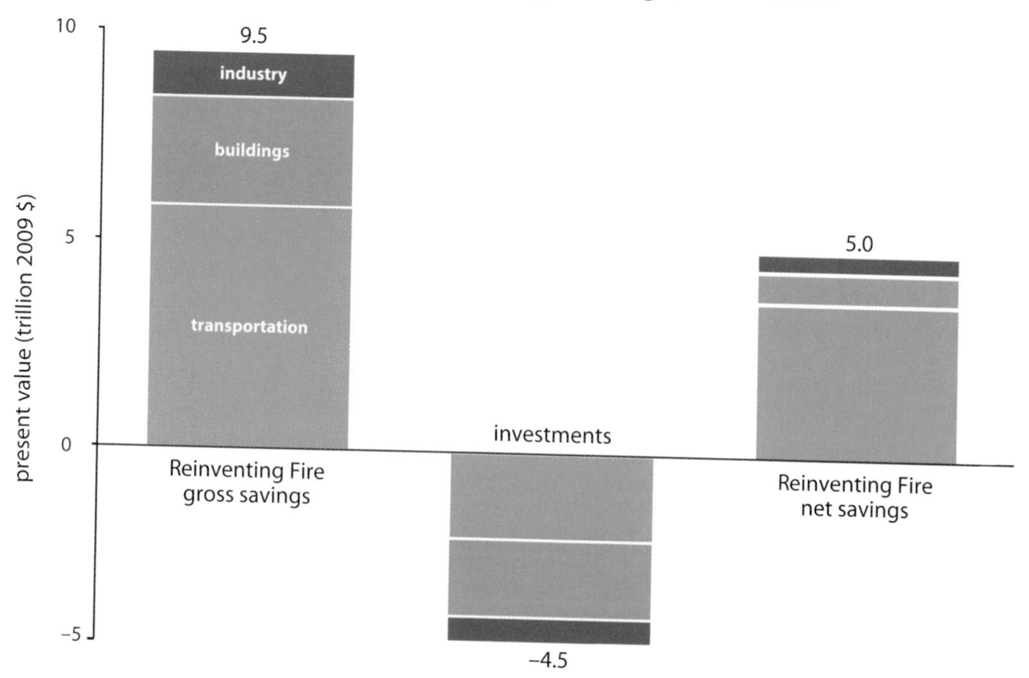

FIG. 6-3. Discounted to 2010 present value at a 3%/y real discount rate, the strategy outlined in chapters 2 through 6 would require $4.5 trillion of cumulative extra investment (beyond business-as-usual) but return $9.5 trillion in fuel savings, creating $5 trillion of cumulative net wealth.[696] This excludes integrative design and non-energy benefits.

expect the Reinventing Fire investments to be carried out by businesses for returns commensurate with their respective industry norms.

Not all the fuel savings are captured by businesses, though. For example, although automakers will make an industry-commensurate return on selling Revolutionary+ vehicles, the oil will be saved by drivers, mainly individuals. This is why, when aggregating all these savings, we used a lower discount rate of 3% when computing their present value. This also seems appropriate because getting off oil and coal has such strategic value for the whole society due to huge avoided capital investments,[697] quicker paybacks for reinvestment, faster learning that amplifies capital savings, stable prices, localization of energy production (avoiding big delivery investments), respending and reinvestment of saved energy dollars (as stimulative as abating a pervasive and regressive tax on society), stronger international competitive advantage and trade balance, healthier people[698] with lower healthcare costs, lower homeland-security and military costs,[699] avoided costs of major mishaps, and abated climate risks. Some of these gains, along with saved government energy costs, could directly and indirectly reduce federal budget deficits.

How Will Reinventing Fire Affect Jobs?

Jobs are where the economic rubber meets the road; they are a tangible way that many people will experience, and benefit from, transitioning away from fossil fuels. Our analysis anticipates the net jobs impact to be at worst neutral, but more likely appreciably positive. While job computations are complex and uncertain—employment depends less on shifts between more and less job-intensive sectors than on national fiscal and monetary policy—some encouraging conclusions seem clear.

Reinventing Fire creates direct jobs in two main ways: through energy efficiency (such as workers insulating a home) and fuel substitution (such as workers building or operating a wind farm). Less tangibly, saved energy expenditures are freed up for respending through the broad economy, supporting further job growth. Finally, the biggest effects, but the least certain and hardest to quantify, are job gains induced by more competitive industrial production, potentially higher share of the global vehicle market, and a more competitive national economy. The fossil-fuel industries will lose jobs, but fortunately, these industries (from fuel extraction and refining to electricity generation) have a very low labor intensity, while efficiency and renewables have above-average labor intensity, so job gains should more than offset losses.

At the sectoral level, lower energy costs and price volatility should create net jobs in buildings and industry, though in relatively modest numbers. There are currently about one million direct and indirect jobs in buildings' energy efficiency sector alone.[700] The transportation and electricity sectors should be relatively net-job-neutral.

In transportation, as Americans drive nearly 50% less to get the same access, and as autos get far simpler and more reliable, it's hard to imagine a scenario in which auto-related jobs are not reduced, chiefly in parts, maintenance, and repair. (Which way vehicle manufacturing and its supply chain evolve will depend on whether U.S. automakers gain or lose market share.) But meanwhile, greater demand for public transit should provide an estimated half-million stable, middle-wage urban jobs. In addition, making the second- and third-generation biofuels for heavy trucks and aerospace would probably support 300,000 jobs—more than the oil industry now employs for supplying today's entire mobility-fuel demand. Two-thirds of the biofuel workers would grow biofuel crops on farms, manage forests, or transport the biomass to nearby biorefineries (or, with portable units, vice versa), benefiting economically hard-hit rural areas. And finally, because the oil industry has a very low labor intensity, the billions of dollars Americans will save annually at the pump by 2050 will be freed up to be respent elsewhere throughout a more labor-intensive general economy, inducing between a half-million and one million jobs.

In the buildings and industry sectors, energy efficiency creates jobs for the installers, designers, auditors, and manufacturers who provide it—their wages are paid by the investment in this efficiency. By 2010, the United States had more insulation installers (about 112,000 full-time home efficiency providers[701]) and renewable energy installers (over 100,000[702] full-time for solar and wind, both rising rapidly) than coal miners (87,000 and falling[703]).

In the electricity sector, the public debate on green jobs has been the most intense. The key point is that renewable energy has more jobs per kWh than fossil-fuel-based generation. Despite producing almost half of U.S. electricity, coal (mining, transportation, and combustion) directly provides only one-eighth of 1% of U.S. jobs. Already windpower alone employs more Americans.[704] Choosing a transformative electricity-production path yields about twice as many jobs as maintaining conventional generation. The U.S. is already starting to see the kinds of significant job creation in clean energy that buoy the economies of such countries as Germany and Denmark.[705] And many of the new jobs in electricity, buildings, and industry are open not just to existing specialists but also to many lower-income and lower-skill workers now being trained in weatherizing houses, renewable energy installation, and similar new occupations.

Maturing renewables sectors will doubtless wring out some jobs, but it's encouraging that of recent new U.S. jobs in wind and solar power, up to 80% are in manufacturing and R&D—both generally high-paying. Although installation will lose jobs as it becomes more efficient, it is necessarily local. In buildings and industry, efficiency providers, mainly small businesses, also do hands-on installations, making these new jobs highly resistant to offshoring. On-the-job training often suffices for older workers (who are especially prevalent in construction), while unskilled and entry-level younger building workers' opportunities expand.

What about workers who will lose their job in the transition? Diversifying coal jobs into new clean-energy jobs alone probably won't suffice, and distressed Appalachian coal communities deserve a fair transition. But coal miners' total payroll, about $5 billion a year, is a fraction of 1% of the nation's energy bills, much less than spontaneous efficiency gains save in a normal year, leaving much flexibility in finding just solutions. Climate policies threaten miners' jobs much less than do the coal companies themselves, which halved mining jobs since 1983 while output rose 37%.[706]

It's not just the number of the jobs that's important, of course, but also their quality. Here the predictions are encouraging as well. We expect a wider spread of occupations, skills, and salaries in the workforce, partly due to increased domestic manufacturing (97% of new U.S. jobs since 1990 were nonmanufacturing, reducing exports and polarizing wages). The jobs should be more stable and much less likely to go offshore. There will be more opportunities for entrepreneurs, small businesses, rural communities, and—since every community needs efficiency—disadvantaged regions and populations. Poor communities can especially benefit from efficiency, because energy dollars that formerly almost all left town, never to return, can instead recirculate around Main Street, supporting local jobs and multipliers.[707]

How Will Reinventing Fire Affect Climate Change?

For you to welcome this book's thesis and embrace its recommendations, you needn't accept the global scientific consensus on the reality and severity of the risks of climate change. Throughout this book, we have highlighted many economic, security, and environmental benefits of moving away from fossil fuel. But if you do agree it's good to make money, you should like protecting the climate by using energy in a way that saves money, because saving fuel is cheaper than buying it. That makes climate protection not costly but profitable—a very convenient truth.

As we saw above, Reinventing Fire saves many trillions of net dollars—while cutting U.S. CO_2 emissions more than 80% from the 2000 level. That slightly surpasses international goals (fig. 6-4), reflected in the U.S.-ratified 1992 Rio Treaty, that were set to avoid "dangerous" climate change.

The December 2009 Copenhagen climate conference proved again how pricing carbon and winning international collaboration are hard if policymakers, pundits, and most citizens *assume* climate protection will be costly. But chapters 2 through 5 show that assumption is wrong. Changing the conversation to wealth creation, jobs, and competitive advantage sweetens the politics so much that any remaining resistance can melt faster than the glaciers.

Since the Kyoto conference in 1997, most efforts to hedge climate risks have made four main errors: assuming solutions will be costly rather than (at

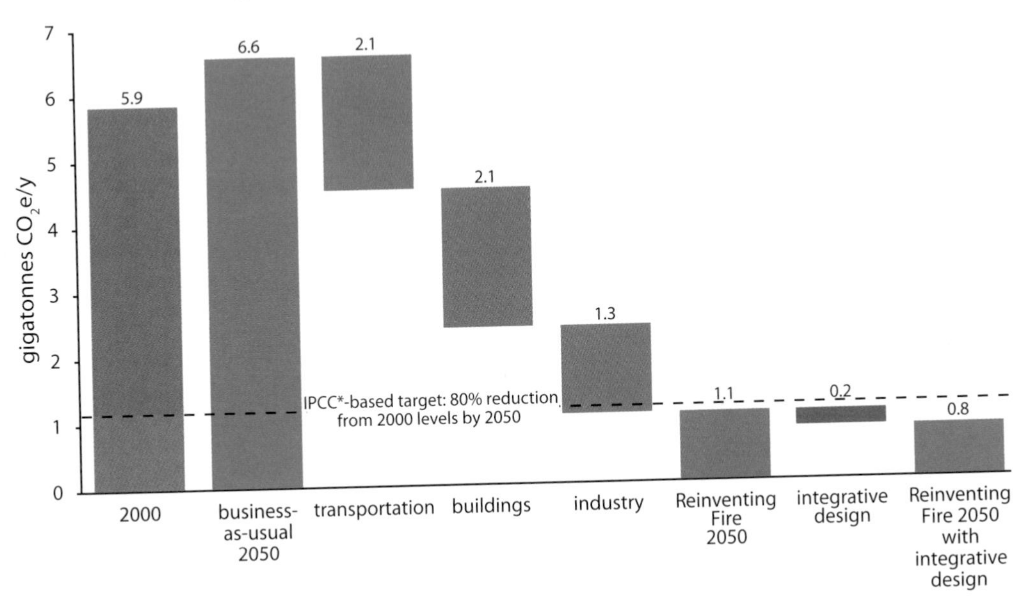

FIG. 6-4. The previous chapters' energy savings and supply shifts can reduce U.S. fossil-fuel carbon emissions by more than 80% from 2000 levels—the nominal reduction needed to hold atmospheric CO_2 concentrations below 450 ppm. (In 2011, the actual level was 390 ppm and lately has been rising by about 2 ppm per year; indeed, it was already equivalent to about 450 ppm when other trace gases' effects are included.) The baseline year formerly used, 1990 instead of 2000, gives equivalent results; the slight difference, at most 5 ppm, would be more than covered by reductions in non-fossil-fuel CO_2 emissions.[708]

least mainly) profitable; insisting they be motivated by concerns about climate rather than about security, profit, or economic development; assuming they require a global treaty;[709] and assuming U.S. businesses can do little or nothing before carbon is priced. As these errors are gradually realized, climate protection is changing course. It will be led more by countries and companies than by international treaties and organizations, more by the private sector and civil society than by governments, more by leading developing economies than by mature developed ones, and more by efficiency and clean energy's economic fundamentals than by possible future carbon pricing of unknown (but not zero) likelihood and price. (These energy benefits will also be augmented by carbon and trace-gas savings from biologically informed agriculture, ranging from perennial polyculture[710] to beef-system reforms[711] to new ways to restore devastated tropical rain forests and their impoverished rural communities[712] while reversing the huge greenhouse-gas emissions of countries like Indonesia and producing abundant biofuel.[713] In short, though it's beyond our scope here, there's as much good news about advantageous ways to abate non-fossil-fuel greenhouse gas emissions as fossil-fuel ones.)

Making these energy changes—and thus getting the needed emissions reductions—quickly enough is a challenge. But it is manageable, just as it was in 1977–1985 when U.S. oil intensity fell 5.2% per year. Today, based on standard economic-growth and decarbonization forecasts, cutting global *energy* intensity (primary energy used per dollar of real GDP) by about 3–4% a year, versus the historic 1%, could more than offset net carbon growth and rapidly abate further climate damage. This looks feasible. The U.S. has long achieved annual intensity reductions of 2–4% without national focus or concerted effort, while China achieved more than 5% for a quarter century through 2001 and 4–5% in the past few years. Some firms have even achieved 6–16%. So why should 3–4% be hard—especially when most of the growth is in countries like China and India that can make their new infrastructure efficient the first time rather than fixing it later as we must do? And since virtually everyone who does energy efficiency makes money at it, why should this be costly?

Sustained effort pays off. Using figures from before the Great Recession of the late 2000s to avoid distortions, in 1990–2006, California shrank greenhouse-gas emissions per dollar of GDP by 30%. In 1980–2006, Denmark shrank its energy intensity 39% and its carbon intensity 50%, made its electricity 28% renewable and three-fourths micropower, and created a world-class renewables industry.[714] Now it's heading for 100% renewable energy to enhance its economy and security.

Impossible? Henry Ford said, "Whatever is worthy and right is never impossible." And, he added, "Whether you think you can, or whether you think you can't, you're right."

Will Reinventing Fire Affect Our Nation's Global Competitiveness and Power?

There's an oddly persistent view that a country can best compete in the global economy by keeping energy prices low. According to this view, if prices rose in the U.S., as they might in a transition away from fossil fuels (though energy bills would be lower, since we'd use energy much more efficiently), then companies would flee to places with cheaper energy.

But that's a myth.[715] If it were true, America, the current world capital of cheap-energy policy,[716] would have no trouble beating countries like Germany and Japan, whose energy prices have long been more than twice those in the U.S. The facts show that such countries' higher energy prices (especially their motor-fuel taxes) have made their economies more efficient, more productive, and more innovative and even led to smarter settlement patterns. Shell's study of 20 countries found that the main reason for Europe's more than halved per-capita driving was land-use patterns shaped by higher energy prices.[717] Of course energy prices, though important, aren't the whole story; ability to respond to price matters even more.[718]

The Reinventing Fire path would tap into America's greatest strength—its mighty engine of innovation, which hasn't failed the nation yet. Remember, for instance, the obituaries lamenting the end of U.S. competitiveness in the late 1980s, when Japan seemed poised to ride VCRs, Walkmans, and other products to world economic dominance? Then along came the Internet, with its vast new business opportunities, and suddenly America was again the world leader.

Once it's clear that market forces, aided by nudges from judicious policies, are aligning behind energy efficiency and clean energy, the trillions of dollars to be gained will kick off another explosion of innovation. Just as has occurred with past market-based environmental regulations, gains would come at lower-than-predicted costs (see Competition and Trading Do Work sidebar). And the resulting creative solutions, invented and developed in the U.S., would help make America a global leader in new energy and efficiency technologies. Moreover, market-based global carbon-trading policies would let U.S. firms buy emission reductions in developing countries first whenever that's cheaper. Such best-buys-first flexibility would further boost U.S. companies' profits while speeding global economic development.

The benefits to the U.S. go far beyond economic gains. A serious commitment and effort to shift off fossil fuels would bring intangible but huge boosts to America's leadership. Right now, the U.S. is being heavily criticized for being the second-biggest emitter of greenhouse gases yet unwilling to lead the way to global reductions. The U.S. has also been forced to maintain good diplomatic relations with sometimes unsavory or unfriendly governments, simply because we need their oil. We're in thrall to outmoded energy sources in much the same way Britain was in 1807, when that era's great source of cheap energy, slavery, drove one-fourth of the economy—far too important to meddle with. Yet the anti-abolitionists were wrong not

COMPETITION AND TRADING DO WORK

A strange irony lurks beneath the climate debate. Why do the same people who favor competitive markets in other contexts argue against their use to cut fossil-fuel consumption? The fact is that markets work. Remember the old debate over the 1990 cap-and-trade system to reduce acid-rain-causing sulfur dioxide emissions? Environmentalists predicted that reductions would cost about $350 a ton. Government economic models predicted $500–$750. Industry claimed the cost would be $1,000–$1,500 per ton. They were all wrong, because the market found clever, cheap, and previously unforeseen ways to cut emissions (including shipping more low-sulfur coal from the West). The sulfur-allowance market opened in 1992 at about $250 a ton; in 1995, it dropped to $130 a ton; in 1996, it was $66. Emissions fell 60%, and faster than required because simple incentives rewarded early achievers. Benefits totaled about 40 times costs. Exceeding the hopes of the mainly Republican lawmakers who had crafted this emissions allowance market, the genius of private enterprise found a way that was over $20 billion a year cheaper than would have happened with command-and-control regulation. It would do so again if we competed to save the most carbon in the cheapest ways, as Europe now does for over 10,000 big industrial plants. In fact, an environmental double bonus for business would emerge: we'd automatically and profitably meet most of the stringent new ozone and fine-particle standards too, via the same reduced combustion that protects the climate and cuts our energy bills.

only morally but economically: Abolishing slavery was a huge boost to the British economy.[719]

Reinventing Fire would resolve that ethical dilemma and thus help restore America's moral authority and standing. To prevail in current struggles and prevent future ones, we must steer by faithful stars: freedom, democracy, justice, enterprise, transparency, diversity, tolerance, and humility. We must strive to be the sort of society others admire, emulate, and want to do business with. What better way to be a moral beacon and reinvigorate our own aspirations than helping lead the world to the cleaner, safer future this book envisions?

How Will Reinventing Fire Affect National Security?

It's easy to see how ending our dependence on oil will make the nation safer and stronger.[720] As former CIA director R. James Woolsey points out, nations once warred over salt, but refrigeration made salt obsolete for food preservation. Now we no longer worry about how we'll fill the saltshakers on our tables. We don't make unholy alliances with salt-producing countries. We don't spend trillions of dollars on a military presence in regions of the world with salt reserves. A clear trajectory off oil will likewise turn oil from scarcity to obsolescence, from black gold to low-value sludge. It would strip power from the oil cartel and ease global tensions and instabilities.

Rivalries over oil—most of all with a giant like China—are a problem we don't need to have, and it's cheaper not to. Imagine the shock and awe if America treated Middle Eastern oil exporters as if they had no oil—and gave the world no reason to believe U.S. actions were motivated by oil.[721]

The demise of oil would have myriad beneficial rippling effects. Ironically, oil is one reason why we now worry about nuclear proliferation, widely agreed to be the gravest threat to our security. The 1970s–1980s boom in nuclear power came substantially from the desire to displace oil-fired power plants. Ultimately this facilitated and disguised[722] the entrepreneurial A. Q. Khan network's one-stop shopping mall for clandestine nuclear weapons programs—funded by the sale of oil. The U.S. invaded oil-rich Iraq over claims of such oil-funded weapons of mass destruction; tried unsuccessfully to buy off North Korea's bomb program with oil; and has failed to stop Iran's oil-funded bomb program because Iran could easily choke oil shipments through the Strait of Hormuz.

Meanwhile, the U.S. military realizes that replacing oil with efficiency and renewables in its own operations can protect warriors, win wars, and help prevent some wars. And, as we've already seen, eliminating the use of even domestically produced oil insulates us from price shocks that cause the economy to stumble, and from brittle infrastructure at home.

So it's obvious that freeing the nation from oil will bring huge national-security benefits. But why not keep using coal and natural gas? They're both abundant domestic resources, and natural gas also offers a smaller carbon footprint.

Eliminating oil use creates only an illusion of security. True energy security requires shifting away from an electricity grid made vulnerable by its centralized power plants and thousands of miles of transmission lines. Even normally reliable coal power plants are not immune to major failures: track and bridge failures can block major rail corridors, for example, while cold snaps can make coal unshippable or unburnable. Not only is such a system prone to expensive blackouts, it also offers many tempting targets.

Will Reinventing Fire Destroy the Fossil-Fuel Companies?

Probably not, but they'll have to change as their bulwarks are pounded by waves of radical innovations.

Take oil, which could be fully obsolete by 2050.[723] Oil companies are nearly all diversified energy companies. Many have shifted or are shifting to natural gas or renewables. But traditionally about two-thirds of their equity value comes from their hydrocarbon reserves. That value could drop dramatically if we stop using oil—and there are already strong signals that oil *demand* is peaking. In October 2009, for instance, Deutsche Bank forecast *world* oil use will peak around 2016, then by 2030 drop to 40% below the consensus forecast or 8% below the 2009 level[724]—just due to electrified autos—and concluded: "We believe that falling oil demand will end the oil age."

Thus the oil industry faces what economist Joseph Schumpeter called "creative destruction"—wrenching change that comes when entirely new businesses replace the old ones. Oil companies no longer control their own destiny.[725] Many major oil companies have seen this coming for more than a decade and have diversified their activities and option portfolios accordingly. Shell, for example, is already the world's leading biofuel distributor. The industry can also use its drilling and reservoir mastery to offer carbon sequestration services—whose success might ultimately require many times the annual cubic meters of throughput that oil now does—and to exploit advanced geothermal energy.

Perhaps the biggest opportunity for the oil industry, however, is the potential hydrogen market, which the industry already knows well and can serve through its retail channels. Hydrogen could earn oil owners higher prices in a world that ultimately buys no oil, because the hydrogen may be worth more without the carbon than with it, even if nobody pays to keep carbon out of the air. Owners may thus make more money taking hydrogen *out* of their hydrocarbons in a reformer than putting hydrogen *into* their hydrocarbons in a refinery. That's because fuel cells' very high efficiency in vehicles, and even higher system efficiency in stationary cogenerators, makes hydrogen more valuable than hydrocarbon per unit of end-use service delivered.

A shift to hydrogen fueling shouldn't significantly increase total U.S. use of natural gas and may, as GM found, reduce it,[726] but hydrogen should earn higher margins than gasoline. By moving toward hydrogen, biofuels, and "green" chemicals, oil companies could continue to use some of their refineries, pipelines, storage facilities, and other infrastructure. Their broad technical know-how, diverse infrastructure, market knowledge, strong balance sheets, and cash flows, global brands, customer relationships, and talented people can help them play and win across a wide range of new games. But they will need to get good at doing many small things, not just a few big things, play as adeptly in efficiency and electricity as in fuels, and probably align better with customers by leasing services (like access and mobility) rather than selling commodities.

Many conventional investments that oil companies are making, especially costly and high-risk frontier exploration and production, look unwise today. As in any other disruptive time, success depends on diversifying options, staying nimble, and timely optioning and acquiring the capabilities needed to compete in the new world. In this respect, at least, many oil companies seem ahead of their counterparts in the older and more traditional coal industry, which is already losing to natural gas for electricity generation and could be entirely replaced by renewables and gas in the U.S. by 2050. Coal owners must either migrate to another business or figure out how to use coal not as a low-value, high-volume boiler fuel but as a high-value, low-volume reagent and reductant. Some, it seems, are beginning that journey.

Will America Have Partners in Reinventing Fire, or Would Others Negate Its Efforts?

Some claim that if America strove to defossilize fuels, the benefits would be offset by others' (chiefly China's) recalcitrance, rendering its labors futile. The truth is just the opposite. The entire industrialized world *except*, at this writing, the U.S. and Australia is being joined by key developing countries in this common quest, with Europe long in the lead[727] but China now surging ahead.

In the 19th century, American innovators and firms laid the foundations for the modern electrical system. Today, the U.S., with less than 5% of the world's population, still produces nearly a fourth of the world's electricity. But by 2050 that share will drop to one-eighth or less as countries like China and India boom, and as electricity finally reaches the last billion and a half people in the world. China and India together burn half the world's coal and in 2008 were building 75% of the world's new coal-fired generating capacity. By 2050 they're expected to spend $5 trillion to install about twice as much generating capacity as the U.S. and Europe will add—but much of that new capacity will be renewable. Asian renewable investments surpassed the Americas in 2009 and outpaced Europe in 2010. India outinvested Japan and Britain, is firmly in the top 10 renewable nations, just quadrupled its renewables target, and aims to install 20 GW of PV by 2022[728]—all displacing coal.

China has surpassed the U.S. in energy and electricity use and in carbon emissions, but also in coal-plant efficiency, installed windpower, and even national climate policy. It's executing the $0.77 trillion (RMB 3 trillion) decarbonization effort it announced in 2010. It has targeted another 18% cut in energy and carbon intensity by 2015 (beyond the 16% cut during 2006–2010), capped coal-mining capacity at 3.8 GT in 2015, effectively capped carbon emissions, embraced market-based emissions trading along with resource consumption fees and environmental taxes, and set up its first dozen environmental exchanges to start the trading. Unlike the U.S., China has laid the foundation for its carbon emissions to peak before 2030,[729] consonant with a 450 ppm world, as its energy demand coasts to saturation.[730]

During 2006–2010, China shut down 71 GW of inefficient coal-fired plants and halved net annual additions of coal-fired capacity (two-thirds of which during 2005–2007 were bootleg projects unauthorized by Beijing). Chinese coal's hidden costs add up to an estimated 7% of GDP, not counting clogged rail networks that delay valuable exports from reaching southern seaports. Coal's air pollution is believed to cause over a million Chinese deaths per year. Coal's habits and bureaucracies remain powerful in China, but its star is starting to wane. Its prices are rising, coal-fired utilities are losing money,[731] environmental rules are tightening, and Chinese solar power is planned and likely to approach coal's cost by 2015.[732] In fact, China plans about 33–39% of its generating capacity and 15% of its electricity to be renewable by 2020.[733] This transition has already begun: China's 2010 net capacity additions were only 59% coal, 38% renewable, and 2% nuclear.[734]

In 2006, China's renewables other than big hydro had seven times the capacity of its nuclear power and were growing seven times faster; since then, the gap has widened.[735]

Overall, China invested $54 billion in clean energy in 2010—60% more than the U.S. (and 139% more in relation to GDP), and more than the world's entire clean energy investment in 2004. The country has also increased spending on research and development, set to rise to 2.2% of GDP.[736] China already dominates patent applications in clean energy technologies,[737] is now the world's leading manufacturer of five renewable technologies (wind, photovoltaics, small hydro, biogas, and solar water heaters), and is closing in on the rest. As the U.S. again slashes its R&D budgets,[738] technological development will be driven increasingly by China and its rivals.

So will prices. China has already driven steep drops in global photovoltaic prices in 2009–2011 and is starting to affect the wind-turbine business.[739]

China's leaders are also terrified of falling into the trap of being dependent on oil, so they're scaling back policies that had encouraged soaring car ownership and use. Their standards have made Chinese new cars surprisingly efficient, and they've put strong policy support and a $15 billion budget behind electric vehicles, with 5–10 million (not counting ordinary hybrids) expected to be on the road by 2010. That's an order of magnitude more than needed to descend the steep part of the learning curves and become a global market force. But even though China is the world's biggest car market, its per-capita car ownership is comparable to that of the U.S. in 1915. Approaching today's U.S. levels seems unwise, simply because China has no place to put 600 million cars.[740] Smart Chinese customers have already responded by buying over 120 million cheap, efficient, space-frugal, and easily parked electric bicycles, triggering a sudden $11-billion-a-year world market.[741] That may prove a harbinger of automotive market disruptions too.

Less visible but just as important as China's new dominance in renewables is the nation's impressive progress with energy efficiency. Between 1980 and 2001, reduced energy intensity cut China's growth in energy demand by about 70%. Then, in 2004, energy efficiency became the top strategic goal. Leaders like Wen Jiabao understand that China can't afford to develop without efficiency, as the needed energy supply would eat up the capital budget. Facing a one-percentage-point shortfall in its 2010 efficiency goal, the central government even shuttered 2,000 factories that had missed their energy-saving targets—a surefire way to encourage the others—but still fell slightly short, and it doesn't intend to do that again. With its long-run efficiency and renewable prospects, the Chinese economy could even grow an order of magnitude bigger than it is now, using no more energy and releasing one-fourth the carbon.[742]

Balancing all these impressive achievements and plans are equally daunting challenges. Urbanization, lagging rural development, and subsidized residential electricity remain major problems. Serious water and farmland shortages loom, and coal shortages are on the horizon.[743] What Wen Jiabao in 2011 called an "irrational industrial structure" will long distort energy use.[744] But for the technocrats who lead China—as a group, they're probably better versed in energy than the leaders of any other nation—China's renewable power and efficiency revolution can help tackle those problems, bringing better lives in a richer China and a safer world. The gains in renewable power also mean thriving exports, as China continues to surge to the front of the pack in clean energy technologies.

For the United States, two lessons about China are clear. First, as American scientists, business leaders, and policymakers engage constructively and respectfully with a vital partner that's making an unprecedented effort to solve extraordinary problems, we should draw inspiration from China's efforts and strive to emulate and surpass them in our very different society. Our two great countries must combine our best talents, share our successes and mistakes, and learn faster together. Though competitive interests are at stake too, and will shape the giant renewable energy industries of the future, shared interests are even more durable and vital.

The second lesson is starker. If we don't move at the speed of heat in the global marketplace, others will. China already has. India is on the move. Korea and Brazil are not far behind. America's private-sector dynamism and powerful capacity to innovate give it the potential to catch up or even jump ahead in capturing the greatest business opportunity of the age. Will we realize that potential? Our own prosperity and the world's hang in the balance.

Will Reinventing Fire Hurt the World's Poor?

If America frees itself from fossil fuels, we will see our economy grow and we can pat ourselves on the back for our leadership. But will that leave the rest of the world's 6.7 billion people, mainly in developing countries, to struggle along on those same fuels they can't yet afford?

The good news is that Reinventing Fire helps create prosperity, most of all for the dispossessed, in ways fully consistent with our highest values and adaptable to diverse cultures and conditions. And it frees up so many wasted energy dollars that a tiny fraction—some $40 billion a year according to the U.N. Development Programme—could, if properly reinvested, provide clean water, sanitation, basic health, nutrition, education, and reproductive healthcare[745] to every deprived man, woman, and child on Earth.

The best proof that this isn't just a pretty theory is what some developing countries actually did. By the time of the 2007 Kyoto climate conference, they were desubsidizing fossil fuels twice as fast as in the rich countries and saving carbon about twice as fast as the rich countries had committed to do—indeed, they were probably saving more carbon in absolute terms than the rich countries were expecting to do—all for *economic* reasons. They were boosting their own growth and quietly reaping the incidental environmental benefits. And by now, developing countries also own the majority of the world's renewable power sources, again for sound economic reasons.

Some say only rich people and nations can afford energy efficiency. That's exactly backward. Poor countries tend to have the *lowest* energy efficiency, averaging about threefold worse than rich countries. They get stuck with the least efficient equipment, including obsolete castoffs from rich countries. Poor people use little energy but waste more of what they do use, with greater economic burden: the poorest quintile may pay 30% or more of their disposable income for energy, versus less than 5% for the general population.[746]

Yet poor countries have the greatest *opportunity* to become efficient as they build their infrastructure from scratch. Building it right is far cheaper than fixing it later. And poor countries have the greatest *need* for energy efficiency. Its lack is an important cause of their poverty. Being trapped in pervasive inefficiency perpetuates crippling fuel expenditures on both personal and national levels. Oil purchases underlie much of the developing world's debt. The inflated costs of fuel delivered to rural and remote areas further heighten its economic burden even as it makes efficiency more valuable.

Bringing efficiency to the developing world frees up an extraordinary amount of capital.[747] Building factories to make superwindows or efficient LED lights, instead of constructing power stations and grids to deliver the same increased cooling and light, requires nearly a thousandfold less capital[748] and recycles that capital cost about 10 times faster. The saved capital could then flow to other critical development needs, such as sanitation, health, and agricultural productivity.

Moreover, some 1.6 billion people lack access to electricity, and another billion have unaffordable or unreliable energy services. In all, two-fifths of all people on Earth live in energy poverty.[749] This leaves many basic needs unmet, economic potentials unfulfilled, and educational aspirations frustrated, especially among women and girls. Building conventional thermal power plants and grids to serve those now lacking reliable or any electricity is not the answer. Not only is such an approach slow and unaffordable, but even if it succeeded, it would lock in decades of unnecessary user cost, fuel dependence, and carbon emissions.

It would be faster, cheaper, and far better for the world's poor to leapfrog the centralized model in both the technology and the delivery of energy. And it's already happening. When a South Indian village switched from kerosene[750] to fluorescent lamps, illumination rose 19-fold, energy input decreased ninefold, and household lighting expenditure was halved.[751] Also in India, whose large-scale solar policy is less helpful to entrepreneurs serving the off-grid poor, Sanjit "Bunker" Roy's Barefoot College trains illiterate African grandmothers as solar engineers so they can return home to build and support solar power in their villages. Efforts like the Lumina Project are bringing efficient solar-powered LED lights to millions across Africa, its takeoff led by private enterprise as customers discover that solar light is cheaper than kerosene lamps, helping girls learn to read at night and freeing up cash flow for mosquito nets, drip irrigation, clean water, and other basic tools for rising from poverty.

In 1947, General George Marshall said, "There can be no political stability and no assured peace without economic security," so U.S. policy must be "directed not against any country or doctrine, but against hunger, poverty, desperation, and chaos." He was right then and even more right today. Weapons and warriors cannot keep us safe when billions of our fellow human beings lack the essentials of a decent life and a reasonable prospect of a better life for their children. Reinventing Fire can help build that world where others live better and we all sleep better.

HOW DO WE SEIZE THE 2050 PRIZE?

There's a conundrum running through this book like a scarlet question mark. If the Reinventing Fire vision is so compelling, if the prize is so large, if the technologies are already available, then why haven't more companies, entrepreneurs, industries, government agencies, and ordinary people leapt to embrace this vision?

Barriers to Reinventing Fire

It is not because the prize doesn't exist. Instead, the reason is that this just-cleared path is strewn with a list of barriers so long that it can become overwhelming: unnerving with complexities, paralyzing with choices. Nine barriers dominate:

- ***Active or passive resistance by incumbents.*** Most organizations have an energy-using and generating asset base based on fossil fuels. Moving away from it is or looks risky and costly.
- ***Economics and technology.*** Economic and technology barriers are shrinking as high-value efficiency and renewable solutions are brought to market, but hurdles persist in some sectors.
- ***Knowledge and culture.*** Energy is not a priority for most organizations, so many lack the knowledge, willingness, or capabilities to move away from fossil fuels.
- ***Financing.*** Energy investments, often with sizable initial and relatively long paybacks, vie with others nearer a company's core priorities.
- ***Value-chain complexity.*** Energy investments often link multiple parties across long value chains, with sometimes misaligned incentives, routing costs and benefits to different parties.
- ***Unclear value proposition.*** Energy is energy, clean or not, so selling efficiency and renewables to undiscriminating customers can be hard.
- ***Lack of long-term leadership.*** Changing the energy strategy of a company, state, or nation requires planning and stewardship over decades—a far longer time period than profit or election horizons.

- ***Policy and regulatory structures.*** Some existing policies and regulatory structures impede energy transformation, so they must be changed or replaced to enable and accelerate it.
- ***Entanglement with partisan politics.*** For example, congressional vacillation has severely damaged the U.S. windpower industry four times and killed every sizable domestic wind-turbine maker except GE, so China, Denmark, and Germany now lead the industry that Charles Brush of Ohio invented in 1888—part of a long and dismal history of such losses in renewable energy.[752] Politicians apparently can't resist the temptation to hold short-term tax policies hostage to other issues, even if the result severely damages some of America's most vibrant and fastest-growing high-tech industries.

Overcoming these barriers is no easy task. It will require education, innovation, leadership, and changes in policy and regulation. For instance, organizations across the economy will need to train their employees systematically on energy issues and incentivize behaviors that move them beyond fossil fuels. Smart entrepreneurs will need to create solutions that minimize the hassle and maximize the benefits of efficiency and renewables. States will need to end-run federal gridlock.

One central conclusion rises above all the details: *Reinventing Fire*'s transformation is based on technologies that have been demonstrated to work economically or are on a clear and steep path to do so. *Therefore the real barrier is slow adoption rates, not inadequate technologies.*

The needed dynamics are simple. As figure 6-5 shows, the needed technologies are adopted only slowly in the first four years as companies gear up their knowledge, strategy, physical capacity, and training. The electricity sector's relatively mature technologies and players give it a useful head start. We don't expect the transportation sector to sell lots of Revolutionary+ autos so soon, and even by 2030, Revolutionary+ autos still won't dominate

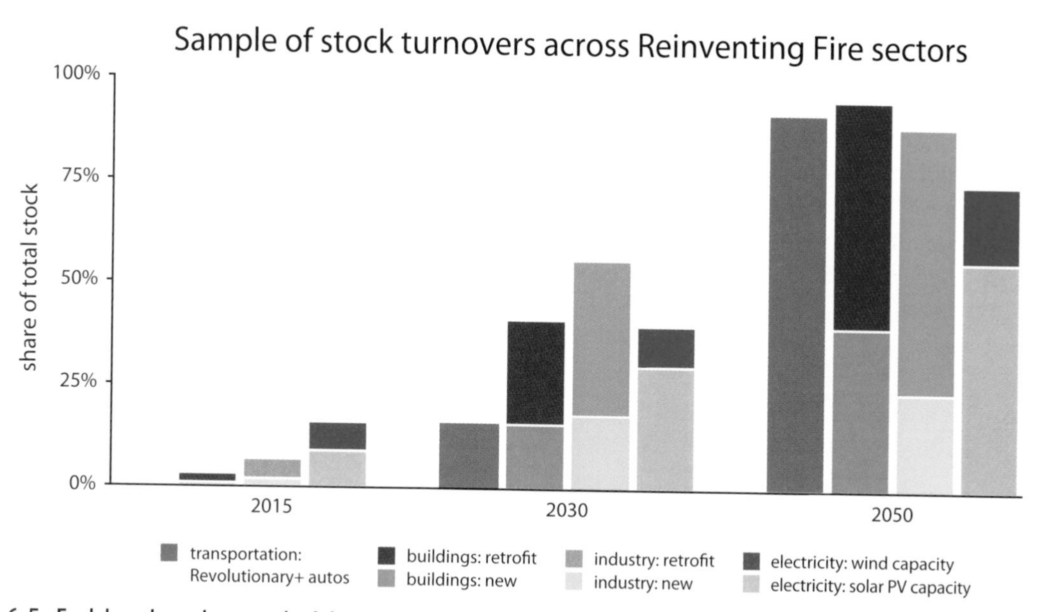

FIG. 6-5. Each bar shows how much of the total stock of vehicles, buildings, factories, or generating capacity will turn over, by the years shown, to the levels of efficiency or renewable supply described in chapters 2 through 5.[753]

the market, but by 2050, nearly three 15-year stock turnovers from now, the tortoise wins the race. The big issue is in slow-turnover sectors like buildings, industry, and electricity. If these sectors don't reach roughly 40% adoption rates by 2030, later years' adoption is unlikely to make up for those early delays.

So how can we light a fire under millions of people, thousands of businesses, and scores of government agencies to adopt efficiency and renewables dramatically faster and transform our energy system?

Framing the Vision

The answer starts with a powerful vision. Henry Ford would never have revolutionized transportation without his simple, compelling idea: "I will build a car for the great multitude," he vowed. John F. Kennedy said, "We choose to go to the moon." And we did.

Reinventing Fire's vision is equally transformational. We can free ourselves from our Faustian bargain with fossil fuels and create a safer, more efficient, cleaner world. Without such a clear goal in sight, chances are slim that the country will stumble its way into a great energy solution or that we will serendipitously avoid more major missteps. Just look at how every U.S. president since Richard Nixon has vowed to end dependence on imported oil. In actuality, net imports stopped rising only in 2005-2006 and by 2010 were still nearly half the total.

We need, and have sought to create in this book, a clear articulation of what a better energy future looks like, what its benefits are, and how we can overcome the challenges on the way. We must not set our sights too low and end up with a 20-year plan instead of a 21st-century goal. And we must not put on blinders. Software wizard Bill Joy, who invented Unix and Java, says, "If you can't solve the problem, make it bigger." Most people try to make tough problems tractable by splitting and narrowing them. But the reason those problems seemed insoluble is that they lacked a big enough design space to create the needed degrees of freedom, so the solution lay unseen outside the system boundary. As we've seen, it's much easier to solve, say, the oil-free-autos and renewable-electricity problems together than separately. Lasting energy solutions need a sharp focus but a wide-angle lens.

Leading the Change

Once we have the vision, the next requirement is leadership. If our elected representatives continue to make decisions based on shifting political winds, if special interests keep blocking meaningful reform, if captains of industry shrink from bold steps out of caution, then we'll get the mediocre energy system we deserve. But we know that America still is a land of leaders; politicians who understand the need for transformative change, CEOs who bet their company's future on risky but potentially revolutionary products, and regulators gutsy enough to take on powerful lobbies. We call upon those leaders in business and in government to embrace the vision of Reinventing Fire, steer a steady course, and make it happen.

We've seen exactly what these leaders must do first. As we've described in this book, business is the engine with the power, speed, scope, and scale to drive this energy transformation. But unleashing the might of business requires policies that allow and reward what we want, not the opposite. We also should rethink our tax system, because it's not only baroque verging on rococo, but fundamentally misdirected. If we taxed bad things like pollution rather than good things like jobs and income, as some European countries already do, we could fire the unproductive tons, gallons, and kilowatt-hours and keep employing more people, who then will have more and better work to do.

The previous chapters have laid out the needed reforms and incentives. Now it's time for our nation's leaders, at whatever level it takes, to put them in place. And when we say leaders we don't just mean lawmakers and regulators, nor presidents, tribal leaders, governors, mayors, county commissioners, and city councilors. Business leaders need to embrace the fact that this great prize will remain beyond their reach and ours unless we unlock the opportunity through public-private coordination and effort. For once, both sectors can get what they want: politicians can take credit for trans-ideological appeal and strong results, while business leaders can create enduring value by pursuing private advantage—rather than, as so often in the past, being forced by government mandates to commit unnatural acts in the marketplace.

Once those incentives to choose sensibly exist, watch out! The powerful American engine of innovation, which the 2010 *National Security Strategy* called a foundation of our nation's power and leadership. will shift into high gear. It will not only bring new technologies to market but also reshape our ideas of how we use energy. Most astonishing will be what happens as IT meets energy, pervading the energy system with distributed intelligence, ubiquitous sensors, and current information. IT-enriched energy will choreograph the convergence between vehicles, buildings, factories, and electricity sources. It will "transform every individual energy using device from a stand alone single purpose entity into a multipurpose interconnected grid asset that will ultimately optimize the efficiency of the entire energy system," as FERC chairman Jon Wellinghoff explains. "It is," he adds, "a revolution that is coming and it will change everything."[754]

We know where we need to end up. To get there, we need to master six critical challenges. We must together:

- ***Transform the auto industry.*** No more obese gas-guzzlers. Our cleaner, safer, oil-free world—and the health of this vital sector—depends upon the industry's ability to produce far fitter vehicles at roughly the same cost, before their old and new competitors do.
- ***Dramatically reduce the distances traveled by our autos and the haul length, weight, and volume of cargo carried by our trucks.*** This won't be achieved through sacrifice, deprivation, or diminished economic vitality. Instead, it will come from gradually redesigning our communities and businesses to increase choices, save time and money, and make life easier, healthier, and safer.
- ***Build efficient buildings and retrofit existing ones on a tremendous scale.*** Just one energy-efficient building makes a difference. Multiply that by 120 million buildings, and we have a revolution. It's a challenging task, of course, and will require everything from education and retreading designers to stricter building codes and innovative financing.
- ***Sustain and accelerate energy savings and cogeneration in industry.*** Both, and increasingly on-site renewable supplies of heat and electricity, are keys to a durably competitive and resilient future for this engine of the national economy.
- ***Keep slashing the cost of renewable energy.*** Wind, solar, and other renewables are already traveling down a steep learning curve. But the further they go—and much more is possible—the more rapidly we can reinvent fire in the sector with arguably the greatest economic, security, and climate leverage.
- ***Revamp utilities' rules and operating models.*** We will never create a future free of fossil fuels if utilities' profits depend on how much electricity they sell, or if distributed renewable sources can't feed electricity into the grid.

> Public utility commissions must change the rules, enabling and rewarding rapid adoption of best buys (efficiency, CHP, and renewables), closing obsolete plants that deserve a graceful retirement, and not imposing on customers any dubious projects' risks that shareholders won't invest in.

With the right leadership, Reinventing Fire will trigger wave upon wave of innovation and entrepreneurship unlike anything that's happened in the energy sector for more than a century. We hope—in fact, we expect—that this wave will carry us much further than our cautious analysis foresees is possible. We can then stand back and marvel as companies and entrepreneurs in mindful markets work their magic. There will be failures and errors, risks and uncertainties. Markets are unpredictable, much as Deng Xiaoping found the monumental transition toward a socialist market economy felt when he said, "We are crossing the river by feeling for stones."[755] But in the end, markets do tend to get it right. As Churchill said of democracy, they're the worst system we know—except for all the rest.

KINDLING THE NEW FIRE

The Reinventing Fire path holds more than just the promise of freeing America and humanity from our dependence on fossil fuels. It also holds the promise of fulfilling our long-held dream of energy without end for the nation and world. Its better path solves many of our most pressing problems, doesn't worsen the rest, makes us wealthier and happier, creates more options and opportunities, and can serve people everywhere, safely and abundantly, so long as we all shall live on Earth.

So how do we get started turning this vision into reality? We can write this book and give speeches, but we can't do anything alone. We need society to throw all its weight behind efficiency and renewables, and into the shifts of design, strategy, technology, and policy that get them done.

So we issue a call to action, a challenge to you—the worker on the assembly line, the CEO in the corner office, the entrepreneur in her garage, the regulator on the hot seat, the engineer struggling to keep the traffic flowing, the auto designer, the rancher, the grandmother, the member of Congress, and, yes, the president. Together you all represent our hope for a world where energy has turned from worrisome to worry-free, from risk to reward, and from cost to profit.

With your ideas, your determination, and your courage, we can make this vision real. This can be the moment when we finally take charge of our energy destiny, powering our future civilization with the inexhaustible resources of sun, wind, water, green plants, and Earth's heat.

This is not a green or a left-wing idea. It was Rupert Murdoch who recently reported that the News Corporation has gone carbon-neutral and thereby "saved millions of dollars."[756] It was his son James who called for a Red-Blue-Green energy agenda based on national security, economic strength, new employment with lower long-term costs, cleaner and healthier communities, and competition trumping regulation.[757] This approach plays just as well in deeply conservative Salina, Kansas,[758] and in the 11 largely Republican states with renewable electricity portfolio standards,[759] as it does in blue states.

Indeed, the fundamental approach we most need in energy is not directive but permissive. We need to allow and require all ways to save or supply energy to compete fairly, at honest prices, regardless of their type, technology, size, location, or ownership. What advocate of private enterprise and free markets wouldn't be in favor of that? Why don't we find out?

Amid the energy cacophony, the swirl of conflicting opinions and agendas, this simple approach bears emphasis: if together we focus on outcomes, not motives, that's good enough, because the actual things to be done are the same no matter whether we each care most about national security, profits and jobs, climate and environment, or something else. We just need to stop insisting that others embrace our differing *why* and together get on with doing the shared *what*. And if we vigorously do the *what*s we agree about, then those we might not agree about may well become superfluous. Together we can rediscover the truth of the African proverb, "If you want to walk fast, walk alone. If you want to walk far, walk together."

Just as shared underlying intent can knit together the factions of our vast and diverse nation, so America is not alone in this gathering global transition. Smart, resourceful people are everywhere, at all levels in every society, and the emerging global nervous system is helping good ideas spread faster. There are now more than a billion transistors for every person on the planet. Facebook has more users than America has people. Revolutions happen on Twitter and YouTube. Paul Hawken's *Blessed Unrest* chronicles the astonishing rise of millions of nonprofit citizen organizations around the world—the greatest social movement in history. As central institutions become more gridlocked and moribund, a new vitality is beginning to spread renewal from the grassroots through the stem to the flower. The search for intelligent life on Earth continues, but many promising specimens are turning up everywhere, just when they're most needed.

So apply your native intelligence, that amazing human gift that hasn't yet failed our species in its entire history, to this question: Shall we continue down the path we're on, toward economic stagnation, rising costs, unpleasant risks, social upheaval, and an ever more dangerous world, or shall we make a bold break and start laying the energy foundations of a world without waste, want, or war?

We get to choose just once. Choose well.

ACKNOWLEDGMENTS

Research and analysis with the scope and depth of *Reinventing Fire* necessarily builds on the work of many others, published in thousands of references, of which nearly 800 basic ones are cited starting on page 289. Yet much of the knowledge we needed hadn't previously been published. Some came from our own three-plus decades of research and practice with a wide range of partners in all four sectors described in chapters 2 through 5. More came from a diverse group of colleagues around the world who graciously shared their learning and their networks, corrected our errors, and provided splendid companionship as we all wondered in the bewilderness. The RMI team owes a deep debt to these many contributors, including many who asked not to be named or whose forgiveness we must beg for having inadvertently overlooked them. Of course, none of them is responsible for this book's content, which reflects solely the views of the authors.

RMI's Board of Trustees authorized this effort and provided helpful guidance and critiques at several stages of research and writing. Its astute counsel makes all our work possible, and we cherish our trustees' indispensible service and support.

DONORS

The research underlying this volume—the most ambitious effort in Rocky Mountain Institute's 30 years, occupying most of our professional staff for about a year and a half—and its release campaign were made possible by the generous philanthropy of Alice and Fred Stanback, Nationale Postcode Loterij, Robertson Foundation, RMI's Board of Trustees, Rachel and Adam Albright, Anonymous (1), Foster and Coco Stanback, Reuben and Mindy Munger, Peter Boyer and Terry Gamble Boyer, Mary Caulkins and Karl Kister, Ayrshire Foundation, Schmidt Family Foundation, RMI National Solutions Council, Sidney E. Frank Foundation, Tom Dinwoodie, Caulkins, Family Foundation, Eleanor N. Caulkins, George P. Caulkins III, Mary I. Caulkins, Maxwell Caulkins, John N. Caulkins, David I. Caulkins, Mac and Leslie McQuown, Sharman and David Altshuler, Pat and Ray C. Anderson, Wiancko Family Advised Donor Fund of the Community Foundation of Jackson Hole, the Concordia Foundation, the Libra Foundation, Interface Environmental Foundation, Inc., Earth-Share, Markell Brooks, Craigslist, Inc., Anonymous (2), Arkay Foundation, the Bunting Family Foundation, Unifi, Inc., Amory B. and Judy Hill Lovins, Suzanne Farver and Clint Van Zee, Argosy

Foundation, Sue and Jim Woolsey, Pacific Gas & Electric Company, William Laney Thornton and Pasha Dritt Thornton, Arntz Family Foundation, the "Anonymous" Trust, Arjun Gupta, the Moses Feldman Family Foundation, MAP Royalty, Inc., Richard D. Kaplan, AIA, the Moore Charitable Foundation, Sandra Pierson Endy, Judith and C. Frederick Buechner, David and Patricia Atkinson, Peter H. and E. Lucille Gaass Kuyper Foundation, Chris Sawyer and Julie Ferguson Sawyer, Energy Future Holdings, Ralph Cavanagh, Esq., Family Capital Corp., Elaine and John French, and Martin and Margaret Zankel, along with the significant collective support of many other donors. We are greatly indebted to them all.

REVIEWERS

Rocky Mountain Institute is deeply grateful to the following experts (and to many others who preferred to remain anonymous) for generously reviewing various drafts of this book or of its four core chapters. The book benefited greatly from their diverse perspectives and insights. Any remaining errors and all content are the sole responsibility of the authors. Reviewers are listed alphabetically in each group. Current affiliations are shown for purposes of identification only, and listings imply no endorsement.

Entire book: Emily Grubert (University of Texas), Bill Joy (Kleiner Perkins Caufield & Byers), Dr. Larry Keeley (Doblin, Inc.), Phil Klein (TEDxRainier), Jane and Bill Knapp, Katherine Lorenz (Mitchell Foundation), William F. Martin (Washington Policy and Analysis, Inc.), Reuben Munger (Bright Automotive), Dr. Arthur H. Rosenfeld (University of California at Berkeley), the Shift Project team, Dr. Paul Sullivan (Georgetown University), LTG Michael Vane (U.S. Army Training and Doctrine Command).

Transportation: Dr. Stephen R. Brand (ConocoPhillips), Bill Browning, Hon. AIA (Terrapin Bright Green), Dr. Dennis Bushnell (NASA), John Casesa (Guggenheim Partners), Robin Chase (GoLoco), David R. Cramer (Fiberforge), Dr. Mike Gallagher (Westport Innovations), Billy Glover (Boeing), Dr. Christoph Grote (BMW), Dr. David Hart (EPFL/Lausanne & Imperial College), Dr. Nicholas Lutsey (University of California at Davis), Glenn Mercer (ex-partner and automotive senior practice expert at McKinsey & Company, and currently an independent consultant), Mike Ogburn (Clean Energy Economy for the Region), Clay Phillips (General Motors), Michael Roethe and colleagues (North American Council for Freight Efficiency), Don Runkle (EcoMotors), Dr. Roger Saillant (Case Western University), Mike Simpson (National Renewable Energy Laboratory), Bill White (former mayor of Houston and U.S. deputy secretary of energy), Steve Williams (Maverick Transportation).

Buildings: Jamy Bacchus, PE (Natural Resources Defense Council), Phil Bernstein, FAIA (Autodesk AEC Solutions), Ralph Cavanagh, Esq. (Natural Resources Defense Council), Dr. Henry Cisneros (CityView), Robert Clarke (Serious Energy), Hannah Granade (Advantix Systems), Ronald Herbst, PE (Deutsche Bank AG), Donald Horn, AIA (U.S. General Services Administration), Donna Hostick (Pacific Northwest National Laboratory), Dr. Karl Knapp (MAP), Christian Kornevall (World Business Council for Sustainable Development), John A. "Skip" Laitner (American Council for an Energy-Efficiency Economy), Anthony E. Malkin (Malkin Holdings), Steven Meyers (Rational Energy LLC), Natalie Mims (Southern Alliance for Clean Energy), Scott R. Muldavin, CRE, CMC, FRICS (Green Building Finance Consortium), Clay G. Nesler, PE (Johnson Controls Inc.), Arah Schuur (Clinton Foundation), Gail Sturm (ProTen Realty Group), Terry Temescu

(The Tanager Group), John Weale, PE (Integral Group), R. Peter Wilcox, AIA (Northwest Energy Efficiency Alliance).

Industry: Dr. Mike Bertolucci (Interface), Dr. Ron Brown (Agenda 2020 Technology Alliance), Kevin Fallon (Breosla LLC), Kenneth W. Nelson (DuPont Engineering Polymers Department), Ron Perkins, PE (Supersymmetry USA, Inc.), James K. Rogers, PE, CEM, CEA, Dan Trombley (American Council for an Energy-Efficient Economy), Cindy Voss, Paul Westbrook (Texas Instruments), Dr. Tim Xu, PE (Lawrence Berkeley National Laboratory).

Electricity: Ralph Cavanagh, Esq. (Natural Resources Defense Council), Thomas Dinwoodie (SunPower), Dr. Andrew Ford (Washington State University), TJ Glauthier (TJG Energy Associates), Dr. Ray Gogel (Current Group), Michael Greene, PE (former vice chairman of Energy Future Holdings), Keith Hay (Colorado Public Utilities Commission), David Ismay, Esq. (Farella, Braun and Martel LLP), Koof Kalkstein (formerly founder of the energy practice at Boston Consulting Group), Brendan Kirby, PE (formerly of Oak Ridge National Laboratory), Dr. Karl Knapp (MAP), Reuben Munger (Bright Automotive), Capt. Scott Pugh, USN Ret. (U.S. Department of Homeland Security).

TECHNICAL CONTRIBUTORS

Josh Agenbroad's analysis of long-term fossil-fuel production in figure 1-3—undertaken because, to our surprise, we couldn't find it in the literature—was made possible by the generous advice of Prof. Kjell Aleklett, Dr. Mikael Höök, Jean Laherrère, Prof. Tadeusz Patzek, and especially Prof. David Rutledge. The pioneering work of Dr. David Greene and his colleagues at Oak Ridge National Laboratory underlies our analysis of oil dependence's economic costs. Reuben Munger kindly got us the data on how the market prices oil and gas price volatility. Our many military mentors over decades, both uniformed and civilian, have helped us understand oil's national-security burdens. Our Danish partners over decades, Prof. Niels Meyer and Prof. Jørgen Nørgård, clarified their national data. Much of the detailed research on the whale-oil story was done by Jeff Bannon. Our understanding of energy rebound issues across the transportation, buildings, and industry chapters was greatly sharpened by the late Dr. Lee Schipper, Dr. Jonathan G. Koomey, Danny Cullenward, Steve Sorrell, and Dr. James Barrett. Dr. Schipper's and Dr. Koomey's deep and diverse insights, like those of E. Kyle Datta, reverberate throughout the book.

Our many teachers, clients, and friends across the global automotive industry for two decades contributed more than they know to chapter 2. Glenn Mercer's extraordinary and generously shared knowledge of automotive component cost breakdowns and manufacturing cost structures made possible that economic and business-model analysis. John Casesa, Maryann Keller, Dr. David Cole, and other senior analysts over the years have contributed greatly to our understanding of the business, as have Clay Phillips, Derrick Kuzak, Sandy Munro, Alan Mulally, Bill Ford, Dr. Christoph Grote, Michael Brylawski, JB Straubel, and dozens of other automotive leaders. Our proprietary automaking clients and our friends in the National Automobile Dealers Association have been wonderful teachers and partners. We are greatly indebted to Dave Cramer, Dr. Jon Fox-Rubin, and their Fiberforge team, as well as Dave Taggart and Burt Rutan, for an ongoing and inspiring education in advanced composites, augmented by Martin O'Connor and many others. Carol and Eddie Sturman and Don Runkle enlightened us about advanced engines, and Bill Joy, John Waters, and Mujeeb Izaz about batteries. Dr. Lee Schipper, Charles Komanoff, Dr. Mark Delucchi, Dr. David Greene, Dr. David Goldstein, and Dr. John DeCicco were among our most valuable informants about policy. Dr. Jon Hykawy, Prof. Paul Sullivan, and Walter Benecki enlarged our understanding of rare-earth markets and uses. Yasushi Santoh kept us informed of Japanese news. Nicholas Lutsey shared his lightweighting work. ADEME in Paris kindly shared data on its feebate program; Laura Schewel, on shopping and traffic; Anna Jaffe, on travel IT. Our truck analysis, built on earlier RMI work by Odd-Even Bustnes, Mike Ogburn, Hiroko Kawai, and others, was informed by our partners in the North American Council for Freight Efficiency that RMI helped form in 2010, by Jeff Byrne and his Walmart team, and by Jimmy Ray, Majora Carter, Tom Wieringa, Sharon Banks, John Gustafson, and Dr. Mike Gallagher. NASA's chief scientist Dr. Dennis Bushnell, and Billy Glover and Dave Daggett of Boeing, were invaluable for their insights on airplanes, Prof. Sridhar Kota, on his innovative morphing surfaces, and Prof. Jim Womack, on point-to-point routing. Jigar Shah plugged us into the Carbon War Room's shipping-efficiency efforts. Our military efficiency work was made possible by hundreds of colleagues and briefers in two Defense Science Board panels—especially Tom Morehouse (who staffed the first and reported the second), Scott Badenoch, Jim Woolsey, Chris DiPetto, VADM Denny McGinn (USN Ret.), RADM Phil Cullom (USN), Mike Aimone, Alun Roberts, Drexel Kleber, Drew Sloan, and Ken Krieg (who suggested framing the two new strategic vectors). Dr. Sandy Thomas was our most insightful and data-rich hydrogen informant. Our biofuels tutors were legion, including Graham Williams, our friends at Shell and elsewhere across the industry, and the master of algae, Dr. Walter Adey.

For our buildings analysis in chapter 3, Scott Muldavin, CRE, CMC, FRICS, shared his encyclopedic knowledge of how efficiency affects value in commercial real estate, and Prof. Roger Ulrich, of how green buildings boost human health and performance. Greg Kats, Ron Herbst, PE, Clay Nesler, PE, Rob Watson, and Rick Fedrizzi opened many doors into their analytic and practice worlds. Steven Meyers shared candid insights on what's needed to scale energy efficiency cost-effectively in small commercial buildings. Technological innovators from Dr. Brandon Titianov, Kevin Surace, and Dr. Steve Selkowitz in glazings to Tsutomu Shimomura, John Gage, Nancy Clanton, PE, FIES, and Robert Sardinsky in lighting, Dr. Alan Meier and Dusko Maravic in appliances, and Onno Koelman, Jay Harman, and their PAX colleagues in turbomachinery kept us apprised of startling developments. Prof. Dr. Wolfgang Feist, his colleague Katrin Klingenberg, AIA, and architects Hans Eek and Christer Nordström helped us explore the economics of passive houses. Practitioners from Paul Braese, PE, and Perry Bigelow to Ted Bakewell III and Dr. David Strong, CEng, FCIBSE, FEI, and in the commercial sector from Bill Browning, Hon. AIA (who long ago launched RMI's buildings work), and Bob Fox, AIA, to Gary Christensen, Tara Darrow, Elizabeth Heider, and Rob Keller, shared their inspiring stories. Lee Scott and his Walmart team helped us understand big-box retailing. Norman Crowley illuminated Crowley Carbon's business model. Steve Jungerberg and Eric Walters helped us understand contractor retraining. Our favorite U.S. mechanical-engineering partner, Peter Rumsey, PE, and his Integral Group team unfailingly responded to our data requests, as did his and our teacher Eng Lock Lee in Singapore. Richard G. Kidd IV, John Simpson, PE, and a host of other friends in government helped us understand their unique challenges and achievements. Prof. Robert H. Socolow and his Princeton group have long informed our efforts. Much of RMI's buildings work is a legacy of Greg Franta, FAIA—RMI's gifted buildings leader until his untimely death in 2009.

In industry (chapter 4), some of our long-time partners like Eng Lock Lee, Frigyes Lestak, Dr. Gunnar Hofstadius, Paul Westbrook, Dr.

Ernst Worrell, Michael Shepard, Prof. Thomas B. Johansson, Dr. David Goldstein, Anita Burke, Dale Hoenshell, Ken Nelson, Jim Clarkson, PE, and James K. Rogers, PE, CEM, CEA, generously shared their data and insights. Dr. Tim Xu, PE, helped us interpret his Berkeley Lab team's essential efficiency analyses. Gil McCoy and his Washington State team ran a special motor-cost analysis for us. Dr. Graham Sinden and Prof. Christopher Weber kindly shared their latest data on energy embodied in international trade; Martha Moore, on the American chemical industry; Jay Stein, on fume hoods; Philipp Schmidt-Pathmann, on European recycling; Dr. Bernie Bulkin, on refining; Lars Nilsson, on pumps; Prof. Julian Allwood and his Cambridge team, on process efficiencies and materials. The late Ray C. Anderson, Jim Hartzfeld, and Dr. Mike Bertolucci patiently helped us grasp Interface's efficiency achievements; the late Dr. Levi Leathers, Dr. Claude Fussler, and Ken Nelson, the efficiency achievements at Dow; Dr. Ron Brown, the intricacies of pulp-and-paper efficiency; Ruksana Mirza, Dr. Luiz Carlos R. de Sousa, and Bruno Fux, the challenges of replacing coal burning in cement kilns; and Prof. Hanns Fischer, his inside-out chemistry lab course. Jonathan Knowles and his colleagues at Autodesk, through the courtesy of CEO and RMI trustee Carl Bass, helped frame and inform our discussion of additive manufacturing. Janine Benyus and her team remain our fount of wisdom on biomimicry, and Dr. Paul Anastas on green chemistry. We appreciate Prof. Alan Fuchs's leadership on our joint AIChE design competition that provided several cases.

Our electricity analysis (chapter 5) drew on decades of learning from many partners. Trieu Mai and Dr. Walter Short conducted the modeling for our four cases using NREL's ReEDS model—a core component of our analysis—and contributed important thinking about the dynamics of grid operations. Bill Joy sharpened our thinking throughout about renewable technologies and economics. Michael Liebreich and his team at Bloomberg New Energy Finance generously shared their unique global database on renewable transactions' value, and Dr. Eric Martinot, REN21's equally important data on their capacity. Solar pioneer and RMI trustee Thomas Dinwoodie and Shayle Kann (GTM Research) kept us updated on the fast-dropping cost of solar photovoltaics and on solar market dynamics. Jack Hidary and Aaron Zubaty illuminated renewable energy finance. Joel Moxley introduced FORO's promising geothermal drilling technology; Lord Oxburgh, the Shell/Eindhoven carbon-capture technology. Ralph Cavanagh, Esq., Dr. Peter Fox-Penner, Jon Wellinghoff, Jack Riggs, Prof. Michael Dworkin, and Richard Cowart have long helped us think through the future of utilities and utility business models. J. Wayne Leonard explained low-income customers' needs; Dan Eggers and Kevin Parker, the effects of EPA coal regulation. Dr. Ray Gogel, Rob Pratt, and Graham Hodge provided important insights on the future of the grid and the potential for microgrids. Holger Kley helped us understand microgrid operations and Denmark's grid. Ed Comer, Esq., provided critical feedback on renewable integration and grid security and feedback on how to build a convincing argument. Capt. Scott Pugh (USN Ret.) added important perspectives on grid security. Prof. Mark Jacobson and Prof. Dan Kammen improved our electricity system dispatch model and much else. Prof. Andrew Ford worked closely with us on developing a system dynamics model of utility investment and finance, while Brad Tirpak provided utility financial data and ideas. Dr. Klaus Töpfer helped us understand German nuclear policy; Peter David Pedersen, Prof. Yoichi Kaya, Susumu Yoda, and Takashi Kiuchi, Japanese nuclear policy. Mycle Schneider, Peter Bradford, Dr. Victor Gilinsky, Robert Alvarez, Prof. Mark Cooper, Prof. Frank von Hippel, and Jim Harding were ever helpful with penetrating nuclear analyses and data. Dr. Darrin Magee, along with Dr. Mark Levine and Dr. David Fridley, got us hard-to-find data on China for chapters 5 and 6, supplemented by Wang Shuxiao, Li Yuqi, Barbara Finamore, and Yi Shenshen.

Our concluding chapter 6 borrowed Jeremy Rifkin's term "Intergrid," and drew on the jobs expertise of Dr. Skip Laitner, the appliance-efficiency work of Dr. Amol Phadke and the SEAD project, the lighting innovations of Dr. Evan Mills's LUMINA Project, Dr. Sarah Emerson's data on shale gas's crude-oil by-product, Robert A. Hefner III's tutelage on natural gas, strategic learnings from nearly four decades of conversation with Shell group planners, the perennial polyculture breakthroughs of Dr. Wes Jackson's team at the Land Institute, integrated ecological and economic restoration by several gifted practitioners (Dr. Ing. Willie Smits, Paolo Lugari, and John Liu), and the broad inspiration of Thomas L. Friedman, Adam Kahane, and Dr. Eric Rasmussen.

The book's title came from my conversation with RMI's Executive Director, Marty Pickett. I was musing about "Uninventing Fire." She said, "No, '*Re*inventing Fire!'" And so we did.

Finally, this labor could never have been done without the devoted support of the entire Rocky Mountain Institute and all our spouses and partners, above all my wife, Judy.

—Amory B. Lovins

ABOUT THE AUTHORS

Amory B. Lovins, physicist, RMI's cofounder, chairman, and chief scientist, was educated at Harvard and Oxford (where he was a don) and has written 30 earlier books and over 450 papers. He has received the Blue Planet, Zayed, Volvo, Onassis, Nissan, Shingo, and Mitchell prizes, MacArthur and Ashoka fellowships, the Benjamin Franklin and Happold medals, 11 honorary doctorates, an Hon. AIA and FRSA, Foreign Membership of the Royal Swedish Academy of Engineering Sciences, and the Heinz, Lindbergh, Right Livelihood ("Alternative Nobel"), National Design, and World Technology awards. His most recent of nine visiting chairs was in 2007 at Stanford's School of Engineering. For nearly 40 years he has advised governments (including the U.S. DOE and DoD) and major firms worldwide, chiefly on advanced energy and resource efficiency, strategy, security, and integrative design. In 2009, *Time* named him one of the world's 100 most influential people, and *Foreign Policy* named him among 100 top global thinkers.

Robert "Hutch" Hutchinson is managing director of RMI's research and consulting practice, emphasizing buildings and transportation. An 18-year consultant, director, and partner of Boston Consulting Group, he holds a BEng summa cum laude in mechanical engineering and mathematics (Vanderbilt), was a Churchill Scholar at Cambridge, and earned master's degrees at Stanford in mechanical engineering and in business (he was an Arjay Miller Scholar). He has consulted worldwide, started a clean-tech venture fund, and led programs at the U.S. DOE's Pacific Northwest Lab.

Stephen Doig, senior fellow, was a program director at RMI from 2007 to 2011, where he led work in electricity, industrial collaborations, and the built environment. He played a senior advisory role on content and key messages across the *Reinventing Fire* effort. Prior to joining RMI, Stephen held leadership positions at the U.S. Air Force and McKinsey & Company and was an adjunct professor at the Wharton School of Business. He holds a chemistry degree from Dartmouth College and a PhD in physical chemistry from the University of California at Berkeley, and he held postdoctoral positions at the Mayo Clinic and Caltech.

James Newcomb, program director at RMI, leads the institute's electricity practice. He is a writer, consultant, and scenario thinker with 25 years' experience in managing cutting-edge research and advisory organizations. Prior to joining

RMI, James served as group manager for markets, policy, and impacts analysis at the National Renewable Energy Laboratory's Strategic Energy Analysis Center. Previously, he was managing director and cofounder of Bio Economic Research Associates (bio-era), was the founding president and CEO of E Source, a for-profit spinoff of RMI, and founded and led the natural gas practice at Cambridge Energy Research Associates. Mr. Newcomb holds a BA in economics from Harvard University and an MA in energy and resources from the University of California at Berkeley.

Lionel Bony is the former director of the office of the chief scientist at RMI, where he managed the Reinventing Fire initiative, coordinated the organization's strategy process, and led the solar photovoltaic balance of system work. Earlier he had cofounded RMI's transportation practice, focusing on platform efficiency and electrification. Before joining RMI in 2006, he worked for L'Oréal as a financial analyst in Madrid and a product manager in Paris, and at Conservation International in Bolivia. He earned a master's degree in finance and marketing from SciencesPo (Paris) and an MBA from Harvard Business School. In September 2011 he returned to France to continue in this line of work.

Other key contributors, in alphabetical order, included:

Mathias Bell is a consultant at RMI, where he has worked on energy efficiency in buildings, focusing on utility program design and delivery. His main responsibilities in this publication included economic modeling and analysis for the building sector, as well as investigating the business and policy approaches to accelerate the adoption of energy efficiency. He has been a researcher on the World Resources Institute's Climate and Energy team and has advised firms and institutions on their energy and sustainability strategy. He holds a BA in international relations with a concentration in environmental and technology studies from Carleton College.

Albert Chan is a consultant in RMI's electricity and industry practices, and his core expertise is in solar energy as well as industrial energy efficiency. He helped lead the analysis and recommendations for *Reinventing Fire*'s industry chapter. He previously worked at SunPower, where he optimized manufacturing processes and introduced next-generation solar modules. Albert graduated from Stanford University with a BS and MS in materials science and engineering, and he has held research positions in organic photovoltaics, solid-oxide fuel cells, and superconducting wires.

Nate Glasgow is a former principal at RMI and led the transportation research for *Reinventing Fire*. He is a coauthor of RMI's *Winning the Oil Endgame* (2004) and managed its implementation in the automotive, airplane, and heavy-truck sectors. Prior to his tenure at RMI, he was a project manager for an energy-efficient construction company in the Napa Valley and a venture capital intern focused on green building technologies. He holds an MBA and bachelor's degree in human biology from Stanford University and an MA in economics from the University of California at Santa Barbara. In April 2011 he took up a position with SunEdison.

Lena Hansen coauthored *Reinventing Fire*'s electricity chapter and drove the sector's analysis and research. She is a principal with RMI's electricity practice, where she leads work focusing on resource planning, carbon strategy, demand-side resources, and variable renewables' integration. Lena holds a bachelor's degree in astrophysics from the University of North Carolina at Chapel Hill and an energy-focused master's degree in environmental management in economics and policy from Duke University.

Virginia Lacy coauthored and co-led the analysis to support *Reinventing Fire*'s electricity chapter. A senior consultant at RMI, Virginia advises utilities, corporations, and governments on electricity strategy, including grid integration of variable renewable energy sources, energy supply and commodity market trends, and financial valuation of energy efficiency and renewable energy sources. She was previously a research analyst with a socially responsible investment research firm, where she analyzed the environmental performance of S&P 1500 companies for some of the largest institutional investors in the U.S. She holds a BA in government from the University of Virginia and a master's degree in environmental management from Yale University.

Eric Maurer is a consultant at RMI, where he has participated in research and consulting projects in the electricity and buildings sectors. He has assisted in the research and writing of the electricity chapter of *Reinventing Fire* and, in succession to Lionel Bony, coordinated the postsubmission production of *Reinventing Fire*. Eric has a bachelor's degree in finance from Miami University and a master's degree in environmental management from Duke University.

Jesse Morris is an analyst with RMI's transportation practice, applying whole-system design to transportation systems by focusing on when, where, why, and how we use vehicles. He coauthored and conducted research for *Reinventing Fire*'s transportation chapter. He previously interned with environmental nonprofit groups and worked in the private sector for a photovoltaics manufacturer after receiving his bachelor's degree in international environmental policy from Colorado College.

Greg Rucks is a consultant at RMI, where he has worked to develop and implement energy-efficient design solutions in transportation and industry. He coauthored the transportation chapter of *Reinventing Fire* and contributed to its underlying research and analysis. Prior to joining RMI, Greg completed a volunteer placement in Ghana and worked for Boeing on structural optimization and lightweight design for the 787 program. He has bachelor's degrees in English from Colorado College and in mechanical engineering from Columbia University.

Caroline (Fluhrer) Traube is a former senior consultant at RMI, where she focused on energy-efficient building design, building energy modeling, lifecycle cost analysis, and green certifications for new and existing buildings, including the Empire State Building retrofit project. Caroline led the buildings-sector research for *Reinventing Fire*, identifying the key business and policy levers required to deepen and scale efficiency. She holds bachelor's and master's degrees in engineering from Stanford University.

Reinventing Fire relied heavily on the work of two RMI modelers:

Mark Dyson is a former analyst with Rocky Mountain Institute's electricity practice, where he helped develop and apply computational tools to model the technical issues and commercial implications of integrating renewable energy into regional electricity systems. Mark and the RMI team have applied these tools to helping utility partners develop low-carbon strategies and to inform RMI's research into the future renewables-dominated systems. Mark received a bachelor's degree in computer science and geology from Carleton College and worked as a geophysics researcher at Carleton and Harvard University. In mid-2011 he left RMI for graduate studies in the Energy and Resources Group at the University of California at Berkeley.

Mark Gately is an analyst at RMI, specializing in quantitative modeling and analysis. He led the development of *Reinventing Fire*'s automotive, buildings, and industry models, as well as the overall integration of models across all sectors. He holds a master's degree in applied economics from the University of Vermont and a bachelor's degree in computer science from Cornell University.

Reinventing Fire also benefited greatly from the contributions of four external consultants and advisors:

John Carey is a freelance science writer and editor. For the two decades prior to 2010, he was a senior correspondent for *BusinessWeek* magazine, covering everything from energy and global warming to cholesterol-lowering drugs and the human genome. Before his tenure at *BusinessWeek*, John was an editor at *The Scientist* and a reporter at *Newsweek*. His stories have won awards from the American Association for the Advancement of Science, the Wistar Institute, and other organizations. He was also a National Magazine Award finalist. John has a bachelor's degree in biochemistry from Yale University and master's degrees in marine biology from the University College of North Wales and in forest ecology from the Yale School of Forestry & Environmental Studies.

E. Kyle Datta, RMI senior fellow, is general partner of Ulupono Initiative, a Hawai'i-focused social investment organization that is dedicated to improving the quality of life for island residents by creating renewable energy, increasing locally produced food, and reducing waste. Formerly, Kyle was the CEO of U.S. Biodiesel Group. He served as managing director of research and consulting at Rocky Mountain Institute and coauthored RMI's *Winning the Oil Endgame* and *Small Is Profitable*. He was a vice president at Booz Allen Hamilton and served as managing partner of the firm's energy practice in Singapore, and as leader of the U.S. Utility practice. Kyle received a master's degree in public and private management from the Yale School of Organization and Management and a master's degree in environmental science in resource economics from the Yale School of Forestry & Environmental Studies.

Jason Denner has 14 years of professional experience across many aspects of sustainable industrial and mechanical engineering, including industrial utility system analysis and improvement, energy efficiency project development, distributed and renewable energy supply, production operations, and distribution logistics. He has used this experience to lead ambitious energy efficiency projects for international companies in food and beverage, manufacturing, pulp and paper, mining, oil and gas, and metallurgical industries. Jason was formerly the director of engineering for DOMANI Sustainability Consulting LLC and a senior project engineer in Rocky Mountain Institute's research and consulting group. He has led projects to improve the energy and water efficiency of numerous industrial facilities on four continents.

Jonathan G. Koomey, RMI senior fellow (www.koomey.com), is a consulting professor at Stanford University, worked for more than two decades at Lawrence Berkeley National Laboratory, and has been a visiting professor at Yale University and Stanford University. He'll be a visiting professor at the University of California at Berkeley for Fall 2011. Dr. Koomey holds MS and PhD degrees from the Energy and Resources Group at UC–Berkeley and an AB in the history of science from Harvard University. He is the author or coauthor of eight books and more than 150 articles and reports, and he is one of the leading international experts on the economics of reducing greenhouse gas emissions and the effects of information technology on resource use. His latest solo book is the second edition of *Turning Numbers into Knowledge: Mastering the Art of Problem Solving* (www.analyticspress.com).

The book received thoughtful internal reviews and guidance from two RMI leaders:

Brad Mushovic, RMI's vice president of marketing and communications, leads branding and messaging for the institute. He most recently served for five years as vice president of marketing at Outward Bound USA, and in leadership positions at First Advantage SafeRent. His other marketing and brand management experience includes Qwest Communications and Proctor & Gamble, where he began as a process engineer and operations manager. He holds a bachelor of science degree in mechanical engineering (summa cum laude) and an MBA with High Distinction from the University of Michigan. He is a rock climber and competitive bicycle racer.

Michael Potts, RMI's president and CEO, was an active trustee before he joined the RMI management team in early 2007 after more than 25 years as a growth-oriented high-technology executive. His most recent commercial position was as CEO of American Fundware, which nearly tripled in revenues during his tenure and produced the award-winning Fundware software suite for non-profit and governmental organizations. AFW was acquired in 2002 by Intuit, maker of Quickbooks and TurboTax. Previously, Potts worked in sales, marketing, and general management positions for IBM, BancTec, and Recognition International. He spent many years working internationally and served for a time as managing director of an Australian subsidiary. He currently serves on commercial and nonprofit boards that align with his passion for innovation and spiritual growth.

Finally, two senior energy executives graciously wrote Forewords to this book:

Marvin E. Odum is President of Shell Oil Company and Director Upstream of Royal Dutch Shell's subsidiary companies in the Americas. Odum holds positions of board leadership and participation in the Business Roundtable, U.S. Climate Action Partnership and the American Petroleum Institute. In addition, he is a member of the Dean's Council of the John F. Kennedy School of Government at Harvard University and the Advisory Board of the Cockrell School of Engineering at The University of Texas at Austin. He also serves on the boards of several Houston-area charities. Odum earned a bachelor's degree in mechanical engineering from the University of Texas and a master's degree in business administration from the University of Houston. He began his Shell career as an engineer in 1982, and has since served in a number of management positions of increasing responsibility in both technical and commercial aspects of energy.

John W. Rowe is the chairman and chief executive officer of Exelon Corporation, a utility holding company headquartered in Chicago. Exelon has the largest market capitalization in the U.S. electric utility industry. Its retail affiliates serve 5.4 million customers in Illinois and Pennsylvania, and its generation affiliate operates the largest fleet of nuclear power plants in the U.S. utility industry. Rowe is the senior CEO in the U.S. utility industry, having served in such positions since 1984, and has led Exelon since its formation in 2000. Rowe previously held CEO positions at the New England Electric System and Central Maine Power Company, served as general counsel of Consolidated Rail Corporation, and was a partner in the law firm of Isham, Lincoln and Beale. Rowe is the past chairman of the Nuclear Energy Institute and the Edison Electric Institute. He is co-chairman of the National Commission on Energy Policy, an industry and environmental organization dealing with climate change. He is a member of the boards of directors of Sunoco, the Northern Trust Company, and UChicago Argonne LLC, the governing body of Argonne National Laboratory. In both 2008 and 2009, Institutional Investor named Rowe the best electric utilities CEO in America.

ABOUT ROCKY MOUNTAIN INSTITUTE

Rocky Mountain Institute is an independent, entrepreneurial, public-benefit think-and-do tank, organized as a §501(c)(3)/§509(a)(1) tax-exempt nonprofit. It was cofounded in 1982 by Amory B. Lovins, who remains its very active thought leader, chairman, and chief scientist. Its leadership, 80+ staff, and board of trustees are described on RMI's website at www.rmi.org, which posts the institute's extensive publications and news.

To achieve its vision of a world thriving, verdant, and secure, for all, for ever, RMI's mission is to drive the efficient and restorative use of resources. RMI shows how to wring far more benefit from energy, water, materials, and other resources by innovative design, technology, business strategy, and public policy—artfully combined, rigorously applied, and vigorously promoted. This effort takes three main forms: transforming design, busting barriers, and spreading innovation. RMI's 2010–2015 strategic focus is Reinventing Fire: mapping the business-led transition from oil, coal, and ultimately natural gas to efficiency and renewable energy, as this book describes. RMI is also implementing that transition through focused, market-driven initiatives in the transportation, building, industrial, and electricity sectors. RMI's work often applies integrative design and the framework of "natural capitalism" (to learn more about that concept, go to www.natcap.org).

Based in Old Snowmass and Boulder, Colorado, RMI works chiefly in the United States but with global context and reach, spanning over 60 countries. Its transdisciplinary research and practice have earned it a reputation for integrity, quality, originality, scholarship, clarity, and effectiveness.

RMI does solutions, not problems; practice, not theory; transformation, not incrementalism. Its style is nonadversarial, nonpartisan, and transideological, embracing diverse partners and collaborators in the private, nongovernmental, and public sectors. The institute is apolitical and does not lobby or litigate.

RMI pioneered nonprofit enterprise via 11 revenue models—all successful, 10 entrepreneurial, and most still operating. Its hybrid approach typically first uses philanthropy-funded innovation to create new solutions to tough old problems. Its integrated team of researcher/practitioners then tests and refines that ideation (and leverages the philanthropic investment into revenue to support further research) via targeted consulting engagements that support the mission. That is, RMI's experts collaborate with skilled and motivated private-sector partners, typically large and influential firms, to turn innovative concepts into rapid mutual learning, practical proofs, business insight and credibility, teachable cases, competitive pressure for emulation, and scaling by powerful partners. This model typically yields 30–70% of RMI's roughly $13 million annual revenue and has so far spawned five for-profit (three formal, two casual) and three nonprofit spinoffs.

OTHER PUBLICATIONS BY ROCKY MOUNTAIN INSTITUTE

Hundreds of free publications on diverse topics, ranging from popular to highly technical, are available for free from the institute's website at www.rmi.org. They date back to RMI's formation in 1982, and in some cases to the 1970s. A search engine is provided. RMI's print and online *Solutions Journal* and *Spark* newsletter are also available through the website. Earlier books include:

P. G. Hawken, A. B. Lovins, and L. H. Lovins, *Natural Capitalism: Creating the Next Industrial Revolution* (Little Brown, 1999); www.natcap.org.

A. B. Lovins, E. K. Datta, T. Feiler, K. R. Rábago, J. N. Swisher, A. Lehmann, and K. Wicker, *Small Is Profitable: The Hidden Economic Benefits of Making Electrical Resources the Right Size* (RMI 2002), an *Economist* book of the year, available from www.smallisprofitable.org.

A. B. Lovins, E. K. Datta, O.-E. Bustnes, J. G. Koomey, and N. J. Glasgow, *Winning the Oil Endgame: Innovation for Profits, Jobs, and Security* (RMI, 2004); www.rmi.org/WTOE.

A. B. Lovins and ed. C. W. Burns, *The Essential Amory Lovins* (London: Earthscan, July 2011); a 1962–2010 anthology; www.earthscan.co.uk/?tabid=102749.

To inquire about translating any of the above publications into other languages, please contact outreach@rmi.org.

As explained in "About This Book" on p. ix, this book's technical backup is posted, and related sectoral initiatives to implement its thesis are described, at www.reinventingfire.com/rfknowledgecenter.

FIGURE CREDITS

The kind permission of the following sources of graphics to reproduce their copyrighted images here is gratefully acknowledged.

CHAPTER 1

Fig. 1-1, Photo by Laurin Rinder/Shutterstock.com.

CHAPTER 2

Fig. 2-1, Photo by ricardoazoury/istockphoto.com; Fig. 2-4, Photo and image courtesy of RMI with thanks to Fiberforge Corporation; Fig. 2-5, Image courtesy of Toyota Motor Corporation; fig. 2-6, Image courtesy of Volkswagen; fig. 2-9, Image courtesy of BMW NA; fig. 2-10, Image courtesy of Bright Automotive; fig. 2-16, Images courtesy of Daimler Trucks North America (left) and Renault Trucks (right); fig. 2-20, Images courtesy of NASA and The Boeing Company (2-20a), NASA and NIA (2-20b), and NASA, Massachusetts Institute of Technology and Aurora Flight Sciences (2-20c); fig. 2-22, Images courtesy of U.S. Air Force (2-22a), Karem Aircraft (2-22b), © U.S. Army (2-22c), Badenoch LLC (2-22d); fig. 2-27, © Fietsersbond.

CHAPTER 3

Fig. 3-1, Photo by Doubleclicks, Inc./iStockphoto.com; fig. 3-9, Image courtesy of NASA/JPL-Caltech. Reproduced by permission from Merrilee Fellows (NASA Policy/Program Management), April 4, 2011; fig. 3-10, Image courtesy of Sony Electronics Inc.; fig. 3-11, Reproduced with permission from Kasey Arnold-Ince, April 11, 2011. Image copyright, PAX Scientific. All rights reserved; fig. 3-12, Reproduced by permission from Heather Lammers, National Renewable Energy Laboratory, March 16, 2011; fig. 3-13, Image courtesy of RMI.

CHAPTER 4

Fig. 4-1, Photo by Oknebulog/Dreamstime.com; fig. 4-7, Graphic by Stanford Kay Studio; fig. 4-8, Graphic by Stanford Kay Studio; fig. 4-9, Image courtesy of Eng Lock Lee; fig. 4-10, Graphic by Stanford Kay Studio.

CHAPTER 5

Fig. 5-1, Photo by Fedorov Oleksiy/Shutterstock.com; fig. 5-2, Graphic by Stanford Kay Studio; fig. 5-8, U.S. Department of Homeland Security, concept by RADM Jay M. Cohen (USN Ret.), courtesy of CAPT Scott Pugh (USN Ret.); fig. 5-26, Graphic by Stanford Kay Studio.

CHAPTER 6

Fig. 6-1, Photo by James Group Studios/istockphoto.com.

NOTES

ABOUT THIS BOOK

1 *Present value* is a standard way to express a series of future costs or benefits as a lump sum now (in this case in 2010), adjusted for the time value of money. For this purpose we discount future costs and benefits at 3% real (after inflation) per year. Net present value subtracts costs from benefits, all expressed in that same way. Business decisions typically use higher discount rates than this low "social" rate, and so do we when analyzing whether an investment in energy efficiency or renewable energy can meet a given sector's financial criteria, as explained on page x.

PREFACE

2 Temple 1986.
3 Lovins 2010a.
4 Mullen 2011.
5 Owen, Inderwildi, and King 2010. Sir David King was the chief scientific advisor to the British government from 2000 to 2007.
6 U.S. Energy Information Administration 2010a, 6-7.
7 U.S. Joint Forces Command 2010, 29, 3; Parthemore and Nagl 2010
8 Rutledge 2010; Rutledge 2011; Patzek and Croft 2010.
9 Nearly 90% if we exclude traditionally scavenged biomass such as wood and dung.

CHAPTER 1. DEFOSSILIZING FUELS

10 McNeill 2011.
11 Koplow 2007, 93–110, table 1.
12 Lovins et al. 2004, notes 137–44.
13 U.S. Energy Information Administration 2010d; 2011a, table 3.5. As an approximation, $3.098/gal average refined-product price times 19.498 Mbbl/d petroleum products supplied yields $929 billion. EIA won't publish its more exact consumer expenditure estimates for energy in 2008 until the August 2011 *Annual Energy Review.*
14 Goldstein 2010, 73.
15 Greene 2010; Greene and Hopson 2010.
16 Greene and Tishchishyna 2000. The model is described in Greene and Leiby 2006.
17 Engemann, Kliesen, and Owyang 2010. The St. Louis Federal Reserve Bank concluded in 2010 that with an average oil shock, with real prices just 13 percentage points over the previous three-year maximum price, the probability of a recession rose nearly 50 percentage points within one year, nearly 90 points within two years, and over 100 points after ten quarters, all else being equal.
18 Munger 2011, analyzing that day's Bloomberg option valuations of the straddle for crude oil and RBOB gasoline (reformulated gasoline blendstock for oxygen blending). These volatility values were respectively 40% and 47% of those commodities' spot price over a five-year horizon, or 23% and 22% over one year.
19 Conservatively valuing all nongasoline petroleum consumption at the crude-oil volatility value adds $0.15 trillion.
20 Greene and Hopson 2010.
21 Based on a conservative reading of the Orszag 2007 analysis summarized in Reuters 2007, but including only federal budgetary costs. The Orszag report comments on differences from Stiglitz and Blimes 2010, a macroeconomic estimate of over $3 trillion.
22 Lovins et al. 2004, note 132.
23 Ibid., notes 132–33.
24 Stern 2010.
25 The U.S. Office of Management and Budget calculates the nominal-dollar national defense function's cost as $661 billion in 2009, $694 billion in 2010, and an estimated $768 billion in 2011 (Office of Management

and Budget 2011). Including defense-related outlays by agencies other than the Department of Defense raises the 2010 total to $0.9 trillion, and including defense-related interest on the national debt raises that to roughly $1.2±0.2 trillion in 2009 dollars (Higgs 2010).

26 Lovins and Lovins 1982.

27 Sullivan 2011a, 2011b.

28 Woolsey 2001; Woolsey, Lovins, and Lovins 2002; Lovins and Lovins 2001, 73.

29 RMI analysis comparing Freedom House 2011 with British Petroleum 2010a.

30 Arezki and Bruckner 2011; Kolstad, Wiig, and Williams 2008.

31 *The Economist* 2005.

32 This hotbed of rebellion and terrorism against Russia lacks its own oil but is a key transit point for Caspian pipelines.

33 Epstein et al. 2011.

34 World Resources Institute 2005. The last one-fifth comes from directly used coal and from natural gas burning. Carbon emissions from burning fossil fuel total about 65% of greenhouse gases; another 12% comes from land-use changes like deforestation, and the last 23% from such trace gas emissions as methane and chlorofluorocarbons (CFCs).

35 U.S. Department of Defense 2010.

36 Mullen 2011.

37 **Assumed ultimately recovered resources** (total area under curves): coal, about 2,500 Gboe (680 gigatonnes or Pg of mixed rank); liquids, 3,000 Gboe (1 boe is about 308 GJ); gas, 3,700 Gboe (about 19,000 Tcf from MIT 2010 report, larger than Laherrère 2010's 13,000 Tcf). Coal converted to Gboe for 1850–2010 using annual conversion factors from Laherrère, and for 2010 onward using the recent typical value of 0.5 toe per coal tonne. (Gboe = gigabarrels oil equivalent; Pg = petagrams or 10^{15} g; GJ = gigajoules; Tcf = trillion cubic feet; toe = tonne oil equivalent; tonne = metric ton = 1,000 kg = 2,240 lb.)
Data sources: Coal data and projection: Rutledge 2010 and Rutledge 2011; see also Patzek and Croft 2010. **Liquids and natural gas data:** Laherrère 2010. **Liquids projection:** Laherrère 2010, approximating his ultimately recovered resource of 3 Tboe. **Gas projection:** Massachusetts Institute of Technology 2010. **Methodological summary:** Please see Knowledge Center at www.reinventingfire.com.

38 Greene 2010.

39 Using the official adjustments for weather and for electricity trade: Danish Energy Agency 2010 and 2009; Energi Styrelsen 2009. GDP updated from Danmarks Statistik 2010.

40 Renewable Energy Policy Network for the 21st Century 2011, 21, 28, 98 n 61.

41 Danish Energy Association 2009, 46, 23, 25..

42 Regeringen 2011; ClimateWire 2011; Richardson et al. 2011.

43 Friedman 2011.

44 Renewable Energy Policy Network for the 21st Century 2010; Pew Charitable Trusts 2010, 2011.

45 U.S. Energy Information Administration 2010b.

46 U.S. Energy Information Administration 2011b.

47 RMI analysis using data from Ibid.

48 RMI analysis.

49 Butti and Perlin 1980.

50 Lovins et al. 2004, 4–5.

CHAPTER 2. TRANSPORTATION: FITTER VEHICLES, SMARTER USE

51 If federal law didn't ultimately harmonize federal-highway weight limits with state ones, about 0.26 Mbbl/d of truck-fuel savings could be lost, but it could easily come from biofuels or natural gas.

52 U.S. Energy Information Administration 2010b.

53 U.S. Energy Information Administration 2010b.

54 Edmunds 2011. See also Lutsey 2010.

55 Taggart 2000.

56 Like an airframe, the structure was suspended from rings, making it stiff, strong, and light. It was also in the style of a monocoque, though stick-on thermoplastic cosmetic panels ensured Class A finish and reparability. In a true monocoque, like a lobster shell, the shell is the structure.

57 The first known Hypercar-class concept cars built by major automakers were even earlier—GM's 1991 Ultralite and a similar unannounced effort by a Japanese competitor several years before.

58 RMI analysis.

59 White 2010.

60 Lovins and Cramer 2004. An Oak Ridge National Laboratory study (Boeman and Johnson 2002) found an 18-part composite design could cut body-in-black weight 60%, versus 57% for a Hypercar SUV with 14 parts, both with superior stiffness.

61 Fuchs et al. 2008.

62 Learning curves are a simple mathematical way to describe how manufactured products and many services get cheaper as you make more of them. The underlying theory, dating back to the 19th century, was elaborated and popularized in the 1960s for business leaders by Bruce Henderson of the Boston Consulting Group. The theory closely fits empirical data for hundreds of diverse manufactured goods and has been observed in virtually all industries from semiconductors to hospital operations, so it's generally accepted as summarizing rather accurately how innovation and scaling up production make things cheaper. The typical metric is how many percentage points cheaper a unit becomes—typically about 15–30% for most technological products—for each doubling of the cumulative number produced.

63 Our carbon-fiber material cost learning curve assumes a current price of $16/lb for large-tow (about 50k), standard modulus, automotive-grade creel fiber and a 2050 price of $5/lb based on assumed adoption of process improvements and precursor alternatives indicated by Department of Energy/ Oak Ridge National Laboratory research on low-cost carbon fiber (Warren 2010). We interpolated between these two price points and expressed the result as a function of cumulative production volume. Price drops below $10/lb with the first half-million vehicles produced, $8/lb with 5 million, and ultimately $5/ lb with 193 million vehicles produced by 2050. U.S. auto production has recently ranged from about 9 million/y to 18 million/y.

64 Warren 2010.

65 RMI analysis based on Fuchs et al. 2008 and Boeman and Johnson 2002.

66 Kahane 2003.

67 Van Auken and Zellner 2003.

68 Ross and Wenzel 2001.

69 Chan-Lizardo, Lovins, Schewel, and Simpson 2008; Schewel 2008.

70 Mitchell, Borroni-Bird, and Burns 2010.

71 Kromer and Heywood 2007. The battery curve assumes a 2015 price of $400/kWh, dropping to $200/ kWh with 22 million packs produced, then ultimately to $165/kWh with 193 million packs produced by 2050. The fuel-cell curve assumes a 2015 price of $200/kW, dropping to $100/kW with 1.2 million units made, ultimately to $48/kW (consistent with longstanding GM analyses) with 193 million units made by 2050. Details are at www.reinventingfire.com.

72 Lovins 2010b. American Physical Society/Materials Research Society 2011 offers some sensible general recommendations but tends to understate the efficiency and substitution potentials.

73 Long et al. 2010.

74 Fickling 2011.

75 McCoy 2010.

76 Tibbits 2010; Loveday 2011.

77 Shirouzu 2010; Bradsher 2010.

78 RMI analysis.

79 Feebate design principles attractive to industry and across the political spectrum were worked out in RMI's industry workshop in 2007 (Mims and Hauenstein 2008). Feebates repair the market failure that makes auto buyers' interests diverge from society's goals because private buyers have a far higher time value of money (discount rate) than society does. By broadening the price spread between more and less efficient autos, feebates can bridge this gap, so buyers can choose the autos that most benefit them *and* society. Being charged or paid at the time of purchase, feebates are a far stronger motivator than an auto's 15-year fuel costs or savings, which amount to only a few dollars' difference per day, a small fraction of the total cost of ownership, and are heavily discounted.

80 Callonec and Blanc 2009. The Austrian feebate (*Normverbrauchsabgabe* or NoVA) differs somewhat.

81 Bunch and Greene 2010 is a valuable analysis of mild feebate options for California.

82 Feebates enable all auto buyers to value long-term savings as fully as society does, so we needn't speculate about what their implicit discount rate is. However, following conventional economic theory, we conservatively assume *all* automobile buyers insist on a three-year or shorter payback on fuel savings, even though a 2010 Consumer Federation of America poll (Cooper 2011) found 64% would be satisfied with a five-year payback on a 60 mpg auto. Our high assumed implicit discount rate is also conservative in three other ways. First, it assumes that car buyers act as if they were paying cash even though in fact most purchases are financed at often-concessionary interest rates. Second, many auto buyers pay up to $9,000 for "trimline features"—the difference between basic and fully loaded versions. That's much more than the premium associated with switching

to one of today's efficient hybrids, and those features don't pay back over time as a hybrid does, so price is clearly not all-important. Third, Revolutionary+ autos' initial premium would also be partly offset by important new value propositions and marketing "exciters." An auto that perfectly does everything expected of it, but no more, will generally undersell a model that adequately does what's expected but also implements an "exciter" even in half-baked fashion.

83 RMI analysis.

84 General Services Administration 2010.

85 Electrification Coalition 2009. We examined how fleet buys could trigger a plausible, though unproven, retooling acceleration of five years in start-up and four in duration.

86 Hybrids are currently the best buy for New York taxis, partly because of their low maintenance costs.

87 Lovins et al. 2004, 194–95.

88 That's already done for retooling to help meet recently raised federal efficiency standards. The 1979–1980 Chrysler loan ultimately profited taxpayers (via equity warrants), and the 2008 Chrysler and GM loans were repaid with interest, though larger elements of their bailout packages persist whose recovery will depend on future stock prices.

89 Automaking, like airlines, is a great industry but a bad business. Its thin profits have long constrained investment and recruitment and made the culture more risk-averse, disheartening its most innovative engineers. The global auto market became overmature: convergent products fought for ever-smaller niches in saturated core markets at cutthroat commodity prices with one-third overcapacity and with stagnating basic innovation and cultures. By the time of the Big Three's 2008–2009 warning heart attack, it was time for something completely different.

90 Schipper 2010.

91 Tabuchi 2010.

92 A lightweight electrified vehicle will in theory be driven more than a normal vehicle because of its low fuel cost per mile. While this rebound effect from increased driving doesn't negate Revolutionary+ autos' efficiency gains, it also can't be ignored: the latest evidence suggests that switching to highly efficient vehicles can boost driving by roughly 3–22% (Small and Van Dender 2010; Hymel, Small, and Van Dender 2010). However, efficient autos' lower fuel cost per mile would probably be partly offset by the need, as we'll see, to shift highway funding from gasoline taxes to driving fees. Two other factors also offset the rebound effect. First, rebound diminishes as median income rises relative to the fuel cost of driving. Despite growing income disparity, median real U.S. income is projected to grow over the next several decades, shrinking historically observed rebounds. An even bigger factor is how much time we actually spend behind the wheel, which already averages about 1.1 hours a day and is about saturated. More autos driving more miles would further increase congestion and traffic. This would actually compel people to drive less, not more: people don't care about fuel savings if the cost is wasted time. Such trends are already observable (Millard-Ball and Schipper 2011). Accordingly, our transportation analysis considers increased auto fuel usage from rebound to be negligible.

93 Delucchi and McCubbin 2010.

94 Chart based on RMI analysis of Delucchi and McCubbin 2010.

95 Cairns et al. 2004.

96 The potential for national VMT reductions is based on RMI's analysis of the following: Oregon DOT 2007, Litman 2005, Urban Land Institute and Cambridge Systematics 2009, Urban Land Institute 2007, TIAX 2006, and Shaheen and Cohen 2006.

97 Ibid.

98 It's therefore common in Europe to signal vehicle-related costs through registration fees based on weight, engine displacement, or fuel intensity.

99 Congressional Budget Office 2011. The EU is exploring such road fees: Rosenthal 2011.

100 Oregon DOT 2007.

101 Bordoff and Noell 2008; Edlin 2002.

102 Goldstein 2010. Progressive offer PAYD.

103 For more information on parking policies, vehicle use, and VMT see Shoup 2005.

104 Luckily, some organizations have encountered little resistance in persuading local governments to exempt even commercial carsharing from the rental-car tax, as they have for Zipcar in Texas and Chicago.

105 Urban Land Institute 2007.

106 Burchell 1998.

107 Goldstein 2010.

108 Goldstein 2010, 114.

109 Schewel and Schipper 2011.

110 Center for Neighborhood Technology 2010.

111 Goldstein 2010, 75–79.

112 Texas Transportation Institute 2009.

113 In this analysis, "heavy trucks" are Class 8 trucks with a gross vehicle weight rating (GVWR) over 33,000 pounds. GWVR is the maximum allowable total weight of a road vehicle when loaded.

114 Based on RMI's analysis of Bockholt and Kromer 2009 and Northeast States Center for a Clean Air Future et al. 2009. The TIAX analysis is the basis for National Research Council 2010, NRC's most recent analysis of heavy-duty truck technology in the U.S.

115 Truck designer/owner Tom Wieringa has so lightened his "Spud Lite" trucks that he can haul in three loads the potatoes his competitors haul in four.

116 Bockholt and Kromer 2009; Vyas, Saricks, and Stodolsky 2002.

117 Curve based on RMI analysis of Bockholt and Kromer 2009, Northeast States Center for a Clean Air Future et al. 2009, Vyas, Saricks, and Stodolsky 2002, and U.S. Energy Information Administration 2010c.

118 Ogburn, Ramroth, and Lovins 2008.

119 Harmonizing a welter of conflicting state and even county rules would most cleanly be done at the federal level, where federal highway rules also need updating (failure to do so would sacrifice a quarter-million-barrel-a-day oil saving: n.51 supra). Trucking-industry groups like the nearly 300-firm North American Council for Freight Efficiency (NACFE), which RMI cofounded in 2010, are starting to coalesce around this agenda. NACFE will share technical evaluations of fuel-saving technologies, giving all operators a trusted source of reliable data measured independently by and for their peers.

120 Walmart 2010.

121 Shpitsberg 2010.

122 VICS Empty Miles 2011; SmartWay Transport Partnership 2004.

123 BestLog 2008.

124 Holland Container 2011.

125 U.S. Energy Information Administration 2010b; U.S. Bureau of Transportation Statistics 2007.

126 Based on RMI's analysis of Lanigan et al. 2006, Global Insight 2006, and Walmart 2010.

127 RMI analysis.

128 Daggett 2003. For a double-aisle airplane, it's about 111 pounds a year. Both numbers depend on many details of airplane and mission profile. We assume the $3.08/U.S. gallon Jet A spot price at May 23, 2011, and a 3%/y real discount rate.

129 Newhouse 2007.

130 Thus the chief designer of Honda's original Insight hybrid car led its carbon-fiber business-jet venture—the top-mounted engines boosted fuel savings to 20%—then returned to cars, while the designer of the machine that filament-wound the 787's carbon-fiber fuselage moved to Fiberforge.

131 To see the Javan cucumber in flight, visit http://news.bbc.co.uk/earth/hi/earth_news/newsid_8391000/8391345.stm. Such airplane designs have been extensively studied since their use in a U.S. Air Force bomber in the late 1940s. Boeing has lately flight-tested a scale model called the X-48B.

132 Kota 2011.

133 Ball 2007.

134 Norris and Wagner 2009.

135 Center for Clean Air Policy and Center for Neighborhood Technology 2006.

136 A drop-in biofuel is a non-petroleum-based fuel that can be "dropped in" to existing fuel tanks and propulsion systems without any modification to infrastructure or components.

137 Electrification Coalition 2010.

138 RMI analysis.

139 Norfolk Southern 2009.

140 Carbon War Room 2011.

141 Anderson 2011; Synovision Solutions 2010.

142 Defense Science Board 2008.

143 Lovins 2010a.

144 At this writing, General Motors is seeking to trademark this widely used term.

145 E Source 2011.

146 The world deployment-speed record may be held by Portugal, which has installed over 1,100 charging stations in just two years—less than 80 miles apart across the whole country—and plans to electrify 10% of all the cars on its roads by 2020 (Socrates 2011).

147 Lovins and Cramer 2004; Lovins et al. 2004, 230-42.

148 Lovins 2003.

149 Lovins et al. 2004, 237.

150 Lovins 2003.

151 Lovins et al. 2004, 237; Thomas 2004–2010.

152 McKinsey & Company 2010a. Hydrogen is a mature technology; indeed, two-thirds of the fossil-fuel atoms burned today are hydrogen. The world already produces enough hydrogen, chiefly for refining and other chemical applications, to fuel a very efficient fleet of U.S. highway vehicles, although such centralized production would usually have a higher delivered cost than decentralized production (Lovins 2003).

153 *New York Times* 2010; Lovins and Cramer 2004, 50–85; Lovins et al. 2004, 230–42.

154 Lovins et al. 2004, 239.
155 Krupnick 2010.
156 Lovins and Lovins 1982, 87–99.
157 Renewable Fuels Association 2010.
158 Notwithstanding such valuable by-products as dried distillers grains, a livestock feed left over from ethanol production.
159 Many automakers are already making cars "total flex"—able to burn anything from pure ethanol to pure gasoline—for foreign markets, such as Brazil, whose governments have ethanol adoption requirements in place. This increases competition among all fuels, because nobody has captive customers who can't switch.
160 International Energy Agency 2010b.
161 RMI analysis.
162 Perlack, Wright, and Turhollow 2005.
163 RMI analysis of International Energy Agency 2010b and Perlack, Wright, and Turhollow 2005.
164 DuPont Danisco Cellulosic Ethanol LLC 2010.
165 Korosec 2010; British Petroleum 2010b; Foroohar 2010.
166 The Smithsonian Institution's Curator of Algae, Dr. Walter Adey, has developed an even cheaper way to make algal fuels while cleaning up polluted lakes and rivers (Adey 2008–2009).
167 But not necessarily sugar palm (*Arenga pinnata*), mentioned in chapter 6; note 712.
168 RMI analysis.
169 RMI analysis.

CHAPTER 3. BUILDINGS: DESIGNS FOR BETTER LIVING

170 U.S. Energy Information Administration 2010b, 1999, 2006a.
171 U.S. Energy Information Administration 2010d.
172 Ibid.
173 Ibid.
174 Ibid. 2008a.
175 U.S. Energy Information Administration 2008a, 2008c, 2010d.
176 D&R International Ltd. 2010.
177 Ibid.
178 Ibid.
179 Klepeis et al. 2001, 231–52.
180 U.S. Energy Information Administration 2010d.
181 Ibid.; GPO 2011.
182 U.S. Energy Information Administration 2008b, 2010d.
183 U.S. Energy Information Administration 2010b.
184 RMI analysis using data from U.S. Energy Information Administration 2010b and 2010d.
185 National Association of Home Builders 2006.
186 Ross and Meier 2000.
187 Meier 2011.
188 U.S. Energy Information Administration 2010d.
189 RMI analysis using data from U.S. Energy Information Administration 2010b and D&R International, Ltd. 2010.
190 U.S. Energy Information Administration 2010d, 2008a.
191 U.S. Energy Information Administration 2010d.
192 RMI analysis using data from U.S. Energy Information Administration 2010b.
193 U.S. Environmental Protection Agency 2010.
194 Milne and Reardon 2008. Research by the Commonwealth Scientific and Industrial Research Organisation has found that the embodied energy used to construct the average household is equivalent to about 15 years of normal operational energy use.
195 RMI analysis using data from U.S. Energy Information Administration 2008a, 2008c, 2010b, and 2010d.
196 RMI analysis using data from U.S. Energy Information Administration 2010b, National Academy of Sciences 2010, and Ehrhardt-Martinez, Donnelly, and Laitner 2010.
197 U.S. Census Bureau 2010.
198 RMI analysis using data from U.S. Energy Information Administration 2010b, National Academy of Sciences 2010, and Ehrhardt-Martinez, Donnelly, and Laitner 2010.
199 Northwest Energy Efficiency Alliance 2009.
200 RMI analysis using data from U.S. Energy Information Administration 2010b, National Academy of Sciences 2010, and Ehrhardt-Martinez, Donnelly, and Laitner 2010.
201 U.S. Green Building Council 2006a.
202 Keller 2011.
203 U.S. Department of Energy 2010a; Brooke Hammett, personal communication with author, March 11, 2011.
204 Singapore National Energy Efficiency Committee 2010.
205 Rosenfeld et al. 1997.
206 Akbari and Rosenfeld 2008.
207 Neubaurer et al. 2009; Goldstein 2010, 132–137. Goldstein makes a convincing case that the lower cost was *because of*, not despite, the efficiency gains.
208 U.S. Energy Information Administration 2008a.
209 Goldstein 2010, 41.

210 Desroches and Garbesi, 2011.
211 PPG Aerospace 2010; RavenBrick LLC 2009, Pleotint 2007.
212 Tinianov 2011; Apte et al. 2003.
213 Tinianov 2011; Apte et al. 2003.
214 Scanlon 2010. Lawrence Berkeley National Laboratory says the world-average room air conditioner provides 2.8 units of cooling per unit of electricity; the best Japanese unit, 7.1; its feasible conventional improvement, 8.4 (SEAD Group 2011).
215 Aspen Aerogels 2011; Proctor Group Ltd. 2010; Cabot Corporation 2011; NanoPore Inc. 2008.
216 National Gypsum 2011.
217 Navigant Consulting 2010.
218 GTM Research 2010.
219 U.S. Department of Energy 2010b.
220 Kanellos 2009.
221 Shimomura and Gage 2009–2011; Beyer 2011. More information can be found at www.neofocal.com.
222 Rose 2010.
223 PAX Scientific 2008; Belko and Smith 2008.
224 PAX Water Technologies 2011.
225 Koelman 2011.
226 Gossamer Ceiling Fans 2008; Eley Associates 1996.
227 WIPO 1999; Nuveco AG 2011.
228 For example, Belkraft 2005; Kuhn Rikon 2011; the Newey & Bloomer "Simplex" kettle; Thermal Cookware 2008. Simplest of all: just put a lid on an ordinary pot to stop four-fifths of the heat from escaping as steam (Cullen, Allwood, and Borgstein, 2011, S24).
229 Wyssen et al. 2010.
230 Browning 2011.
231 Cooney 2011.
232 Ehrhardt-Martinez, Donnelly, and Laitner 2010.
233 Laitner 2011.
234 Xu 2009.
235 Kats 2009.
236 Sandborg 2010; W&H Properties 2009; ENR New York 2009; Schneider 2011.
237 Herman Miller 2008.
238 Goldstein 2010, 103. Goldstein achieved roughly 95% savings in lighting energy in his kitchen and 85% in his bedroom—with better aesthetics and visibility—by combining efficient luminaires with the sort of direct-indirect lighting design normally used in commercial buildings.
239 Zeller 2010.
240 Hawken, Lovins, and Lovins 1999.
241 Lovins 1995, 79–81.
242 Lovins 1992.
243 Empire State Building Company LLC 2010; Lockwood 2009; Empire State Building Company LLC and Rocky Mountain Institute 2009.
244 Tommerup and Nørgård 2007.
245 Desroches and Garbesi 2011.
246 LeClaire 2011.
247 National Renewable Energy Laboratory 2010a; Kiatreungwattana and Salasovich 2010; National Renewable Energy Laboratory 2009.
248 Rocky Mountain Institute 2008.
249 Interface Engineering 2009.
250 Aster 2010; Deutsche Bank Corporate Real Estate & Services 2011.
251 U.S. Green Building Council 2006b; City of Fort Collins Utilities 2004.
252 Parker 2009; Aspen Publishers 2004; Klingenberg 2011.
253 Feist 2011.
254 Eek 2011.
255 Bakewell 2011.
256 National Academy of Sciences 2010.
257 Kats 2009.
258 D&R International 2010.
259 Heschong Mahone Group 2003a.
260 Heschong Mahone Group 1999, 2001, and 2003b.
261 Ulrich 2011.
262 Eichholtz, Kok, and Quigley 2010.
263 Malin 2010.
264 Cochran and Muldavin 2010.
265 LaSalle Investment Management 2010.
266 Barrett 2010.
267 Contradicting this dictum, however, Nobel economist Ken Arrow has told me that he had once indeed found, and picked up, a $20 bill on the sidewalk.
268 Goldstein 2010, 49.
269 D&R International 2010.
270 Leonard 2009.
271 U.S. Energy Information Administration 2006a.
272 Goldstein 2010, 51–53.
273 Norton 2010. Separately, another military base found its least efficient home had the lowest energy use while its most efficient had the highest, because the occupants' behavior overwhelmed technical factors.
274 BetterBricks 2006; U.S. Green Building Council 2006c; Idaho Business Review 2007; Christensen 2011.
275 Johnson Controls 2010.
276 Carbon Disclosure Project 2011.

277 Darrow 2011.
278 Mills 2009.
279 Bendewald and Franta 2010; Edmonson and Rashid 2009.
280 Eubank and Browning 2004.
281 Westbrook 2011.
282 BetterBricks 2011.
283 Gohring 2008.
284 Flex Your Power 2011.
285 Braese 2011.
286 RMI analysis with support from U.S. Energy Information Administration 2010b, National Academy of Sciences 2010, and Ehrhardt-Martinez, Donnelly, and Laitner 2010.
287 U.S. Energy Information Administration 2006a.
288 Bell 2008.
289 Crowley 2010.
290 Meyers 2011.
291 Kinsley 1997.
292 Segall 2011.
293 Bonneville Power Administration 1992.
294 Lovins 1992.
295 Bigelow 2011.
296 Jungerberg 2010.
297 Walters 2010.
298 Hawken, Lovins, and Lovins 1999.
299 Bundesministerium der Justiz 2009.
300 Mayerowitz 2009; Novak 2010.
301 Attari et al. 2010.
302 Nadel and Goldstein 1996; Neubaurer et al. 2009.
303 Gold et al. 2011.
304 Rohmund et al. 2011.
305 California Energy Commission, 2007. This and other smart policies helped California consumers avoid at least $60 billion of utility investments (Bernstein et al. 2000; Roland-Holst 2008) for which their economy doubtless found more productive uses—investing in infrastructure and entrepreneurship rather than in the capacity to make and waste more electricity. The 2008 Roland-Holst study notes that California households saved $56 billion through energy efficiency during 1972–2006, helping create about 1.5 million full-time-equivalent jobs.
306 Business Wire 2000.
307 Northwest Power and Conservation Council 2010a.
308 Online Code Environment and Advocacy Network 2011a and 2011b.
309 U.S. Department of Energy 2010c; Laustsen 2008; Thomsen et al. 2008.
310 European Union 2010.
311 Architecture 2030 2011; Stand-To! 2011.
312 City of Sioux Falls Building Services 2010.
313 Reed et al. 2004.
314 U.S. General Services Administration 2011.
315 City of Boulder 2011.
316 Carey 2009.
317 Pew Center on Global Climate Change 2011a.
318 American Gas Association 2010; Institute for Electric Efficiency 2010.
319 Northwest Power and Conservation Council, 2010b.
320 Efficiency Vermont 2010.
321 Database of State Incentive for Renewables & Efficiency (DSIRE) 2011; PACEnow 2011.
322 SMUD reportedly collateralizes these using a Uniform Commercial Code fixture lien.
323 RMI analysis.
324 Danish Energy Agency 2009, 33. For example, Denmark cut its oil-fired home heating sixfold in the past 30 years.

CHAPTER 4. INDUSTRY: REMAKING HOW WE MAKE THINGS

325 Terborgh 1992.
326 Indigo Development 2003.
327 Biddle 2011.
328 U.S. Energy Information Administration 2005.
329 RMI analysis based on data from U.S. Energy Information Administration 2010b, National Academy of Sciences 2010, Xu, Slaa, and Sathaye 2010, Martin et al. 2000, and Bailey and Worrell 2005.
330 U.S. Energy Information Administration 2010a.
331 Another option: adding rubber crumbs from old tires to road paving can save half the asphalt or an eighth of the feedstock oil, at strongly negative marginal cost and with many advantages. See Lovins et al. 2004, 93–94.
332 Committee on Biobased Industrial Products 2000.
333 RMI analysis based on data from U.S. Energy Information Administration 2010b.
334 Ibid.
335 American Chemistry Council 2010a.
336 U.S. Energy Information Administration 2010e; American Chemistry Council 2010b; American Chemistry Council staff 2011.
337 U.S. Energy Information Administration 2006b.

338 Williams, Larson, and Ross 1987; Larson 1991; Wernick and Ausubel 1995a. Per Larson 1991, U.S. steel consumption per capita was back to its ~1880 level by 1990.

339 Weber 2008, quoted in National Academy of Sciences 2010.

340 The Carbon Trust's 2011 analysis of 2004 data found 25%, implying that U.S. fossil carbon emissions were 13.5% higher on a consumption than on a production basis (Sinden 2011).

341 If the products are made in a more energy-intensive fashion abroad than they would be in the U.S., as is often the case with current Chinese techniques, then offshoring can increase energy use. For example, Xu, Allenby, and Chen 2009 estimate that producing 2002–2007 Chinese exports to the U.S. consumed about 12–17% of China's energy use.

342 Weber and Matthews 2007, excluding up to about 10% from bunker oil. Transferring "gray energy" to the U.S. account would reduce it for trading partners, such as China.

343 Weber 2008 describes many of the sources of uncertainty. Some others are methodological; for example, Xu, Williams, and Allenby 2010 analyze the energy needed to support Chinese workers. In net-energy analysis this is normally not counted because the workers would need to carry on their lives whether they were making exports to the U.S. or not.

344 Our analysis also includes the shift to carbon fiber for ultralight autos (chapter 2) and the large increases in photovoltaic-power and wind-turbine production (chapter 5), but their net effect on industrial energy use is small.

345 Dow 2009.

346 United Technologies Corporation 2010.

347 U.S. Department of Energy 2010d.

348 Fletcher et al. 2002 is a useful early summary. This area is moving very rapidly and modular microreactors are commercially available, yet few results are published because of their breakthrough value if kept proprietary.

349 Tucker 2009.

350 Bruggink, Schoevaart, and Kieboom 2003.

351 Wirth 2008.

352 This proposition was proven by Dr. Sandy Thomas's firm H2Gen: his forecourt reformers' tighter thermal integration offset their less favorable geometrical scaling (Thomas 2004–2010).

353 Burns 2011.

354 Mills and Sartor 2005.

355 Rumsey 2011, illustrated by a 175,000-square-foot teaching lab his firm designed at Cal Poly in San Luis Obispo, where the sniffer system allowed 30–40% less ventilation with the same or better indoor air quality, downsizing HVAC and hence saving about 15–20% of the lab's total energy at lower total cost.

356 Howe 1993.

357 U.S. Department of Energy 2003.

358 U.S. Environmental Protection Agency 1998.

359 Galitsky and Worrell 2008.

360 Scheihing 2009.

361 E Source 2002.

362 McCoy 2011.

363 McCoy 2010.

364 Desroches and Garbesi 2011.

365 Europump 2003.

366 Elliott et al. 2002, fig. 7-1.

367 A late-1980s Swedish study found average new industrial pumps were about 3–5 percentage points better than those in use, and the best new ones were 3–10 points better still—but researchers checking one size found 8 points' gain at the same or lower price (courtesy of Larson and Nilsson 1990, fig. 7). This is to say nothing of specialized types like ITT Flygt's sewage pumps, which often save about 40–50% because they're nonclogging and self-cleaning (Flygt 2008).

368 Sator, Krepchin, and Horsey 2010.

369 Oak Ridge National Lab 2008.

370 Groscurth and Kümmel 1989.

371 Worrell, Laitner, and Michael 2003; Worrell, Price, and Martin 2001. Thus when Greenville Tube Corporation demonstrated new drivesystems under the U.S. Department of Energy's Motor Challenge program, energy efficiency rose 30%, but productivity also rose 15% while scrap fell 15%, cutting the payback to five months (Romm 1997).

372 Lawrence Berkeley National Laboratory, formerly called Lawrence Berkeley Laboratory and now often shortened to Berkeley Lab, is the lead national laboratory for energy efficiency and collaborates with other national labs responsible for specific parts of that opportunity.

373 KEMA, Inc. 2006.

374 Itron Inc. and KEMA, Inc. 2008.

375 California Public Utilities Commission 2010.

376 Students and faculty from 26 universities visit small and medium-size manufacturing plants and

survey opportunities for saving energy, increasing productivity, and reducing waste and pollution. During 1976–2010, the program saved manufacturers more than $700 million, created or maintained more than 1.5 million U.S. manufacturing jobs, and trained legions of students in energy engineering.

377 Shipley and Elliott 2006.

378 U.S. Energy Information Administration 1997 and 2006b; U.S. Census Bureau 2005.

379 U.S. Energy Information Administration 2011b.

380 Viswanathan, Davies, and Holbery 2006.

381 Bailey and Worrell 2005.

382 Gemmer et al. 2010.

383 Chan-Lizardo et al. 2011.

384 Glasser et al. 2009.

385 Enterprise Lane 2010.

386 Westbrook 2011.

387 E Source 1999; Lovins 1989. See also Elliott et al. 2002. Our analysis eliminates potential double-counting within the EIA- and LBNL-based previous tranches of efficiency.

388 Howe and Shepard 1993. Our 1989 analysis conservatively omitted this term, but it can be important because pump or fan energy rises as the cube of speed.

389 Martin et al. 2000; Xu, Slaa, and J. Sathaye 2010.

390 Our estimate of biofuel process energy is conservatively high because it assumes the National Renewable Energy Laboratory (NREL) standard corn-ethanol plant design. However, a 2007 RMI/NREL/industry charrette found that redesigning NREL's best corn stover–ethanol plant design could save roughly 50% of its heat, 60% of its electricity, and 30% of its capital cost while increasing its yield. Also, many advanced biofuels like algal fuels may need little or no processing.

391 Wherever biofuels aren't suitable to replace oil in industrial mobility, any combination of natural gas, electricity, and hydrogen could be substituted. Some mining trucks have already begun switching from diesel to LNG (InfoMine—Mining Equipment and Supplier News 2010).

392 U.S. Energy Information Administration 2010c. Values are extrapolated to 2050.

393 Greening, Greene, and Difiglio 2000.

394 Sorrell 2007.

395 Ayres and Ayres 2009.

396 Ertesvag 2001.

397 Cullen, Allwood, and Borgstein 2011.

398 Ayres, Ayres, and Warr 2003.

399 Cullen, Allwood, and Borgstein 2011.

400 Blok 2004.

401 Choate 2003; U.S. Department of Energy 2004, 2007, 2006b (see also 2006a); Institute of Paper Science and Technology 2006.

402 Remember that we don't include the chemical-reagent (reductant) part of the coal because that's a feedstock use, not a fuel.

403 International Energy Agency 2007.

404 Smith, T. 2011.

405 LaFarge 2010.

406 Energetics Incorporated 2006.

407 Holcim AG 2010.

408 Lee et al. 2007.

409 Criscione 2011; U.S. Department of Energy 2008a; Apogee Technology, Inc. 2010; Foundry Management & Technology 2008.

410 Assumes capex of $4 per watt electric, 0.26 capacity factor, 20–30% electric conversion efficiency, 95% heat exchanger efficiency, 30% investment tax credit, 3% O&M costs, and 20-year system life. The levelized cost is then $5.25 to $7.90 per million BTU.

411 Heschong Mahone Group 2007.

412 Over a one- or five-year time horizon, the May 2, 2011, straddle (the price of simultaneously sold put and call options) for Henry Hub wholesale natural gas was respectively $1.03 and $2.15 (the figures come from Munger 2011). In a perfect market, that's how much you'd have to pay a zero-profit trader to assume the price risk and sell you gas at a constant price—making it no more risky than the constant-price solar steam.

413 SunChips 2011.

414 Mujumdar 2006.

415 Greenberg 2010.

416 Metcalfe et al. 2006.

417 Lester, Rudman, and Metcalfe 2009.

418 U.S. Department of Energy 2010e.

419 Benyus 2002.

420 Ziemlewski 2010.

421 American Institute of Chemical Engineers 2010. These two cases and the RAM mixer won recognition in AIChE's 2010 10xE competition, which I cochaired.

422 Lovins 2007.

423 Allwood et al. 2011.

424 Despite the materials intensity of microchips, which a decade ago each consumed about 1.7 kg of fossil-fuel and chemical inputs, 32 kg of water, and 0.7 kg of

separated gases (Williams, Ayres, and Heller 2002). Modern microchip fabrication plants are far more efficient, and some approach 99% water recycling.

425 Material Flows 2011.

426 Wernick and Ausubel 1995b.

427 Holtzclaw 2004, quoted by Goldstein 2010, 125–26. Goldstein also notes a 300-fold shorter route for the average mail carrier. The comparison is between a suburban development and a dense urban development with three-story houses.

428 Hawken, Lovins, and Lovins 1999.

429 Eurostat 2011.

430 Hageluken and Corti 2010.

431 Fischer 1997; Hawken, Lovins, and Lovins 1999, ch. 4, note 27. We thank Claude Fussler, then of Dow, for this example.

432 Allwood 2009.

433 Bögle and Cachola Schmal 2005.

434 Knowles 2011.

435 *The Economist* 2011a.

436 Drexler 1987.

437 McKinsey & Company, 2009a, 2009b, 2010b.

438 Goldstein, 2010, 122–23.

439 U.S. Department of Energy 2010e.

440 Nelson 1998.

441 Green Manufacturer 2010.

442 RMI analysis based on data from U.S. Energy Information Administration 2010b, National Academy of Sciences 2010, Xu, Slaa, and Sathaye 2010, Martin et al. 2000, and Bailey and Worrell 2005.

443 Ibid.

444 Anderson 1999, 2009.

445 When commercial construction, Interface's main market, crashed in 2008–2009, Interface survived and even thrived because its loyal customers knew its products matched both their needs and their values. Furthermore, Interface's biomimicry-informed design wizards have long been developing products and service-leasing business models that may ultimately reduce by 97%—with eventual recycling, 99.9%—the amount of material per year needed to carpet a floor beautifully (Hawken, Lovins, and Lovins 1999, 139–41).

CHAPTER 5. ELECTRICITY: REPOWERING PROSPERITY

446 Grant and Mitton 2010, 40.

447 Hilbert and Lopez 2011, 60–65.

448 U.S. Energy Information Administration 2000.

449 Constable and Somerville 2003.

450 RMI analysis of the estimated undiscounted capital investment in the U.S. electricity system during 2010–2050 in the *Maintain* case.

451 A rigorous assessment of electric system reliability for the contiguous United States in 2050 with 80% renewable generation is not possible using existing tools and models. The models used for *Reinventing Fire* use simplifying assumptions to balance supply and demand at hourly timescales, taking into consideration transmission system constraints. Future work is needed, however, to ensure the reliability of the electric power system at subhourly timescales and at the level of distribution networks. North America's overseer of electricity reliability, the North American Electric Reliability Council (NERC), defines electric system reliability as "the ability to meet the electricity needs of end-use customers, even when unexpected equipment failures or other factors reduce the amount of available electricity." NERC sets reliability standards to ensure adequacy (adequate resources are available to provide customers with a continuous supply of electricity) and security (the ability to withstand sudden unexpected disturbances to the electric system). Power system reliability must be measured from several different aspects that span time frames from long-term planning to real-time operations. A complete analysis of power system reliability would require extensive additional efforts, including system adequacy analyses, high-resolution production modeling, power flow analyses, power system stability studies, and contingency analyses.

452 U.S. Department of Energy 2006c.

453 RMI analysis based on ReEDS modeling.

454 Ibid.

455 U.S. Energy Information Administration 2011d.

456 U.S. Energy Information Administration 2011b.

457 Electric Energy Market Competition Task Force 2005.

458 U.S. Energy Information Administration 2007.

459 Standard & Poor's 2008.

460 Ironically, Edison may have the last word: now that most of the loads in buildings, which use most of our electricity, do or readily can run on DC, there's a growing movement to distribute low-voltage DC within buildings and even outside them, eliminating the costs and losses of individual DC power supplies in our lights, electronics, and DC motors. The most

efficient data centers and most telephone switching centers already run on DC.

461 Lovins et al. 2002.

462 An anomaly annoying to economic purists is that some 17 U.S. cities have dual distribution systems owned by competing private and public utilities. The lights stay on, and prices appear to be lower.

463 Kahn 1988.

464 U.S. Energy Information Administration 2011e.

465 Hirsh 1999.

466 U.S. Energy Information Administration 2010b.

467 For example, in 1990, New England Electric System captured 90% of a small-commercial-customer pilot retrofit market in two months. That same year, PG&E signed up 25% of its entire new-commercial-construction market for a design-improvement program—150% of the year's target—in three months, so it then raised its 1991 target . . . and got all of it in the first nine days of January. Utilities from the sprawling Bonneville Power Administration to PG&E to the tiny Osage (Iowa) Municipal Utility showed they could capture about 70–90+% of particular efficiency micromarkets—from efficient showerheads to the large-scale superinsulation home retrofits of the Hood River experiment (see p. 111)—in just a year or two.

468 Eggers 2010.

469 Lovins et al. 2002.

470 U.S. Energy Information Administration 2010f.

471 U.S. Energy Information Administration 2010g.

472 Gupta, Tirpak, and Burger 2007.

473 U.S. Climate Action Partnership 2009.

474 National Academy of Sciences 2009.

475 Ibid.

476 Schneider and Banks 2010.

477 For comparison, by 2008, over 95% of Chinese coal plants had electrostatic precipitators; by 2009, 71% (460 GW) had installed flue-gas desulfurization; and Chinese plants' sulfur emissions per kWh probably fell below U.S. levels starting in 2008 (Wang 2011).

478 Wang 2011; North American Electric Reliability Corporation 2010a.

479 Chupka et al. 2008; North American Electric Reliability Corporation 2010b.

480 Celebi et al. 2010.

481 Krauss 2010.

482 Shuster 2009.

483 Wilbanks 2009, 45–80.

484 Kenny et al. 2009; Torcellini et al. 2003.

485 California State Water Resources Control Board 2011a and 2011b; California Energy Commission 2010.

486 Mufson 2011.

487 Ibid.

488 Ibid.

489 U.S. Energy Information Administration 2010b.

490 World Energy Council 2010.

491 Luppens et al. 2008.

492 Rutledge 2010, 23–24.

493 Database of State Incentives for Renewables and Efficiency (DSIRE) 2010.

494 RMI analysis using data from Database of State Incentives for Renewables and Efficiency (DSIRE) 2010 and U.S. Energy Information Administration 2009.

495 Wiser and Bolinger 2010; Sherwood 2010.

496 Renewable Energy Policy Network for the 21st Century; 2011.

497 U.S. Energy Information Administration 2010b.

498 Earth Policy Institute 2007; Wiser and Bolinger 2010, 2011; Mints 2010; Liebreich 2010.

499 Bindewald 2004; Electricity Consumers Resource Council (ELCON) 2004.

500 Burns, Potter, and Witkind-Davis 2004, 11–15.

501 Bindewald 2004; Electricity Consumers Resource Council (ELCON) 2004.

502 LaCommare, Eto, and Hamachi 2004.

503 Ibid.

504 Lovins and Lovins 1982.

505 Pugh 2010.

506 National Research Council 2008; Fountain 2010.

507 LogLogic Inc. 2009.

508 RMI analysis using data from North American Electric Reliability Corporation 2011.

509 RMI analysis using data from U.S. Nuclear Regulatory Commission 2011a. North American Electric Reliability Council data show that during 2002–2009, U.S. reactors average 0.55 such unplanned sudden shutdowns (SCRAM events) per year, but some plants had more than two per year.

510 U.S. Energy Information Administration 2010d.

511 U.S. Nuclear Regulatory Commission 2011b.

512 Massachusetts Institute of Technology 2011.

513 RMI analysis based on ReEDS modeling.

514 Ibid.

515 Koplow 2010; Lovins 2010c.

516 International Atomic Energy Agency 2011.

517 Lochbaum 2008.

518 Koomey and Hultman 2007.

519 Moody's Investors Service 2009.
520 Schneider, Froggatt, and Thomas 2011; Härkönen 2011; World Nuclear News 2011.
521 Grübler 2009 and 2010.
522 Koomey and Hultman 2007.
523 GPO 2011.
524 U.S. Department of Energy 2011a; Wald 2011a. Another $2 billion was guaranteed for AREVA, 93% owned by the French state, to build a uranium enrichment plant competing with American ones. The Georgia project's loan guarantees are conditional on NRC licensing, recently delayed by safety questions.
525 Wald 2011a; Wald 2011b; Lovins 2010c.
526 Henry 2011.
527 Sturgis 2011.
528 Downey 2011.
529 Murawski 2011.
530 Smith, M. 2011.
531 Reuters 2011b; Tabuchi 2011.
532 U.S. Nuclear Regulatory Commission 2011c.
533 O'Grady 2011.
534 Buckley 2011.
535 Dempsey 2011; Kraemer 2011. Chanceller Merkel wants "to end the use of nuclear energy and reach the age of renewable energy as fast as possible," 85 percent of Germans polled want that too, and no political party at this writing opposes it. The nearly 9 GW initial shutdown had no material effects on electricity prices, reliability, or power imports (Töpfer 2011). The basic bet, so far looking sound, is that doubling renewables' 17% electricity share will continue to lead Germany's small-business-centric economic development strategy, buying Germans' labor rather than Russians' gas.
536 Fackler 2011.
537 Tabuchi; 2011b.
538 Yasu 2011.
539 Mijuk and Germann 2011.
540 Kanter 2009.
541 *The Economist* 2011b.
542 Alvarez et al. 2003, 1–51 and 213–23; Beyea, Lyman, and von Hippel 2004; National Research Council 2005.
543 Biello 2009.
544 Ibid.
545 American Electric Power 2011; Wald and Broder 2011.
546 Theunissen et al. 2011; Brouwers and Kemenade 2010; van Benthum et al. 2010; Willems et al. 2010. Expanding and cooling the flue gas causes the CO_2 to condense out as fine droplets that can be mechanically separated. The compact equipment could be widely retrofittable and is estimated at scale to cost about one-tenth as much as chemical processing, cutting total CCS cost by roughly two-thirds. If successful, this could be useful for retrofitting old coal plants, especially in China, but it may still raise costs enough to make many coal plants no longer worth operating, and it would certainly not create a business case for building more. However, it could help clean up big industrial plants, or any gas-fired generators still desired to operate under stringent long-term carbon limits.
547 Herzog 2010.
548 National Energy Technology Laboratory 2011.
549 California's nuclear plants, rather than cycling like French ones when in surplus, have a special "must-run" status dispatching them before cheaper windpower—penalizing wind operators.
550 Töpfer 2011.
551 Pöyry 2010; Redpoint Energy 2011; Nicholson, Rogers, and Porter 2010; Bode and Groscurth 2010; Tirpak 2010.
552 Paley 1952; Makhijani and Saleska 1999.
553 Newton 1989, 31.
554 International Energy Agency 2010a, 8–9.
555 EconPost 2011.
556 Wiser and Bollinger 2010, 2011.
557 Ibid.
558 RMI analysis based on ReEDS modeling. Well-sited wind turbines typically achieve capacity factors around 35–45% (the best can exceed 50%), but utilities refuse to take some output and rapid growth leaves many new turbines operating for only part of a year, so the 2008 global average was 26%. U.S. PVs average around 17% without or 30% with single-axis tracking (the latter in sunny sites). In an average weather year, the recent capacity factor of global renewables other than big hydro was about 40% and rising, but it was about 60% for the sum of geothermal, small hydro, and biomass and waste combustion. These, like cogeneration (at about 83%), run quite steadily. "Micropower"—cogeneration plus renewables minus big hydro—averages around 66%. U.S. nuclear plants have lately averaged around 90%. The global nuclear average reported by the IAEA, 79% in 2009, omits all 44 plants performing below 50%, 13 of which produced 0%.
559 RMI analysis based on ReEDS modeling.
560 *Popular Mechanics* 2010.

561 RMI analysis using the following sources: National Renewable Energy Laboratory 2010b and U.S. Department of Energy 2008b.
562 RMI analysis using the following sources: National Renewable Energy Laboratory 2011a, Navigant Consulting 2008, U.S. Census Bureau 2004, Mehos, Kabel, and Smithers 2009, and Paidipati et al. 2008.
563 Turner 1999, 687, adjusted for economic growth and for modern PVs, which are nearly twice as efficient as Turner assumed.
564 National Renewable Energy Laboratory and AWS True Wind 2010; U.S. Energy Information Administration 2011b. The NREL assessment counts wind potential on available land (net of land-use restrictions) in the Lower 48 states at an 80-meter hub height. Modern turbines typically use a 100-meter hub height, capturing even more windpower.
565 Solar recovery conservatively assumes existing commercial and residential roofspace for distributed PV applications and 1% of land area for centralized PV applications. Available roofspace takes into account estimates for shading, structural soundness, and orientation, and assumes 15–22% capacity factor.
566 The forward cost projections in blue are RMI analysis. The forward cost projections in green are from a range of experts including the National Renewable Energy Laboratory and Electric Power Research Institute and are detailed in Tidball et al. 2010. The 2010–11 windpower costs are updated from Wiser and Bolinger 2011.
567 RMI learning curves analysis.
568 Kanellos 2011; Mints 2010; Dinwoodie 2011.
569 Hidary 2011. The network is sambaenergy.com, with over 110 commercial-building solar installers.
570 Wingfield 2011.
571 Dinwoodie and Shugar 2011.
572 Martin, D. 2010. An iPod using transistors at the 1976 price would cost a billion dollars (and be the size of a building).
573 European Photovoltaic Industry Association 2011.
574 Kann 2011.
575 First Solar 2010.
576 Schwede et al. 2010.
577 Mints 2010.
578 Bony, Newman, and Doig 2010.
579 ZepSolar 2011.
580 Wiser and Bolinger 2010.
581 Terra Magnetica 2010.
582 Kota 2011. The calculated payback is two years, excluding the value of reduced blade and gearbox stresses.
583 Marsh 2009.
584 Moxley 2011.
585 Burgermeister 2008.
586 Bony, Newman, and Doig 2010.
587 North American Electric Reliability Corporation 2010c.
588 RMI analysis and stylized chart adapted from the following sources: Electric Reliability Council of Texas 2004, GE Energy 2010, and National Renewable Energy Laboratory 2011a.
589 RMI analysis.
590 RMI analysis using data from U.S. Energy Information Administration 2010h and 2011e.
591 U.S. Energy Information Administration 2010.
592 Swisher and Orans 1995.
593 RMI analysis.
594 Straub and Behr 2009.
595 Not solar in general, since concentrating solar thermal electric plants typically have many hours of thermal storage so they can run into evening peak periods. One has even run continuously for days.
596 Palmintier, Hansen, and Levine 2008.
597 In May 2011, GE launched a fast-ramp gas turbine that can go from a standstill to its 510 MW full power in just 10 minutes, twice as fast as its nearest competitor. It's designed specifically to complement variable renewables.
598 North American Electric Reliability Corporation 2010b.
599 Kim and Powell 2010.
600 ERCOT 2008.
601 Federal Energy Regulatory Commission 2011a.
602 Ibid.
603 U.S. Energy Information Administration 2010b. Pumped storage uses cheap offpeak power, typically at night, to pump water up to a high reservoir from which it runs back down through a turbine to generate about one-fourth less (but far more valuable) power at peak periods.
604 Succar and Williams 2008. As of June 2011, there is one compressed-air energy storage (CAES) facility in the United States, in McIntosh, Alabama, with 110 MW of storage. At least two other CAES plants are under development, including FirstEnergy's 2.7 GW facility in Norton, Ohio, and a 150 MW pilot project in New York that uses an existing depleted salt cavern (Business Wire 2011; Messenger Post 2010).
605 American Electric Power 2007.
606 RMI analysis and stylized chart adapted from the following sources: Electric Reliability Council of Texas 2004, GE Energy 2010, and National Renewable Energy Laboratory 2011a.

607 Skea et al. 2008.
608 EnerNex 2011; GE Energy 2010.
609 International Energy Agency 2011.
610 Gross 2006; Skea et al. 2008. Further findings are reviewed in Lovins et al. 2002, 172–200.
611 Molly 2010.
612 Danish Energy Agency 2009; Renewable Energy Policy Network for the 21st Century 2011.
613 Red Eléctrica de España 2010.
614 Rosenthal 2010c. During the same period, the U.S. rose from 9.2% to 10.5% renewable electricity.
615 Wiser and Bolinger 2010c.
616 RMI analysis based on ReEDS modeling.
617 Wald 2010.
618 Jha 2010; Global Transmission report 2010.
619 RMI analysis based on ReEDS modeling.
620 Lovins 2009; Hansen and Lovins 2010.
621 Gladwell 2011.
622 Mick 2011.
623 Gregor 2011.
624 Sunverge Energy 2011.
625 Tecogen 2011.
626 RMI analysis based on ReEDS modeling.
627 Ibid.
628 Ibid.
629 Morales 2011.
630 Rocky Mountain Institute analysis from RMI 2010.
631 Denholm and Margolis 2008. Adjusted from 13.5% to the 2011 market-leading 19.8% module efficiency.
632 RMI analysis based on ReEDS modeling. These costs are model outputs and show approximate composite values for our four cases. Direct costs of construction, finance, interconnection, and operation are included (plus growth-induced bottlenecks and constraints, administrative costs, and regional multipliers; for example, onshore windpower costs rise as less favorable sites are exploited). Grid costs and losses, spilled surplus renewable power, and reserves are excluded. All costs are levelized over 20 years at a 5.7%/y real weighted average cost of capital (WACC). Capital markets would actually differentiate technologies markedly in both expected economic life and financial risk. New U.S. nuclear build has been offered major financial subsidies not explicitly shown. For further details, such as learning curves (some shown in Fig. 5-14), please see the Knowledge Center at www.reinventingfire.com.
633 RMI analysis using data from North American Electric Reliability Corporation 2011.
634 Institute of Electrical and Electronics Engineers 2011.
635 Lovins et al. 2002.
636 ABB 2007.
637 Lovins et al. 2002.
638 Electric Power Research Institute 2007.
639 Bindewald 2004; Electricity Consumers Resource Council (ELCON) 2004.
640 Lidula and Rajapakse 2011.
641 Washom 2009; Reitenbach 2010.
642 U.S. Department of Energy 2011b.
643 Lund 2011; Danish Energy Agency 2010; van der Vleuten and Raven 2006. More than three-fifths of Danish homes pipe their heat from a central plant (district heating), and more than four-fifths of that heat is a previously wasted by-product of power generation. The Danish electricity-and-heat sector releases 42% less carbon per dollar of GDP than in the U.S.
644 Lund 2011.
645 Galvin Electricity Initiative 2010.
646 Barringer 2011.
647 Cho 2010.
648 Defense Science Board 2008; U.S. Government Accountability Office 2009.
649 Torres and Hightower 2010; Ka'iliwai and Jost 2009.
650 Torres and Hightower 2010; Stockton 2011.
651 RMI analysis based on ReEDS modeling.
652 Ibid; Renewable Energy Policy Network for the 21st Century 2011.
653 Pew Charitable Trusts 2011.
654 The last bits of oil used to make U.S. electricity are typically in isolated diesel-powered systems in places like Hawai'i. That state's utilities can now profit from efficiency and renewables, which are now being rapidly installed to relieve that dependence. In general, any remaining engine-generator needs could be met at reasonable cost by substituting renewables, biofuels, or both. By 2050, this issue should be long gone. On the contrary, electrified autos will be an important way of *saving oil*, as described in chapter 2.
655 An impressive example from a very different setting is Lovins 2010d. I have not been to Cuba but have met with the main architect of this remarkable story.
656 RMI analysis and stylized chart adapted from the following sources: Electric Reliability Council of Texas 2004, GE Energy 2010, and National Renewable Energy Laboratory 2011a.
657 Renewables International 2011. Some reports indicate a similar cost gap for 2010 windpower.
658 Cooper 2010.

659 Maslin 2009.
660 Wiser and Bolinger 2010.
661 NextEra Energy Resources 2010.
662 Cogenra 2011.
663 Wesoff 2011.
664 SolarCity 2011; SunRun 2010; Sungevity 2011; SunPower 2011.
665 This worked in 2001–2002 at the Santa Rita Jail in Alameda County, California. The facility first cut demand 0.7 MW with efficiency, saving 1 GWh/y, then added three acres (1.18 MW) of rooftop PVs producing 1.5 GWh/y. Operating the jail to shave peak load freed up solar power to sell back to PG&E on hot afternoons at a premium price. This $9 million package, shaved to $4 million by state subsidies, was expected to earn a 10.3% IRR by saving $15 million over 25 years at 2002 electricity prices, which later soared. PV and efficiency prices have since plummeted, so today the case would be even stronger.
666 Austin Energy had previously demonstrated that "green power" sold at a premium when gas prices were low but undercut "brown power" when gas prices were high. More broadly, a least-cost electricity portfolio should include a large renewable component for the same reason that a financial portfolio should include investments like Treasury debt: the riskless part of the portfolio improves the overall risk/reward performance of the whole (Lovins et al. 2002).
667 EnerNOC Inc. 2011.
668 PJM News 2011.
669 Network for New Energy Choices 2010.
670 Bony, Newman, and Doig 2010.
671 Pew Center on Global Climate Change 2011b.
672 A potential alternative approach is to treat the money that utilities invest to help reduce their customers' electricity consumption like a capital investment that can be included in the rate base to boost revenues. That can be a win-win for utilities and consumers, since cutting electricity use enough to eliminate the need for a new power plant is usually cheaper. However, rate-basing efficiency investments rewards the utility not for the desired outcome—saving electricity and money—but for spending money. It also doesn't decouple revenues from sales and may not make the utility whole due to regulatory lag.
673 *The Economist* 2008.
674 American Council for an Energy-Efficient Economy 2011.
675 American Gas Association 2010, NRDC 2011.
676 Patel 2006.
677 Frantzis 2010, 15.
678 U.S. Department of Energy 2011c.
679 Federal Energy Regulatory Commission 2011b; St. John 2011.
680 At PG&E and Ontario Hydro, all led by John C. Fox (RMI's former chairman).

CHAPTER 6. MANY CHOICES, ONE FUTURE

681 Butti and Perlin (1980) chronicle how, from northern China's earth-sheltered homes to the Anasazi pueblos of the American Southwest, from classical Greek and Roman buildings to Persian draft towers and Bedouin goat-hair tents, passively comfortable structures were mastered over thousands of years. New England's 18th-century industry and Aspen's silver mines throve on hydropower and fuelwood. Most water-heating in Florida was solar before natural gas. Henry Ford built biofueled cars made of soybeans. Photovoltaics were invented and fuel cells demonstrated in 1839. My grandmother drove one of the electric cars that outnumbered gasoline cars in early 20th-century America. Fossil fuels until lately had only repeatedly interrupted and derailed the development and use of quite sophisticated renewables.
682 Diamond 1999 explains how local circumstances led some civilizations to flourish and others to fade even though all began with people of similar abilities. In hands like his, a "virtual history" of how humanity might have evolved without fossil fuels could make a great book.
683 Friedman 2011.
684 For example, the 4,300 Danish inhabitants of Samsø in 1997–2007 cut their island's carbon emissions 140%—their island became a net exporter of renewable energy—often through local and cooperative investment (Edin Energy 2011); the 80,000 natives of Kristianstad, Sweden (home of Absolut vodka), in 1991–2010 switched all their heating from fossil fuels to local surplus biomass and wastes (Rosenthal 2010b); Växsjö and several other Swedish municipalities aim to eliminate fossil fuel use; the thousand-year-old German village of Jühnde switched to a biogas-CHP plant (Harsch 2010; Babits 2010); other German villages and American tinkerers

are competing to get off fossil fuels fastest (Deutsche Welle 2008).

685 RMI analysis.

686 Gas infrastructure does share some pipeline and control-system vulnerabilities with oil and has special system-restoration issues if interrupted supply drops retail distribution pressure enough to extinguish pilot lights. See Lovins and Lovins 1982, chapter 9.

687 The average U.S. gas well in 2009 was 76% less productive than at its 1971 peak. The U.S. had nearly twice as many producing wells in 2010 as in 1990, for just 21% higher output. However, learning is rapid: service companies halved their drilling time per well during 2009–2011.

688 Newell 2011.

689 Or oil in the big new Eagle Ford play in Texas (Krauss 2011). Many fracked shale-gas wells coproduce not only gas and natural-gas liquids but also a significant amount of light crude oil that substantially improves the economics (Emerson 2011).

690 For skeptical views, see Berman 2009, Urbina 2011a, 2011b. The industry disagrees.

691 Massachusetts Institute of Technology 2010.

692 Methane, the main component of natural gas, is continually produced by living organisms feeding on biomass. The transportation sector becomes fossil-fuel-free by 2050 using only half of the 12-quad U.S. supply of nonedible biofeedstocks—mainly farming and forestry residues, energy crops not displacing food crops, and municipal waste (see "Pumping Biofuels" in chapter 2). The other half could be fed to methane-emitting microbes in industrial-scale digesters, or it could be converted thermochemically and then upgraded to pipeline quality and put into existing natural-gas infrastructure to supplement and diversify the U.S. natural gas supply. Such "biomethane" would increase competition among natural gas sources and hedge against environmental and regulatory uncertainties.

693 About 40% of world natural gas is still sold at administered prices and 35% at contractual prices indexed to oil (Emerson 2011). Shale gas's most important contribution to world gas markets could be its tendency to move those artificial price regimes toward market-based prices, especially in Europe and Japan.

694 Massachusetts Institute of Technology 2010 estimated that the long-term wholesale price of U.S. natural gas could range from roughly $6 to $8 (per million BTU, thousand cubic feet, or gigajoules, all roughly equivalent) depending on how hard all this turns out to be. Gas converted from biomass would cost roughly $7–$12 depending on the size of the conversion facility and type of feedstock (according to an RMI analysis using data from Chen et al. 2010 and Electrigaz 2010). Over half the conventional onshore gas reserves come in below $4, and perhaps 80% of the shale and coalbed methane below $6; some of the cheaper shale gas plays can even beat the costlier sorts of conventional gas, though most shale gas would sell above 2011's low gas prices.

695 For example, Jacobson and Delucchi 2011 and Delucchi and Jacobson 2011. The European Climate Foundation's Roadmap 2050 concluded in 2010 that "by developing technologies already commercial today or in the late development stage, Europe could reduce greenhouse gases emissions by 80% by 2050 compared to 1990 and still provide the same level of reliability as the existing system" at "comparable" electricity cost (European Climate Foundation 2010). The European Renewable Energy Council in 2010 found that 96% renewable European energy was feasible by 2050 with a $4t investment (European Renewable Energy Council 2010). A July 2010 study by the German Federal Environmental Agency (UBA), a linked consortium of research teams, showed that a 100%-renewable national energy system was feasible using already available technologies (ForschungsVerbund Erneuerbare Energien 2010). Similar futures have been outlined for New Zealand and the United States (Sovacool and Watts 2009). A 2010 UK report mapped the potential to meet national electricity needs up to six times over from offshore resources, making Britain a net energy exporter with a six-year payback (Offshore Valuation Group 2010). A noted British consulting engineering firm described 80% UK greenhouse-gas reductions by 2050 (Parsons Brinckerhoff 2009). Sweden adopted (until a change of government) a 2006 commission's strategy for eliminating oil use by 2020 (Swedish Government Commission on Oil Independence 2006). And the International Energy Agency's *World Energy Outlook 2010* also found a 450-ppm-CO_2 global scenario viable. Of course, official studies have confirmed similar or better potentials for the United States, like the five-national-labs studies in 1990 (Solar Energy Research Institute 1990) and in 1997 (Brown and Levine, eds., 1997).

696 RMI analysis.

697 The International Energy Agency's 2010 New Policies Scenario implies 2010–2035 global investments in energy supply equivalent to about one-sixth of global net fixed capital formation. But meanwhile, global trends look set to include increased investment needs in other sectors, decreasing savings, and hence higher interest rates, so business-as-usual energy policies would encounter and exacerbate a global capital crunch (McKinsey Global Institute 2010). This could be more acute in countries with rapid growth, like China, or old equipment, like the U.S.

698 Reasons include cleaner outdoor and indoor air, less-polluted water and land, more livable and walkable or bikeable communities, and less car-centric and less stressful living.

699 Over time, the fuel efficiencies summarized in chapter 2 could let the U.S. Department of Defense realign multiple divisions away from fuel logistics and protection, leaning out the force structure to save tens of billions of dollars per year while making warfighting more effective. *Winning the Oil Endgame* (Lovins et al. 2004) offers an illustrative estimate on p. 267, but current savings estimates are both larger and more empirical.

700 Ehrhardt-Martinez, Donnelly, and Laitner 2010. Lawrence Berkeley National Laboratory estimates there were 380,000 people directly engaged in the energy efficiency services sector in 2008 (excluding product manufacturers and distributors) and that this could double or even quadruple by 2020. See also Goldman et al. 2010.

701 Goldman et al. 2010.

702 In 2010 the U.S. had about 5,000 solar installation firms averaging eight employees each; to this must be added manufacturing, monitoring, et cetera, and the same for windpower. The solar and wind trade associations (the Solar Energy Industries Association and American Wind Energy Association) cite respectively 93,000 and 85,000 jobs, but some of those are part-time. The 100,000 total full-time-equivalent estimate is from Hidary 2011.

703 SourceWatch 2011.

704 Bode 2010.

705 Danish Wind Energy Association 2011. Thanks substantially to its booming renewable-energy equipment exports, for example, Germany now has fuller employment than before the so-called Great Recession, while wind turbines were 8.5% of Denmark's total 2010 exports.

706 National Mining Association 2010.

707 Kinsley 1997.

708 RMI analysis.

709 China's 10th Five-Year Plan set energy efficiency via leapfrog technologies as its top priority for national development, not because a treaty required it, but because China's leaders wisely realized China couldn't otherwise afford to develop. Conversely, the most effective climate-protection treaty yet has been the Montréal Protocol, whose primary aim, largely successful, was to protect stratospheric ozone.

710 Lynd 2006.

711 Tweaking the feedstuff system to feed people efficiently could even replace 100% of U.S. gasoline (Dale et al. 2009). This doesn't count other worthwhile reforms like switching from feedlot to range beef or partially from beef to poultry.

712 Brilliant work by Dr. Willie Smits of the Masarang Foundation has been demonstrated in six Indonesian communities, spurring demand from hundreds more. One of the more than 1,400 trees he uses to restore devastated tropical rainforests is the sugar palm, *Arenga pinnata*. It provides more than 60 forms of value—health-promoting sugar, the strongest known wood and fiber, medicinals, fuels, food, and more. Its sap outyields sugarcane 6–12:1 and it is extremely hardy and low-input. It provides 20–50 times the jobs of palm-oil plantations and needs diverse forest (not monoculture) to thrive, so it's uninteresting to industrial monoculturists but perfect for the economic and ecological needs of much of Southeast Asia. Similarly integrated ecological and economic restoration has been demonstrated in such places as Las Gaviotas in eastern Colombia (Weisman 1998) and the Loess Plateau of China (Liu 2011).

713 Smits has devised and demonstrated—and is now miniaturizing into air-droppable containers—a village-scale processing plant to produce more than two dozen valuable outputs from the sugar palm tree alone, including fuel ethanol (for local use and export) competitive with Brazilian sugarcane ethanol and, if adapted across much of the world's tropical region, potentially able to displace the world's entire oil production.

714 Danish Ministry of Climate and Energy 2009.

715 Electricity prices would be the most sensitive to potential carbon pricing, yet in 2009, electricity bills averaged only 1.04% of U.S. industries' value of shipments, ranging from 0.5% for the oil, coal, and transportation equipment subsectors to nearly 3% for primary metals (Edison Electric Institute 2009,

116). Even a considerable change in such a small factor cost would hardly induce a factory to move abroad, any more than European and Japanese firms moved to the U.S. to cut their high electricity prices; instead, they saved more electricity, stayed home, and prospered. Often the same people who claim higher electricity prices would force their factories to move overseas are the same ones who say in the next breath that the reason they've saved so little electricity is that it's such a small factor cost.

716 Except for a few even cheaper-energy countries like Saudi Arabia and Venezuela to which U.S. industries don't generally seem eager to flee.

717 Shell International 2011.

718 Thus DuPont found in the 1990s that its European chemical plants, despite having long paid twice the energy prices of its U.S. plants, were no more efficient, because they were all designed and built similarly. People in Seattle during 1990–1996 paid half as much for a kilowatt-hour as people in Chicago. But they cut electricity use 3,640 times as fast and peak load 12 times as fast as in Chicago (Lovins and Lovins 1997), mainly because Seattle City Light helped them save while Commonwealth Edison encouraged greater use (to protect its revenues and profits under the usual perverse regulatory incentive). Thus people and firms can save energy faster if they have extensive ability to respond to a weak price signal than if they have little ability to respond to a strong one. This means that comprehensive barrier-busting can sustain brisk energy savings even if energy prices stay low—and thus can help keep them low.

719 As Lord Puttnam told the House of Lords' 2007 climate debate (Hansard 2007), abolishing slavery removed "a colossal impediment . . . [that had hindered] the development of more efficient business formations, leading to the generation of many new forms of wealth and success" as the British economy, freed from this "metaphorical ball and chain," leapt forward.

720 Lovins 2006; Brown and Sovacool 2010. What energy security is and how to get it are more complex than meets the eye.

721 America today has normal relations with countries on which it once critically depended, e.g., for naval masts in the 1820s (Norway), naval coal a century ago (various Pacific island nations), and natural rubber in the 1920s–1940s (Southeast Asian countries).

722 Lovins 2010e; Lovins, Lovins, and Ross 1980.

723 Chapter 2 showed how we can obviate oil's use for transportation through efficient vehicles, their more productive use, and a flexible mix of electricity, hydrogen (typically from natural gas with today's technologies but perhaps renewable with tomorrow's), advanced biofuels, and optional natural gas. This will cost trillions of net dollars less than business-as-usual oil burning. For buildings, chapter 3 mentioned that the 0.04 Mb/d of oil use projected for 2050 (96% less than in 2010) could readily be eliminated: all the oil furnaces will have expired by then. They can be replaced with any combination of efficiency, passive and active solar heat, wood (generally pelletized), wastes, or electricity—the same techniques Denmark used to cut its oil-fired home heating sixfold in the past 30 years (chapter 3, note 324). (Tiny amounts of LPG coproduced with natural gas—otherwise destined for feedstock use—could be used in especially difficult cases but should not be necessary.) And in industry, chapter 4 showed that efficiency, natural gas, and some solar process heat offer ample replacements for oil in process heat. (Oil is already less used for industrial process heat than for industrial feedstocks because as a fuel it's so uncompetitive with natural gas.) All these substitutions have a strong economic case even if all the hidden costs of buying and burning oil (chapter 1) were worth nothing.

724 Deutsche Bank 2009.

725 Oil has some great upside potential but, like airlines and automaking, some discouraging fundamentals. It is extremely capital-intensive, with decadal lead times; geologically and technologically risky (as illustrated by the 2010 Macondo blowout); owning only about 6% of its resource base (governments own the rest and can confiscate or tax away what's left); often unpopular; politically fraught and interfered with; utterly at the mercy of vehicle technology revolutions it can't control; and a price-taker in volatile commodity markets whose liquidity is largely and increasingly focused in the world's most political volatile region, currently undergoing multiple revolutions. What a recipe for headaches! These fundamentals are why many industry strategists have been seeking graceful paths beyond oil since the 1970s, as illustrated by Royal Dutch/Shell Group's often prescient scenario-planning efforts. Those of 2001 and 2011 (Shell International 2011) are of particular value.

726 All hydrocarbons' value should rise as hydrogen end uses are widely deployed and hydrogen's market premium emerges. The interplay of fuel-cell

applications between buildings, factories, and vehicles will fundamentally change how the value of mobility, power, and heat is ultimately delivered and monetized. The hydrogen transition is thus an important piece of the oil puzzle. It can widen the portfolio of oil displacements in transportation, better use oil companies' fuel and plant assets and skills, offer complete primary-energy flexibility and climate-safe mobility, increase autos' versatility as extended-duration power plants on wheels, and thus speed and deepen adoption of renewable electricity sources, improve electrical resilience, and help preserve vehicle industries on which the oil industry's health depends. These shifts can also sustain or enhance the value of gasfields (full or empty), pipelines, storage facilities, and other midstream infrastructure; possibly of some refineries' methane steam reformers; of higher-quality and lower-cost retail stores and brands; and of oil companies' technical and organizational skills. Just as refineries have become integrated centers that convert both energy and molecules across the oil, gas, electricity, and chemical value chains, so oil companies must start re-visioning their whole range of assets through the lens of the emerging hydrogen, biofuel, green-chemicals, and electrified-mobility value chains.

727 EU countries are striving to meet 20% of their primary energy from renewables by 2020 (and raise efficiency 20%), saving about €50 billion in annual oil and gas costs (http://ec.europa.eu/clima/policies/brief/eu/package_en.htm); with better-coordinated investments, they may beat that target (Reuters 2011a). During 2000–2010, Europe added 111 GW of renewable generation (74 GW wind, 26 GW PVs, 3 GW each big hydro and biofuel, et cetera) and 118 GW of gas-fired generation while decreasing capacity by 8 GW for nuclear, 10 for coal, and 13 for oil. Nor is leadership confined to the EU: by early 2010, more than 100 countries had a renewable-energy policy target or promotional policy, versus 55 countries in early 2005 (Renewable Energy Policy Network for the 21st Century 2010).

728 Pew Charitable Trusts 2011. However, India also plans to double its 2008 coal-fired generation by 2017.

729 Climate Group 2011.

730 Zhou et al. 2011.

731 Xinhua 2011.

732 Bloomberg New Energy Finance 2011 estimates that continued price drops in China could bring installed system costs for utility-scale crystalline-silicon PV to $1.12/W in 2015, down from $2.60 in 2010, with balance-of-system costs 40% below the world level.

733 Renewable Energy Policy Network for the 21st Century 2010; Magee 2011.

734 Another 195 GW of renewables is to be added to today's 1,600 GW capacity by 2015 and 520–630 GW during 2010–2020. China's spectacular renewables progress reflects formidable industrial capabilities backed by consistent policies, including mandatory purchase of all renewable power produced. (Grid companies late in hooking it up must pay for it anyway, plus a 100% penalty. Enforcement of that late-2010 law is reportedly beginning in 2011.) China beat its 2010 windpower target in 2007 and its original 2020 windpower target in 2010, when it installed 19 (CWEA 2011) of the world's 40 GW of new windpower. The Chinese wind industry will relish bringing installed windpower capacity to 240+ GW by 2020—60% above the official target, or 13 times the capacity of the troubled (Wines 2011) Three Gorges dam—and to generate 465 TWh/y, displacing 200 coal-fired plants (Global Times 2011). China's practical and cost-effective windpower potential has been conservatively measured at twice its total electricity use today (McElroy et al. 2009). China also plans interregional ultra-high-voltage (UHV) and other transmission projects ($77 billion worth) with supporting smart-grid intelligence. China may now lead the world in UHV and storage technologies, and it plans grid investment for 2011–2015 totaling $391 billion, explicitly to help enable large-scale renewables.

735 The ambitious but unofficial pre-Fukushima nuclear target of 70–80 GW in 2020 has probably slipped back to its previous level of 40 GW, if that, as overheated nuclear ambitions cool and suspended approvals and preconstruction work give way to sober assessments of increased precautions and their costs, and as efficiency, renewables, and the nascent natural-gas industry continue to advance.

736 Of the 12th Five-Year Plan's seven strategic emerging industries to receive major investment ($1.5 trillion is being discussed), three are entirely and two partly in energy.

737 Renewable Energy Policy Network for the 21st Century 2010, 52.

738 AAAS 2011; U.S. Department of Energy 2011d. Amidst political maneuvering over the FY2012 budget, DOE's applied energy R&D budget was cut by about 22% with only half the year left.

739 Bradsher 2010.

740 The world has nearly a billion cars, about one for every seven people, and is only half as good at birth control for cars as for people. More want a car. But even if there were enough fuel, air, climate, and money for all those cars, most countries have no space for them, especially now that the majority of the world's people live in cities. Instead of running out of air, oil, and climate, we'd run out of land, roads, and patience.

741 Goodman 2010. In 2009, Chinese people bought 9.4 million cars and 21 million electric bikes costing around $300. Up to two-thirds of them are in Beijing, choked by 4.7 million cars plus 2,000 more each day. China has nearly tripled its cars in three years, and by 2015, Beijing rush-hour traffic is projected to slow from 15 to 9 mph, making electric bikes about as fast, especially in wide bike lanes. (Many people remove their 12 mph speed limiter, and 30 mph, 50-mile-range electric scooters are popular.) Hydrogen-fuel-cell bikes are also emerging. Human-electric hybrid bikes are selling hundreds of thousands a year in the U.S. and 300,000 a year in Japan. Electric vehicle analyst Frank Jamerson, when at GM, suggested GM give away an electric bike with every new car to get people used to a vehicle they recharge at night. Not a bad idea.

742 China has nearly three times the energy intensity and (as of 2006) 11 times the carbon intensity of Japan, which in turn can potentially double or triple its own energy productivity and cut its carbon emissions 70% from the 1990 level (2050 Japan Low-Carbon Society Scenario Team 2007, 2009; Matsuoka 2006) while providing projected 2050 energy services with 1–2%/y GDP growth per capita, and without using integrative design. The estimated direct marginal cost is about 0.1% of 2050 GDP. The technical assumptions appear conservative compared to those in chapters 2 through 5 of this book. This comparison assumes long-run convergence of the general composition of economic output.

743 China's longstanding shortages of arable land and water are colliding with urbanization (Beijing now drills for water five times deeper than a decade ago), affluence, an increasing taste for meat, paving farmland for cars, and severe northwest desertification (akin to the U.S. Dust Bowl era) to set the stage for major dependence on U.S. grain production. Already, China has outsourced four-fifths of its soybean production—a crop it originated—thereby restructuring agriculture across the Americas. Some knowledgeable agronomists fear that China will inevitably need to import more grain than America can sell without seriously raising its historically low food prices—and that such a need cannot be refused now that China finances U.S. deficits and oil imports and holds nearly $1 trillion of U.S. Treasury debt. (A controversial but sobering assessment of the emerging U.S./Chinese codependency in food and its embodied water, as in finance, is Brown 2011.) Climate change could make this codependency far more fragile and dangerous as Himalayan glaciers dwindle at the headwaters of the great Asian rivers, and drought stalks many nations' breadbaskets. Even in spring 2011, food prices averaging 136% above those of the 1990s are causing great hardship and noticeable political instabilities, including in the Middle East.

744 By 2006, a binge on energy-intensive basic materials industries, chiefly large state-owned enterprises with seriously distorted factor costs, had created a supersized and inefficient industrial sector using 54% of national energy, up from 39% just five years before. China used three-fifths more energy for iron and steel than for households, more for chemicals than for transportation, and more for aluminum than for the commercial sector (Rosen and Houser 2006). The 12th Five-Year Plan shifts back toward consumption-led investment. But then more, bigger, better-equipped households will also make electric loads peakier—urban air-conditioner ownership just quintupled in a decade—so Beijing is promoting high-efficiency air conditioners and a smart grid and trying to catch up on efficient building standards and practices. (One-fourth of China's total fixed investment is slated for real estate, most still inefficient.) Urbanization, lagging rural development, and subsidized residential electricity remain major challenges. Policy remains more vigorous and consistently applied for new supply than for better efficiency. China's renewable power revolution will come none too soon for her domestic needs as 400 million urban and 900 million rural customers start to live better.

745 Engleman 2011: "If all births resulted from women actively intending to conceive, fertility would immediately fall slightly below the replacement level; world population would peak within a few decades and subsequently decline."

746 Clarke and Wallsten 2002.

747 For example, around 2005 a standard Brazilian electric showerhead (*chuveiro elétrico*) costing R$20 used on-peak hydro and grid capacity costing about R$1,800–3,000—two orders of magnitude more. Replacing chuveiros with solar, butane, or natural-gas water heaters (plus a permanent fuse to prevent reinstallation) would free up at least 8 GW of national generating capacity, equivalent to several years' demand growth, at comparatively trivial cost.

748 Gadgil et al. 1991, 4–5. Such a shift can therefore cut by roughly 10,000-fold the capital needs of the most capital-intensive sector, electricity, which traditionally consumes about one-fourth of global development capital. The first saved trillion dollars or so could come from the Super-Efficient Equipment and Appliance Deployment (SEAD) initiative approved at the 2011 Clean Energy Ministerial in Abu Dhabi. Its full execution could save by 2030 1,800 TWh/y—the output of 300 GW of coal plants—plus 20 quadrillion BTU/y of primary energy and about $150 billion a year. SEAD targets four kinds of appliances (lighting, refrigeration, air-conditioning, and TVs) that use about 60% of residential electricity but most of which haven't yet been manufactured or bought. The program aims at the U.S., EU, China, and India (which together account for nearly three-fourths of those appliances' global electricity use) and the 15 manufacturers that control three-fourths of their market. For example, four firms make 60% of the world's TVs, and 15 make 75% of white goods (Phadke 2011).

749 One of the commonest issues is cooking. Much good work has been done on more-efficient and solar cookers, but virtually none on how to apply up to five established technologies for making *pots* about two- to sixfold more efficient, without the cultural problems that may accompany stove innovations. See chapter 3, note 228, and Kelly Kettle 2011, plus selective surfaces that better retain or radiate heat.

750 Fuel-based (chiefly kerosene) lighting is a source not only of substantial carbon emissions but also of major health and economic burdens ($38 billion/y) (Mills 2005).

751 Reddy, Williams and Johansson 1997, 70.

752 Bell Labs invented crystalline silicon PVs in 1954, and the U.S. made 45% of world production in 1995, but by 2010 that was 6% and China's share was 60%: the U.S. captured less than 6% of 1995–2008 production growth. Similarly, U.S. inventors developed compact fluorescent lamps (1976), LEDs (1962), and three-way catalytic converters (1973), but China now dominates those markets. As the sun rises in the East, must it set in the West? Regaining leadership will take different signals than of late: U.S. clean-energy investment slid from #1 in 2008 to #2 in 2009 to #3 in 2010, thanks to Washington's erratic policy environment (Pew Charitable Trusts 2011; Renewable Energy Policy Network for the 21st Century 2010).

753 RMI analysis.

754 Makower 2011, Comments.

755 Kahane 2010, 120.

756 Rudolf 2011.

757 Murdoch 2009.

758 Kaufman 2010.

759 Usher 2010.

REFERENCES

2050 Japan Low-Carbon Society Scenario Team. 2007. *Japan Scenarios towards Low-Carbon Society (LCS): Feasibility Study for 70% CO_2 Emission Reduction by 2050 Below 1990 Level*. National Institute for Environmental Studies (NIES) at Kyoto University, February. 2050.nies.go.jp/press/070215/file/20070215_report_e.pdf.

———. 2009. "Japan Roadmaps towards Low-Carbon Societies (LCSs)." http://2050.nies.go.jp/report/file/lcs_japan/20090814_japanroadmap_e.pdf.

AAAS. 2011. *AAAS Report XXXVI: Research and Development FY 2012*. AAAS. http://www.aaas.org/spp/rd/rdreport2012/tablelist.shtml.

ABB. 2007. *Energy Efficiency in the Power Grid*. ABB. http://tinyurl.com/yd2deqm.

Adey, Walter. 2008–2009. Personal communications with author.

Akbari, Hashem, and A. Rosenfeld. 2008. *White Roofs Cool the World, Directly Offset CO_2 and Delay Global Warming*. Lawrence Berkeley National Laboratory, Heat Island Group, November 10. http://coolcolors.lbl.gov/assets/docs/fact-sheets/Global-cooling-2pp.pdf.

Allwood, J. M. 2009. "Steel, Aluminium and Carbon Targets: Alternative Strategies for Meeting the 2050 Carbon Emission Targets." In *R'09 Twin World Congress: Resource Management and Technology for Material and Energy Efficiency, 14–16 September 2009, Davos, Switzerland*. Davos, Switzerland: R'09 World Congress. http://publications.eng.cam.ac.uk/17014/.

Allwood, Julian M., Michael F. Ashby, Timothy G. Gutowski, and Ernst Worrell. 2011. "Material Efficiency: A White Paper." *Resources, Conservation and Recycling* 55 (3): 362–381. doi:10.1016/j.resconrec.2010.11.002.

Alvarez, Robert, Jan Beyea, Klaus Janberg, Jungmin Kang, Ed Lyman, Allison Macfarlane, Gordon Thompson, and Frank von Hippel. 2003. "Reducing the Hazards from Stored Spent Power-Reactor Fuel in the United States." *Science & Global Security* 11 (1): 1. http://www.informaworld.com/10.1080/08929880309006.

American Chemistry Council. 2010a. *Industry Fact Sheet*. American Chemistry Council. http://209.190.243.167/chemistry-industry-facts.

———. 2010b. *Guide to the Business of Chemistry. American Chemistry Council*. http://store.americanchemistry.com/detail.aspx?ID=261.

American Chemistry Council staff. 2011. Personal communication with author, March 4.

American Council for an Energy-Efficient Economy. 2011. "Incentivizing Utility-Led Efficiency Programs: Lost Margin Recovery." http://www.aceee.org/sector/state-policy/toolkit/utility-programs/lost-margin-recovery.

American Electric Power. 2007. Press release. Newsroom, September 11. http://www.aep.com/newsroom/newsreleases/?id=1397.

———. 2011. "AEP to Receive Funds From Global CCS Institute for Commercial Scale Carbon Dioxide Capture and Storage Project." Newsroom, February 16. http://www.aep.com/newsroom/newsreleases/?id=1673.

American Gas Association. 2010. "Revenue Decoupling Resources." http://www.aga.org/our-issues/RatesRegulatoryIssues/ratesregpolicy/Issues/Decoupling/Pages/default.aspx.

American Institute of Chemical Engineers. 2010. "10xE Challenge." http://www.aiche.org/Energy/GetInvolved/10xEChallenge.aspx.

American Physical Society and the Materials Research Society. 2011. *Energy Critical Elements: Securing Materials for Emerging Technologies*. College Park, MD: American Physical Society and the Materials Research Society.

Anderson, Ray C. 1999. *Mid-Course Correction: Toward a Sustainable Enterprise: The Interface Model*. Peregrinzilla Press.

——— 2009. *Confessions of a Radical Industrialist*. New York: St. Martin's Press.

Anderson, S. 2011. *Why We Need DOD Policy to Require Energy Efficient Expeditionary Structures*. Brief to Army Science Board, April 20.

Apogee Technology, Inc. 2010. "Isothermal Melting." http://www.apogeetechinc.com/products/advanced-heating-2/isothermal-melting/.

Apte, J., D. Arasteh, and J. Huang. 2003. "Future Advanced Windows for Zero-Energy Homes." ASHRAE Transactions 109 (2).

Architecture 2030. 2011. "Architecture 2030: Adopters." http://architecture2030.org/2030_challenge/adopters.

Arezki, Rabah, and Markus Bruckner. 2011. "Oil Rents, Corruption, and State Stability: Evidence from Panel Data Regressions." School of Economics Working Papers 2011-07. Adelaide, Australia: University of Adelaide School of Economics (January). http://ideas.repec.org/p/adl/wpaper/2011-07.html.

Aspen Aerogels. 2011. Aspen Aerogels. www.aerogel.com.

Aspen Publishers. 2004. "An Illinois 'PassivHaus.'" *Energy Design Update* 24 (5). http://www.passivehouse.us/passiveHouse/Articles_files/EDU%20May2004%20Postable.pdf.

Aster, Nick. 2010. "Deutsche Bank: The Payoff in Building Green." Triple Pundit, April 20. http://www.triplepundit.com/2010/04/deutsche-bank-green-building-leed/.

Attari, Shahzeen Z., Michael L. DeKay, Cliff I. Davidson, and Wändi Bruine de Bruin. 2010. "Public Perceptions of Energy Consumption and Savings." *Proceedings of the National Academy of Sciences* 107 (37): 16054–59. doi:10.1073/pnas.1001509107. http://www.pnas.org/content/107/37/16054.abstract.

Ayres, R. U., and E. H. Ayres. 2009. *Crossing the Energy Divide: Moving from Fossil Fuel Dependence to a Clean-Energy Future*. 1st ed. Upper Saddle River, NJ: Pearson Prentice Hall.

Ayres, R. U., L. W. Ayres, and Benjamin Warr. 2003. "Exergy, Power and Work in the U.S. Economy, 1990–1998. q." *Energy* 28 (3): 219–273. doi:10.1016/S0360-5442(02)00089-0.

Babits, Sadie. 2010. "Germany's Green Lead." *Boise Weekly*, March 3. http://www.boiseweekly.com/boise/umweltschutz/Content?oid=1507039.

Bailey, Owen, and Ernst Worrell. 2005. *Clean Energy Technologies: A Preliminary Inventory of the Potential for Electricity Generation*. Lawrence Berkeley National Laboratory, April. http://www.recycled-energy.com/_documents/news/LBNL_clean_energy.pdf.

Bakewell, Ted (Bakewell Corporation). 2011. Personal communication with author, February 9.

Ball, Christopher P. 2007. "Rethinking Hub versus Point-to-Point Competition: A Simple Circular Airline Model." *Journal of Business & Economic Studies* 13 (1).

Barrett, James. 2010. "Horse, Meet Water: Getting Efficiency to Market." *Great Energy Challenge* (blog for *National Geographic*), November 8. http://www.greatenergychallengeblog.com/blog/2010/11/08/horse-meet-water/.

Barringer, Felicity. 2011. "Smart Meters Draw Fire from Left and Right in California." *New York Times*, January 30. http://www.nytimes.com/2011/01/31/science/earth/31meters.html.

Belko, John, and A. O. Smith. 2008. *Novel Fan Design Offers Energy Savings to Refrigeration Market*. International Appliance Manufacturing. http://www.appliancedesign.com/AM/Home/Files/PDFs/30_AO%20Smith.pdf.

Belkraft. 2005. "Vacumatic Waterless Cookware." http://www.belkraft.com/Waterless_cookware.html.

Bell, Geoffrey. 2008. *Aerosol Ductwork Sealing in Laboratory Facilities*. U.S. Environmental Protection Agency, July 29. http://www.epa.gov/lab21gov/pdf/bulletin_lab_duct_seal_508.pdf.

Bendewald, Mike, and Lindsay Franta. 2010. *Autodesk AEC Headquarters and Integrated Project Delivery: Factor Ten Engineering Case Study*. Snowmass, CO: Rocky Mountain Institute. http://www.rmi.org/rmi/Library%2F2010-16_AutodeskCaseStudy.

Benyus, Janine M. 2002. *Biomimicry: Innovation Inspired by Nature*. New York: Harper Perennial.

Berman, Arthur. 2009. "Facts Are Stubborn Things." Posted on the website of the Association for the Study of Peak Oil & Gas, November 5. http://www.aspousa.org/index.php/2009/11/facts-are-stubborn-things-arthur-e-berman-november-2009/.

Bernstein, Mark A., Robert J. Lempert, David S. Loughran, and David S. Ortiz. 2000. *The Public Benefit of California's Investments in Energy Efficiency*. RAND Monograph

Report 1212.0. Santa Monica, CA: RAND Corporation. http://www.rand.org/pubs/monograph_reports/MR1212z0.html.

BestLog. 2008. *IKEA: Increased Transport Efficiency by Product and Packaging Redesign*. European Commission. http://green4pl.com/blog/bestLog_best_practice_Ikea_transport_efficiency_redesign.pdf.

BetterBricks. 2006. *Banner Bank Building*. BetterBricks. http://www.betterbricks.com/sites/default/files/casestudies/pdf/betterbricks-case-study-banner-bank.pdf.

———. 2011. *PeaceHealth Commits to Energy Efficiency through SEMP*. BetterBricks. http://www.betterbricks.com/healthcare/peacehealth-commits-energy-efficiency-through-semp.

Beyea, Jan, Ed Lyman, and F. von Hippel. 2004. "Damages from a Major Release of ^{137}Cs into the Atmosphere of the United States." *Science & Global Security* 12: 125–36.

Beyer, Dale. 2011. Personal communications with author, June 27–28.

Biddle, Dr. Thomas B. (MBA Polymers). 2011. Personal communications with author, July 3.

Biello, David. 2009. "First Look at Carbon Capture and Storage in a West Virginia Coal-Fired Power Plant." Slide show. *Scientific American* website, November 6. http://www.scientificamerican.com/article.cfm?id=first-look-at-carbon-capture-and-storage.

Bigelow, Perry. 2011. Personal communication with author, February 24.

Bindewald, G. 2004. *Transforming the Grid to Revolutionize Electric Power in North America*. Report prepared for the U.S. Department of Energy, September.

Blok, Kornelis. 2004. "Improving Energy Efficiency by Five Percent and More per Year?" *Journal of Industrial Ecology* 8 (4): 87–99. doi:10.1162/1088198043630478.

Bloomberg New Energy Finance. 2011. "China Profits From Solar-Power Strategy as Europeans Backpedal." February 14. http://bnef.com/News/43833.

Bockholt, Wendy, and Matthew Kromer. 2009. *Assessment of Fuel Economy Technologies for Medium- and Heavy-Duty Vehicles*. TIAX LLC.

Bode, D. 2010. "Statement from Denise Bode, CEO American Wind Energy Association." American Wind Energy Association, December 9. http://www.awea.org/newsroom/inthenews/release_120910.cfm.

Bode, S., and H. M. Groscurth. 2010. "The Impact of PV on the German Power Market." *Zeitschrift für Energiewirtschaft* 35 (2): 105–115. doi:10.1007/s12398-010-0041-x. http://www.springerlink.com/content/u8828j3182vw8n27/.

Boeman, Raymond G., and N. L. Johnson. 2002. *Development of a Cost Competitive, Composite Intensive, Body-in-White*. Publication 2002-01-1905. Oak Ridge National Laboratory. http://www.ornl.gov/~webworks/cppr/y2002/pres/113371.pdf.

Bögle, A., and P. Cachola Schmal, eds. 2005. *Leicht weit / Light Structures: Jörg Schlaich Rudolf Bergermann*. New York: Prestel.

Bonneville Power Administration. 1992. *Hood River Conservation Project*. Portland, OR: Bonneville Power Administration.

———. 2009. *Energy Efficiency Emerging Technologies (E3T) Overview*. Bonneville Power Administration. http://www.bpa.gov/energy/n/emerging_technology/index.cfm.

Bony, Lionel, Sam Newman, and Stephen Doig. 2010. *Achieving Low-Cost Solar PV: Industry Workshop Recommendations for Near-Term Balance of System Cost Reductions*. Rocky Mountain Institute. http://www.rmi.org/rmi/Library/2010-19_BalanceOfSystemReport.

Bordoff, Jason, and Pascal Noell. 2008. "The Impact of Pay-As-You-Drive Auto Insurance in California." Brookings Institution. http://www.brookings.edu/papers/2008/07_payd_california_bordoffnoel.aspx.

Bradsher, Keith. 2010. "To Conquer Wind Power, China Writes the Rules." *New York Times*, December 14. http://www.nytimes.com/2010/12/15/business/global/15chinawind.html.

Braese, Paul. 2011. Personal communication with author, March 21.

British Petroleum. 2010a. *BP Statistical Review of World Energy*. British Petroleum. http://www.bp.com/liveassets/bp_internet/globalbp/globalbp_uk_english/reports_and_publications/statistical_energy_review_2008/STAGING/local_assets/2010_downloads/statistical_review_of_world_energy_full_report_2010.pdf .

———. 2010b. "BP and Verenium Announce Pivotal Biofuels Agreement." Press release, July 15. http://

www.bp.com/genericarticle.do?categoryId=2012968&contentId=7063758.

Brouwers, B., and E. V. Kemenade. 2010. *Condensed Rotational Separation for CO_2 Capture in Coal Gasification Processes*. Eindhoven University of Technology. http://www.gasification-freiberg.org/PortalData/1/Resources/documents/paper/14-3_Brouwers.pdf.

Brown, Lester. 2011. "Can the United States Feed China?" Earth Policy Institute, March 23. http://www.earth-policy.org/plan_b_updates?2011/update93.

Brown, M. A., and M. D. Levine, eds. 1997. *Scenarios of U.S. Carbon Reductions: Potential Impacts of Energy Technologies by 2010 and Beyond*. LBNL-40533. Interlaboratory Working Group on Energy-Efficient and Low-Carbon Technologies, Lawrence Berkeley National Laboratory, September 25.

Brown, Marilyn A., and Benjamin K. Sovacool. 2010. "Competing Dimensions of Energy Security: An International Perspective." *Annual Review of Energy and the Environment* 35: 77–108.

Browning, Bill. 2011. Personal communication with author, February 14.

Bruggink, Alle, Rob Schoevaart, and Tom Kieboom. 2003. "Concepts of Nature in Organic Synthesis: Cascade Catalysis and Multistep Conversions in Concert." *Organic Process Research & Development* 7 (5): 622–40. doi:10.1021/op0340311.

Buckley, Chris. 2011. "China Freezes Nuclear Approvals after Japan Crisis." Reuters, May 16. http://tinyurl.com/4bc76fl.

Bunch, David, and David L. Greene. 2010. *Potential Design, Implementation, and Benefits of a Feebate Program for New Passenger Vehicles in California: Interim Statement of Research Findings*. University of California at Davis, Institute of Transportation Studies, April. http://www.arb.ca.gov/research/apr/past/08-312main.pdf.

Bundesministerium der Justiz. 2009. *Energieausweis nach Energieeinsparverordnung* [German Energy Saving Ordinance]. Section 16, Abs. 4.

Burchell, Robert. 1998. *The Costs of Sprawl—Revisited*. Transit Cooperative Research Program. http://gulliver.trb.org/publications/tcrp/tcrp_rpt_39-a.pdf.

Burgermeister, Jane. 2008. "Geothermal Electricity Booming in Germany." RenewableEnergyWorld.com, June 2. http://www.renewableenergyworld.com/rea/news/article/2008/06/geothermal-electricity-booming-in-germany-52588.

Burns, C. 2011. "Caltech's Linde + Robinson Laboratory to be First LEED Platinum Lab." *RMI Solutions Journal* 4 (2).

Burns, R. E., S. Potter, and V. Witkind-Davis. 2004. "After the Lights Went Out." *Electricity Journal* 17 (1): 11–15.

Business Wire. 2000. "Whirlpool Corporation Endorses Landmark Agreement Among Appliance Manufacturers, Energy Advocates and Government; Agreement Will Save Consumers Billions." Business Wire, May 23. http://findarticles.com/p/articles/mi_m0EIN/is_2000_May_23/ai_62262436/.

———. 2011. "Fitch Affirms Ratings of FirstEnergy Corp.; Revises Outlook to Stable." Business Wire, May 27. http://www.businesswire.com/news/home/20110527005626/en/Fitch-Affirms-Ratings-FirstEnergy-Corp.-Revises-Outlook..

Butti, Ken, and John Perlin. 1980. *A Golden Thread: 2,500 Years of Solar Architecture and Technology*. 1st ed. New York: Cheshire Books / Van Nostrand Reinhold.

Cabot Corporation. 2011. "Cabot Aerogel: Aerogel for Insulation, Daylighting, Additives." http://www.cabot-corp.com/Aerogel.

Cairns, Sloman, Anable Newson, A. Kirkbride, and P. Goodwin. 2004. *Smarter Choices—Changing the Way We Travel*. London, UK: The Robert Gordon University and Eco-Logica.

California Energy Commission. 2007. "2007 Integrated Energy Policy Report." http://www.energy.ca.gov/2007_energypolicy/.

California Energy Commission. 2010. *The Role of Aging and Once-Through-Cooled Power Plants in California—An Update*. CEC-200-2009-018. www.energy.ca.gov/2009publications/CEC-200-2009-018/CEC-200-2009-018.PDF.

California Public Utilities Commission. 2010. *Draft 2006–2008 Energy Efficiency Evaluation Report*. California Public Utilities Commission. http://www.dra.ca.gov/NR/rdonlyres/07ED3986-1D4F-455B-B3FF-B1AA96482022/0/200608DraftFinalEDEvaluationReport.pdf.

California State Water Resources Control Board. 2011. *Once-Through Cooling Policy Protects Marine Life and Insures Electric Grid Reliability*. California State Water Resources Control Board, May 4. http://www.waterboardsca.gov/publications_forms/publications/factsheets/docs/oncethroughcooling0811.pdf.

California State Water Resources Control Board. 2011b. "Thermal Discharges—Cooling Water Intake Structures, Once-Through Cooling." August 4. www.swrcb.ca.gov/water_issues/programs/ocean/cwa316/.

Callonec, Gael, and Nicolas Blanc. 2009. *Evaluation of the Economic and Ecological Effects of the French 'Bonus Malus' for New Cars.* French Environment and Energy Management Agency (ADEME). http://www.eceee.org/conference_proceedings/eceee/2009/Panel_2/2.273/Paper/.

Carbon Disclosure Project. 2011. *Water and Carbon Continue to Rise as Investment Issues.* Carbon Disclosure Project. https://www.cdproject.net/en-US/WhatWeDo/CDPNewsArticlePages/water-and-carbon-continue-to-rise-as-investment-issues.aspx.

Carbon War Room. 2011. "Shipping Efficiency." Shippingefficiency.org: Information For a More Efficient Market. http://www.shippingefficiency.org/.

Carey, John. 2009. "Wanted: A New Biz Model for Electric Power." *Bloomberg Businessweek,* August 6. http://tinyurl.com/42esqaa.

Casillas, C. E., and D. M. Kammen. 2010. "The Energy-Poverty-Climate Nexus." *Science* 330 (6008): 1181. http://esmap.org/esmap/sites/esmap.org/files/1126PolicyForumD.pdf.

Celebi, Metin, Frank C. Graves, Gunjan Bathla, and Lucas Bressan. 2010. Potential Coal Plant Retirements Under Emerging Environmental Regulations. The Brattle Group, Inc. December 8. http://www.brattle.com/_documents/uploadlibrary/upload898.pdf.

Center for Clean Air Policy and Center for Neighborhood Technology. 2006. *High Speed Rail and Greenhouse Gas Emissions in the U.S.* Center for Clean Air Policy and Center for Neighborhood Technology. http://www.cnt.org/repository/HighSpeedRailEmissions.pdf.

Center for Neighborhood Technology. 2010. *Penny Wise, Pound Fuelish: New Measures of Housing + Transportation Affordability.* Center for Neighborhood Technology. http://www.cnt.org/repository/pwpf.pdf.

Chan-Lizardo, Kristine, Davis Lindsey, John Carey, and Elliot Harry. 2011. *Big Pipes, Small Pumps: Interface, Inc. Factor Ten Engineering Case Study.* Rocky Mountain Institute. http://www.rmi.org/rmi/Library/2011-04_BigPipesSmallPumps.

Chan-Lizardo, Kristine, Amory Lovins, Laura Schewel, and Mike Simpson. 2008. *Ultralight Vehicles: Non-Linear Correlations between Weight and Safety.* Rocky Mountain Institute. http://www.rmi.org/rmi/Library%2F2008-24_UltralightVehicles.

Chen, Patrick, Astrid Overholt, Brad Rutledge, and Tomic Jasna. 2010. *Economic Assessment of Biogas and Biomethane Production from Manure.* CalStart, March 20.

Cheung, W. H, K. K. H Choy, D. C. W. Hui, J. F. Porter, and G. Mckay. 2006. "Use of Municipal Solid Waste for Integrated Cement Production." *Developments in Chemical Engineering and Mineral Processing* 14 (1–2): 193–202. doi:10.1002/apj.5500140117. http://onlinelibrary.wiley.com/doi/10.1002/apj.5500140117/abstract.

Cho, Hanah. 2010. "BGE to Move Ahead with 'Smart Grid' Plan." *Baltimore Sun,* August 16. http://articles.baltimoresun.com/2010-08-16/business/bs-bz-bge-smart-grid-response-20100816_1_regular-rate-increase-requests-bge-estimates-smart-grid-plan.

Choate, William T. 2003. *Energy and Emission Reduction Opportunities in the Cement Industry.* Columbia, MD: BCS, Inc. report to U.S. Department of Energy. Dec. 29. http://www1.eere.energy.gov/industry/imf/pdfs/eeroci_dec03a.pdf.

Christensen, Gary. 2011. Personal communication with author, March 18.

Chupka, Marc, Robert Earle and Peter Fox-Penner, and Ryan Hledik. 2008. *Transforming America's Power Industry: The Investment Challenge 2010–2030.* Washington DC: The Edison Foundation.

City of Boulder. 2011. "SmartRegs: Smart Regulation for Sustainable Places." http://www.bouldercolorado.gov/index.php?option=com_content&task=view&id=13982&Itemid=22%29.

City of Fort Collins Utilities. 2004. *The Next Generation: Fossil Ridge High School.* City of Fort Collins Utilities. http://www.fcgov.com/utilities/img/site_specific/uploads/cs-fossilridge.pdf.

City of Sioux Falls Building Services. 2010. "ISO's Building Code Effectiveness Grading Schedule Results for Sioux Falls." City of Sioux Falls Building Services, June 14. http://www.siouxfalls.org/News/2010/June/14/iso_grading_results.

Clarke, G. R. G., and S. J. Wallsten. 2002. *Universal(ly Bad) Service: Providing Infrastructure Services to Rural and Poor Urban Consumers*. The World Bank Development Research Group. http://www-wds.worldbank.org/servlet/WDSContentServer/WDSP/IB/2002/08/23/000094946_02081004010494/Rendered/PDF/multi0page.pdf.

Climate Group. 2011. *Delivering Low Carbon Growth: A Guide to China's 12th Five Year Plan*. The Climate Group, March. http://www.theclimategroup.org/_assets/files/FINAL_14Mar11_-TCG_DELIVERING-LOW-CARBON-GROWTH-V3.pdf.

ClimateWire. 2011. "Denmark Reveals a Plan to End Reliance on Fossil Fuels by 2050." ClimateWire, February 25. http://www.eenews.net/cw/2011/02/25.

Cochran, Maura M., and Scott R. Muldavin. 2010. *Value Beyond Cost Savings: How to Underwrite Sustainable Properties*. The Counselors of Real Estate, March 22. http://www.greenbuildingfc.com/Home/ValueBeyondCostSavings.aspx.

Cogenra. 2011. "Solar Hot Water & Electricity in a Single Module." http://www.cogenra.com/products-services/.

Committee on Biobased Industrial Products, National Research Council. 2000. *Biobased Industrial Products: Research and Commercialization Priorities*. Washington DC: The National Academies Press.

Congressional Budget Office. 2011. *Alternative Approaches to Funding Highways*. Congressional Budget Office, pub. no. 4090, March. http://thehill.com/images/stories/blogs/flooraction/Jan2011/cboreport.pdf.

Constable, G., and B. Somerville. 2003. "Greatest Engineering Achievements of the 20th Century." National Academy of Engineering. http://www.greatachievements.org/.

Cooke, Stephanie. 2009. *In Mortal Hands: A Cautionary History of the Nuclear Age*. New York: Bloomsbury USA.

Cooney, Kevin. 2011. *Evaluation Report: OPower SMUD Pilot Year 2*. Navigant Consulting, Inc., February 20. http://www.opower.com/LinkClick.aspx?fileticket=sSraBTDAtSA%3D&tabid=72.

Cooper, Mark. 2010. *Policy Challenges of Nuclear Reactor Construction, Cost Escalation and Crowding Out Alternatives*. Vermont Law School. http://www.vermontlaw.edu/Documents/IEE/20100909_cooperStudy.pdf.

Cooper, Mark. 2011. *Rising Gasoline Prices and Record Household Expenditures: Will Policymakers Get Serious About Ending Our "Addiction to Oil" by Supporting a 60 Mile Per Gallon Fuel Economy Standard?* Consumer Federation of America, May 16. http://www.consumerfed.org/pdfs/CFA-Auto-Standard-Report-May-16-2011.pdf.

Criscione, Peter. 2011. *Melting Away Energy Waste in Foundries*. E Source, March 15. http://www.esource.com/members/TAS-RB-33/Research_Brief/EnergyWaste_Foundries.

Crowley, Norman. 2010. Personal communication with author, November 16.

Cullen, Jonathan M., Julian M. Allwood, and Edward H. Borgstein. 2011. "Reducing Energy Demand: What Are the Practical Limits?" *Environmental Science & Technology* 45 (4): 1711–18. doi:10.1021/es102641n. http://dx.doi.org/10.1021/es102641n.

CWEA. 2011. *2010 China Statistics: Installed Capacity of Windpower. China Renewable Energy Society (Wind Division)*. In Chinese. CWEA. http://www.cwea.org.cn/upload/2010年风电装机容量统it.pdf.

D&R International, Ltd. 2010. *2010 Buildings Energy Data Book*. U.S. Department of Energy. http://buildingsdatabook.eren.doe.gov/docs/DataBooks/2010_BEDB.pdf.

Daggett, Dave (Boeing). 2003. Personal communication with author, August 29.

Dale, B. E., M. S. Allen, M. Laser, and L. R. Lynd. 2009. "Protein Feed Coproduction in Biomass Conversion to Fuels and Chemicals." *Biofuels, Bioproducts, and Biorefining* 3: 219–30. doi: 10.1002/bbb.132.

Dale, Bruce E., Lee R. Lynd, Thomas L. Richard, Robert P. Anex, Mark S. Laser, and Bryan D. Bals. 2009. *Reimagining Agriculture to Accommodate Large Scale Energy Production*. Presented at 2009 AIChE Annual Meeting, (Nashville), November 10.

Danish Energy Agency. 2009. "Energy Statistics 2009." http://www.ens.dk/en-US/Info/FactsAndFigures/Energy_statistics_and_indicators/Annual%20Statistics/Documents/Tables2009.xls.

———. 2010. *Danish Energy Policy 1970–2010: Vision: 100% Independence of Fossil Fuels*. http://www.ens.dk/en-US/Info/news/Factsheet/Documents/DKEpol.pdf%20engelsk%20til%20web.pdf.

Danish Energy Association. 2009. "Danish Electricity Supply '09 Statistical Survey." http://www.danishenergyassociation.com/~/media/Energi_i_tal/Statistik_09_UK.ppt.ashx.

Danish Ministry of Climate and Energy. 2009. "The Danish Example: The Way to an Energy Efficient and Energy Friendly Economy." http://www.kemin.dk/documents/publikationer%20html/the%20danish%20example/html/kap01.html.

Danish Wind Energy Association. 2011. "New All Time High Record in Exports." April 4. http://www.windpower.org/en/news/news.html#719.

Danmarks Statistik. 2010. *Statistical Yearbook 2010*. Danmarks Statistik. http://www.dst.dk/pukora/epub/upload/15198/15nuk.pdf.

Darrow, Tara. 2011. Personal communication with author, February 15.

Database of State Incentives for Renewables & Efficiency (DSIRE). 2010. "RPS Data Spreadsheet." http://www.dsireusa.org/rpsdata/index.cfm.

Database of State Incentives for Renewables & Efficiency (DSIRE). 2011. "PACE Financing." http://www.dsireusa.org/solar/solarpolicyguide/?id=26.

Defense Science Board. 2008. *More Fight—Less Fuel*. Defense Science Board. http://www.acq.osd.mil/dsb/reports/ADA477619.pdf.

Delucchi, M. A., and M. Z. Jacobson. 2011. "Providing All Global Energy with Wind, Water, and Solar Power, Part II: Reliability, System and Transmission Costs, and Policies." *Energy Policy* 39 (3): 1170–90.

Delucchi, Mark, and Donald McCubbin. 2010. *External Costs of Transport in the U.S.* University of California at Davis Institute of Transportation Studies. http://www.its.ucdavis.edu/publications/2010/UCD-ITS-RP-10-10.pdf.

Dempsey, Judy. 2011. "Panel Urges Germany to Close Nuclear Plants by 2021." *New York Times*, May 11. http://www.nytimes.com/2011/05/12/business/energy-environment/12energy.html.

Denholm, Paul, and Robert Margolis. 2008. *Impacts of Array Configuration on Land-Use Requirements for Large-Scale Photovoltaic Deployment in the United States*. National Renewable Energy Laboratory. http://www.nrel.gov/analysis/pdfs/42971.pdf.

Desroches, Louis-Benoit, and Karina Garbesi. 2011. *Max Tech and Beyond*. Berkeley, CA: Lawrence Berkeley National Laboratory, April 22. http://ees.ead.lbl.gov/bibliography/max_tech_and_beyond.

Deutsche Bank. 2009. *The Peak Oil Market: Price Dynamics at the End of the Oil Age*. Deutsche Bank, October 4. http://www.petrocapita.com/attachments/128_Deutsche%20Bank%20-%20The%20Peak%20Oil%20Market.pdf.

Deutsche Bank Corporate Real Estate & Services. 2011. *Greentowers: Setting a New Global Standard Transforming Deutsche Bank's Head Office into a Green Building*. http://www.greenprintfoundation.org/Libraries/Public_Library_-_Case_Study_-_3/Deutsche_Bank_makes_sustainable_improvements_at_headquarters.sflb.ashx?download=true.

Deutsche Welle. 2008. "Eco-Village: A Bavarian Village Goes It Alone in Cooperation with Handelsblatt." http://www.dw-world.de/dw/article/0,,2338265,00.html.

Diamond, Jared M. 1999. *Guns, Germs, and Steel: The Fates of Human Societies*. New York: W. W. Norton.

Dinwoodie, Thomas. 2011. Personal communication with author, June 5.

Dinwoodie, Thomas (CTO, SunPower), and Dan Shugar (CEO, Solaria). 2011. Personal communications with author, June 5–6.

Dow. 2009. "New BASF and Dow HPPO Plant in Antwerp Completes Start-Up Phase." News. http://www.dow.com/news/corporate/2009/20090305a.htm.

Downey, John. 2011. "N.C.'s Customer Advocate Now Opposes Nuclear Legislation." *Charlotte Business Journal*, April 12. http://www.bizjournals.com/charlotte/blog/power_city/2011/04/ncs-customer-advocate-opposes.html.

Drayton, B., and V. Budinich. 2010. "A New Alliance for Global Change." *Harvard Business Review* 88 (9): 56–64. http://hbr.org/products/10353/10353p4.pdf.

Drexler, Eric. 1987. *Engines of Creation: The Coming Era of Nanotechnology*. New York: Anchor.

DuPont Danisco Cellulosic Ethanol LLC. 2010. "DDCE: DuPont Danisco Cellulosic Ethanol." http://www.ddce.com/.

E Source. 1999. *E Source Technology Atlas Series: Drivepower Atlas*. Boulder, CO: E Source.

———. 2002. *Industrial Machinery*. Boulder, CO: E Source.

———. 2011. "E Source Study Finds 85% of U.S. Consumers Want to Buy an Electric Vehicle." Press release, May 3. http://www.esource.com/esource/getpub/public/pdf/press_releases/ES-PR-EVData-4-11.pdf.

Earth Policy Institute. 2007. "World Average Photovoltaic Module Cost per Watt, 1975–2006." http://www.earth-policy.org/datacenter/xls/indicator12_2007_7.xls.

The Economist. 2005. "A Survey of Oil: Oil in Troubled Waters." *The Economist*, April 28. http://www.economist.com/node/3884623.

———. 2008. "Energy Efficiency: The Elusive Negawatt." *The Economist*, May 8. http://www.economist.com/node/11326549.

———. 2010. "Power to the People." *The Economist*, September 2. http://www.economist.com/node/16909923.

———. 2011a. "Print Me a Stradivarius." *The Economist*, February 10. http://www.economist.com/node/18114327.

———. 2011b. "Another blow for Berlusconi." *The Economist*, June 13. http://www.economist.com/blogs/newsbook/2011/06/italys-referendums.

EconPost. 2011. "Texas Economy Ranking in the World." February 03. http://econpost.com/texaseconomy/texas-economy-ranking-world.

Edin Energy. 2011. "Samsø, Denmark, Strives to Become a Carbon-Neutral Island." Edin Energy. http://www.edinenergy.org/samso.html.

Edison Electric Institute. 2009. *Statistical Yearbook of the Electric Power Industry*. Washington, DC: Edison Electric Institute.

Edlin, Aaron. 2002. "Per-Mile Premiums for Auto Insurance." Working Paper no. E02-318. University of California at Berkeley, June 2. http://repositories.cdlib.org/iber/econ/E02-318.

Edmonson, Amy, and Faaiza Rashid. 2009. *Integrated Project Delivery at Autodesk, Inc.* Harvard Business School, September 24. http://hbr.org/product/integrated-project-delivery-at-autodesk-inc-b/an/610017-PDF-ENG.

Edmunds. 2011. "New Cars, Used Cars, Car Reviews and Pricing." http://www.edmunds.com/.

Eek, Hans. 2011. Personal communications with author, March.

Efficiency Vermont. 2010. *Efficiency Vermont 2009 Annual Report*. Efficiency Vermont, November. http://www.efficiencyvermont.com/stella/filelib/FINAL2009AnnualReport.pdf.

Eggers, Dan. 2010. *Credit Suisse*. Brief to Aspen Energy Forum in Aspen, Colorado, July 3.

Ehrhardt-Martinez, K., K. A. Donnelly, and S. Laitner. 2010. *Advanced Metering Initiatives and Residential Feedback Programs: A Meta-Review for Household Electricity-Saving Opportunities*. American Council for an Energy-Efficient Economy, June. http://www.aceee.org/research-report/e105.

Eichholtz, Piet, Nils Kok, and John Quigley. 2010. *The Economics of Green Building*. Self-published. http://nilskok.typepad.com/EKQ3/EKQ_Economics.pdf.

Electric Energy Market Competition Task Force. 2005. *Report to Congress on Competition in Wholesale and Retail Markets for Electric Energy Pursuant to Section 1815 of the Energy Policy Act of 2005*. Electric Energy Market Competition Task Force. http://www.ferc.gov/legal/fed-sta/ene-pol-act/epact-final-rpt.pdf.

Electric Power Research Institute. 2007. *The Galvin Path to Perfect Power: A Technical Assessment*. Galvin Electricity Initiative, January. http://www.galvinpower.org/sites/default/files/documents/Perfect_Power_Technical_Assessment.pdf.

Electric Reliability Council of Texas. 2004. "FERC Form No. 714-ERCOT." http://www.ferc.gov/docs-filing/forms/form-714/data.asp.

Electricity Consumers Resource Council (ELCON). 2004. *The Economic Impacts of the August 2003 Blackout*. Washington DC: ELCON, February 9. http://www.elcon.org/Documents/EconomicImpactsOfAugust2003Blackout.pdf.

Electrification Coalition. 2009. *Electrification Roadmap*. Electrification Coalition. http://www.electrificationcoalition.org/policy/electrification-roadmap.

———. 2010. *Fleet Electrification Roadmap: Revolutionizing Transportation and Achieving Energy Security*. Electrification Coalition. http://www.electrificationcoalition.org/reports/EC-Fleet-Roadmap-screen.pdf.

Electrigaz. 2010. "Biogas FAQ." http://www.electrigaz.com/faq_en.htm.

Eley Associates. 1996. *ACT² Davis Residential Site EEM Impact Analysis.* Davis, CA: PG&E.

Elliott, RN, Michael Shepard, S. Greenberg, Gail Katz, Anibal T. De Almeida, and Steven Nadel. 2002. *Energy-Efficient Motor Systems: A Handbook on Technology, Program, and Policy Opportunities.* 2nd ed. Washington, DC: American Council for an Energy-Efficiency Economy.

Emerson, Sarah. 2011. Personal communication with author, July 7.

Empire State Building Company LLC. 2010. "Sustainability & Energy Efficiency." http://www.esbnyc.com/sustainability_energy_efficiency.asp.

Empire State Building Company LLC and Rocky Mountain Institute. 2009. *Empire State Building Case Study: Cost-Effective Greenhouse Gas Reductions via Whole-Building Retrofits: Process, Outcomes, and What Is Needed Next.* Empire State Building Company LLC and Rocky Mountain Institute. http://www.esbnyc.com/documents/sustainability/ESBOverviewDeck.pdf.

Energetics Incorporated. 2006. "Manufacturing Energy and Carbon Footprint: Cement." http://www1.eere.energy.gov/industry/rd/footprints.html.

Energi Styrelsen. 2009. *The Danish Example.* Energi Styrelsen. http://tinyurl.com/437vmxe.

EnerNex. 2011. *Eastern Wind Integration and Transmission Study.* Report prepared for National Renewable Energy Laboratory.

EnerNOC Inc. 2011. *Annual Report on Form 10-K for the Fiscal Year Ended December 31, 2010.* Boston, MA: EnerNOC.

Engemann, Kristie M., Kevin L. Kliesen, and Michael T. Owyang. 2010. "Do Oil Shocks Drive Business Cycles? Some U.S. and International Evidence." Federal Reserve Bank of St. Louis Working Papers, no. 2010-007. http://ideas.repec.org/p/fip/fedlwp/2010-007.html.

Engleman, Robert. 2011. "An End to Population Growth: Why Family Planning Is Key to a Sustainable Future." *Solutions* 2 (3).

ENR New York. 2009. "Skanska USA New York Office, Empire State Building." December. http://newyork.construction.com/features/2009/1201_EmpireStateBuilding.asp.

Enterprise Lane. 2010. "ISG Data Centre Presentation." February 17. http://www.youtube.com/watch?v=QsIyzdva780&feature=youtube_gdata_player.

Epstein, Paul R., Jonathan J. Buonocore, Kevin Eckerle, Michael Hendryx, Benjamin M. Stout III, Richard Heinberg, Richard W. Clapp, et al. 2011. "Full Cost Accounting for the Life Cycle of Coal." *Annals of the New York Academy of Sciences* 1219 (1): 73–98. doi:10.1111/j.1749-6632.2010.05890.x.

ERCOT. 2008. "ERCOT Demand Response Program Helps Restore Frequency Following Tuesday Evening Grid Event." Press release, February 27. http://www.ercot.com/news/press_releases/2008/nr02-27-08.

Ertesvag, I. S. 2001. "Society Exergy Analysis: A Comparison of Different Societies." *Energy* 26: 253–70.

Eubank, Huston, and William Browning. 2004. *Energy Performance Contracting for New Buildings.* Eley Associates. http://www.rmi.org/rmi/Library%2FD04-23_EnergyPerformanceNewBuildings.

European Climate Foundation. 2010. *Roadmap 2050: A Practical Guide to a Prosperous, Low Carbon Europe.* European Climate Foundation. http://www.roadmap2050.eu/.

European Photovoltaic Industry Association. 2011. *Global Market Outlook for Photovoltaics until 2015.* http://www.epia.org/publications/photovoltaic-publications-global-market-outlook/global-market-outlook-for-photovoltaics-until-2015.html.

European Renewable Energy Council. 2010. *Re-thinking 2050: A 100% Renewable Energy Vision for the European Union.* European Renewable Energy Council. http://www.rethinking2050.eu/.

European Union. 2010. "Directive 2010/31/EU of the European Parliament and the Council of 19 May 2010 on the Energy Performance of Buildings." *Official Journal of the European Union* 53. http://eur-lex.europa.eu/JOHtml.do?uri=OJ:L:2010:153:SOM:EN:HTML.

Europump. 2003. "European Guide to Pump Efficiency for Single Stage Centrifugal Pumps." http://work.sitedirect.se/sites/europump/europump/index.php?show=226_SWE&&page_anchor=http://work.sitedirect.se/sites/europump/europump/p226/p226_swe.php.

Eurostat. 2011. "Recycling Accounted for a Quarter of Total Municipal Waste Treated in 2009." *Eurostat News Release*, March 8.

Exeter Associates, K. 2007. *Review of International Experience Integrating Variable Renewable Energy Generation*. Sacramento, CA: California Energy Commission.

Fackler, Martin. 2011. "Japan's Leader Cancels Plan for New Nuclear Plants." *New York Times*, May 10. http://www.nytimes.com/2011/05/11/world/asia/11japan.html.

Federal Energy Regulatory Commission. 2004. *Form No. 714—Annual Electric Balancing Authority Area and Planning Area Report—ERCOT 2004*. Federal Energy Regulatory Commission. http://www.ferc.gov/docs-filing/forms/form-714/data.asp.

———. 2011a. *Assessment of Demand Response and Advanced Metering*. Federal Energy Regulatory Commission. http://www.ferc.gov/legal/staff-reports/demand-response.pdf.

———. 2011b. "FERC Approves Market-Based Demand Response Compensation Rule." Press release, March 15. http://www.ferc.gov/media/news-releases/2011/2011-1/03-15-11.asp.

Feist, Wolfgang. 2011. Personal communication with author, March.

Ferguson, Will. 2010. "Texas Wind Industry's Rapid Growth Creates New Challenges." *Texas Business Review* (University of Texas at Austin), February.

Fickling, David. 2011. "A Warning on Rare Earth Elements." *Wall Street Journal*, May 6. http://online.wsj.com/article/SB10001424052748703992704576304712512256774.html.

First Solar. 2010. *First Solar Corporate Overview*, March 1, 2010. Tempe, Arizona: First Solar. http://files.shareholder.com/downloads/FSLR/1301877449x0x477649/205c17cb-c816-4045-949f-700e7c1a109f/FSLR_CorpOverview.pdf.

Fischer, Hanns. 1997. Personal communication with author, December 4.

Fletcher, Paul D. I., Stephen J. Haswell, Esteban Pombo-Villar, Brian H. Warrington, Paul Watts, Stephanie Y. F. Wong, and Xunli Zhang. 2002. "Micro Reactors: Principles and Applications in Organic Synthesis." *Tetrahedron* 58 (24): 4735–57. doi:10.1016/S0040-4020(02)00432-5.

Flex Your Power. 2011. "Best Practice Guide: Adobe Systems Incorporated and Cushman & Wakefield." http://www.fypower.org/bpg/case_study.html.

Flygt. 2008. *N-Pumps 3153, 3171, 3202 & 3301: A New Generation of Submersible Wastewater Pumps*. Flygt. http://www.ittwww.com/n/1396800.pdf.

Foroohar, Kambiz. 2010. "Exxon $600 Million Algae Investment Makes Khosla See Pipe Dream." Bloomberg Markets Magazine, June 3. http://www.bloomberg.com/news/2010-06-03/exxon-600-million-algae-investment-spurs-khosla-to-dismiss-as-pipe-dream.html.

ForschungsVerbund Erneuerbare Energien. 2010. *Energiekonzept 2050*. June. www.fvee.de/fileadmin/politik/kurzfassung_energiekonzept.pdf.

Foundry Management & Technology. 2008. "Isothermal Melting Research Now into New Stage." Posted on the website of *Foundry Management & Technology*, January 3. http://www.foundrymag.com/frontpage/feature/77480/isothermal_melting_research_now_into_new_stage.

Fountain, Henry. 2010. "Solar Storms Force Electric Utilities to Plan for the Worst." *New York Times*, November 16. http://www.nytimes.com/2010/11/17/business/energy-environment/17GRID.html.

Frantzis, L. 2010. "Renewable Energy Global and Domestic Market Drivers." Presented at the Renewable Energy Finance Forum, New York, June 29.

Freedom House. 2011. "Map of Freedom." http://www.freedomhouse.org/images/File/fiw/FIW_2011_MOF_Final.pdf.

Friedman, Thomas L. 2010. "Their Moon Shot and Ours." *New York Times*. September 25.

Friedman, Thomas L. 2011. "Washington vs. the Merciless." *New York Times*. March 19.

Froggatt, Antony, and Glada Lahn. 2010. *Sustainable Energy Security: Strategic Risks and Opportunities for Business*. Chatham House, June. http://www.chathamhouse.org.uk/publications/papers/view/-/id/891/.

Fuchs, Erica R. H., Frank R. Field, Richard Roth, and Randolph E. Kirchain. 2008. "Strategic Materials Selection in the Automobile Body: Economic Opportunities for Polymer Composite Design." *Composites Science and Technology* 68 (9): 1989–2002. doi:10.1016/j.compscitech.2008.01.015.

Gadgil, A., A. H. Rosenfeld, D. Arasteh, and E. Ward. 1991. "Advanced Lighting and Window Technologies for Reducing Electricity Consumption and Peak

Demand: Overseas Manufacturing and Marketing Opportunities." In *Proceedings of the IEA/ENEL Conference on Advanced Technologies for Electric Demand-Side Management, April*. (Paris: OECD Publishing), 4–5.

Galitsky, C., and E. Worrell. 2008. *Energy Efficiency Improvement and Cost Saving Opportunities for the Vehicle Assembly Industry*. Lawrence Berkeley National Laboratory. http://ies.lbl.gov/drupal.files/ies.lbl.gov.sandbox/Vehicle%20Assembly_0.pdf.

Galvin Electricity Initiative. 2010. *The Value of Smart Distribution and Microgrids*. The Galvin Electricity Initiative, January. http://www.galvinpower.org/sites/default/files/ValuesRpt_Microgrids0113%20(2).pdf.

GE Energy. 2010. *Western Wind and Solar Integration Study*. Report prepared for the National Renewable Energy Laboratory. http://www.nrel.gov/wind/systemsintegration/pdfs/2010/wwsis_final_report.pdf.

Gemmer, Bob, Ted Bronson, John Cuttica, and Tommi Makila. 2010. "US Department of Energy Pushes for CHP in Industry." *Cogeneration & On-site Power Production* 11 (5). http://www.cospp.com/articles/print/volume-11/issue-5/features/us-department-of-energy-pushes-for-chp-in-industry.html.

General Services Administration. 2010. *Federal Fleet Report, Fiscal Year 2009*. General Services Administration, January 29.

German Federal Environmental Agency (UBA). 2010. *Energieziel 2050: 100% Strom aus erneuerbaren Quellen*. http://www.uba.de/uba-info-medien/3997.html.

Gladwell, Malcom. 2011. "What Is the Tipping Point?" http://www.gladwell.com/tippingpoint/index.html.

Glasser, D., Diane Hildebrandt, Brendon Hausberger, B. Patel, and B. J. Glasser. 2009. "Systems Approach to Reducing Energy Usage and Carbon Dioxide Emissions." *AIChE Journal* 55 (9): 2202–2207. doi:10.1002/aic.12009.

Global Insight. 2006. *Four Corridor Case Studies of Short-Sea Shipping Services: Short-Sea Shipping Business Case Analysis*. Transportation Research Board. http://www.marad.dot.gov/documents/USDOT_-_Four_Corridors_Case_Study_(15-Aug-06).pdf.

Global Times. 2011. "China Dethrones U.S. as Largest Wind Power Installer." *Global Times*, January 13. http://business.globaltimes.cn/industries/2011-01/612386.html.

Global Transmission Report. 2010. "France seeks to build transmission lines under Mediterranean." March 28. http://www.globaltransmission.info/archive.php?id=4110.

Gohring, Nancy. 2008. "Good Incentives Boost Data-Center Energy Efficiency." July 9. http://www.networkworld.com/news/2008/070908-good-incentives-boost-data-center-energy.html?page=1.

Gold, Rachel, Steven Nadel, J. A. Laitner, and Andrew de Laski. 2011. *Appliance and Equipment Efficiency Standards: A Money Maker and Job Creator*. Washington DC: American Council for an Energy-Efficient Economy. http://www.aceee.org/research-report/a111.

Goldman, Charles, Merrian Fuller, Elizabeth Smart, Jane S. Peters, et al. 2010. *Energy Efficiency Services Sector: Workforce Size and Expectations for Growth*. LBNL-3987E. Lawrence Berkeley National Laboratory, September.

Goldstein, D. B. 2010. *Invisible Energy: Strategies to Rescue the Economy and Save the Planet*. Pt. Richmond, CA: Bay Tree Publishing.

Goodman, J. David. 2010. "An Electric Boost for Bicylists," *New York Times*, January 31.

Gossamer Ceiling Fans. 2008. "What Is So Special About These Fans?" http://www.gossamerwind.com/content/what-so-special-about-these-fans-0.

GPO. 2011. *Budget of the United States Government, Fiscal Year 2011*. GPOAccess (website of the U.S. Government Printing Office). http://www.gpoaccess.gov/usbudget/fy11/index.html.

Grant, M., and J. Mitton. 2010. "Case Study: The Glorious, Golden, and Gigantic Quaking Aspen." *Nature Education Knowledge* 1 (8): 40.

Green Manufacturer. 2010. "Manufacturer Finds Lighting Energy Efficiency Convenient, Truthfully." March 1. http://www.greenmanufacturer.net/article/facilities/manufacturer-finds-lighting-energy-efficiency-convenient-truthfully.

Greenberg, D. 2010. "Pint-Size Electron Beams Portend Prodigious Savings." E Source, December 21. http://www.esource.com/node/36659.

Greene, David L. 1990. "CAFE or Price? An Analysis of the Effects of Federal Fuel Econmoy Regulations adn Gasoline Price on New Car MPG, 1978-89. Energy J. 11(3)-37-57.

Greene, David L. 2010. "Measuring Energy Security: Can the United States Achieve Oil Independence?" *Energy Policy* 38 (4): 1614–21. doi:10.1016/j.enpol.2009.01.041.

Greene, David L., and Janet Hopson. 2010. "The Costs of Oil Dependence." Oak Ridge National Laboratory Vehicle Technology Program Fact of the Week, fact #632, July 19. http://www1.eere.energy.gov/vehiclesandfuels/facts/m/2010_fotw632.html.

Greene, David L., and Paul N. Leiby. 2006. *The Oil Security Metrics Model: A Tool for Evaluating the Prospective Oil Security Benefits of DOE's Energy Efficiency and Renewable Energy R&D Programs*. Oak Ridge National Laboratory. http://www-cta.ornl.gov/cta/Publications/Reports/ORNL_TM_2006_505.pdf.

Greene, David L., and Nataliya Tishchishyna. 2000. *Costs of Oil Dependence: A 2000 Update*. Oak Ridge National Laboratory. http://www-cta.ornl.gov/cta/Publications/Reports/ORNL_TM_2000_152.pdf.

Greening, Lorna, David L. Greene, and Carmen Difiglio. 2000. "Energy Efficiency and Consumption—The Rebound Effect—A Survey." *Energy Policy* 28 (6–7): 389–401. doi:10.1016/S0301-4215(00)00021-5.

Gregor, Alison. 2011. "Solar Farms Put Vacant Land to Work." *New York Times*, March 22. http://www.nytimes.com/2011/03/23/realestate/commercial/23solar.html.

Groscurth, H. M., and R. Kümmel. 1989. "The Cost of Energy Optimization: A Thermoeconomic Analysis of National Energy System." *Energy* 14 (11): 685–96. doi:10.1016/0360-5442(89)90002-9.

Gross, Robert. 2006. *The Costs and Impacts of Intermittency*. UK Energy Research Centre. http://www.ukerc.ac.uk/Downloads/PDF/06/0604Intermittency/0604IntermittencyReport.pdf.

Grübler, Arnulf. 2009. *An Assessment of the Costs of the French Nuclear PWR Program 1970–2000*. Laxenburg, Austria: International Institute for Applied Systems Analysis, October 6. http://www.iiasa.ac.at/Admin/PUB/Documents/IR-09-036.pdf.

———. 2010. "The Costs of the French Nuclear Scale-Up: A Case of Negative Learning by Doing." *Energy Policy* 38 (9): 5174–88. doi:10.1016/j.enpol.2010.05.003.

GTM Research. 2010. *Enterprise LED Lighting: Commercial and Industrial Market Trends, Opportunities & Leading Companies*. Greentech Research, December 1. http://www.gtmresearch.com/report/enterprise-led-lighting.

Gupta, S., D. A. Tirpak, and N. Burger. 2007. "Policies, Instruments and Co-Operative Arrangements." In *Climate Change 2007: Mitigation*. Contribution of Working Group III to the Fourth Assessment Report of the Intergovernmental Panel on Climate Change. Cambridge: Cambridge University Press.

Hageluken, Christian, and Christopher Corti. 2010. "Recycling of Gold from Electronics: Cost-Effective Use through 'Design for Recycling.'" *Gold Bulletin* 43 (3). http://cat.inist.fr/?aModele=afficheN&cpsidt=23966564.

Hansard. 2007. *Parliamentary Debates*, House of Lords. Column 1138, November 27. http://www.publications.parliament.uk/pa/ld200708/ldhansrd/text/71127-0004.htm.

Hansen, Lena, and Lovins. 2010. "Keeping the Lights on While Transforming Electric Utilities." *Rocky Mountain Institute Solutions Journal* 3 (1). http://www.rmi.org/rmi/Transforming+Electric+Utilities.

Härkönen, Jehki. 2011. "New problems in Olkiluoto." Greenpeace International. July 21.

Harsch, John. 2009. *1,059-Year-Old German Village Has Created the Bioenergy Future*. Agri-Pulse Communications. http://www.agri-pulse.com/uploaded/20091014H9.pdf.

Hawken, Paul, Amory B. Lovins, and L. Hunter Lovins. 1999. *Natural Capitalism: Creating the Next Industrial Revolution*. Boston: Little, Brown.

Henry, Ray. 2011. "Georgia Power Says No Deal on Nuclear Costs." *Bloomberg Businessweek*, March 31. http://www.businessweek.com/ap/financialnews/D9MA6FF80.htm.

Herman Miller. 2008. *The Attributes of Thermal Comfort*. Herman Miller. http://www.hermanmiller.com/MarketFacingTech/hmc/solution_essays/assets/se_Attributes_of_Thermal_Comfort.pdf.

Herzog, Howard. 2010. "Carbon Dioxide Capture and Storage." In *The Economics and Politics of Climate Change*, by Dieter Helm and Cameron Hepburn, eds. New York: Oxford University Press.

Heschong Mahone Group. 1999. *Daylighting in Schools: An Investigation into the Relationship between Daylighting and Human Performance*. Fair Oaks, CA: California Board for Energy Efficiency.

———. 2001. *Re-Analysis Report, Daylighting in Schools.* Fair Oaks, CA: California Board for Energy Efficiency.

———. 2003a. *Daylight and Retail Sales.* Fair Oaks, CA: California Board for Energy Efficiency.

———. 2003b. *Windows and Classrooms: A Study of Student Performance and the Indoor Environment.* Fair Oaks, CA: California Board for Energy Efficiency.

———. 2007. *Performance Study of a Mechanical Vapor Recompression (MVR) Evaporation System.* Emerging Technologies Coordinating Council, December. http://www.etcc-ca.com/component/content/article/20/2408-mechanical-vapor-recompression-tomato.

Hidary, Jack. 2011. Personal communication to author, May 21.

Higgs, Robert. 2010. "Defense Spending Is Much Greater than You Think." The Independent Institute *Beacon* (blog), April 17. http://blog.independent.org/2010/04/17/defense-spending-is-much-greater-than-you-think/.

Hilbert, M., and P. López. 2011. "The World's Technological Capacity to Store, Communicate, and Compute Information." *Science* 332 (6025): 60.

Hirsh, R. 1999. *Power Loss: The Origins of Deregulation and Restructuring in the American Electric Utility System.* Cambridge, MA: MIT Press.

Holcim AG. 2010. "Pressemitteilung." http://www.holcim.de/de/kommunikation/pressemitteilungen/pressemitteilung/article/holcim-investiert-7-millionen-euro-in-den-standort-laegerdorf.html.

Holland Container. 2011. "Holland Container Innovations." http://www.hcinnovations.nl/.

Holtzclaw, John. 2004. "A Vision of Energy Efficiency." In *2004 ACEEE Summer Study on Energy Efficiency in Buildings.* Washington, DC: American Council for an Energy-Efficient Economy.

Howe, B. 1993. *Distribution Transformers: A Growing Energy Savings Opportunity.* Boulder, CO: E Source.

Howe, B., and M. Shepard. 1993. *Balancing Speed and Efficiency in Motor Selection.* Boulder, CO: E Source.

Hymel, Kent, Kenneth Small, and Kurt Van Dender. 2010. "Induced Demand and Rebound Effects in Road Transport." *Transportation Research Part B* 44 (February): 1220–41.

Idaho Business Review. 2007. "Boise Developer Gary Christensen Plans Zero Net Energy Buildings." *Idaho Business Review*, April 19. http://idahobusinessreview.com/2007/04/19/boise-developer-gary-christensen-plans-zero-net-energy-buildings/.

Indigo Development. 2003. "The Industrial Symbiosis at Kalundborg, Denmark." http://www.indigodev.com/Kal.html.

InfoMine—Mining Equipment and Supplier News. 2010. "First LNG Mining Truck in U.S.: GFS Corp. Converts CAT 777C to 60% Liquefied Natural Gas." Mining.com, Equipment and Suppliers, October 28. http://suppliersandequipment.mining.com/2010/10/28/first-lng-mining-truck-in-us-gfs-corp-converts-cat-777c-to-60-liquefied-natural-gas/.

Institute for Electric Efficiency. 2010. *Changes in State Regulatory Frameworks for Utility Administered Energy Efficiency Programs.*

Institute of Electrical and Electronics Engineers. 2011. *IEEE Draft Guide for Design, Operation, and Integration of Distributed Resource Island Systems with Electric Power Systems.* April. IEEE. http://grouper.ieee.org/groups/scc21/1547.4/1547.4_index.html.

Institute of Paper Science and Technology at Georgia Institute of Technology. 2006. *Pulp and Paper Industry: Energy Bandwidth Study.* American Institute of Chemical Engineers. http://www1.eere.energy.gov/industry/forest/bandwidth.html.

Interface Engineering. 2009. *Engineering a Sustainable World: Design Process and Engineering Innovation for the Center for Health & Healing at Oregon Health & Science University.* Portland, OR: Interface Engineering.

International Atomic Energy Agency. 2011. "Power Reactor Information System." http://www.iaea.or.at/programmes/a2/.

International Energy Agency. 2007. *Tracking Industrial Energy Efficiency and CO_2 Emissions.* Paris: OECD Publishing.

———. 2010. *Sustainable Production of Second-Generation Biofuels.* International Energy Agency, February. http://www.europeanclimate.org/documents/Power_trains_for_Europe.pdf.

———. 2010a. World Energy Outlook. http://www.worldenergyoutlook.org/docs/weo2010/WEO2010_ES_English.pdf .

———. 2011. *Harnessing Variable Renewables—A Guide to the Balance Challenge.* www.iea.org/publications/free_new_Desc.asp?PUBS_ID=2403.

Itron Inc. and KEMA Inc. 2008. *California Energy Efficiency Potential Study.* Pacific Gas & Electric Company. http://www.nwcouncil.org/dropbox/6th%20Plan%20Industrial/Industrial%20Conservation%20Data%20Catalogue/ISC%20Document%20Catalogue_Public%20Version-5%20June%202009/Documents/Tier%201/California_PotentialStudy_Vol1_May%202006.pdf.

Jacobson, Mark Z., and Mark A. Delucchi. 2011. "Providing All Global Energy with Wind, Water, and Solar Power, Part I: Technologies, Energy Resources, Quantities and Areas of Infrastructure, and Materials." *Energy Policy* 39 (3) (March): 1154–69.

Jha, Alok. 2010. "Sun, Wind and Wave-Powered: Europe Unites to Build Renewable Energy 'Supergrid'." January 3. http://www.guardian.co.uk/environment/2010/jan/03/european-unites-renewable-energy-supergrid.

Johnson Controls. 2010. *2010 Energy Efficiency Indicator.* Institute for Building Efficiency. http://www.institutebe.com/Energy-Efficiency-Indicator/global-energy-efficiency-indicator-results.aspx.

Jungerberg, Steven. 2010. Personal communication with author, December 15.

Kahane, Adam. 2010. *Power and Love: A Theory and Practice of Social Change.* 1st ed. San Francisco: Berrett-Koehler Publishers.

Kahane, C. 2003. "Vehicle Weight, Fatality Risk, and Crash Compatibility of Model Year 1991–99 Passenger Cars and Light Trucks." National Highway Traffic Safety Administration Technical Report DOT HS 809 662. NHTSA, October. http://www.nhtsa.gov/cars/rules/regrev/evaluate/pdf/809662.pdf.

Kahn, Alfred E. 1988. "The Rationale of Regulation and the Proper Role of Economics." In *The Economics of Regulation: Principles and Institutions.* Cambridge, MA: MIT Press.

Ka'iliwai, George, and Wade Jost. 2009. *SPIDERS: Energy Security JCTD Proposal.* U.S. Department of Defense. http://www1.eere.energy.gov/femp/pdfs/fupwg_fall2009_jost.pdf.

Kanellos, Michael. 2009. "Starbucks Goes Bonkers for LED Lights." Greentech Enterprise, November 12. http://www.greentechmedia.com/articles/read/starbucks-goes-bonkers-for-led-lights/.

———. 2011. "How to Drop Solar to $1 a Watt? Try Diamond Saws, Says Dick Swanson." Greentech Solar, March 17. http://www.greentechmedia.com/articles/read/how-to-drop-solar-to-1-a-watt-try-diamond-saws-says-dick-swanson.

Kann, Shayle (GTM Research). 2011. Personal communication with author, June 21.

Kanter, James. 2009. "Siemens Pulls Out of Nuclear Venture with Areva." *New York Times*, February 6. http://www.nytimes.com/2009/01/26/business/worldbusiness/26iht-26nuclear.19695526.html.

Kats, Greg. 2009. *Greening Our Built World: Costs, Benefits, and Strategies.* Washington, DC: Island Press.

Kaufman, Leslie. 2010. "In Kansas, Climate Skeptics Embrace Cleaner Energy." *New York Times*, October 18. http://www.nytimes.com/2010/10/19/science/earth/19fossil.html.

Keller, Rob (JCPenney). 2011. Personal communication with author, February 7.

Kelly Kettle. 2011. Kelly Kettle. http://en.wikipedia.org/wiki/Kelly_Kettle

KEMA, Inc. 2006. *California Industrial Existing Construction Energy Efficiency Potential Study.* Pacific Gas & Electric Company. http://tinyurl.com/3oyqpt5.

Kenny, J. F., N. L. Barber, S. S. Hutson, K. S. Linsey, J. K. Lovelace, and M. A. Maupin. 2009. *Estimated Use of Water in the United States in 2005.* U.S. Geological Survey. http://pubs.usgs.gov/circ/1344/.

Kiatreungwattana, Kosol, and Jimmy Salasovich. 2010. *Performance Based Design-Build Process Research Support Facility.* Golden, CO: National Renewable Energy Laboratory.

Kim, J. H., and W. B. Powell. 2010. "An Hourly Electricity Price Model, a Robust Trading Policy, and the Value of Storage, for Heavy Tailed Markets." Preprint, submitted November 3. http://www.castlelab.princeton.edu/Papers/Kim-Heavy%20tailed%20pricing%20and%20storageNov042010.pdf.

Kinsley, Michael J. 1997. *Economic Renewal Guide: A Collaborative Process for Sustainable Community Development.* 3rd ed. Snowmass, CO: Rocky Mountain Institute.

Klepeis, Neil E., W. C. Nelson, Wayne R. Ott, and John P. Robinson. 2001. "The National Human Activity Pattern Survey: A Resource for Assessing Exposure to Environmental Pollutants." *Journal of Exposure Analysis and Environmental Epidemiology* 11: 231–52.

Klingenberg, Katrin. 2011. Personal communication with author, March 28.

Knowles, Jonathan. 2011. Personal communication with author and colleagues, April 25.

Koelman, Onno. (PAX Scientific). 2011. Personal communication with author, May 25.

Kolstad, Ivar, Arne Wiig, and D. A. Williams. 2008. *Tackling Corruption in Oil Rich Countries: The Role of Transparency.* Chr. Michelsen Institute. http://www.cmi.no/publications/publication/?2938=tackling-corruption-in-oil-rich-countries.

Koomey, J., and N. E. Hultman. 2007. "A Reactor-Level Analysis of Busbar Costs for US Nuclear Plants, 1970–2005." *Energy Policy* 35 (11): 5630–42.

Koplow, Doug. 2007. "Energy." In *Subsidy Reform and Sustainable Development: Political Economy Aspects.* Paris: OECD Publishing. http://www.earthtrack.net/files/uploaded_files/OECD_Reform2007.pdf.

———. 2010. *Nuclear Power in the United States: Still Not Viable without Subsidies.* Union of Concerned Scientists. http://www.ucsusa.org/nuclear_power/nuclear_power_and_global_warming/nuclear-power-subsidies-report.html.

Korosec, Kirsten. 2010. "Ethanol Love-In: Valero Energy Snaps Up Another Plant." BNET, February 4. http://www.bnet.com/blog/clean-energy/ethanol-love-in-valero-energy-snaps-up-another-plant/1150.

Kota, Sridhar. 2011. Personal communication with author, January 31.

Krauss, Clifford. 2010. "Utilities Shift to Gas-Based Plants as Alternative to Coal." *New York Times*, November 29.

———. 2011. "Oil in Shale Sets Off a Boom in Texas." *New York Times*, May 27.

Kraemer, R. Andreas. 2011. "The Nuclear Power Endgame in Germany." American Institute for Contemporary German Studies. www.aicgs.org/analysis/c/kraemer063011.aspx.

Kromer, Matthew, and John Heywood. 2007. *Electric Powertrains: Opportunities and Challenges in the U.S. Light-Duty Vehicle Fleet.* Laboratory for Energy and the Environment. http://web.mit.edu/sloan-auto-lab/research/beforeh2/files/kromer_electric_powertrains.pdf.

Krupnick, Alan. 2010. *Energy, Greenhouse Gas, and Economic Implications of Natural Gas Trucks.* Resources for the Future; National Energy Policy Institute. http://www.rff.org/RFF/Documents/RFF-BCK-Krupnick-NaturalGasTrucks.pdf.

Kuhn Rikon. 2011. "Durotherm Thermal Cookware. http://www.kuhnrikon.com/products/duro/group.php3?id=3.

LaCommare, Kristina, Joseph Eto, and Kristina Hamachi. 2004. *Understanding the Cost of Power Interruptions to U.S. Electricity Consumers.* Lawrence Berkeley National Laboratory. http://certs.lbl.gov/pdf/55718.pdf.

LaFarge. 2010. *Annual Report Registration Document 2010.* Lafarge. http://www.lafarge.com/03222011-press_publication-2010_annual_report-uk.pdf.

Laherrère, Jean. 2010. Personal communications with author, December.

Laitner, Skip (American Council for an Energy-Efficient Economy). 2011. Personal communication with author, March 3.

Lanigan, Jack, John Zumerchik, Jean-Paul Rodrigue, Randall Guensler, and Michael Rodgers. 2006. "Shared Intermodal Terminals and the Potential for Improving the Efficiency of Rail-Rail Interchange." Presented to the Transportation Research Board Committee on Intermodal Freight Terminal Design and Operations at the 86th Annual Meeting of the Transportation Research Board, Washington DC, January 21–25. http://people.hofstra.edu/Jean-paul_Rodrigue/downloads/TRB_JPR_2007.ppt.

Lankarani, Nazanin. 2011. "Generating the Unlikeliest of Heroes." *New York Times*, April 18. http://www.nytimes.com/2011/04/18/business/global/18iht-rbog-barefoot-18.html.

Larson, Eric. D. 1991. *Trends in the Consumption of Energy-Intensive Basic Materials in Industrialized Countries and Implications for Developing Regions.* http://www.princeton.edu/pei/energy/publications/texts/Larson_91_Trends_Consumption_Basic_Materials.pdf.

Larson, Eric D., and Lars J. Nilsson. 1990. *A System-Oriented Assessment of Electricity Use and Efficiency in Pumping and Air-handling*. PU/CEES Report no. 253; Department of Environmental and Energy Systems Studies Report no. 1990:2. Lund, Sweden: Lund University, September. http://www.princeton.edu/pei/energy/publications/reports/No.253.pdf.

LaSalle Investment Management. 2010. *Investment Strategy Annual 2011*. LaSalle Investment Management. http://www.lasalle.com/Research/ResearchPublications/2011%20ISA%20Cover.pdf.

Laustsen, Jens. 2008. *Energy Efficiency Requirements in Building Energy Codes, Energy Efficiency Policies for New Buildings*. International Energy Agency, March. http://www.iea.org/g8/2008/Building_Codes.pdf.

Lawrence Berkeley National Laboratory. 2011. "The Lumina Project." http://light.lbl.gov/.

LeClaire, Nicole. 2011. Personal communication with author, February 7.

Lee, V. K. C., K. C. M. Kwok, W. H. Cheung, and G. McKay. 2007. "Operation of a Municipal Solid Waste Co Combustion Pilot Plant." *Asia Pacific Journal of Chemical Engineering* 2 (6): 631–39. doi:10.1002/apj.77.

Leonard, J. Wayne (Entergy). 2009. Personal communication with author, April 8.

Lester, D. R., M. Rudman, and G. Metcalfe. 2009. "Low Reynolds Number Scalar Transport Enhancement in Viscous and Non-Newtonian Fluids." *International Journal of Heat and Mass Transfer* 52 (3–4): 655–64. doi:10.1016/j.ijheatmasstransfer.2008.06.039.

Lidula, N. W. A., and A. D. Rajapakse. 2011. "Microgrids Research: A Review of Experimental Microgrids and Test Systems." *Renewable and Sustainable Energy Reviews* 15 (1): 186–202. doi:10.1016/j.rser.2010.09.041.

Liebreich, Michael. 2010. Personal communication with author, November 21.

Litman, Todd. 2005. "Pay-As-You-Drive Pricing and Insurance Regulatory Objectives." *Journal of Insurance Regulation* 23 (3). http://www.vtpi.org/jir_payd.pdf.

Liu, John. 2011. http://www.earthshope.org.

Lochbaum, David A. 2008. Testimony to the Select Committee on Energy Independence and Global Warming, U.S. House of Representatives, March 12. http://www.ucsusa.org/assets/documents/nuclear_power/20080312-ucs-house-nuclear-climate-testimony.pdf.

Lockwood, Charles. 2009. "Building Retrofits." *Urban Land*, November/December. http://www.esbnyc.com/documents/sustainability/uli_building_retro_fits.pdf.

LogLogic Inc. 2009. *Securing U.S. Critical Infrastructure from Cyber Attacks*. LogLogic. http://loglogic.com/resources/white-papers/securing-critical-infrastructure.

Long, Keith, Bradley Van Gosen, Nora Foley, and Daniel Cordier. 2010. *The Principal Rare Earth Elements Deposits of the United States: A Summary of Domestic Deposits and a Global Perspective*. U.S. Geological Survey. http://pubs.usgs.gov/sir/2010/5220/downloads/SIR10-5220.pdf.

Loveday, Eric. 2011. "Report: BMW: 3 Uses Carbon Fiber to . . . Cut Costs?" 27 July. http://green.autoblog.com/2011/07/27/report-bmw=:3uses-carbon-fiber-to-cut-costs/.

Lovins, Amory B. 1989. *The State of the Art: Drivepower*. Snowmass, CO: Competitek.

———. 1992. *Energy-Efficient Buildings: Institutional Barriers and Opportunities*. E Source Strategic Issues Paper #2. Boulder, CO: E Source. http://www.rmi.org/rmi/Library%2F1992-02_EnergyEfficientBuildingsBarriersOpportunities.

———. 1995. "The Super-Efficient Passive Building Frontier." *ASHRAE Journal* 37 (6): 79–81. http://www.rmi.org/rmi/Library/E95-28_SuperEfficientPassiveBuilding.

———. 2003. *Twenty Hydrogen Myths*. Snowmass, CO: Rocky Mountain Institute. http://www.rmi.org/rmi/Library%2FE03-05_TwentyHydrogenMyths.

———. 2005. "More Profit with Less Carbon." *Scientific American* 293 (3): 52–61. http://www.scientificamerican.com/article.cfm?id=more-profit-with-less-car.

———. 2006. "How Innovative Technologies, Business Strategies, and Policies Can Dramatically Enhance Energy Security and Prosperity." Invited Testimony to the U.S. Senate Committee on Energy and Natural Resources, March 7, 2006. http://www.rmi.org/rmi/Library/E06-02_SenateEnergyTestimony.

———. 2007. "Janine Benyus: Heroes of the Environment." *Time Magazine*, October 17, 2007. http://www.time.com/time/specials/2007/article/0,28804,1663317_1663319_1669888,00.html.

———. 2009. "Does a Big Economy Need Big Power Plants?" Freakonomics (blog), February 9. *New York Times*. http://www.rmi.org/rmi/Library/2009-06_FreakonomicsBlog.

———. 2010a. "DOD's Energy Challenge as Strategic Opportunity." *Joint Force Quarterly* 57: 33–42. http://www.ndu.edu/press/ifg_pages/editions/i57/lovins.pdf.

———. 2010b. National Defense University Press: Letters to the Editor. http://www.ndu.edu/press/lovins.html.

———. 2010c. "Nuclear Socialism." *Weekly Standard*, October 25. http://www.weeklystandard.com/articles/nuclear-socialism_508830.html.

———. 2010d. *Efficiency and Micropower for Reliable and Resilient Electricity Service: An Intriguing Case-Study from Cuba*. Snowmass, CO: Rocky Mountain Institute. http://www.rmi.org/rmi/Library/2010-23_CubaElectricity.

———. 2010e. "On Proliferation, Climate, and Oil: Solving for Pattern." *Foreign Policy*, January 21. http://www.rmi.org/rmi/Library/2010-03_ForeignPolicyProliferationOilClimatePattern (unabridged version at www.rmi.org/rmi/Library%2F2010-02_ProliferationOilClimatePattern).

Lovins, Amory B., and David Cramer. 2004. "Hypercars, Hydrogen, and the Automotive Transition." *International Journal of Vehicle Design* 35 (1): 50–85. http://www.rmi.org/rmi/Library/T04-01_HypercarsHydrogenAutomotiveTransition.

Lovins, Amory B., E. Kyle Datta, Odd-Even Bustnes, J. G. Koomey, and Nathan J. Glasgow. 2004. *Winning the Oil Endgame*. Snowmass, CO: Rocky Mountain Institute. http://www.oilendgame.com.

Lovins, Amory B., E. K. Datta, J. Swisher, A. Lehmann, T. Feiler, K. R. Rábago, and K. Wicker. 2002. *Small Is Profitable: The Hidden Economic Benefits of Making Electrical Resources the Right Size*. Snowmass, CO: Rocky Mountain Institute. http://www.smallisprofitable.org.

Lovins, Amory B., and A. Gadgil. 1991. "The Negawatt Revolution: Electric Efficiency and Asian Development." Published in abridged version in *Far Eastern Economic Review*, August. Full version at http://old.rmi.org/images/PDFs/Energy/E91-23_NegawattRevolution.pdf.

Lovins, Amory B. and L. H. Lovins. 1982. *Brittle Power: Energy Strategy for National Security*. Andover, MA: Brick House Pub. Co.. http://www.rmi.org/rmi/Library%2FS82-03_BrittlePowerEnergyStrategy.

———. 2001. "Fool's Gold in Alaska (Annotated Version)." *Foreign Affairs*, July/August. http://www.rmi.org/rmi/Library/E01-04_FoolsGoldAlaskaAnnotated.

Lovins, Amory B., L. H. Lovins. 1997. *Climate: Making Sense and Making Money*. Snowmass, CO: Rocky Mountain Institute. http://www.rmi.org/rmi/Library/C97-13_ClimateSenseMoney.

Lovins, Amory B., L. H. Lovins and L. Ross. 1980. "Nuclear Power and Nuclear Bombs." *Foreign Affairs* 58: 1137–77; 59: 172.

Lund, Per. 2011. "Running the Cell Controller Pilot Project as a Virtual Power Plant." Presented at the 4th International Conference on Integration of Renewable and Distributed Energy Resources, Albuquerque, NM, December 9.

Luppens, J. A, D. C Scott, J. E. Haacke, L. M. Osmonson, T. J. Rohrbacher, and M. S. Ellis. 2008. *Assessment of Coal Geology, Resources, and Reserves in the Gillette Coalfield, Powder River Basin, Wyoming*. U.S. Geological Survey. http://pubs.usgs.gov/of/2008/1202/pdf/ofr2008-1202.pdf.

Lutsey, Nicholas P. 2010. "Review of technical literature and trends related to automobile mass-reduction technology." Institute of Transportation Studies. University of California at Davis. UCD-ITS-RR-10-10. http://escholarship.org/uc/item/9t04t94w.

Lynd, Lee. 2006. *Biomass Energy Systems of the Future*. August 28. http://www.bioeconomyconference.org/images/Lynd,%20Lee—Keynote.pdf. Presented at Growing the Bioeconomy Conference, Ames, IA, August 28.

Magee, Darin. 2011. Personal communication with author, May 13.

Makhijani, Arjun, and Scott Saleska. 1999. *The Nuclear Power Deception*. New York: Apex Press.

Makower, Joel. 2011. "The Emergence of VERGE." GreenBiz, April 18. http://www.greenbiz.com/blog/2011/04/18/emergence-verge/?src-int.

Malin, Nadav. 2010. "Non-Green Office Buildings Sacrifice 8% in Rent Revenues." BuildingGreen.com, November 9. http://www.buildinggreen.com/auth/article.cfm/2010/11/9/Non-Green-Office-Buildings-Sacrifice-8-in-Rent-Revenues/.

Marsh, George. 2009. "Turbine Innovation at BWEA30." *Renewable Energy Focus,* January/February. http://www.insix.files.wordpress.com/2009/02/focus-magazine.pdf.

Martin, D. 2010. "A Steeper Learning Curve for PV." EXPO Solar, July 30. http://www.exposolar.org/2012/eng/center/contents.asp?idx=94&page=5&search=&searchstring=&news_type=C.

Martin, N., E. Worrell, M. Ruth, L. Price, R. Elliott, and A. Shipley. 2000. *Emerging Energy-Efficient Industrial Technologies.* Lawrence Berkeley National Laboratory. http://ies.lbl.gov/iespubs/46990.pdf

Maslin, Thomas. 2009. "US Solar Market Continues Evolution." In *Emerging Energy Research Breakfast Briefing Solar Power International.* Anaheim, CA: North America Solar Power Advisory.

Massachusetts Institute of Technology. 2010. *The Future of Natural Gas: An Interdisciplinary MIT Study.* Massachusetts Institute of Technology. http://web.mit.edu/mitei/research/studies/naturalgas.html.

———. 2011. "Power Plant Carbon Dioxide Capture and Storage Projects." CCS Project Database. http://sequestration.mit.edu/tools/projects/index.html.

Material Flows. 2011. "Trends in Global Resource Extraction, GDP and Material Intensity 1980–2007." http://www.materialflows.net database, maintained by Sustainable Europe Research Institute. http://www.materialflows.net/index.php?option=com_content&task=view&id=32&Itemid=48.

Matsuoka, Yuzuru. 2006. "Modeling Activity to Support Japan: LCS Toward 2050." In Japanese. http://www.nies.go.jp/gaiyo/media_kit/12.200606workshop/.../matsuoka.pdf.

Mayerowitz, Scott. 2009. "Austin Forces Home Sellers to Pay for Energy Audits." ABC News. http://abcnews.go.com/Business/Economy/story?id=7797595&page=1.

McCoy, Gilbert. 2010. "'Super Premium' Efficiency Motors Are Now Available." Washington State University Extension Energy Program. http://www.energy.wsu.edu/Documents/EEFactsheet-Motors-Dec22.pdf.

———. 2011. Personal communication with author, March 10.

McElroy, M. B., X. Lu, C. P. Nielsen, and Y. Wang. 2009. "Potential for Wind-Generated Electricity in China." *Science* 325 (5946): 1378. http://www.sciencemag.org/content/325/5946/1378.short.

McKinsey & Company. 2009a. *Unlocking Energy Efficiency in the US Economy.* McKinsey & Company. http://www.mckinsey.com/en/Client_Service/Electric_Power_and_Natural_Gas/Latest_thinking/Unlocking_energy_efficiency_in_the_US_economy.aspx.

———. 2009b. *Pathways to a Low-Carbon Economy.* McKinsey & Company, January. https://solutions.mckinsey.com/ClimateDesk/default.aspx.

———. 2010a. *A Portfolio of Power-Trains for Europe: A Fact-Based Analysis.* McKinsey & Company. http://www.europeanclimate.org/documents/Power_trains_for_Europe.pdf.

———. 2010b. *Energy Efficiency: A Compelling Global Resource.* McKinsey & Company, 209.172.180.115/clientservice/.../pdf/A_Compelling_Global_Resource.pdf.

McKinsey Global Institute. 2010. *Farewell to Cheap Capital? The Implications of Long-Term Shifts in Global Investment and Saving.* McKinsey Global Institute, December. http://www.mckinsey.com/mgi/publications/farewell_cheap_capital/pdfs/MGI_Farewell_to_cheap_capital_full_report.pdf.

McNeill, John. 2011. "Global Environmental History in the Age of Fossil Fuels (1800–2007)." Mapping the World. http://www.cartografareilpresente.org/article254.html.

Mehos, M., D. Kabel, and P. Smithers. 2009. "Planting the Seed: Greening the Grid with Concentrating Solar Power." *IEEE Power and Energy Magazine* 7 (May/June 2009).

Meier, A. 2011. Personal communication with author, May 19.

Messenger Post. 2010. "NYSEG to Study Compressed Air Energy Storage." MPNnow.com, December 6.

http://www.mpnnow.com/business/x1817614900/NYSEG-to-study-compressed-air-energy-storage.

Metcalfe, Guy, M. Rudman, A. Brydon, L. J. W. Graham, and R. Hamilton. 2006. "Composing Chaos: An Experimental and Numerical Study of an Open Duct Mixing Flow." *AIChE Journal* 52 (1): 9–28. doi:10.1002/aic.10640.

Meyers, Steve (Rational Energy). 2011. Personal communication with author, February.

Mick, Jason. 2011. "Company Poised to Blanket Former Sears Tower with 2 MW of Solar Panels." DailyTech, Blogs, March 22. http://www.dailytech.com/Company+Poised+to+Blanket+Former+Sears+Tower+With+2+MW+of+Solar+Panels/article21194.htm.

Mijuk, Goran, and Markus Germann. 2011. "Swiss To Exit Nuclear Power After Fukushima Disaster." May 25. http://online.wsj.com/article/BT-CO-20110525-710147.html.

Millard-Ball, Adam, and Lee Schipper. 2011. "Are We Reaching Peak Travel? Trends in Passenger Transport in Eight Industrialized Countries." *Transport Reviews* 31 (3): 357–78.

Mills, Evan. 2005. "The Specter of Fuel-Based Lighting." *Science* 308 (5726): 1263.

———. 2009. "Building Commissioning: A Golden Opportunity for Reducing Energy Costs and Greenhouse-Gas Emissions." Lawrence Berkeley National Laboratory. http://cx.lbl.gov/2009-assessment.html.

Mills, Evan, and Dale Sartor. 2005. "Energy Use and Savings Potential for Laboratory Fume Hoods." *Energy* 30 (10): 1859–64. doi:10.1016/j.energy.2004.11.008.

Milne, G., and C. Reardon. 2008. *Your Home Technical Manual* 5.2 "Embodied Energy." Section 5.2 of *Your Home Technical Manual*, 4th ed. Commonwealth of Australia. http://www.yourhome.gov.au/technical/fs52.html.

Mims, Natalie, and Heidi Hauenstein. 2008. *Feebates: A Legislative Option to Encourage Continuous Improvements to Automobile Efficiency.* Snowmass, CO: Rocky Mountain Institute. http://www.rmi.org/rmi/Library%2FT08-09_FeebatesLegislativeOption.

Mints, Paula. 2010. *Photovoltaic Manufacturer Shipments, Capacity & Competitive Analysis 2009/2010.* Palo Alto, CA: Navigant Consulting.

Mitchell, William J., Chris E. Borroni-Bird, and L. D. Burns. 2010. *Reinventing the Automobile: Personal Urban Mobility for the 21st Century.* Cambridge, MA: MIT Press.

Moody's Investors Service. 2009. *New Nuclear Generation: Ratings Pressure Increasing.* Moody's Investors Service, June. http://tinyurl.com/3cbplsd.

Molly, J.P. 2010. "Status der Windenergienutzung in Deutschland—Stand 31.12.2010," p. 7, DEWI GmbH. www.dewi.de/dewi/fileadmin/pdf/publications/Statistics%20Pressemitteilungen/31.12.10/Foliensatz_2010.pdf.

Morales, Alex. 2011. "Low-Carbon Energy Investment Hit a Record $243 Billion in 2010, BNEF Says." Bloomberg, January 11. http://www.bloomberg.com/news/2011-01-11/low-carbon-energy-investment-hit-a-record-243-billion-in-2010-bnef-says.html.

Moxley, Joel. 2011. Personal communication with author, February 11.

Mufson, Steven. 2011. "Coal's Burnout: Have Investors Moved On to Cleaner Energy Sources?" *Washington Post*, January 1. http://www.washingtonpost.com/wp-dyn/content/article/2011/01/01/AR2011010102146.html.

Mujumdar, Arun S. 2006. *Handbook of Industrial Drying.* 3rd ed. Boca Raton, FL: CRC Press.

Mullen, M. G. 2011. "From the Chairman." *Joint Force Quarterly* 60 (first quarter). http://www.ndu.edu/press/from-the-chairman-60.html.

Munger, Reuben. 2011. Personal communication with author, May 2.

Murawski, John. 2011. "Utilities Won't Get Nuclear Cost Break This Year." *Charlotte Observer*, May 4. http://www.charlotteobserver.com/2011/05/04/2269975/utilities-wont-get-nuclear-cost.html#ixzz1Lyr4kGoR.

Murdoch, James. "Clean Energy Conservatives Can Embrace." 2009. *Washington Post*, December 4. http://www.washingtonpost.com/wp-dyn/content-article/2009/12/03/AR2009120303698.html.

Nadel, Steven, and D. Goldstein. 1996. *Appliance and Equipment Efficiency Standards: History, Impacts,*

Current Status, and Future Directions. Washington DC: American Council for an Energy-Efficient Economy, June. http://www.aceee.org/sites/default/files/publications/researchreports/A963.pdf.

Nan Zhou, David Fridley, Michael McNeil, Nina Zheng, Jing Ke, and Mark Levine. 2011. *China's Energy and Carbon Emissions Outlook to 2050*. LBNL-4472E. Lawrence Berkeley National Laboratory, April. http://china.lbl.gov/publications/Energy-and-Carbon-Emissions-Outlook-of-China-in-2050.

NanoPore Inc. 2008. "NanoPore™ Thermal Insulation." http://www.nanopore.com/thermal.html.

National Academy of Sciences. 2009. *Hidden Costs of Energy: Unpriced Consequences of Energy Production and Use*. Washington DC: The National Academies Press. www.nap.edu/catalog.php?record_id=12794.

———. 2010. *Real Prospects for Energy Efficiency in the United States*. Washington DC: The National Academies Press.

National Association of Home Builders. 2006. *Housing Facts, Figures and Trends*. National Association of Home Builders. http://www.soflo.fau.edu/report/NAHBhousingfactsMarch2006.pdf.

National Energy Technology Laboratory. 2011. "Clean Coal Power Initiative (CCPI)." NETL, Coal & Power Systems, Major Demonstrations. http://www.netl.doe.gov/technologies/coalpower/cctc/ccpi/index.html.

National Gypsum. 2011. "ThermalCORE PCM Panel." http://www.thermalcore.info/ThermalCore.pdf.

National Mining Association. 2010. *Trends in U.S. Coal Mining 1923–2009*. National Mining Association. http://www.nma.org/pdf/c_trends_mining.pdf.

National Renewable Energy Laboratory. 2009. "NREL Sets the Bar for Office Building Energy Use." NREL Newsroom, December 7. http://www.nrel.gov/features/20091207_rsf.html.

———. 2010a. *Research Support Facility: A Model of Super Efficiency*. National Renewable Energy Laboratory. http://www.nrel.gov/docs/fy10osti/48943.pdf.

———. 2010b. *Wind Maps and Wind Resource Potential Estimates*. Resource materials provided by the Wind Powering America project of the National Renewable Energy Laboratory. http://www.windpoweringamerica.gov/windmaps/.

———. 2011a. *National Solar Radiation Data Base 1991–2005 Update*. National Renewable Energy Laboratory. http://rredc.nrel.gov/solar/old_data/nsrdb/1991-2005/.

———. 2011b. *Western Wind and Solar Integration Study 2004 Wind Data*. National Renewable Energy Laboratory. http://wind.nrel.gov/public/WWIS/.

National Renewable Energy Laboratory and AWS True Wind. 2010. *Estimates of Windy Land Area and Wind Energy Potential by State for Areas ≥30% Capacity Factor at 80m*. NREL and AWS True Wind, February 4. http://www.windpoweringamerica.gov/pdfs/wind_maps/wind_potential_80m_30percent.pdf.

National Research Council. 2005. *Safety and Security of Commercial Spent Nuclear Fuel Storage*. Washington DC: The National Academies Press.

———. 2008. *Severe Space Weather Events—Understanding Societal and Economic Impacts: A Workshop Report*. Washington DC: The National Academies Press.

———. 2010. *Technologies and Approaches to Reducing the Fuel Consumption of Medium- and Heavy-Duty Vehicles*. Washington DC: The National Academies Press.

Navigant Consulting. 2008. *Florida Renewable Energy Potential Assessment*. Navigant Consulting. http://www.psc.state.fl.us/utilities/electricgas/RenewableEnergy/FL_Final_Report_2008_12_29.pdf.

———. 2010. *Energy Savings Potential of Solid-State Lighting in General Illumination Applications 2010 to 2030*. Washington DC: U.S. Department of Energy.

Nelson, K.E. 1998. "Finding Process Improvements: The Six Places to Look!" In *Proceedings from the Twentieth National Industrial Energy Technology Conference, Houston, TX, April 22–23, 1998*. Houston, TX. http://repository.tamu.edu/handle/1969.1/91180?show=full.

Network for New Energy Choices. 2010. *Freeing the Grid: Best Practices in State Net Metering Policies and Interconnection Procedures*. Network for New Energy Choices. http://www.newenergychoices.org/index.php?page=publications&sd=no.

Neubaurer, Max, Andrew deLaski, Marianne DiMascio, and Steven Nadel. 2009. *Ka-Boom! The Power of Appliance Standards*. Report no. ASAP-7/ACEEE-A091. Washington DC: American Council for an Energy-Efficient Economy; Boston: Appliances Standards Awareness Project. http://www.aceee.org/sites/default/files/publications/researchreports/a091.pdf.

New York Times. 2010. "The Future of Cars." *New York Times*, June 24.

Newell, Richard. 2011. "The Long-Term Outlook for Natural Gas." Presented at the United States Energy Consultation, Washington DC, February 2. http://www.eia.doe.gov/neic/speeches/newell_aeo_ng.pdf.

Newhouse, John. 2007. *Boeing versus Airbus: The Inside Story of the Greatest International Competition in Business*. New York: Knopf.

Newton, James. 1989. *Uncommon Friends: Life with Thomas Edison, Henry Ford, Harvey Firestone, Alexis Carrel, and Charles Lindbergh*. New York: Mariner Books.

NextEra Energy Resources. 2010. "NextEra Energy Resources Remains Nation's No. 1 Wind Energy Owner, According to New AWEA Report." April 8. http://www.nexteraenergyresources.com/news/contents/2010/040810a.shtml.

Nicholson, E., J. Rogers, and K. Porter. 2010. *The Relationship between Wind Generation and Balancing-Energy Market Prices in ERCOT: 2007–2009*. U.S. National Renewable Energy Laboratory. http://www.nrel.gov/wind/systemsintegration/pdfs/2010/nicholson_balancing_energy_market.pdf.

Norfolk Southern. 2009. "Batteries Are Included: Norfolk Southern Unveils Experimental Electric Locomotive." Press release, September 28. http://www.nscorp.com/nscportal/nscorp/Media/News%20Releases/2009/batteries.html.

Norris, Guy, and Mark Wagner. 2009. *Boeing 787 Dreamliner*. Minneapolis, MN: Zenith Press.

North American Electric Reliability Corporation. 2010a. *2010 Special Reliability Scenario Assessment: Resource Adequacy Impacts of Potential U.S. Environmental Regulations*. North American Electric Reliability Corporation, October. http://www.nerc.com/files/EPA_Scenario_Final.pdf.

———. 2010b. *2010 Long-Term Reliability Assessment*. North American Electric Reliability Corporation. October. http://www.nerc.com/files/2010%20LTRA.pdf.

———. 2010c. *2005–2009 Generating Unit Statistical Brochure—All Units Reporting*. North American Electric Reliability Corporation. http://www.nerc.com/page.php?cid=4|43|47.

———. 2011. *System Disturbance Reports*. North American Electric Reliability Corporation. http://www.nerc.com/page.php?cid=5|66.

Northeast States Center for a Clean Air Future, International Council on Clean Transportation, Southwest Research Institute, and TIAX. 2009. *Reducing Heavy-Duty Long Haul Combination Truck Fuel Consumption and CO_2 Emissions*. Boston: NESCCF; Washington DC and San Francisco: ICCT. http://www.nescaum.org/documents/heavy-duty-truck-ghg_report_final-200910.pdf.

Northwest Energy Efficiency Alliance. 2009. *CFL Update*. Northwest Energy Efficiency Alliance, November 5. http://neea.org/research/documents/CFL_Slide_Presentation_111009.pdf.

Northwest Power and Conservation Council. 2010a. "Energy Efficiency in the Future: The Sixth Northwest Power Plan" and "Northwest Energy Efficiency Achievements 1980–2008." http://www.nwcouncil.org/library/2010/2010-08.htm.

———. 2010b. *Sixth Northwest Power Plan*. Northwest Power and Conservation Council. http://www.nwcouncil.org/energy/powerplan/6/default.htm.

Norton, Paul. 2010. "Utility Data Analysis: Program Evaluation and Tools Improvement." Presented at Affordable Comfort, Inc., Austin, TX, April 21, on behalf of the National Renewable Energy Laboratory. http://www.affordablecomfort.org/images/Events/46/Courses/1622/MODL2_Norton.pdf.

Novak, Shonda. 2010. "Impact of Home Energy Audit Rule Less Than Expected." *American Statesman* (Austin, TX), July 16. http://www.statesman.com/business/impact-of-home-energy-audit-rule-less-than-807664.html.

NRDC. 2011. "Gas and Electric Decoupling in the US" map set. July. San Francisco: Natural Resources Defense Council.

Nuveco AG. 2011. "The New Generation of Cooking." http://www.lifepr.de/attachment/35202/PRESENTATION_PDF.pdf. The firm has been renamed and is at conduction.ch.

Oak Ridge National Lab (ORNL). 2008. *Combined Heat and Power: Effective Energy Solutions for a Sustainable Future*. Oak Ridge National Laboratory, December 1.

http://www1.eere.energy.gov/industry/distributedenergy/pdfs/chp_report_12-08.pdf

Office of Management and Budget. 2011. "Table 3.2—Outlays by Function and Subfunction: 1962–2016." Historical Tables. http://www.whitehouse.gov/omb/budget/historicals.

Offshore Valuation Group. 2010. *The Offshore Valuation*. Y Plas, Machynlleth, Wales: Public Interest Research Centre. http://www.offshorevaluation.org/downloads/offshore_vaulation_full.pdf.

Ogburn, Michael, Laurie Ramroth, and Amory Lovins. 2008. *Transformational Trucks: Determining the Energy Efficiency Limits of a Class-8 Tractor-Trailer.* Snowmass, CO: Rocky Mountain Institute, July. http://www.rmi.org/rmi/Library/T08-08_TransformationalTrucksEnergyEfficiency.

O'Grady, Eileen. 2011. "NRG Energy Abandons Texas Nuclear Expansion Plan." Reuters, April 19. http://www.reuters.com/article/2011/04/19/us-nuclear-nrg-idUSTRE73I7E620110419.

Online Code Environment and Advocacy Network. 2011a. *Code Status: Commercial*. Online Code Environment and Advocacy Network. http://bcap-ocean.org/code-status-commercial.

———. 2011b. *Code Status: Residential*. Online Code Environment and Advocacy Network. http://bcap-ocean.org/code-status-residential.

Oregon DOT. 2007. *Oregon's Mileage Fee Concept and Road User Fee Pilot Program*. Oregon Department of Transportation. http://www.oregon.gov/ODOT/HWY/RUFPP/docs/RUFPP_finalreport.pdf.

Orszag, Peter. 2007. *Estimated Costs of U.S. Operations in Iraq and Afghanistan and of Other Activities Related to the War on Terrorism*. Washington DC: Congressional Budget Office, October 24.

Owen, Nick A., Oliver R. Inderwildi, and David A. King. 2010. "The Status of Conventional World Oil Reserves—Hype or Cause for Concern?" *Energy Policy* 38 (8): 4743–49. doi:16/j.enpol.2010.02.026.

PACEnow. 2011. PACE blog and portal. www.pacenow.org.

Paidipati, J., L. Frantzis, H. Sawyer, and A. Kurrasch. 2008. *Rooftop Photovoltaics Market Penetration Scenarios*. National Renewable Energy Laboratory. http://www.nrel.gov/docs/fy08osti/42306.pdf.

Paley, William. 1952. *Resources for Freedom: Report of the President's Materials Commission*. Washington DC: U.S. Government Printing Office.

Palmintier, Bryan, Lena Hansen, and Jonah Levine. 2008. "Spatial and Temporal Interactions of Solar and Wind Resources in the Next Generation Utility." Presented at SOLAR 2008, San Diego, CA, May 3-8. http://www.rmi.org/rmi/Library/2008-21_SolarWindNGU.

Parker, D. S. 2009. "Very Low Energy Homes in the United States: Perspectives on Performance from Measured Data." *Energy and Buildings* 41 (5): 512–20.

Parsons Brinckerhoff. 2009. *Powering the Future: Mapping Our Low Carbon Path to 2050.* Newcastle upon Tyne, UK: Parsons Brinckherhoff, December. http://www.pbpoweringthefuture.com.

Parthemore, Christine, and John Nagl. 2010. *Fueling the Future Force: Preparing the Department of Defense for a Post-Petroleum Era*. Center for a New American Security. http://www.cnas.org/files/documents/publications/CNAS_Fueling%20the%20Future%20Force_NaglParthemore.pdf.

Patel, M. R. 2006. *Wind and Solar Power Systems: Design, Analysis, and Operation*. Boca Raton, FL: CRC Press.

Patton, K. J., and M. A. Gonzales. 2010. "Development of High-Efficiency Clean Combustion Engines: Designs for SI and CI Engines." Presented at GM Powertrain Advanced Engineering, June 11.

Patzek, T. W., and G. D. Croft. 2010. "A Global Coal Production Forecast with Multi-Hubbert Cycle Analysis." *Energy* 35 (8): 3109–22. doi:10.1016/j.energy.2010.02.009.

PAX Scientific. 2008. "A Fascination with Flow." PAX Scientific Technology. http://www.paxscientific.com/tech.html.

PAX Water Technologies. 2011. *Inside America's Water Storage Tanks There's a Revolution Going on . . .* http://www.paxwater.com/Default.aspx?app=LeadgenDownload&shortpath=docs%2fData_Sheet_PAX_Mixer.pdf.

Perlack, Robert, Lynn Wright, and Anthony Turhollow. 2005. *Biomass as Feedstock for a Bioenergy and Bioproducts Industry: Technical Feasibility of a Billion-Ton Annual Supply*. U.S. Department of Agriculture. http://www.scag.ca.gov/rcp/pdf/summit/billion_ton_vision.pdf.

Pernick, Ron, and Clint Wilder. 2011. *Clean Energy Trends 2011*. Clean Edge, March. http://www.cleanedge.com/reports/pdf/Trends2011.pdf.

Pew Center on Global Climate Change. 2011a. *Energy Efficiency Standards and Targets*. Pew Center on Global Climate Change, February 10. http://www.pewclimate.org/what_s_being_done/in_the_states/efficiency_resource.cfm.

———. 2011b. *Decoupling Policies*. Pew Center on Global Climate Change, February 10. http://www.pewclimate.org/what_s_being_done/in_the_states/decoupling.

Pew Charitable Trusts. 2010. *Who's Winning the Clean Energy Race?: Growth, Competition, and Opportunity in the World's Largest Economies*. Washington DC: Pew Charitable Trusts. http://www.pewtrusts.org/uploadedFiles/wwwpewtrustsorg/Reports/Global_warming/G-20%20Report.pdf.

Pew Charitable Trusts. 2011. *Who's Winning the Clean Energy Race?* 2010 Edition. G-20 Investment Powering Forward. March 29. www.pewenvironment.org/uploadedFiles/PEG/Publications/Report/G-20Report-LOWRes-FINAL.pdf.

Phadke, Amol (Lawrence Berkeley National Laboratory). 2011. Personal communication with author, May 5.

PJM News. 2011. "Demand Response and Energy Efficiency Continue to Grow in PJM's RPM Auction Renewable Resources Offer More Capacity." *Electric Energy Online*, May 16. http://www.electricenergyonline.com/?page=show_news&id=155034&cat=5.

Pleotint. 2007. "Pleotint LLC Sunlight Responsive Thermochromic (SRT)." http://www.pleotint.com/default.asp.

Plumber, Bradford. 2007. "Wind Vain." *New Republic*, November 5.

Pollard, S. 1980. "A New Estimate of British Coal Production, 1750–1850." *Economic History Review* 33 (2): 212–34.

Popular Mechanics. 2010. "The New Wildcatters." *Popular Mechanics*, January 1. http://www.popularmechanics.co.za/article/the-new-wildcatters-2010-01-01.

Pöyry. 2010. *Wind Energy and Electricity Prices: Exploring the "Merit Order Effect."* The European Wind Energy Association. http://www.ewea.org/fileadmin/ewea_documents/documents/publications/reports/MeritOrder.pdf.

PPG Aerospace. 2010. "Boeing 787 Flies with PPG Aerospace Transparencies, Coatings, Sealants." http://www.ppg.com/coatings/aerospace/newsroom/news/Pages/2010-02-04.aspx.

Proctor Group LTD. 2010. "Spacetherm®." http://www.proctorgroup.com/Products/ThermalInsulation/Spacetherm.aspx.

Pugh, Scott. 2010. "Protecting America's Electric Grid." Presentation at the Electric Power Grid Resiliency Workshop, Monterey, CA, May 19–21.

RavenBrick, LLC. 2009. "The Technology." http://www.ravenbrick.com/.

Red Eléctrica de España. 2010. *El sistema eléctrico español 2010*. Red Eléctrica de España, December 21. http://www.ree.es/sistema_electrico/pdf/infosis/Avance_REE_2010.pdf.

Reddy, A. K. N., R. H. Williams, and T. B. Johansson. 1997. *Energy After Rio: Prospects and Challenges*. New York: UN Development Programme.

Redpoint Energy. 2011. *The Impact of Wind on Pricing within the Single Electricity Market*. Irish Wind Energy Association. http://www.iwea.com/contentFiles/Documents%20for%20Download/Publications/News%20Items/Impact_of_Wind_on_Electricity_Prices.pdf?uid=1298912434703.

Reece, Erik. 2009. "Hell Yeah, We Want Windmills." *Orion Magazine*, July/August. http://www.orionmagazine.org/index.php/articles/article/4809/.

Reed, J. H., K. Johnson, J. Riggert, and A. Oh. 2004. *Who Plays and Who Decides: The Structure and Operation of the Commercial Building Market*. Report prepared for the U.S. Department of Energy Office of Building Technology, State and Community Programs, March. Rockland, MD: Innovologie, LLC.

Regeringen (Government of Denmark). 2011. *Energistrategi 2050—fra kul, olie og gas til grøn energi*. Regeringen, February 24. http://www.ens.dk/Documents/Netboghandel%20-%20publikationer/2010/Energistrategi_2050_-_final-print_A4.pdf.

Reitenbach, Gail. 2010. "Smart Power Generation at UCSD." *Power Magazine*, November 1. http://www.slideshare.net/UCSD-Strategic-Energy/

smart-power-generation-at-ucsd-power-magazine-nov-2010.

Renewable Energy Policy Network for the 21st Century (REN21). 2010. *Renewables 2010 Global Status Report*. Renewable Energy Policy Network for the 21st Century (REN21), July 15. http://www.ren21.net/REN21Activities/Publications/GlobalStatusReport/GSR2010/tabid/5824/Default.aspx.

Renewable Fuels Association. 2010. "Statistics." Renewable Fuels Association. http://www.ethanolrfa.org/pages/statistics.

Renewables International. 2011. "Cost of Turnkey PV in Germany Drops." *Renewables International—The Magazine*, January 14. http://www.renewablesinternational.net/cost-of-turnkey-pv-in-germany-drops/150/452/29911/.

Reuters. 2007. "U.S. CBO Estimates $2.4 Trillion Long-Term War Costs." October 24.

———. 2011a. "EU Will Surpass 20 pct Green Energy Goal." January 4.

———. 2011b. "Japan's Tepco Reports Record Loss after Quake." *Sydney Morning Herald*, May 20. http://tinyurl.com/3o67jfk.

Richardson, Katherine et al. 2011. "Denmark's Road Map for Fossil Fuel Independence." Solutions 2(4). July 21.

Rocky Mountain Institute. 2008. "State of Missouri Department of Natural Resources Lewis & Clark State Office Building." A case study from the Rocky Mountain Institute film *High Performance Building: Perspective and Practice*. http://bet.rmi.org/files/case-studies/mo-dnr/State_of_Missouri.pdf.

———. 2010. Micropower Database. September. http://www.rmi.org/rmi/Library/2010-14_MicropowerDatabaseSeptember2010.

Rohmund, Ingrid, Anthony Duer, Sharon Yoshida, Jan Borstein, Lisa Wood, and A. Cooper. 2011. *Assessment of Electricity Savings in the U.S. Achievable through New Appliance/Equipment Efficiency Standards and Building Efficiency Codes (2010–2025)*. Washington DC: Institute for Energy Efficiency, May. http://www.edisonfoundation.net/IEE/reports/IEE_CodesandStandardsAssessment_2010-2025_UPDATE.pdf.

Roland-Holst, David. 2008. *Energy Efficiency, Innovation, and Job Creation in California*. Berkeley, CA: Center for Energy, Resources, and Economic Sustainability (CERES), October. http://areweb.berkeley.edu/~dwrh/CERES_Web/Docs/UCB%20Energy%20Innovation%20and%20Job%20Creation%2010-20-08.pdf.

Romm, J. 1997. *Technology Scenarios of U.S. Carbon Reduction*. Briefing paper, U.S. Department of Energy, July 25.

Rose, Melinda. 2010. "Displays Add a Dimension, Durability." *Photonics Spectra*, July. http://www.photonics.com/Article.aspx?AID=42944.

Rosen, D. H., and T. Houser. 2006. *What Drives China's Demand for Energy (and What It Means for the Rest of Us)*. Center for Strategic and International Studies. http://csis.org/files/media/csis/pubs/090212_02what_drives_china_demand.pdf.

Rosenfeld, A., J. J. Romm, Hashem Akbari, and Alan C. Lloyd. 1997. "Painting the Town White—and Green." *MIT Technology Review*, February/March 1997. http://heatisland.lbl.gov/PUBS/PAINTING/.

Rosenthal, E. 2010a. "African Huts far from the Grid Glow with Renewable Power." *New York Times*, December 25. http://www.nytimes.com/2010/12/25/science/earth/25fossil.html.

———. 2010b. "Using Waste, Swedish City Shrinks Its Fossil Fuel Use." *New York Times*, December 10.

———. 2010c. "Portugal gives itself a clean-energy makeover." *New York Times*, Aug. 11. http://www.nytimes.com/2010/08/10/science/earth/10portugal.html.

———. 2011. "In Auto Test in Europe, Meter Ticks Off Miles, and Fee to Driver." *New York Times*. 10 August. http://www.nytimes.com/2011/08/11/science/earth/11meter.html.

Ross, J. P., and A. Meier. 2000. *Whole House Measurements of Stand-by Power Consumption*. LBNL-45967. Lawrence Berkeley National Laboratory. http://eetd.lbl.gov/ea/reports/45967.pdf.

Ross, Marc, and Tom Wenzel. 2001. "Losing Weight to Save Lives: A Review of the Role of Automobile Weight and Size in Traffic Fatalities." American Council for an Energy-Efficient Economy Report ACEEE-T013. July. http://eetd.lbl.gov/ea/teepa/pdf/LBNL-48009.pdf.

Rudolf, John Collins. 2011. "News Corp. Is Carbon-Neutral, Murdoch Declares." *New York Times*, March 4. http://green.blogs.nytimes.com/2011/03/04/news-corp-is-carbon-neutral-murdoch-declares/?hpw.

Rumsey, Peter. 2011. Personal communication with author, March 7.

Rutledge, David. 2010. Personal communications with author, December.

———. 2011. "Estimating Long-Term World Coal Production with Logit and Probit Transforms." *International Journal of Coal Geology* 85 (1): 23–33. doi:10.1016/j.coal.2010.10.012.

Sandborg, Hans. 2010. "Taking Green to the Platinum Level at the Empire State Building." *Currents* (published by the Swedish-American Chambers of Commerce), August 12. http://sacc-usa.org/currents/people/taking-green-to-the-platinum-level-at-the-empire-state-building/.

Sator, Spencer. 2008. *Managing Office Plug Loads*. Boulder, CO: E Source.

Sator, Spencer, I. Krepchin, and M. Horsey. 2010. "Standout Industrial Programs and Technologies." Presented at E Source, Boulder, CO, June 30, 2010.

Scanlon, Bill. 2010. "Energy Saving A/C Conquers All Climates." National Renewable Energy Laboratory, June 11. http://www.nrel.gov/features/20100611_ac.html.

Scheihing, Paul. 2009. *United States Industrial Motor-Driven Systems Market Assessment: Charting a Roadmap to Energy Savings for Industry*. U.S. Department of Energy. http://www1.eere.energy.gov/industry/bestpractices/m/us_industrial_motor_driven.html.

Schewel, Laura, and Lee Schipper. 2011. *Shop Till We Drop: Historical and policy analysis of driving-for-shopping and freight energy use in the U.S.* Presented at the European Council for an Energy-Efficient Economy 2011 Summer Study, June 7, France.

Schipper, Lee. 2010. "Automobile Use, Fuel Economy and CO_2 Emissions in Industrialized Countries: Encouraging Trends Through 2008?" *Transport Policy* 18 (2): 358–72.

Schmidt-Pathmann, Philipp. 2011. Personal communication with author, April 27.

Schneider, C., and Jonathan Banks. 2010. *The Toll from Coal: An Updated Assessment of Death and Disease from America's Dirtiest Energy Source*. Clean Air Task Force, September. http://www.catf.us/resources/publications/files/The_Toll_from_Coal.pdf.

Schneider, Dana. 2011. Personal communication with author, March 18.

Schneider, M., Antony Froggatt, and Steve Thomas. 2011. *The World Nuclear Industry Status Report 2010–2011: Nuclear Power in a Post-Fukushima World, 25 Years after the Chernobyl Accident*. Worldwatch Institute and Mycle Schneider Consulting. http://download.www.arte.tv/permanent/u1/tchernobyl/report2011.pdf.

Schwede, Jared W., Igor Bargatin, Daniel C. Riley, Brian E. Hardin, S. J. Rosenthal, Yun Sun, Felix Schmitt, et al. 2010. "Photon-Enhanced Thermionic Emission for Solar Concentrator Systems." *Nature Materials* 9: 762–67. doi:10.1038/nmat2814.

Schewels, Laura 2008. "Triple Safety: Lightweighting Automobiles to Improve Occupant, Highways, and Global Safety," SAE 2008-01-1282. http://www.rmi.org.rmi/Library/2008-23_Triple Safety Lightweighting.

SEAD Group. 2011. Personal communication with author, May 5.

Segall, Justin (Simple Energy). 2011. Personal communication with author, March 25.

Shaheen, Susan, and Adam Cohen. 2006. "Carsharing in North America: Market Growth, Current Developments, and Future Potential." *Transportation Research Record: Journal of the Transportation Research Board of the National Academies No. 1986*:9. http://pubs.its.ucdavis.edu/publication_detail.php?id=1080.

Shell International. 2011. *Shell Energy Scenarios to 2050: Signals and Signposts*. The Hague: Shell International. http://www.shell.com/scenarios/.

Sherwood, Larry. 2010. *U.S. Solar Market Trends 2009*. Interstate Renewable Energy Council, July. http://irecusa.org/wp-content/uploads/2010/07/IREC-Solar-Market-Trends-Report-2010_7-27-10_web1.pdf.

Shimomura, Tsutomu, and John Gage. 2009–2011. Personal communications with author.

Shipley, A., and R. Elliott. 2006. *Ripe for the Picking: Have We Exhausted the Low-Hanging Fruit in the Industrial Sector?* Report no. IE061. Washington DC: American Council for an Energy-Efficient Economy, April. http://www.aceee.org/research-report/ie061.

Shirouzu, Norihiko. 2010. "Train Makers Rail Against China's High-Speed Designs." *Wall Street Journal*, November 17. http://online.wsj.com/article/SB10001424052748704814204575507353221141616.html.

Short, W., N. Blair, P. Sullivan, and T. Mai. 2009. *ReEDS Model Documentation: Base Case Data and Model Description*. National Renewable Energy Laboratory. http://www.nrel.gov/analysis/reeds/pdfs/reeds_full_report.pdf.

Shoup, Donald. 2005. *The High Cost of Free Parking*. Washington DC: American Planning Association.

Shpitsberg, Anna. 2010. *Federal Freight Efficiency Authority*. Rocky Mountain Institute. http://www.rmi.org/rmi/Library/2010-28_FederalFreightEfficiencyAuthority.

Shuster, Erik. 2009. *Estimating Freshwater Needs to Meet Future Thermoelectric Generation Requirements*. U.S. Department of Energy National Energy Technology Laboratory, September 30. http://www.netl.doe.gov/energy-analyses/refshelf/detail.asp?pubID=278.

Sinden, Graham. 2011. Personal communication with author, May 11.

Singapore National Energy Efficiency Committee. 2010. *Green Energy Management (GEM) at Grand Hyatt Singapore*.

Skea, J., D. Anderson, T. Green, R. Gross, P. Heptonstall, and M. Leach. 2008. "Intermittent Renewable Generation and the Cost of Maintaining Power System Reliability." *IET Generation, Transmission & Distribution* 2 (1): 82–89. doi:10.1049/iet-gtd:20070023.

Small, Kenneth, and Kurt Van Dender. 2006. "Fuel Efficiency and Motor Vehicle Travel: The Declining Rebound Effect." *Energy Journal* 28 (1): 25–51.

SmartWay Transport Partnership. 2004. *A Glance at Clean Freight Strategies: Intermodal Shipping*. U.S. Environmental Protection Agency. http://www.epa.gov/smartwaytransport/transport/documents/tech/intermodal-shipping.pdf.

Smith, M. 2011. "Japan Faces Lengthy Recovery from Fukushima Accident." CNN, April 22. http://articles.cnn.com/2011-04-22/world/japan.fukushima.future_1_nuclear-power-plant-plutonium-plant-nuclear-accident?_s=PM:WORLD.

Smith, T. 2011. "Shale Gas: A Renaissance for DRI Production in USA?" *Steel Times International*, March.

Sócrates, José. 2011. Address to World Energy Summit, Abu Dhabi, January 17.

Solar Energy Research Institute. 1990. *The Potential of Renewable Energy*. Interlaboratory White Paper SERI/TP-260-3674. Golden, CO: Solar Energy Research Institute (now National Renewable Energy Laboratory).

SolarCity. 2011. "SolarCity Solar Lease: SolarLease Lets Homeowners Install Solar Power for $0 Down and Save Money Every Month with Lower Electricity Costs." http://www.solarcity.com/residential/solar-lease.aspx.

Sony. 2011. "OLED Specifications." http://www.sonystyle.com/webapp/wcs/stores/servlet/CategoryDisplay?catalogId=10551&storeId=10151&langId=-1&categoryId=8198552921644579396#.

Sorrell, Steve. 2007. *The Rebound Effect: an assessment of the evidence for economy-wide energy savings from improved energy efficiency*. UK Energy Research Centre. Main report and five Technical Reports. www.ukerc.ac.uk/support/tiki-index.php?page=ReboundEffect.

Sorrell, Steve. 2009. "The Evidence for Direct Rebound Effects." In *Energy Efficiency and Sustainable Consumption: The Rebound Effect*. New York: Palgrave Macmillan.

SourceWatch. 2011. *Coal and Jobs in the United States*. SourceWatch, April. http://www.sourcewatch.org/index.php?title=Coal_and_jobs_in_the_United_States.

Sovacool, Benjamin K., and Charmaine Watts. 2009. "Going Completely Renewable: Is It Possible (Let Alone Desirable)?" *The Electricity Journal* 22 (4): 95–111. doi:16/j.tej.2009.03.011.

St. John, Jeff. 2011. "Demand Response 'Negawatts' Getting a Pay Day." GigaOM, March 15. http://gigaom.com/cleantech/demand-response-%E2%80%9Cnegawatts%E2%80%9D-getting-a-pay-day/.

Stand-To! 2011. "Today's Focus: Net Zero." *Stand To!* January 31. http://www.army.mil/standto/archive/2011/01/31/?s_cid=email.

Standard & Poor's. 2008. *Key Credit Factors: Business and Financial Risks in the Investor-Owned Utilities Industry*. Standard & Poor's, November 26.

Stern, Roger J. 2010. "United States Cost of Military Force Projection in the Persian Gulf, 1976-2007." *Energy Policy* 38 (6): 2816–25. http://ideas.repec.org/a/eee/enepol/v38y2010i6p2816-2825.html.

Stiglitz, Joseph E., and Linda Blimes. 2010. "The True Cost of the Iraq War: $3 Trillion and Beyond." *Washington Post*, September 5. http://www.washingtonpost.com/wp-dyn/content/article/2010/09/03/AR2010090302200.html.

Stockton, Hon. Paul (Asst. Sec. of Def.). 2011. Testimony to Subcommittee on Energy and Power. USHR Committee on Energy and Commerce. May 31.

Straub, N., and P. Behr. 2009. "Energy Regulatory Chief Says New Coal, Nuclear Plants May Be Unnecessary." *New York Times*, April 22. http://www.nytimes.com/gwire/2009/04/22/22greenwire-no-need-to-build-new-us-coal-or-nuclear-plants-10630.html.

Sturgis, Sue. 2011. "Duke Energy's Nuclear Ambitions Face Fallout from Japanese Crisis." *Online Magazine of the Institute for Southern Studies*, March 16. http://www.southernstudies.org/2011/03/duke-energys-nuclear-ambitions-face-fallout-from-japanese-crisis.html.

Succar, Samir, and Robert H. Williams. 2008. *Compressed Air Energy Storage: Theory, Resources, and Applications for Wind Power.* Princeton, NJ: Princeton Environmental Institute, April 8.

Sullivan, Paul. 2011a. "Threats to the Global Oilconomy." Al Arabiya.net, April 28. http://english.alarabiya.net/articles/2011/04/28/147058.html.

———. 2011b. *Written Testimony in Support of the Oral Testimony of Professor Paul Sullivan, National Defense University and Georgetown University for the Western Hemisphere Subcommittee of the Foreign Affairs Committee, US House of Representatives Regarding the Need for Canadian Oil As We Face Increasing Turmoil in the Middle East, Increasing Competition for Energy Resources, Peak Conventional Oil, and an Increasingly Complex Geostrategic Environment.* March 31. http://www.foreignaffairs.house.gov/112/sul033111.pdf.

SunChips. 2011. "SunChips®—Healthier Planet." http://www.sunchips.com/healthier_planet.shtml.

Sungevity. 2011. "The Solar Lease." http://www.sungevity.com/solar-lease

SunPower. 2011. "Financing Your SunPower® Residential Solar System." http://us.sunpowercorp.com/homes/how-to-buy/financing/

SunRun. 2010. "Home Solar Power Systems: Installation, Financing, & Leasing." http://www.sunrunhome.com/.

Sunverge Energy. 2011. "First Micro-Grid, Distributed Energy Resource Community Coming to California." PRWeb, February 1. http://www.prweb.com/releases/2011/02/prweb5023374.htm.

Sustainable Homes. 1999. *Embodied Energy in Residential Property Development: A Guide for Registered Social Landlords.* Teddington, Middlesex, UK: Sustainable Homes. http://www.sustainablehomes.co.uk/upload/publication/Embodied%20Energy.pdf.

Swedish Government Commission on Oil Independence. 2006. *Making Sweden an OIL-FREE Society.* http://www.sweden.gov.se/content/1/c6/06/70/96/7f04f437.pdf.

Swisher, J., and R. Orans. 1995. "The Use of Area-Specific Utility Costs to Target Intensive DSM Campaigns." *Utilities Policy* 5 (3–4): 185–97.

Synovision Solutions. 2010. "Our Strategic Opportunity in Afghanistan: Fixing Our Energy Inefficiencies to Save $Billions . . . and BLOOD." Briefing prepared by Synovision Solutions, February 13. http://www.gaco.com/press/Anderson_Amadee_Afghanistan_Assmt_021310.pdf.

Tabuchi, Hiroko. 2010. "Electric Cars Make Japan's Gas Engine Industry Anxious." *New York Times*, November 2. http://www.nytimes.com/2010/11/03/business/global/03japancar.html.

———. 2011. "Head of Japanese Utility Steps Down After Nuclear Crisis." *New York Times*, May 20. http://www.nytimes.com/2011/05/21/business/global/21iht-tepco21.html.

———. 2011b. "Japan Premier Wants Shift Away From Nuclear Power." *New York Times.* July 13.

Tabuchi, Hiroko. 2011b. "Japan Premier Wants Shift Away From Nuclear Power." *New York Times.* July 13.

Taggart, D. F. 2000. "Riding a Bike." Presentation to Defense Science Board Task Force on Improving Fuel Efficiency, September 20. http://www.slideshare.net/lightspeed65/DSB-Weapon-Platforms-092000a-from-CD.

Tecogen. 2011. "Tecogen Supplies Three Ultra-Clean CHP Modules to Sacramento Electric Utility for Microgrid Demonstration Project." *PR Newswire.* March 1. http://www.prnewswire.com/news-releases/tecogen-supplies-three-ultra-clean-chp-modules-to-sacramento-electric-utility-for-microgrid-demonstration-project-117149568.html.

Temple, Robert K. G. 1986. "Petroleum and Natural Gas as Fuel, Fourth Century BC." In *The Genius of China: 3,000 Years of Science, Discovery, and Invention*. New York: Simon and Schuster.

Terborgh, J. 1992. "Table of Data on Tropical Rainforests." Table 1, modified for the class "The Tropical Rain Forest" at the University of Michigan's Global Change Curriculum from J. Terborgh, *Diversity and the Tropical Rainforest* (New York: Scientific American Library, 1992). http://www.globalchange.umich.edu/globalchange1/current/lectures/kling/rainforest/rainforest_table.html.

Terra Magnetica. 2010. "Siemens Launches Permanent Magnet-Based Gearless Wind Turbine." April 25. http://www.terramagnetica.com/2010/04/25/siemens-launches-permanent-magnet-based-gearless-wind-turbine/.

Texas Transportation Institute. 2009. *Urban Mobility Report 2009*. University Transportation Center for Mobility. http://transportationblog.dallasnews.com/UMReport%202009%20WEB%20July%2009_Embargoed.pdf.

Thermal Cookware. 2008. "How It All Works." http://thermalcookware.com/main.php?mod=Dynamic&id=22.

Theunissen, Ton, Mike Golombok, J. J. H. (Bert) Brouwers, Gagan Bansal, and R. van Benthum. 2011. "Liquid CO_2 Droplet Extraction from Gases." *Energy* 36 (5): 2961–67. doi:16/j.energy.2011.02.040.

Thomas, Sandy. 2004–2010. Personal communications with author.

Thomsen, K. E., K. B. Wittchen, Statens Byggeforskningsinstitut, and European Alliance of Companies for Energy Efficiency in Buildings. 2008. *European National Strategies to Move Towards Very Low Energy Buildings*. Danish Building Research Institute. http://www.buildup.eu/publications/1519.

TIAX. 2006. *The Energy and Greenhouse Gas Emissions Impact of Telecommuting and E-Commerce*. Cambridge, MA: TIAX LLC. http://www.ce.org/Energy_and_Greenhouse_Gas_Emissions_Impact_CEA_July_2007.pdf.

Tibbits, George. 2010. "Gregoire: Moses Lake Carbon Fiber Plant to Grow." *Bloomberg Businessweek*, November 5. http://www.businessweek.com/ap/financialnews/D9JA87BO0.htm.

Tidball, Rick, Joel Bluestein, Nick Rodriguez, and Stu Knoke. 2010. *Cost and Performance Assumptions for Modeling Electricity Generation Technologies*. National Renewable Energy Laboratory. www.nrel.gov/docs/fy11osti/48595.pdf.

Tinianov, Brandon (Serious Materials). 2011. Personal communication with author, February 8.

Tirpak, Brad. 2010. Personal communication with author, December 28.

Tommerup, Henrik, and Jørgen Nørgård. 2007. "Proper Sizing of Circulation Pumps." In *ECEEE 2007 Summer Study "Saving Energy—Just Do It!" Conference Proceedings* (Stockholm: European Council for an Energy Efficient Economy).

Töpfer, Klaus. 2011. Personal communication with author, May 25.

Torcellini, P., N. Long, and R. Judkoff. 2003. *Consumptive Water Use for U.S. Power Production*. National Renewable Energy Laboratory. www.nrel.gov/docs/fy04osti/33905.pdf.

Torres, Juan, and Mike Hightower. 2010. *2010 Smart Grid Peer Review Project Summary Form—Energy Surety Microgrids*. U.S. Department of Energy. http://events.energetics.com/SmartGridPeerReview2010/pdfs/summaries/16_Energy_Surety_Microgrids_and_SPIDERS.pdf.

Tucker, Jonathan. 2009. "The Future of Chemical Weapons." *The New Atlantis*, Fall 2009/Winter 2010: 3–29. http://www.thenewatlantis.com/publications/the-future-of-chemical-weapons.

Turner, J. A. 1999. "A Realizable Renewable Energy Future." *Science* 285 (5428): 687.

Ulrich, Roger. 2011. Personal communication with author, March 2.

United Technologies Corporation. 2010. *2010 Annual Report*. United Technologies Corporation. http://utc.com/About+UTC/Company+Reports/2010+Annual+Report+English.

Urban Land Institute. 2007. *Growing Cooler: Evidence on Urban Development and Climate*

Change. Urban Land Institute. http://www.mwcog.org/uploads/committee-documents/u1ZbXlk20070921140031.pdf.

Urban Land Institute and Cambridge Systematics. 2009. *Moving Cooler: An Analysis of Transportation Strategies for Reducing GHG Emissions*. Urban Land Institute. http://www.movingcooler.info/.

Urbina, Ian. 2011a. "Insiders Raise Alarm Amid a Rush for Natural Gas." *New York Times*, p. 1, June 25. http://www.nytimes.com/2011/06/26/us/26gas.html.

———. 2011b. "Behind Veneer, Doubt on Future of Natural Gas." *New York Times*, June 26. http://www.nytimes.com/2011/06/27/us/27gas.html.

U.S. Bureau of Transportation Statistics. 2001. *Highlights of the 2001 National Household Travel Survey*, fig. 10, "Mean Minutes and Miles Spent Driving by Driver Age." Washington DC: U.S. Bureau of Transportation Statistics. http://www.bts.gov/publications/highlights_of_the_2001_national_household_travel_survey/html/figure_10.html.

———. 2007. *National transportation Statistics*, table 1-46b, "U.S. Ton-Miles of Freight." Washington DC: U.S. Bureau of Transportation Statistics. http://www.bts.gov/publications/national_transportation_statistics/html/table_01_46b.html.

U.S. Census Bureau. 2004. "United States Summary: 2000." From the 2000 Census of Population and Housing, Population and Housing Unit Counts. Washington DC: U.S. Census Bureau.

———. 2005. "Value of Product Shipments: 2005." From the 2006 Annual Survey of Manufactures (ASM). Washington DC: U.S. Census Bureau, November 2006. http://www.census.gov/prod/2006pubs/am0531vs1.pdf.

———. 2010. "Annual Value of Construction Put in Place 2002–2010." http://www.census.gov/const/C30/total.pdf.

U.S. Climate Action Partnership. 2009. *A Call for Action*. Washington DC: U.S. Climate Action Partnership. http://us-cap.org/USCAPCallForAction.pdf.

U.S. Department of Defense. 2010. *Quadrennial Defense Review*. Washington DC: U.S. Department of Defense, January 26. http://www.defense.gov/qdr/QDR%20as%20of%2026JAN10%200700.pdf.

U.S. Department of Energy. 2003. *Improving Compressed Air System Performance: A Sourcebook for Industry*. Washington DC: U.S. Department of Energy.

———. 2004. *Steel Industry Energy Bandwidth Study*. Washington DC: U.S. Department of Energy.

———. 2006a. *Energy Bandwidth for Petroleum Refining Processes*. Washington DC: U.S. Department of Energy.

———. 2006b. *Chemical Bandwidth Study: Exergy Analysis: A Powerful Tool for Identifying Process Inefficiencies in the U.S. Chemical Industry*. Washington DC: U.S. Department of Energy.

———. 2006c. *Benefits of Demand Response in Electricity Markets and Recommendations for Achieving Them: A Report to the United States Congress Pursuant to Section 1252 of the Energy Policy Act of 2005*. Washington DC: U.S. Department of Energy. http://eetd.lbl.gov/ea/ems/reports/congress-1252d.pdf.

———. 2007. *U.S. Energy Requirements for Aluminum Production*. Washington DC: U.S. Department of Energy.

———. 2008a. "Isothermal Melting: Reaching for the Peak of Efficiency in Aluminum Melting Operations." *Energy Matters*, Spring 2008, a publication of the U.S. Department of Energy Industrial Technologies Program. http://www1.eere.energy.gov/industry/bestpractices/energymatters/archives/spring2008.html#a27.

———. 2008b. *20% Wind Energy by 2030*. Washington DC: U.S. Department of Energy.

———. 2010a. "Case Study: SMUD's Energy Efficient Remodel Demonstration Project." *Building America: Comprehensive Energy Retrofit: Efficient Solutions for Existing Home*. U.S. Department of Energy. http://apps1.eere.energy.gov/buildings/publications/pdfs/building_america/ba_casestudy_smud_sacramento_hot-dry.pdf.

———. 2010b. *Solid-State Lighting Research and Development: Multi-Year Program Plan*. Washington DC: U.S. Department of Energy.

———. 2010c. *Program: Preliminary Determination Regarding Energy Efficiency Improvements in the Energy Standard for Buildings, Except Low-Rise Residential Buildings, ANSI/ASHRAE/IESNA Standard 90.1-2007*. U.S. Department of Energy. http://federalregister.gov/a/2010-22060.

———. 2010d. *3M's Model Rewards and Recognition Program Engages Employees and Drives Energy Savings Efforts.* Washington DC: U.S. Department of Energy. http://www1.eere.energy.gov/industry/saveenergynow/pdfs/3m.pdf.

———. 2010e. *Impacts: Industrial Technologies Program: Summary of Program Results for CY 2008: Boosting the Productivity and Competitiveness of US Industry.* Washington DC: U.S. Department of Energy.

———. 2011a. *Our Projects: The Financing Force Behind America's Clean Energy Economy.* Washington DC: U.S. Department of Energy. https://lpo.energy.gov/?page_id=45.

———. 2011b. "Smart Grid Project Information." *SmartGrid.* U.S. Department of Energy. http://www.smartgrid.gov/smartgrid_projects?order=field_total_value_value&sort=asc&category=1.

———. 2011c. *The Role of Electricity Market Design in Integrating Solar Generation.* U.S. Department of Energy. http://www.nrel.gov/docs/fy11osti/50058.pdf.

———. 2011d. *Department of Energy FY 2012 Congressional Budget Request*, vol. 2. Washington DC: "U.S. Department of Energy. http://www.cfo.doe.gov/budget/12budget/Content/Volume2.pdf.

U.S. Energy Information Administration. 1997. "U.S. Manufacturing Energy Intensity, 1977 through 1991." In *Changes in Energy Intensity in the Manufacturing Sector 1985–1991.* Washington DC: U.S. Energy Information Administration. http://www.eia.gov/emeu/mecs/mecs91/intensity/mecs2g.html.

———. 1999. Commercial Building Energy Consumption Survey, 1999. Washington DC: U.S. Energy Information Administration. http://www.eia.doe.gov/emeu/cbecs/.

———. 2000. *The Changing Structure of the Electric Power Industry 2000: An Update.* Washington DC: U.S. Energy Information Administration, October. http://www.eia.gov/cneaf/electricity/chg_stru_update/update2000.pdf.

———. 2005. "MECS Definition of Nonfuel (Feedstock)." From the 2005 Manufacturing Energy Consumption Survey. U.S. Energy Information Administration. http://www.eia.doe.gov/emeu/mecs/mecs98/datatables/nonfueldef.html.

———. 2006a. "Energy Use in Commercial Buildings." From the 2003 Commercial Building Energy Consumption Survey. Washington DC: U.S. Energy Information Administration. http://www.eia.doe.gov/emeu/cbecs/cbecs2003/.

———. 2006b. "2006 Manufacturing Energy Consumption by Manufacturers—Data Tables." From the 2006 Manufacturing Energy Consumption Survey. U.S. Energy Information Administration. http://www.eia.doe.gov/emeu/mecs/mecs2006/2006tables.html.

———. 2007. *Electric Power Industry Overview.* Washington DC: U.S. Energy Information Administration. http://www.eia.gov/cneaf/electricity/page/prim2/toc2.html.

———. 2008a. "Consumption and Expenditures in U.S. Households." Data tables from the Residential Energy Consumption Survey. Washington DC: U.S. Energy Information Administration. http://www.eia.doe.gov/emeu/recs/historicaldata/historical_data80_02.html.

———. 2008b *International Energy Statistics.* Washington DC: U.S. Energy Information Administration. http://tonto.eia.doe.gov/cfapps/ipdbproject/IEDIndex3.cfm?tid=44&pid=44&aid=2.

———. 2008c. 2005 Residential Energy Consumption Survey. Washington DC: U.S. Energy Information Administration. http://205.254.135.24/emeu/recs/recs2005/hc2005_tables/detailed_tables2005.html.

———. 2009. "Retail Sales of Electricity by State by Sector by Provider, 1990–2009." Data from the *Electric Power Annual.* Washington DC: U.S. Energy Information Administration. http://www.eia.doe.gov/cneaf/electricity/epa/sales_state.xls.

———. 2010a. "Macroeconomic Indicators, Reference Case." From *Annual Energy Outlook 2010.* Washington DC: U.S. Energy Information Administration. http://www.eia.gov/oiaf/aeo/tablebrowser/#release=AEO2010&subject=14-AEO2010&table=18-AEO2010®ion=0-0&cases=aeo2010r-d111809a.

———. 2010b. *Annual Energy Outlook 2010.* May 11. http://www.eia.gov/oiaf/archive/aeo10/index.html.

———. 2010c. "Petroleum Product Prices, Early Release Reference." From *Annual Energy Outlook 2010.* Washington DC: U.S. Energy Information Administration. http://www.eia.doe.gov/oiaf/

aeo/tablebrowser/#release=AEO2011&subject=0-AEO2011&table=12-AEO2011®ion=0-0&cases=ref2011-d120810c.

———. 2010d. *Annual Energy Review 2009.* August 19, 2010. http://www.eia.doe.gov/aer/.

———. 2010e. Table 6.8, "Natural Gas Prices by Sector, 1967–2009." From *Annual Energy Review.* Washington DC: U.S. Energy Information Administration, August 19. http://www.eia.doe.gov/totalenergy/data/annual/txt/ptb0608.html.

———. 2010f. Table 8.1, "Electricity Overview, 1949–2009." From *Annual Energy Review.* Washington DC: U.S. Energy Information Administration, August 19. http://www.eia.gov/totalenergy/ data/annual/txt/ptb0801.html.

———. 2010g. "Form EIA-860, Annual Electric Generator Report." http://www.eia.doe.gov/cneaf/electricity/page/capacity/existingunitsbs2008.xls.

———. 2010h. "Form EIA-923." *Electricity Data Files.* U.S. Energy Information Administration. http://www.eia.doe.gov/cneaf/electricity/page/eia906_920.html.

———. 2011a. *January 2011 Monthly Energy Review.* U.S. Energy Information Administration. http://www.eia.gov/totalenergy/reports.cfm.

———. 2011b. *March 2011 Monthly Energy Review.* U.S. Energy Information Administration. http://www.eia.gov/totalenergy/reports.cfm.

———. 2011c. "Existing Net Summer Capacity by Energy Source and Producer Type." From the *Electric Power Annual.* Washington DC: U.S. Energy Information Administration. http://www.eia.doe.gov/cneaf/electricity/epa/epat1p1.html.

———. 2011d. Table 7.2b, "Electricity Net Generation: Electric Power Sector." From *April 2011 Monthly Energy Review.* Washington DC: U.S. Energy Information Administration. http://www.eia.gov/totalenergy/data/monthly/pdf/sec7_6.pdf.

———. 2011e. *Form EIA-860 Annual Electric Generator Report.* U.S. Energy Information Administration, January 4. http://www.eia.doe.gov/cneaf/electricity/page/eia860.html.

U.S. Environmental Protection Agency. 1998. *Role of Technology in Climate Change Policy.* EP 1.2:R 64. Washington DC: U.S. Environmental Protection Agency, Atmospheric Pollution Prevention Division.

———. 2009. *Light-Duty Automotive Technology, Carbon Dioxide Emissions, and Fuel Economy Trends: 1975 through 2009.* Washington DC: U.S. Environmental Protection Agency.

———. 2010a. *ENERGY STAR and Other Climate Protection Partnerships 2009 Annual Report.* Washington DC: U.S. Environmental Protection Agency.

———. 2010b. *Light-Duty Automotive Technology, carbon Dioxide Emissions, and Fuel Economy Trends: 1975 Through 2010.* EPA-420-R-10-023. November. http://www.epa.gov/oms/

U.S. General Services Administration. 2011. "2.2.1.1 Scope 1 & 2 Greenhouse Gas Emissions Reductions in Federal Buildings." http://www.gsa.gov/portal/content/185129.

U.S. Government Accountability Office. 2009. *Defense Critical Infrastructure: Actions Needed to Improve the Identification and Management of Electrical Power Risks and Vulnerabilities to DOD Critical Assets.* Report to Congressional Committees, October.

U.S. Green Building Council. 2006a. *Project Profile: Toyota Motor Sales South Campus Office Development Torrance, California.* http://www.usgbc.org/ShowFile.aspx?DocumentID=3382 .

———. 2006b. *Fossil Ridge High School Fort Collins, Colorado.* http://tinyurl.com/3e8cmk2.

———. 2006c. *Project Profile: Banner Bank Building.* http://www.usgbc.org/ShowFile.aspx?DocumentID=2057.

U.S. Joint Forces Command. 2010. *The Joint Operating Environment (JOE) 2010.* U.S. Joint Forces Command. http://tinyurl.com/3tpey9x.

U.S. Nuclear Regulatory Commission. 2011a. "Power Reactor Status Reports for 2003." *Power Reactor Status Reports.* Nuclear Regulatory Commission. http://www.nrc.gov/reading-rm/doc-collections/event-status/reactor-status/2003/index.html.

———. 2011b. *Operating Nuclear Power Reactors (by Location or Name).* Washington DC: Nuclear Regulatory Commission. http://www.nrc.gov/info-finder/reactor/.

———. 2011c. "NRC Appoints Task Force Members and Approves Charter for Review of Agency's Response to Japan Nuclear Event." Press release, April 1. http://pbadupws.nrc.gov/docs/ML1109/ML110910479.pdf.

Usher, Bruce. 2010. "On Global Warming, Start Small." *New York Times*, November 27.

Van Auken, R. M., and J. W. Zellner. 2003. *A Further Assessment of the Effects of Vehicle Weight and Size Parameters on Fatality Risk in Model Year 1985–98 Passenger Cars and 1985–97 Light Trucks*. Torrance, CA: Dynamic Research, Inc.

van Benthum, R. J., H. P. van Kemenade, J. J. H. Brouwers, and M. Golombok. 2010. "CO_2 Capture by Condensed Rotational Separation." Presented at the International Pittsburgh Coal Conference, Istanbul, October 11–14. http://www.mate.tue.nl/mate/pdfs/12195.pdf.

van der Vleuten, E., and R. Raven. 2006. "Lock-in and Change: Distributed Generation in Denmark in a Long-Term Perspective." *Energy Policy* 34 (18): 3739–48.

VICS Empty Miles. 2011. *Empty Miles Webinar*. A production of the Voluntary Interindustry Commerce Solutions (VICS) Association. https://www.emptymiles.org/.

Viswanathan, V., R. Davies, and J. Holbery. 2006. *Opportunity Analysis for Recovering Energy from Industrial Waste Heat and Emissions*. Pacific Northwest National Laboratory / U.S. Department of Energy. http://www1.eere.energy.gov/industry/imf/pdfs/4_industrialwasteheat.pdf.

Vyas, A., C. Saricks, and F. Stodolsky. 2002. *The Potential Effect of Future Energy-Efficiency and Emissions: Improving Technologies on Fuel Consumption of Heavy Trucks*. Argonne, IL: Center for Transportation Research.

Wald, Matthew L. and John M. Broder. 2011. "Utility Shelves Ambitious Plan to Limit Carbon." *New York Times*. July 13. www.nytimes.com/2011/07/14/business/energy-environment/utility-shelves-plan-to-capture-carbon-dioxide.html.

W&H Properties. 2009. *The Empire State Building Takes Leadership Role In Energy and Cost Savings for Tenants*. http://www.esbnycleasing.com/graphics/ads/1stESB.pdf.

Wald, Matthew L. 2010. "Wind Power Backbone Sought Off Atlantic Coast." *New York Times*, October 12. http://www.nytimes.com/2010/10/12/science/earth/12wind.html.

———. 2011a. "Westinghouse Nuclear Reactor Design Flaw Is Found." *New York Times*, May 20. http://www.nytimes.com/2011/05/21/business/energy-environment/21nuke.html.

———. 2011b. "Despite Bipartisan Support, Nuclear Reactor Projects Falter." *New York Times*, April 28. http://www.nytimes.com/2011/04/29/business/energy-environment/29utility.html.

Walmart. 2010. *Walmart Global Sustainability Report 2010 Progress Update*. Walmart. http://cdn.walmartstores.com/sites/sustainabilityreport/2010/WMT2010GlobalSustainabilityReport.pdf.

Walters, Eric. 2010. Personal communication with author, December 16.

Wang, Shuxiao. 2011. "Mercury Emissions from Coal-Fired Power Plants in China." Presentation to UNEP INC2 Technical Meeting, Tsinghua University, January 23.

Warren, Dave. 2010. *Low Cost Carbon Fiber Overview*. Oak Ridge National Laboratory. http://www1.eere.energy.gov/vehiclesandfuels/pdfs/merit_review_2010/lightweight_materials/lm002_warren_2010_o.pdf.

Washom, Byron. 2009. "Integration of Distributed Energy Resources within a Smart Microgrid." Presented at Smart Grid Rulemaking Distribution Workshop, California Public Utilities Commission, University of California at San Diego, June 5. http://www.cpuc.ca.gov/NR/rdonlyres/FEB2952C-4CF5-4600-8A4F-A6566D94329C/0/UCSD.pdf.

Weber, C. L. 2008. *An Examination of Energy Intensity and Energy Efficiency*. Pittsburgh, PA: Carnegie Mellon University.

Weber, C. L., and H. S. Matthews. 2007. "Embodied Environmental Emissions in US International Trade, 1997–2004." *Environmental Science & Technology* 41 (14): 4875–81.

Weisman, Alan. 1998. *Gaviotas: A Village to Reinvent the World*. White River Junction, VT: Chelsea Green Publishing.

Wernick, I. K., and J. H. Ausubel. 1995a. "National Material Metrics for Industrial Ecology." *Resources Policy* 21 (3): 189–98. doi:10.1016/0301-4207(96)89789-3.

———. 1995b. "National Materials Flows and the Environment." *Annual Review of Energy and the*

Environment 20 (1): 463–92. doi:10.1146/annurev.eg.20.110195.002335.

Wesoff, Eric. 2011. "Clarian: Moving Solar to the Mainstream." Greentech Solar. http://www.greentechmedia.com/articles/read/clarian-lowering-the-entry-price-to-solar/.

Westbrook, Paul (Texas Instruments). 2011. Personal communications with author, February 15 and 17.

White, Joseph. 2010. "Luxury Cars Go on a Diet." *Wall Street Journal*, October 5. http://online.wsj.com/article/SB10001424052748704631504575532162477960150.html.

Wilbanks, Thomas J. 2009. *Effects of Climate Change on Energy Production and Use in the United States*. Derby, PA: DIANE Publishing.

Willems, G. P., J. P. Kroes, M. Golombok, B. P. M. van Esch, H. P. van Kemenade, and J. J. H. Brouwers. 2010. "Performance of a Novel Rotating Gas-Liquid Separator." *Journal of Fluids Engineering* 132. doi:10.1115/1.4001008.

Williams, E. D., R. U. Ayres, and Miriam Heller. 2002. "The 1.7 Kilogram Microchip: Energy and Material Use in the Production of Semiconductor Devices." *Environmental Science & Technology* 36 (24): 5504–10. doi:10.1021/es025643o.

Williams, R. H., E. D. Larson, and M. H. Ross. 1987. "Materials, Affluence, and Industrial Energy Use." *Annual Review of Energy* 12 (1): 99–144. doi:10.1146/annurev.eg.12.110187.000531.

Wines, Michael. 2011. "China Admits Problems with Three Gorges Dam." *New York Times*, May 19.

Wingfield, Brian. 2011. "GE Sees Solar Cheaper Than Fossil Power in Five Years." Bloomberg, May 26. http://www.bloomberg.com/news/2011-05-26/solar-may-be-cheaper-than-fossil-power-in-five-years-ge-says.html.

WIPO. 1999. "(WO/1999/020165) High-Performance Cooking Pot." World Intellectual Property Organization *PatentScope*, a database of patent applications. http://www.wipo.int/pctdb/en/wo.jsp?WO=1999020165.

Wirth, Thomas. 2008. *Microreactors in Organic Synthesis and Catalysis*. Germany: Wiley-VCH.

Wiser, Ryan, and Mark Bolinger. 2010. *2009 Wind Technologies Market Report*. Lawrence Berkeley National Laboratory, August. http://eetd.lbl.gov/ea/emp/reports/lbnl-3716e.pdf.

Wiser, Ryan, and Mark Bolinger. 2011. *2010 Wind Technologies Market Report*. Lawrence Berkeley National Laboratory, June. http://eetd.lbl.gov/ea/emp/reports/lbnl=4820e.pdf

Woolsey, R. James. 2001. Testimony to the Subcommittee on Energy, Committee on Science, U.S. House of Representatives, November 1. http://commdocs.house.gov/committees/science/hsy75842.000/hsy75842_0.HTM.

Woolsey, R. James, Amory B. Lovins, and L. H. Lovins. 2002. "Energy Security: It Takes More than Drilling." *Christian Science Monitor*, March 29. http://www.csmonitor.com/2002/0329/p11s02-coop.html.

World Energy Council. 2010. *Survey of Energy Resources 2010*. London: World Energy Council. http://www.worldenergy.org/publications/3040.asp.

World Nuclear News. 2011. "'New approach' puts back Flamanville 3." World Nuclear News. July 21. http://www.world-nuclear-news.org/NN_New_approach_puts_back_Flamanville_3_2107111.html.

World Resources Institute. 2005. *World Greenhouse Gas Emissions: 2005*. World Resources Institute. http://www.wri.org/chart/world-greenhouse-gas-emissions-2005.

Worrell, Ernst, J. A. Laitner, and Ruth Michael. 2003. "Productivity Benefits of Industrial Energy Efficiency Measures." *Energy* 28 (11).

Worrell, Ernst, Lynn Price, and Nathan Martin. 2001. "Energy Efficiency and Carbon Dioxide Emissions Reduction Opportunities in the US Iron and Steel Sector." *Energy* 26 (5): 513–36. doi:10.1016/S0360-5442(01)00017-2.

Wyssen, Ivan, Lucas Gasser, and M. Meier. 2010. "Chiller with Small Temperature Lift for Efficient Building Cooling." Presented at the 10th REHVA World Congress, Antalya, Turkey, May 9–12.

Xinhua. 2011. "China's Thermal Power Plants to See More Losses." *China Daily*, April 30.

Xu, M., Braden Allenby, and Weiqiang Chen. 2009. "Energy and Air Emissions Embodied in China–U.S. Trade: Eastbound Assessment Using Adjusted

Bilateral Trade Data." *Environmental Science & Technology* 43 (9): 3378–84. doi:10.1021/es803142v.

Xu, M., E. Williams, and Braden Allenby. 2010. "Assessing Environmental Impacts Embodied in Manufacturing and Labor Input for the China–U.S. Trade." *Environmental Science & Technology* 44 (2): 567–73. doi:10.1021/es901167v.

Xu, P. 2009. *Evaluation of Demand Shifting Strategies with Thermal Mass in Two Large Commercial Buildings.* Lawrence Berkeley National Laboratory. http://gundog.lbl.gov/dirpubs/SB06/pengxu.pdf.

Xu, T., J. Slaa, and J. Sathaye. 2010. *Characterizing Costs and Savings Benefits from a Selection of Energy Efficient Emerging Technologies in the United States.* BOA-99-205-P. Lawrence Berkeley National Laboratory. http://ies.lbl.gov/drupal.files/ies.lbl.gov.sandbox/efficient%20tech%20working-%20US.pdf.

Yasu, Mariko. 2011. "Softbank's CEO Wants a Solar-Powered Japan." *Bloomberg Businessweek*, June 23.

Zeller, Tom. 2010. "Can We Build a Brighter Shade of Green?" *New York Times*, September 25. http://www.nytimes.com/2010/09/26/business/energy-environment/26smart.html?_r=4&pagewanted=1.

ZepSolar. 2011. "Zep System II." http://www.zepsolar.com/zepsystem2.html.

Ziemlewski, Joanna. 2010. "Saving Energy in a Polyethylene Facility." JOURNAL. http://findarticles.com/p/articles/mi_qa5350/is_201007/ai_n54718274/.

Zhou, Nan, David Fridley, Michael McNeil, Nina Zheng, Jing Ke, and Mark Levine. 2011. "China's Energy and Carbon Emissions Outlook to 2050." LBNL-4472E. Lawrence Berkeley National Laboratory. http://china.lbl.gov/publications/Energy-and-Carbon-Emissions-Outlook-of-China-in-2050.

INDEX

A

additive manufacturing, 154–55
Adobe, 109
Advantix Systems, 91
aerodynamics, 21, 50, 56–57, 60, 61, 73
AeroVironment, Inc., 63
Afghan War, 60–61
air compressors
  compressed air energy storage, 198
  efficiency of, 135
air conditioning, 84, 90, 91, 104, 111
air pollution, xvi, 6, 43, 175–76
airplanes, 56–59, 61, 70, 132, 231
  biofuel, 69
  hydrogen fuel, 64
  manufacturing process waste, reducing, 154
  military use, 60, 69, 154
algal fuels, 68–69, 132
Alphabet Energy, 138
alternative fuels, xiii, xv, 5, 62–69, 71, 73, 74. *See also* biofuels; *specific fuels*
  Chinese investments in, 243–44
  industrial sector use, 144, 148
  natural gas as vehicle fuel, 59, 64–65
aluminum, 21, 29, 34, 148–49, 153
American Electric Power, 176, 185, 197
Anderson, Ray C., 161–62
appliances, 93, 113–14
AREVA, 182
armored vehicles, ultralight, 62
Atlantic Station (Atlanta, Georgia), 47–48
Audi AG, 21, 33
Austin, Texas, building energy check-up policy, 113
auto dealers, 41
auto industry, x, 73, 249
  China, 21, 35, 40
  history of, 19–20
  imports and exports, 9, 40
  manufacturing methods, 26–28, 41, 153, 154
  Revolutionary+ autos, transition to, 32–36, 40–42, 70–71
  U.S. automakers, handicaps of, 40
auto insurance, 45–46, 47
Automotive X Prize, Progressive Insurance 2010, 21, 40
autos. *See also* fuel efficiency standards; gasoline
  alternatives to, 43–45, 73, 244
  definition, 18
  driving methods, fuel-efficient, 49
  energy use by, 18
  external costs, U.S., 43
  fleet use, 34, 39, 72, 73, 74
  imports and exports, 9, 40
  physics of, 18–22
  productive use of, 42–49
  vehicle-miles traveled (VMT), 18, 44–49
  weight of, 17, 18, 19 (*See also* autos, Revolutionary+)
autos, Revolutionary+, 16–42, 70–71, 229–30, 235–36, 247–49
  aerodynamics, 21
  composites, use of advanced, 23–29, 33, 73, 230, 235
  electrification, 24–26, 29–35, 38, 41, 42
  engines, 25, 30
  feebates for, 37–39, 72, 234
  integrative whole-system design, 22, 24–25, 35, 73
  manufacturing costs/methods, 26–28, 41, 153
  metals, use of, 21, 26, 34
  policies for, 36–41, 72, 234
  price of, 36–40
  rolling resistance, 22
  safety of, 29, 42
  transition to, 32–36, 40–42
  ultralight materials for, 17, 24, 26, 35, 41, 73 (*See also subhead:* composites, use of advanced)
  ultrastrong, 17, 29
  vehicle fitness (*See* vehicle fitness)
  vision for, 22–24
  weight savings, 17, 21–23, 25, 26, 29, 33, 34

B

Baltimore Gas and Electric, 210
Banks, Sharon, 52
Banner Bank Building (Idaho), 106
BASF SE, 91
batteries, lithium, 31, 32
battery-electric vehicle (BEVs), 18, 20, 30, 31, 33, 36, 42
Benyus, Janine, 152
Better Place, 63
bicycles, use of, 47, 75
Bigelow Homes, 111
biofuels, xv, 65–69, 72, 232, 242
 for airplanes, 30, 57, 59, 69, 72
 algal fuels, 68–69, 132
 for autos, 30
 for buses, 59
 drop-in fuel, 59, 66, 72
 ethanol, 65–69
 industrial use, 132, 144
 for trucks, 30, 64, 72
biogas, 144
biomass
 electricity generation, 187, 189–90, 198, 201, 212
 industrial use, 125, 127, 129, 132, 137–38, 144, 159
 waste biomass, use of, 137–38
biomaterial feedstocks, 128
biomimicry, 124–25, 151–52, 155
biopower, 171, 181, 188, 203
BMW AG, 32–33, 41, 65
Boeing Co., 56, 64, 90
boilers, industrial, 136
Boise, Idaho, 106
Boonyatikarn, Suntoorn, 97
Boulder, Colorado, 115
Braese, Paul, 109
Brazil, 147, 205, 244
Bright Automotive, 34
BroadStar Wind Systems, 191–92
building codes, 88–89, 114–15, 118, 121
building sector, energy efficiency in, ix–x, xvi, 76–121, 211, 229–31, 236–37, 247. *See also* commercial buildings; residences
 appliances, 93, 113–14
 behavioral changes, 93–94, 105, 111
 best-practice design, 107–8
 building codes and, 88–89
 business opportunities, 77, 80–81, 86–93, 102–3, 105–12, 117, 118, 120
 capital spending, 104–5, 107
 consumers, energy usage information for, 94, 106, 110–13, 117, 121
 cost-effectiveness of, x, 78, 80–81, 86–90, 93–94, 97–98, 103–5, 108–11, 114
 current status of, 82–86
 dematerialization, 154
 embodied energy, 85–86
 emerging technologies, 90–93
 employee incentives, 109
 financing for, 80, 89, 116–17
 integrative design, 94–96, 98–101
 key parameters, targets for, 108
 maintenance and installation savings, 101–2, 106
 monitoring and feedback systems, 110, 218
 new buildings, 82, 249
 nonenergy benefits, 98–102
 performance standards, 114–15
 policies for, 77, 80–81, 112–18, 121
 retrofitting existing buildings, xvi, 78–79, 82, 89–90, 97, 99–101, 104, 106, 110–11, 114, 120, 234, 249
 right-sizing, 96–98
 right-timing, 89–90, 97, 107
 stock turnover, savings from, 89
 surveys, inspections, and analyses, 110, 116, 121
 terminology, 79
 training, workforce, 111–12, 117–18, 120, 234, 237
 utility regulation, changes in, 115–16
Burke, Anita, 158
buses, 59
Byron G. Rogers Federal Building, 99

C

California, 176, 210
 appliance efficiency standards, 90, 114
 building codes, 88–89, 115
 feebate program, 38
 foreclosed home retrofits, 89–90
 greenhouse emissions, reducing, 239
 industry, electricity savings in, 137
 solar power, use of, 191
 utility efficiency targets, 118
 utility regulation, 220
Canada, 6
capacity, defined, 170
capacity factor of generators, defined, 170
carbon capture and sequestration (CCS), xvii, 171, 180–81, 185–86, 188, 203, 205, 212, 216, 242
carbon dioxide emissions. *See* carbon emissions
carbon dioxide waste, use of, 124
Carbon Disclosure Project, 106
carbon emissions, xvi, 6–8, 238, 243. *See also* carbon capture and sequestration (CCS)

biofuels and, 66
building sector, 78, 90, 106
electricity sector, 172, 175
France, 37
industrial sector, 128, 132, 136, 143, 153
from natural gas, 233
carbon-fiber composites, 71, 73
airplanes, 27, 28, 56–57
autos, 23–29, 33, 230, 235
crash safety, 29
manufacturing process waste reduction, 154
carbon pricing, 175, 238, 240
carbon reduction, 98, 147
carpooling, 46–47
Carrefour Demeter Environment and Logistics Club, 53
carsharing, 47, 72
Carter, Majora, 51
Cascade Sierra Solutions, 52
cash for clunkers program, 39–40, 74
CCS. *See* carbon capture and sequestration (CCS)
cement manufacture, 132, 147, 148, 151
Centre of Material and Process Synthesis, 143
Chalouhi, Olivier, 62
charging stations, 62–63, 96
Chase, Robin, 47
chemical industry, 130
Chevrolet Volt, 20, 33
China, 241, 243–44, 247
auto industry, 21, 35, 40
auto use, controlled, 43, 244
bullet trains, 60
carbon emissions, 175
electricity sector, 167, 177, 184, 187, 191
energy intensity reductions, 239, 243, 244
steelmaking, 147
Choren Group, 68
CHP. *See* combined heat and power (CHP)
Christensen, Gary, 106
Cisco, 109
Clarian Power, 217
Clean Air Act, 126, 147
Clean Water Act, 126
Cleaver-Brooks, Inc., 136
climate change, xvi, 6–7, 237–39, 243
electricity sector and, 175, 176, 180, 216
industrial sector and, 128, 132, 145–47
transportation sector and, 20, 40, 43, 66
coal, xvi, 12, 232, 242–43
"clean" (*See* carbon capture and sequestration (CCS))
costs of, 6–8, 12
electricity generation, 6, 8, 171, 172, 174–75, 180, 181, 188, 195, 201, 203, 205, 212, 214, 222
industrial use of, 127, 129, 131, 137, 144, 147–48, 159–60
liquefication process, 143
mining, 176
peak production, xii, 7–8, 12
rail transport of, 55, 180
cogeneration of heat and power. *See* combined heat and power (CHP)
Cogenra Solar, 217
combined cooling, heating, and power unit (CCHP), 209
combined heat and power (CHP), 126, 127, 136–38, 144, 147–48, 158–59, 205, 207, 209, 211, 231
commercial buildings, 81, 83–84, 88–89, 106–10, 230–31
government entities, owned by, 99, 109, 115
integrative design, 95–96, 99–100
as market for innovations, 110–11
tenants, energy use by, 104, 106
worker productivity, boosting, 102
commissioning, 79, 106
commuting, 44–47
compact fluorescent lamps, 88, 92
composites. *See also* carbon-fiber composites
autos, use in, 23–29, 33, 73, 230, 235
crash safety of, 29
compressed air energy storage, 198
compressed natural gas (CNG), 64–65
concentrating solar power (CSP), 150, 171, 181, 188, 189, 198, 201, 203, 205, 212
congestion pricing, 45
construction work in progress (CWIP) charges, 184
Consumer Assistance to Recycle and Save Act of 2009, 39–40
cooperatives (co-ops), defined, 170
Copper Development Association, 135
cost of saved energy (CSE), defined, 18
Coulomb Technologies, 63
Cramer, David, 22, 26
crash safety, carbon fiber composites, 29, 42
crossover vehicles, 18, 22. *See also* autos
Crowley Carbon, 110
CSP. *See* concentrating solar power (CSP)
Cushman & Wakefield, 109
cyberattacks, electric grid, 179, 214
CyberTran, 60
Czinger, Kevin, 9

D

Danielson, Antje, 47
Darbee, Peter, 115–16

data centers, 138–39, 143
decoupling, 116, 158, 220, 234
delivered energy, 79, 82–85, 88, 99, 100, 108, 131
demand response, 170, 222
demand-side resources, 170, 186, 196–97, 213, 221
dematerialization, 152–55
democracy, 6
Denmark, 8, 60, 114, 125, 186, 199, 209, 237, 239, 247
desiccant enhanced evaporation, 91
Deutsche Bank Twin Towers (Germany), 100
diesel engines, 30, 50, 51, 64, 132
direct-reduced iron process, 147–48
distillation processes, 134, 151
distributed generation, 205, 206, 219, 220–21
distributed resources, 170, 206–9, 216, 219, 221–22
distribution company, defined, 170
distribution system, defined, 170
Dow Chemical Co., 68, 110, 128, 133, 137, 153, 157–58
drivetrain, 18, 25, 33, 50
drying processes, industrial, 150
Duany, Andres, 48
ducts, finding air leaks in, 110
Duke Energy, 184
DuPont company, 68, 128, 154
Durning, Alan, 48
Dutch Cyclists' Union, 75

E

EcoMotors International, 30
economics of fossil fuel use, 3–4
economy, reinventing fire's effect on, xiii, xvi, 235–36
Edison, Thomas, 173, 187
Edison 2 Very Light Car, 21
Electric Reliability Council of Texas, 197
electric vehicles (EVs), 18, 56, 59–63, 71, 73, 96, 132. *See also* autos, Revolutionary+; powertrain
Électricité de France, 182–83
electricity
  buildings, use in, 78–82, 84, 92, 94, 98, 119
  generation of, 6, 8, 171, 172, 174–77, 180–93, 198–201, 203, 205–6, 209–10, 212–14, 216, 217, 219, 220–22
  industrial use, 127, 129, 131, 134–35, 137, 144–45, 148–49, 158, 161, 162
  production costs, 171, 182, 188, 190, 192, 204
  storage systems, 171, 181, 188, 197–99, 203, 206, 212
electricity scenarios, RMI, 168–217
  choosing between, 211–17
  evaluation criteria, 169
  maintain, 168, 169, 171–80, 211–16
  migrate, 169, 180–87, 211–16
  renew, 169, 187–202, 211–16
  transform, 169, 202–11, 212–16
electricity sector, energy efficiency in, ix–x, xvii, 8, 9, 129, 164–225, 237, 243. *See also* electricity scenarios, RMI; grid
  business models, 173, 180, 186, 202, 205, 219–23, 249–50
  business opportunities, 165
  competition, effect of, 167, 169, 186, 216–19
  cost-effectiveness of, x, 165, 189–93, 209–10, 213–14, 216
  customers, role of, 169, 186, 196–97, 202, 210, 220–21, 223–24
  demand for electricity, 12–13, 174, 197
  distributed generation market, 205
  efficiency standards, electricity saved by, 113–14
  environmental constraints, 175–76
  infrastructure, aging, 167, 174–75
  investments, necessary, 167, 183–84, 186, 216
  metering, 218, 219 (*See also* smart meters)
  new technologies, 167, 176–78, 201–2, 218, 221–22
  policies, 165, 177, 183–84, 205, 216–20
  power storage systems, 212
  pricing of electricity, 167, 170, 196–97, 210, 216, 220–22, 249
  public support for, 215–16
  regulation, 172, 173, 176, 180, 202, 216–19, 224–25, 249–50 (*See also* public utility commissions (PUCs), state)
  security and supply disruptions, 169, 178, 214, 229, 241
  terminology, 170
electrified parking spaces (EPS), 51
electron beams, 150
Electronic Data Systems, 143
electronics, recycling, 153
Empire State Building, 78–79, 95, 97–98, 118
employment. *See* job creation
Empty Miles Service, 53
end-use devices, 79
end users, 79
energy, defined, 170
energy efficiency, defined, 170
energy intensity, 239, 243, 244
  building sector, 85, 89, 108, 113, 115
  definition, 79, 126
  industrial sector, 125, 130, 132, 133, 137, 146
energy labeling, 103, 113, 116, 120
energy monitoring and feedback systems, 110, 218
energy performance standards, 114–15
ENERGY STAR, 85, 103, 113, 114
energy surveys/analyses, 110, 116, 121

energy use, U.S., 2, 9–10, 13, 16–17
building sector, 79, 80–87, 92, 94, 114, 119
electricity sector, 174, 204, 211, 232, 235
energy intensity reductions, 239
future, 10–11
industrial sector, 126–33, 137, 144, 159–60
transportation sector, 16–17, 19, 25
EnerNOC, 218
EnerPath, 111
engines, 25, 30, 50–51, 56–57, 60
environmental effects. *See also* air pollution; carbon emissions
coal use, 6
electricity sector, 175–76, 184–85, 214
industrial sector, 125
ethanol, 65–69
Eugster, C. H., 153–54
Europe. *See also specific countries*
building sector, 112, 113, 114
carbon reduction, 98
combined heat and power (CHP), 136
electricity sector, 182–87, 199, 216
energy prices, 240
nuclear power, 182–85
recycling programs, 153
renewables, use of, 243
evaporative cooling, enhanced, 91

F

Fannie Mae, 48, 105
fans, industrial, 136
Federal Energy Regulatory Commission (FERC), 172
FedEx Corporation, 59
feebates, 37–39, 41, 72, 74, 116–17, 234
feedstocks, 9–10, 28, 124, 126, 128
Feist, Wolfgang, 97
Fiberforge Corporation, 26
First Solar, Inc., 191
Fischer, Hanns, 153–54
Ford, Henry, 21, 239, 248
Ford Motor Co., 20, 21, 35
foreclosed homes, energy retrofits of, 89–90
fossil-fuel companies, effect of energy efficiency on, 241–42
fossil fuels, xi–xii, xvi, 228, 232–34. *See also specific fossil fuels*
electricity sector use, 171–72
as feedstock, 9–10, 28, 124, 128
industrial use, 124, 128–29, 133, 144–47, 159–61
price volatility, 28, 145 (*See also* oil, price volatility)
Fossil Ridge High School (Colorado), 100
fracking, 233
France, 37–38, 47, 182–83
Friedman, Thomas L., 35, 228
Frito-Lay (Pepsico), 150
fuel-cell vehicles (FCVs), 18, 30, 31, 34, 36, 63–64
fuel cells, 60, 207, 209
fuel efficiency standards, 19–20, 37, 40, 50, 74
fuel providers, 73. *See also specific fuel companies*
fuel refineries, 132
fuel-switching, 126, 147–50, 160, 161
Fukushima nuclear disaster, 184–85
fume hoods, 134

G

Galvin Electricity Initiative, 210
gas-fired combined-cycle plants, 174, 176, 192, 209
Gas Technology Institute, 136
gasoline, 3, 5–6, 8–9, 19, 37, 40, 132
GDP (gross domestic product), U.S.
cash for clunkers program, effect of, 39
energy use and, 8, 13
industry, generated by, 128
General Electric Co., 63, 92, 191
General Motors, 34, 35, 59
generating company, defined, 170
Georgia Power, 184
geothermal, 171, 181, 187–89, 192, 201, 203, 212–13
Geothermal Solutions LLC, 192
Germany, 100, 237, 247
buildings, energy passport for, 113
electricity sector, 184, 186, 187, 192, 199, 205, 216
energy prices in, 159, 239
recycling in, 153
Gillette Coalfield, 176
Glasser, David, 143
Goldstein, David, 48
Gossamer Wind® ceiling fans, 93
government policies, xiii, 234, 238, 240, 247. *See also* subsidies; tax policy
building sector (*See* building sector, energy efficiency in, subhead: policies)
electricity sector (*See* electricity sector, energy efficiency in, subhead: policies)
industrial sector (*see* industrial sector, energy efficiency in, subhead: policies)
transportation sector (*See* transportation, subhead: policies)
Gramm, Phil, 13
Greene, David, 3, 8

greenhouse gas emissions, 106, 133, 162, 240. *See also* carbon emissions
Greenspan, Alan, 5
grid, 170, 178–80, 193–96, 205–9, 214. *See also* microgrid; smart grid
Grübler, Arnulf, 183
Gustafson, Jon, 52

H

Harman, Jay, 93
health costs
  asthma, diesel-particulate-induced, 51
  coal use, 6
  electricity sector, 175–76, 214
  ozone reduction, 90
  smart grid, 210
heat, electricity production, 172
heat, industrial processes, 138. *See also* combined heat and power (CHP)
  electricity, direct use of, 148–49
  heat pumps, 149
  primary energy used for, 131
  solar process heat, 149–50
  waste heat, reuse of, 124, 127, 136–38, 149, 163, 231
heat exchangers, 151
heat pumps, 93, 149
heating systems, 84, 97, 98, 158
Herman Miller, Inc., 154
high-voltage transmission system, 170
highways and roads, 43, 45, 62–63
Hildebrandt, Diane, 143
Holcim Ltd., 148
Holland Container Innovations, 54
Honda Motor Co., 26, 27, 33–34
Hood River Conservation Project, 111
HVAC systems, xvi, 95, 96
hybrid-electric vehicles, 18, 30, 39, 42, 52, 59
hydrogen, 131, 232, 242
hydrogen-fueled vehicles, 34, 63–64, 71, 132
hydropower, 131, 171, 181, 187–89, 196, 201, 203, 205, 212, 243
Hyman, Leonard S., 223
HyperCar®, 22–24, 26, 29, 73

I

IGCC. *See* integrated gasification combined cycle (IGCC)
IKEA, 54
imports
  autos, 9, 40
  energy embodied in, 132
  oil, 3, 6, 8, 16
India, 35, 147, 177, 187, 239, 243, 244
industrial sector, energy efficiency in, ix–x, 122–63, 211, 231, 236–37, 247, 249. *See also* heat, industrial processes; waste, reuse of industrial
  additive manufacturing, 154–55
  barriers to, 155–57
  biomimicry, 124–25, 151–52, 155
  boilers, efficiency of, 136
  business opportunities, 123, 153
  capital spending for, 143, 156, 158, 160
  competitiveness and, 145, 158–62
  cost-effectiveness of, x, 125–26, 137, 143, 155, 157–58, 160
  customers, role of, 162
  dematerialization, 152–55
  diverse energy uses by, 128–32
  emerging technologies for, 127, 137, 158
  energy-intensive products, decreasing need for, 132
  energy overkill, reducing, 146
  energy services distribution, facility, 134–36
  energy use by sector, 126–33
  fossil fuel use, minimizing, 144–47
  information transparency, 134
  integrative design, 125, 128, 138–44
  international trade, effect of, 132
  investment rate of return and, 123, 156, 158, 160–61
  maintenance savings, 134–36
  management-led innovations, 157–58, 161, 162
  new facilities, 143
  non-energy benefits, 136–37
  policies, 123, 128, 138, 147, 157, 159–63
  primary energy consumption, 126–27, 129
  process redesign, 133–34, 150–51
  product life-cycle, producer responsibility for, 126, 153, 159, 161, 234
  rebound effect, 145
  regulations, effects of, 126
  retrofits, 143–45
  terminology, 126
  training, workforce, 158, 161, 162, 237
  utilities, work with, 158
information technology, 9, 12, 41, 110, 169
  electricity sector, use in, 167, 177–79, 197, 201–2, 205–6, 221, 231
  industrial processes, 134
  transportation systems, 49, 53, 72, 73
Ingomar Packing Co., 149
insulation, 91, 110, 111, 113
Insull, Samuel, 173

integrated gasification combined cycle (IGCC), 171, 181, 188, 203, 205, 212
integrated project delivery (IPD), 107
intelligent transportation systems, 49
Interface, Inc., 161–62
Interface Nederland, 139–41
intermodal freight, 55
internal combustion engine (ICE), 18, 30
Intratec Solutions LLC, 151
investor-owned utilities (IOUs), defined, 170
Iran, 6, 241
Iraq War, 5, 60–61, 241
islanding, 197, 207–10, 214, 217, 219, 222–24

J

Jaffe, Anna, 49
Japan, 114, 135, 153, 159, 184–85, 239
JC Penney, 89
job creation, 236–37
  building sector, 80, 82, 111, 113, 118
  cash for clunkers program, 39
Joint Strike Fighter (JSF), F-35, 22
Jungerberg, Steve, 112

K

kBTU, defined, 79
Kozubal, Eric, 91
kW (kilowatt), defined, 79
kWh (kilowatt-hour), defined, 79

L

Land Rover, 35
Lanni, Dennis, 90
LaSalle Investment Management, 103
Lawrence Berkeley National Laboratory, 110, 137, 138, 178
Lee, Eng Lock, 90, 140–41
LEED ratings, xvi, 85, 103, 114
Lewis and Clark State Office Building (Missouri), 100
light-duty vehicles. *See* autos
light-emitting diodes (LEDs), 92, 101, 245–46
light rail, 60
lighting, xvi, 84, 106, 245–46
  compact fluorescent lamps, 88, 92
  energy used for, 113
  industrial sector, 151, 158
  integrated building systems, 96
  light-emitting diodes, 92, 101, 245–46
  maintenance and installation, 101–2
  natural light, benefit of, 102–3
liquefied natural gas (LNG), 64–65, 143
lithium batteries, 31, 32
load, defined, 170
logistics, 53, 58, 60–61
Lufthansa, 69

M

machines, industrial, 131
MakerBot, 155
Malkin, Anthony, 78–79, 81, 97, 118
Maravic, Dusko, 93
Marshall, George, 246
Marston, Jim, 218
Massachusetts, 118
Mbbl/d, defined, 18
McKinsey & Co., 155
McNeill, John R., 2
mechanical vapor decompression, 149
Mesilla Valley Transportation, 50
metals, auto use of, 21, 26, 29, 34
metering, electricity, 218, 219. *See also* smart meters
microgrid, 203, 206–11, 221–23, 229
micropower, 205
microreactors, 134
Microsoft, 109
Midrex, 131
Minnkota Power Cooperative, 199
mixing processes, industrial, 150–51
mobility industries. *See* transportation
Modern Forge Companies, 135
molecular assembly, 155
Mondi Group, 148
Mori Seiki Co., 135
mortgages
  energy-efficient, 105
  locationally efficient, 48, 72
motors, industrial, 135–36, 143–44
Mulally, Alan, 21
Mullen, Mike, 6–7
municipal-solid-waste incinerators, 148
Münter, Leilani, 65
Munters Corporation, 91
Musk, Elon, 31

N

National Gypsum Co., 91

National Highway Traffic Safety Administration (NHTSA), 29
National Renewable Energy Laboratory, 99, 169, 189
national security, xii, xvi, 5–6, 230, 240–41
  electricity sector issues, 169, 178–80, 214, 229, 241
  transportation sector issues, 16, 43, 62
natural gas, xv, xvii, 3, 8, 13, 231
  buildings, use in, 80, 82, 119
  carbon-fiber production, use in, 28
  electricity generated by, 8, 171, 172, 176, 181, 188, 195, 198, 201, 203, 205, 212, 214, 216
  hydrogen made from, 64, 242
  industry, use in, 127, 129, 130, 144–45, 147–48, 158–60
  liquid fuel, converting to, 143
  peak production, 7
  price volatility, 28, 233
  shale gas, 233
  steel production using, 131
  as transition fuel, 232–33
  utilities, regulatory changes for, 115–16
  as vehicle fuel, 59, 64–65
natural light, benefit of, 102–3
nature, learning from, 124–25, 151–52, 155, 158–59
Nelson, Ken, 157–58, 161
net-zero energy buildings, 106, 114
New Jersey, development in, 48
New Orleans, energy use in, 104
New York City, 39. *See also* Empire State Building
NextEra Energy, Inc., 217
NGOs, 74, 121, 162
Nissan Motor Co., 20, 21, 26, 33, 34, 42, 62
nitrogen oxides, xvi, 6, 175, 176, 233
non-governmental organizations (NGOs), 74, 121, 162
Nordstrom's, 106
Norfolk Southern Corporation, 60
North Carolina, 184
Northwest Power Planning Council, 114
NRG Energy, 63, 184
nuclear power, xvii, 232, 243
  construction loan guarantees, 183–84
  electricity generated by, 171, 172, 180–86, 188, 195, 201, 203, 205, 212, 214, 216, 222
  Fukushima nuclear disaster, 184–85
  waste storage issues, 185
Nucor Corporation, 148

O

Oak Ridge National Laboratory (ORNL), 28, 29
Oakland Museum, 141
oil, ix, 12, 231, 232, 242
  costs of, 3–7, 12, 16
  electricity generated by, 171, 181, 188, 201, 203, 212, 214
  as feedstock, 9–10, 28, 124, 128
  imported, 3, 6, 8, 16
  peak, xii, 7–8, 242
  price volatility, 3–4, 8, 13, 16, 20, 28, 37, 50
  subsidies, 3, 20
  supply chain, vulnerability of, 5, 6
oil use, U.S., 5, 8, 13
  in buildings, 119
  by Defense Department, 60
  by industry, 127, 129, 159–60
  natural gas substituted for, 65
  for transportation, 16–17, 19, 25–26, 49–51, 56, 59, 60, 69, 71
OLEDs (organic light-emitting diodes), 92
Opower company, 94
Oregon Health & Science University Center for Health and Healing, 100
Organization of Petroleum Exporting Countries (OPEC), 3, 8
ozone reduction, 90

P

PACE (Property Assessed Clean Energy) bonds, 116, 117
Pacific Gas & Electric (PG&E), 210, 220
Pacific Northwest, 88, 114, 116
Paley Commission, 187
Panasonic Corporation, 93
Parker, Kevin, 176
parking lots and garages, 73
  auto, 43, 46
  charging stations at, 63, 96
  light-colored surface, energy savings from, 90, 96
  tax policies for, 74
  truck, 51
Passivhaus, 79, 97, 100, 114
PAX Scientific, 92–93
pay-as-you-drive insurance, 45–46
PeaceHealth, 108–9
peak production, fossil fuels, xii, 7–8, 12, 242
Persian Gulf, 5, 8, 62
phase-change materials, 91
photovoltaic systems, 171, 177, 181, 187–93, 196, 198, 201, 203, 205, 211, 212, 216, 217, 220, 244, 247
piping, energy efficiency of, 140–42
PJM Interconnection, 218

Pleotint, Inc., 90
plug and play standards, 219
plug-in hybrid electrical vehicles (PHEVs), 18, 20, 23–24, 30, 62, 197, 198
plug loads, 84, 114
Pool Power, 112
power, defined, 170
Powerhouse Dynamics, 110
Powerplant and Industrial Fuel Use Act, 147
powertrain, 18, 73
  electrified, 24, 29–32, 34, 38, 41, 42, 73
  redesign of, 25, 35
Pratt & Whitney Co., 154
primary energy, 79, 82, 84, 126–27, 129, 131, 239
prize competitions, 21, 40, 74
process energy, defined, 126
Progress Energy, Inc., 176
Progressive Insurance 2010 Automotive X Prize, 21, 40
Project Get Ready, 63
public transportation, 49, 59, 73, 74
public utility commissions (PUCs), state, 172, 173, 184, 250
Public Utility Regulatory Policies Act of 1978, 219
publicly owned utilities (POUs), defined, 170
pulp-and-paper mills, 137
pumps, 112, 136, 139–42
Pythagoras Solar, 202

Q

Qatar, 131, 148
quads, primary energy, 79

R

r-value, 79, 91
rail transport, 55, 59, 60, 180
Rak, Paul, 158
rare-earth materials, supply of, 32
RavenBrick LLC, 90
Ray, Jimmy, 50, 53
reagents, recycling of, 153–54
rebates, 37–39
Recurve, 110
recycling. *See* waste, reuse of industrial
refrigeration, 84, 90, 108
reinventing fire, xi–xvii, 9–13, 226–51
  barriers to, 246–48
  climate change, effect on, 237–39
  economic viability of, xiii, xvi, 234
  economy, effect on, 235–36
  fossil-fuel companies, effect on, 241–42
  global competitiveness, effect on, 239–41
  innovation, effect of, 240
  investments required for, 235–36
  jobs, effect on, 236–37
  leadership for, 246, 248–50
  national security, effect on, 240–41
  partners for America in, 243–45
  technological viability of, 234
  transition to energy efficiency, 232–34
  vision for, 248
  world's poor, effect on, 245–46
Reinventing the Automobile, 29
remanufacturing, 154, 162
renewable portfolio standards (RPS), 177, 218
renewables, xiii, xvii, 8, 218–19, 231, 234, 243, 249
  buildings, use in, 112, 119
  Chinese investments in, 243–44
  costs of, 3, 12, 235
  Defense Department use, 60, 210
  electricity generated by, 9, 162, 174, 176–78, 186–202, 205, 210, 221–22
  locally sited, 178, 207, 211, 220, 229, 231
  public support for, 215–16
  variability of output, 193–96, 214, 221–22
Rentech, Inc., 68
residences, 81–84
  cost *vs.* savings, energy efficiency, 104
  foreclosed homes, energy retrofits of, 89–90
  integrative design, 100–101
  as market for energy efficiency innovations, 110–11
  rental housing energy use, 104, 115
  solar power microgrid, 203
  transportation and cost of, 48
Retro Green Homes, 112
retrocommissioning, 79, 106
Revolutionary+ autos. *See* autos, Revolutionary+
ridesharing, 46–47, 72
roads. *See* highways and roads
Robertson, Ernie, 151
Rocky Mountain Institute (RMI), ix, 263–64. *See also* electricity scenarios, RMI; HyperCar®
  Bright Automotive, 34
  industry retrofits, 143
  Project Get Ready, 63
  pump efficiency, 140
rolling resistance, 22, 50, 73
roofs, 90
Ross, Marc, 29
rotors, efficient, 92–93
Rumsey, Peter, 141

Runkle, Don, 30
Rutan, Burt, 30–31
Rutledge, Dave, 176

S

Sacramento, California, 89, 94
Samba Energy, 111
Samsung Corporation, 92
Sapphire Energy, Inc., 68
Schellnhuber, Hans-Joachim, 47
Schilham, Jan, 139–41
Schlaich, Michael, 154
Schlesinger, James R., 210
Schneider Electric, 63
Scientific Conservation, 110
Scott, Lee, 102, 109
seat-mile, defined, 18
Seibu Giken Co., 91
semiconductor industry, 148
sensors, 110, 134
Serious Energy, Inc., 91
shale gas, 233
Shanghai Automotive, 35
shareholder-owned utilities, 170
Shell Oil Co., 68, 143, 240, 242
Shimomura, Tsutomu, 92
ships, 55, 60–62
Siemens AG, 185, 191
site energy, defined, 79
Skanska USA, 95, 109
SkySails, 60
smart controls, boiler, 136
smart grid, 167, 170, 177–78, 197, 201–3, 205–6, 235
  cost of, 209–10
  health effects of, 210
smart growth, 44, 47–49, 72, 74, 153, 230, 240, 249
smart meters, 94, 110, 170, 216
smart windows, 90–91
smartphones, energy savings reports generated with, 110, 111
social networking, 44, 46–47, 111
software
  auto innovation, 23, 31, 41, 42
  for buildings, 106, 110, 118
solar systems, 111, 237, 249
  electricity generated by, 181, 185, 186, 198–202, 199, 202, 217, 219, 220, 221 (*See also* concentrating solar power (CSP); photovoltaic systems)
  industrial use, 145, 149–50, 161, 163
  manufacture of, 231
  shading parking areas with solar cells, 96
  vertical solar farm, 202
SolarCity, 217
Solazyme, Inc., 68
Sony Corporation, 92
South Carolina, 184
South Korea, 147, 244
Southwest Airlines, 58
sport-utility vehicles (SUVs), 18, 20–24, 62. *See also* autos
Starbucks, 92
state public utility commissions (PUCs). *See* public utility commissions (PUCs), state
steam, 137, 149, 151, 174
steel
  auto use of, 21, 29
  manufacture of, 131, 136, 137, 147–48, 153
  recycling, 153
Steelcase, Inc., 154
Stop & Shop, 102
Straubel, JB, 30–31
Sturman, Eddie, 30
subsidies, 14
  auto efficiency, 29, 34, 36
  biofuels, 66, 67
  developing countries, 245
  industrial energy efficiency and, 126, 159, 163
  oil and gasindustry, 3, 20
  public transportation, 46
  whale oil, 13
substations, 170
sulfur, 6, 176, 233
sulfur dioxide, xvi, 175, 240
supply-side resources, 170, 196–200, 213
SUVs. *See* sport-utility vehicles (SUVs)
Sweden, 131, 153
swimming pool pumps, 112

T

Taggart, David F., 22, 24
Tata, 35
Tavares, Felipe, 151
tax policy
  auto efficiency and, 20, 39, 40, 41
  electricity sector, 205
  fuel prices and, 239
  income taxes, 3
  industrial energy efficiency and, 126, 159, 161
  parking spaces, 74
  vehicle-miles traveled (VMT), taxes based on, 45

Teijin Limited, 26
telecommuting, 44
teleconferencing, 58
tenants, energy use by, 104, 106, 115, 121
Tendril, 110
Tesla Motors, 20, 31, 34
Texaco, Inc., 158
Texas, 176, 184, 186–89, 197, 218
Texas Instruments, 108, 143
thermoplastics, auto use of, 26
third-party logistics, 53
3M Co., 133, 137
tires, 49, 50, 59
Tokyo Electric Power Co., 184–85
ton-mile, defined, 18
Toray Industries, Inc., 26–27
Toyota Motor Corporation, 21, 23–24, 26–27, 32, 33, 42, 88–89
trade. *See also* imports
  energy embodied in international trade, 132
  for oil, 5
Trane Co., 91
Trans-Alaska Pipeline, 6
transmission company, defined, 170
transmission system, 170, 200
transportation, ix–x, xvi, 3, 14–75, 132, 211, 229–30, 236–37, 247, 249. *See also* airplanes; autos; rail transport; trucks, heavy
  business opportunities, 15, 16, 26–27, 43, 47, 49, 52–53, 62–63, 68–72
  commuting, 44–47
  cost to U.S. consumers, 16, 43
  Defense Department, use and design by, 60–62
  efficiency improvements, system-wide, 44–45, 49
  infrastructure, 43
  innovative pricing, 45–46
  intelligent transportation systems, 49
  intermodal freight, 55
  policies, 15, 36–41, 57, 66, 72, 74, 234
  savings available from, 15
  ships, 55, 60–62
  terminology, 18
  transformation of, 16 (*See also* autos, Revolutionary+)
transportation applications (apps), 46–47, 49
trucking industry, logistics and structure of, 52–53
trucks, delivery, 34, 59
trucks, heavy, 50–56, 70, 72, 132, 249
  biofuels for, 64
  containers, 54
  driving methods, fuel-efficient, 53
  fuel-efficient design, 50–52
  idling engines, auxiliary power units to replace, 51
  intermodal freight, 55
  loads, space and weight of, 53, 54
  natural gas-fueled, 64–65
  obstacles to innovation, 52–53
  productive use of, 53–54
  turnpike doubles, 51–52
  weight of, 50, 51, 52, 54
Tsuda, Hiroshi, 42
turnpike doubles, 52

U

ultralight airplanes, 56, 61
ultralight armored vehicles, 62
ultralight autos. *See* autos, Revolutionary+
ultralight rail, 60
underfloor displacement ventilation, 95, 96
United States Department of Defense (DoD), xii, 32, 109, 178
  biofuels, use of, 69
  climate change concerns, 6–7
  costs, effect of fossil fuel use on, 5–6
  microgrids, use of, 210
  military structures, heating and cooling of, 60–61
  remanufacturing by, 154
  renewables, use of, 210
  transportation use and design, 60–62
United States Department of Energy (DOE), 32, 67, 89, 92, 178
  Industrial Assessment Centers energy surveys, 137
  Industrial Technologies Program, 138, 157
  nuclear construction loan guarantees, 183–84
United States Energy Information Administration, 17, 36, 56, 137
United States Environmental Protection Agency (EPA), 176
United States GDP. *See* GDP (gross domestic product), U.S.
United States General Services Administration (GSA), 115
United States government. *See also* government policies; subsidies; *specific agencies*
  auto fleets, purchase of, 39, 74
  budget deficit, 3
  cash for clunkers program, 39–40
  federal buildings, energy efficient design of, 99, 109
  feebate program, proposed, 38–39
United States Nuclear Regulatory Commission, 184

United Technologies, 133
University of California at San Diego, 209
University of North Carolina at Asheville, 109
UPS, 59
uranium, 214
U.S. Climate Action Partnership, 175, 180
U.S. Green Building Council, 85
utilities, 120, 178, 224. *See also* electricity sector
  definition, 170
  efficiency projects funded by, 116, 117
  energy conservation encouraged by, 94, 110–11, 115–16
  industrial sector, work with, 158
  interconnection rules, 158
  regulatory changes, need for, 115–16, 118, 121, 158, 163, 220, 234, 249–50
  state-mandated energy efficiency, 116, 118

V

Valero Energy, 68, 69
vehicle fitness, 20–22, 25, 31, 32, 34, 62, 64, 73
vehicle-miles traveled (VMT), 18, 44–49
vehicles. *See* autos; transportation; trucks, heavy
ventilation, 95, 96
VeriForm Inc., 158
Vermont, 116, 118
Viridity Energy, 218
Volkswagen AG, 23–24, 27, 33
Volvo, 35

W

Walmart, 52, 102, 109
Walters, Eric, 112
Washington, D.C., bikesharing program, 47
waste, reuse of industrial, 125, 126, 136, 137–38, 144, 148, 152–55, 159, 161–62. *See also* heat, industrial processes
  biomass, 137–38
  carbon dioxide, 124
  process waste, reducing, 154
  recycling, 152–54, 159, 162
  remanufacturing, 154, 162
waste combustion, electricity generated by, 189
water heating, 84
water use, 153, 162, 176, 214
Westinghouse, George, 173
whale oil, 13
Whirlpool Corporation, 114
wind power, xv, 171, 177, 181, 186–93, 195–96, 198–201, 203, 205, 209, 212–13, 217, 221, 237, 243–44, 247, 249
windows, 90–91, 97
wiring, improved, 135
Woolsey, R. James, 5, 6, 241
Worrell, Ernst, 136–37

Z

Zhejiang Geely, 35
Zipcar, 47
Zoltek Companies Inc., 26–27